D1376695

Software Defined Radio

WILEY SERIES IN SOFTWARE RADIO

Series Editor: Dr Walter Tuttlebee, Mobile VCE, UK

The Wiley Series in Software Radio aims to present an up-to-date and in-depth picture of the technologies, potential implementations and applications of software radio. Books in the series will reflect the strong and growing interest in this subject. The series is intended to appeal to a global industrial audience within the mobile and personal telecommunications industry, related industries such as broadcasting, satellite communications and wired telecommunications, researchers in academia and insutry, and senior undergraduate and postrgraduate students in computer science and electronic engineering.

Mitola: Software Radio Architecture: Object-Oriented Approaches to Wireless Systems Engineering, 0471384925, 568 Pages, October 2000
Mitola and Zvonar (Editors): Software Radio Technologies: Selected Readings: 0780360222, 496 Pages, May 2001
Tuttlebee: Software Defined Radio: Origins, Drivers and International Perspectives, 0470844647, £65, 350 pages
Tuttlebee: Software Defined Radio: Enabling Technologies, 0470843187, £65, 304 pages
Dillinger, Madani and Alonistioti (Editors): Software Defined Radio: Architectures, Systems and Functions, 0470851643, £85, 456 pages

Software Defined Radio

Architectures, Systems and Functions

Markus Dillinger

Siemens AG, Germany

Kambiz Madani

University of Westminister, United Kingdom

Nancy Alonistioti

University of Athens, Greece

WILEY

Telephone (+44) 1243 779777

Email (for orders and customer service enquiries): cs-books@wiley.co.uk
Visit our Home Page on www.wileyeurope.com or www.wiley.com

Other Wiley Editorial Offices

John Wiley & Sons Inc., 111 River Street, Hoboken, NJ 07030, USA

Jossey-Bass, 989 Market Street, San Francisco, CA 94103-1741, USA

Wiley-VCH Verlag GmbH, Boschstr. 12, D-69469 Weinheim, Germany

John Wiley & Sons Australia Ltd, 33 Park Road, Milton, Queensland 4064, Australia

John Wiley & Sons (Asia) Pte Ltd, 2 Clementi Loop #02-01, Jin Xing Distripark, Singapore 129809

John Wiley & Sons Canada Ltd, 22 Worcester Road, Etobicoke, Ontario, Canada M9W 1L1

Wiley also publishes its books in a variety of electronic formats. Some content that appears in print may not be available in electronic books.

Library of Congress Cataloging-in-Publication Data

Software defined radio : architectures, systems, and functions / [edited by] Markus
Dillinger, Kambiz Madani, Nancy Alonistioti.
p. cm. – (Wiley series in software radio)
Includes bibliographical references and index.
ISBN 0-470-85164-3 (alk. paper)
1. Software radio. I. Dillinger, Markus. II. Madani, Kambiz. III. Alonistioti, Nancy. IV. Series.

TK5103.4875.S57 2003
621.384 – dc21

2003043063

British Library Cataloguing in Publication Data

A catalogue record for this book is available from the British Library

ISBN 0-470-85164-3

Typeset in 10/12pt Times by Laserwords Private Limited, Chennai, India
Printed and bound in Great Britain by Biddles Ltd, Guildford and King's Lynn
This book is printed on acid-free paper responsibly manufactured from sustainable forestry in which at least two trees are planted for each one used for paper production.

Contents

Preface

It is not uncommon for breakthrough concepts in wireless communications to take a considerable time to mature and gain momentum. In this respect, Software Defined Radio (SDR) is no exception, evolving very slowly from its military origin. J. Mitola first defined the area and also coined the term back in 1991 while Steinbrecher made initial investigations on the suitability of Software Radio in implementing multi-standard, flexible, smart base stations. As J. Mitola pointed out in his keynote speech at the 1st European Workshop on Software Radio in May 1997 when he referred to the slow start of SDR in the United States: … *it was not that the technology was not within reach, but rather the fact that commercial applications are quite cost sensitive.* A renewal of interest in the area occurred around 1996 driven by two events, namely the creation of the Modular Multifunction Information Transfer System (MMITS) Forum[1] in March 1996, and the launch of the first European research project on SDR in the context of the Advanced Communications Technologies and Services (ACTS) programme of European Union (EU) funded research and development (R&D).

One could say that the title of the first European research project in this area was prophetic as very becomingly it was entitled FIRST (Flexible Integrated Radio Systems Technology). Observing the rising interest by the international research community in what was still known at that time as S/W radio technologies, the European Commission (EC) organised the 1st European Workshop on Software Radio in May 1997. The objective of this event was twofold. First it provided an excellent opportunity to assess the interest of the European research community in this area while at the same time it aimed to provide valuable feedback that would assist in shaping future research activities in this area.

Without doubt the 1st European Workshop on Software Radio was a major success. The attendance of more than 170 researchers representing academic and industrial centres not only from Europe, clearly showed that European industry was ready to invest in this challenging new area. This was reflected in the 3rd Call for Proposals of the ACTS R&D programme which called for proposals on '… novel technological work in S/W Radio technologies'. As a result of this call, two projects were launched in this area: Smart Universal Beam-forming (SUNBEAM) and Software Radio Technology (SORT), the first addressing the integration of smart array antennas in a 'software radio base station', while the second investigated intermediate frequency (IF) and baseband issues.

The follow-up of the very successful 1997 workshop took place in the summer of 1998, this time jointly organised by the EC and the MMITS Forum. The objective of the 2nd

[1] In December 1998 the MMITS Forum decided to change its name to the SDR (Software Defined Radio) Forum.

Workshop on Software Radio was to foster the exchange of experience in the field and explicitly to promote a broader approach, extending beyond the terminal. However, the most significant moment in this chronology of events was probably March 1999 when the three ACTS research projects active in this area took upon themselves the initiative to organise the First European Colloquium on Reconfigurable Radio Systems and Networks (RRSN). For the first time in the context of that event a different system-oriented and all-encompassing perspective/approach to SDR was put forward that went beyond the mainstream thinking in the area that was viewed SDR as a technology 'restricted' to the baseband and radio frequency (RF) functions of a wireless terminal.

This system-oriented point of view of SDR became the starting point for the 1st Call for Proposals of the IST (Information Society Technologies) R&D programme launched early in 2000. As we are now approaching the end of the IST programme we can safely claim that the projects performing work in the area of SDR have made considerable progress in the direction of the above-mentioned vision. This is manifested by the large number of contributions made to the SDR Forum as well as ITU Working Party (WP) 8F, which is in charge of defining the technical, operational, and spectrum issues related to the future development of IMT-2000 and systems beyond IMT-2000. The purpose of this book is to provide an integrated view of the work performed in this area in the context of the IST programme. In this light, the network functions for managing flexible devices and network elements are presented without ignoring at the same time the users' requirements and expectations from reconfigurable systems reflecting the system-oriented and all-encompassing perspective of the IST programme.

As we approach the end of the IST programme, this book acts as a repository of the ideas and concepts developed and investigated by the projects active in the area of RRSN. Without doubt reconfigurability will continue to play a central role in the overall system concept of wireless systems beyond 3G and also in the context of the 6th Framework Programme of EU funded R&D. In this context, the high-level objectives for the research work to be performed in the next four years could be summarised as follows:

- to achieve wide acceptance at standardisation level especially in what concerns network and service aspects; and
- to obtain to a clear understanding of the regulatory implications;

In this light, work in the context of the 6th Framework Programme will certainly build upon the sound basis established by the IST programme.

In conclusion, we wish to thank our colleagues who participated in the IST projects in the area of wireless communications for their commitment and co-operative spirit that is best reflected in this book. In particular, we thank the editors, Nancy Alonistioti, Markus Dillinger and Kambiz Madani, for their high quality contributions and Parbhu Patel for his enthusiasm and continuous support.

Brussels, 20 November, 2002
Dr Demosthenes Ikonomou
European Commission
DG Information Society

Abbreviations

2G	Second Generation Mobile Network
3GPP	Third Generation Partnership Project
AA	Alternating MAC Frame Utilisation Algorithm
AAA	Authentication, Authorisation and Accounting
AAAARCH	Authorisation, Authentication and Accounting ARCHitecture research group
ACH	Access feedback Channel
ADC	Analogue to Digital Converter
AF	Alternating Factor
AMM	Alternative Mode Monitoring
ANAI	Access Network Added Intelligence
ANFR	Council of the National Frequency Agency
AP	Access Point
APC	Access Point Controller
API	Application Programming Interface
ARIB	Association of Radio Industries and Business
ARQ	Automatic Repeat Request
AS	Application Server
ASIC	Application-specific Integrated Circuit
ASP	Application Services Provider
ATM	Asynchronous Transfer Mode
BCCH	Broadcast Control Channel
BCH	Broadcast Channel
BGCF	Breakout Gateway Control Function
BER	Bit Error Rate
BOV	Book of Vision
BPC	Baseband Processing Cells
BS	Base Station
BSC	Base Station Controller
BST	Base Station Transmitter
CAB	Charging, Accounting and Billing
CAC	Call Admission Control
CAMEL	Customised Applications for Mobile Network Enhanced Logic
CAP	CAMEL Application Part
CAST	Configurable Radio with Advanced Software Technology

CCF	Charging Collection Function
CDF	Cumulative Density Function
CDMA	Code Division Multiple Access
CDR	Charging Data Record
CEPT	Conference Européen des administration des Postes et des Télécommunications
CERP	Comité Européen des Régulateurs Postaux
CGF	Charging Gateway Function
CORBA	Common Object Request Broker Architecture
COPS	Common Open Policy Service protocol
COPS-PR	Common Open Policy protocol for Policy Provisioning
CPU	Central Processing Unit
CN	Core Network
CMM	Configuration Management Module
CSD	Circuit Switched Data
CSEL	Cell Selection
DCA	Dynamic Channel Allocation
DCH	Data Channel
DES	Discrete Event System
DHCP	Dynamic Host Configuration Protocol
DL	Downlink
DLC	Data Link Control
DLCC	DLC Connection
Drive	Dynamic Radio for IP-services in vehicular environments
DSA	Dynamic Spectrum Allocation
DSM	Distributed Shared Memory
DSP	Digital Signal Processor
DTX	Discontinues Transmission
DVB	Digital Video Broadcasting
DVB-T	Digital Video Broadcasting – Terrestrial
ECC	Electronic Communications Committee
ECTRA	European Committee for Regulatory Telecommunications Affairs
ED	Event Driven
EDGE	Enhanced Date Rate for GSM Evolution
ERC	European Radio Communications Committee
ETNO	European Telecommunications Network Operators' Association
ETO	European Telecommunication Office
ETSI	European Telecommunications Standards Institute
EXP	EXport Protocol
FA	False Alarm
FCA	Fixed Channel Allocation
FCC	Federal Communications Commission
FCCH	Frame Control Channel
FCFS	First Come First Serve
FCH	Frame Channel
FDD	Frequency Division Duplex

FEC	Forward Error Correction
FER	Frame Error Rate
FIFO	Fast in First Out
FPGA	Field Programmable Gate Array
FSA	Flexible Spectrum Allocation
FSM	Finite State Machine
FTAM	File Transfer, Access and Management
FTP	File Transfer Protocol
GGSN	Gateway GPRS Support Node
GIRC	Global Intelligent Reconfiguration Component
GMSC	Gateway Mobile Switching Centre
GoS	Grade of Service
GPS	Generic Process Sharing
GPRS	General Packet Radio Service
GRSF	Gateway Reconfiguration Support Function
GSM	Global System for Mobile Communications
GTP	Generic Tunnelling Protocol
GUI	Graphical User Interface
HAVi	Home Audio Video Interoperability
HCA	Hybrid Channel Allocation
HIPERLAN/2	High Performance Radio Local Area Network Type 2
HLR	Home Location Register
HO	Handover
HRM	Home Reconfiguration Manager
HSCSD	High Speed Circuit Switched Data
HSS	Home Subscriber Server
HTML	Hypersett Markup Language
HTTP	Hypersett Transfer Protocol
I-CSCF	Interrogating – Call Session Control Function
ID	Identification
IDL	Interface Definition Language
IEICE	Institute of Electronics, Information and Communications Engineers
IETF	Internet Engineering Task Force
IMEI	International Mobile Equipment Identity
IMS	IP Multimedia Subsystem
IMI	Initial Mode Identification
IMT-2000	International Mobile Telephony (3rd generation networks are referred as IMT-2000 within ITU)
INAP	Intelligent Network Application
I/O	Input/Output
IP	Internet Protocol
IRAC	Inter-Department Radio Advisory Committee
ISP	Internet Service Provider
IST	Information Society Technologies
IT	Information Technology
ITU	International Telecommunication Union

IU	Traffic User Interface
JAIN	Java APIs for Integrated Networks
JCC	Java Call Control
JCAC	Joint Call Admission Control
JOCAC	Joint Call Admission Control
JOLDC	Joint Load Control
JOSAC	Joint Session Admission Control
JOSCH	Joint Resource Scheduler
JRRM	Joint Radio Resource Management
JVM	Java Virtual Machine
LAN	Local Area Network
LCH	Long Transport Channel
LCS	Location Services
LCLS	Last Come Last Served
LFFS	Last Finished First Served
LHP	Last Hop Protocol
LIFO	Last in First Out
LIM	Location Information Manager
LODCL	Load Control
MAC	Medium Access Control
MAP	Mobile Application Part
mAh	milli Ampere hour
MD	Metering Device
MDC	Metering Devices Controller
MExE	Mobile Execution Environment
MGCF	Media Gateway Control Function
MIMM	Mode Identification and Monitoring Module
MM	Mobility Management
MMI	Man–Machine Interface
MNSM	Mode Negotiation and Switching Module
MO	Mobile Operator
MOBIVAS	Downloadable Mobile Value Added Services through Software Radio and Switching Integrated Platforms
MPHPT	Ministry of Public Management, Home Affairs, Posts and Telecommunications
MRFC	Multimedia Resource Function Controller
MSC	Mobile Switching Centre
MT	Mobile Terminal
MTS	Mobile Terminal Software
NTIA	National Telecommunications and Information Administration
NPRM	Notice of Proposed Rulemaking
OC	Occupied Carrier
OLAP	Online Analytical Processing
OMG	Object Management Group
ONS	Open Network Services
OSGi	Open Services Gateway initiative

OSM	Office of Spectrum Management
OSA	Open Service Access
OSI	Open Systems Interconnection
OTA	Over The Air
Overdrive	Over Dynamic multi Radio networks in Vehicular Environments
P2P	Peer to Peer
PCCPCH	Physical Common Control Channel
PCI	Personal Computer Interface
P-CSCF	Proxy-Call Session Control Function
PDA	Personal Digital Assistant
PDM	Packaging and Downloading Module
PDP	Policy Decision Point
PDU	Protocol Data Unit
PEP	Policy Enforcement Point
PHY	Physical Layer
PLC	Physical Layer Controller
PLMN	Public Land Mobile Network
PN	Pseudo Noise
PPP	Point-to-Point Protocol
PRM	Proxy Reconfigurable Manager
PS	Packet Switched
PW	Priority Weight
PV	Priority Vector
PVI	Priority Vector Information
QoS	Quality of Service
R&TTE	Radio Equipment and Telecommunications Terminal Equipment
RACH	Random Access Channel
RAN	Radio Access Network
ReSoA	Remote Socket Architecture
RAT	Radio Access Technology
RBB	Reconfigurable Baseband
RCH	Random Channel
RCM	Reconfiguration Manager
RDF	Resource Description Framework
RegTP	Regulierungsbehörde für Telekommunikation und Post
RESAC	Regular Traffic Session Admission Control
RF	Radio Frequency
RLC	Radio Link Control
RMM	Reconfigurable Baseband Management Module
RMSPP	Reconfiguration Management and Service Provision Platform
RNC	Radio Network Controller
RR	Round Robin
RRM	Radio Resource Management
RRN	Reconfigurable Radio Network
RRSF	RAN Reconfiguration Support Function
RRU	Radio Resource Unit

RS	Resource Scheduling
RSC	Resource Controller
RSCPM	Reconfiguration Control and Service Provision Manager
RSF	Reconfiguration Support Functions
RT	Real Time
SAC	Session Admission Control
SADL	Service Architecture Definition Language
SCF	Service Capabilities Features
SCH	Short Transport Channel
S-CSCF	Serving – Call Session Control Function
SCOUT	Smart ser-Centric cOmmUnication environmenT
SDL	Specification and Description Language
SDR	Software Defined Radio
SDSAC	Software Download Session Admission Control
SDM	Service Deployment Manager
SDU	Service Data Unit
SF	Spreading Factor
SGSN	Service GPRS Support Node
SHO	Soft Handover
SIP	Session Initiation Protocol
SIR	Signal to Interference Ratio
SP&P	Spectrum Plans and Policies
SME	Small and Medium Enterprise
S-MSC	Mobile Switching Center Server
SMS	Short Message Service
SOAP	Simple Object Access
SRM	Serving Reconfiguration Manager
SRSF	Serving Reconfiguration Support Function
SS7	Signalling System No. 7
SSA	Static Spectrum Allocation
SWR	Software Radio
TAP	Transferred Account Procedure
TCAP	Transaction Capabilities Application Protocol
TCP	Transmission Control Protocol
TCAM	Telecommunications Conformity Assessment and Market Surveillance Committee
TDD	Time Division Duplex
TDMA	Time Division Multiple Access
TELEC	Telecom Engineering Centre
TGS	TCAM Group on SDR
TREST	Traffic Estimation
TRSCH	Traffic Scheduler
TRSF	Terminal Reconfiguration Support Function
TRUST	Transparently Reconfigurable Ubiquitous Terminal
TTI	Transmission Time Interval
UDCH	User Data Channel

UE	User Equipment
UIMM	User Interaction Management Module
UL	Uplink
UML	Unified Markup Language
UMTS	Universal Mobile Telecommunications System
UPnP	Universal Plug and Play
UTRA	UMTS Terrestrial Radio Access
UTRAN	Ultra Terrestrial Radio Access Network
VAS	Value-Added Service Provider
VASM	Value-Added Service Manager
VASP	Value-Added Service Provider
VC	Void Carrier
VHE	Virtual Home Environment
VoIP	Voice over IP
VSM	Viable Systems Model
WAP	Wireless Access Protocol
WCDMA	Wideband Code Division Multiple Access
WLAN	Wireless Local Area Network
WML	Wireless Markup Language
WSDL	Web Service Definition Language
WTB	Wireless Telecommunications Bureau
XML	Extensible Markup Language

Contributors' Biographies

Series and Book Editors

Markus Dillinger

Markus Dillinger received his Diplom-Ing. degree in telecommunications in 1990 from the University of Kaiserslautern, Germany. In 1991 he joined the Mobile Network Division at Siemens to develop call processing software for the GSM base stations. Later, he joined the system engineering division where he was responsible for the evaluation software for the Siemens Channel Sounder measurement equipment. In addition, he was responsible for the definition of the technical requirements and specification on the GSM base station line interfaces. From 1995 on he was working on the definition of the third mobile radio generation in the European research project FRAMES and in 1999 he was appointed Technical Manager of FRAMES. Since January 2000 he has been the project leader of the European research project TRUST and the follow-up project SCOUT for Software Radio. He has published many articles in the field of WCDMA and software radio.

Nancy Alonistioti

Nancy Alonistioti has a BSc degree and a PhD degree in Informatics and Telecommunications (University of Athens). She had been working for 7 years at the Institute of Informatics and Telecommunications of NCSR 'Demokritos' in the areas of protocol and service design and testing, mobile systems (UMTS), open architectures, CORBA, Intelligent Networks and SDR systems and networks. She has collaborated for one year as an expert at the Greek regulatory organisation and she is currently working as senior researcher and project manager in the Communication Networks Laboratory (University of Athens). She has participated in several national and European projects, (CTS, SS#7, ACTS RAINBOW, etc.) and is the Technical Manager of the IST-MOBIVAS project. She specialises in mobile communications, reconfigurable systems and networks, adaptable service engineering, formal specification and testing of communication protocol and services, test case design for conformance testing, and communications software engineering. Her current research includes: reconfigurability and adaptability management, protocol/software download and open architectures and platforms, CORBA, OSA/Parlay, JAIN, MEXE, etc.

Kambiz Madani

Kambiz Madani is the Technical Director of the Westminster Wideband research laboratory, University of Westminster, London, and has over 20 years of direct experience in leading many international industrial research and development projects for a variety of communication applications. With a PhD from King's College London (1979), he served as a senior principal design engineer at STC Telecommunications and as Technical Executive at ERA Technology, in charge of all research activities of the RF Technology Division, up until 1992. He then joined the University of Westminster and created the Westminster Wideband Research Laboratory, carrying out contract research in the communication systems area, and establishing the first specialist MSc Course in the UK in Mobile, Personal and Satellite Communications. He has been the technical leader for over 25 funded projects in the telecommunications field including the EC IST FP5 CAST Project. He has over 70 technical publications, and has been an active member of Technical Committees and sessions for many international conferences.

Contributors

Didier Bourse

Didier Bourse (*didier.bourse@motorola.com*) currently holds the position of Team Leader with RASSEL Lab of Motorola Labs–CRM in Paris. He received his diploma degree in telecommunications in 1992 from ENSTBr (Ecole Nationale Supérieure des Télécommunications de Bretagne, France) and obtained his PhD degree in 1997 from IRCOM (Institut de Recherche en Communications Optiques et Micro-ondes, France). In 1997 he joined Thomson–CSF Communications and worked in the field of military tactical SDR. He was a member of the NATO FM^{3}TR Technical Group. In 2000 he was the French Technical Manager of a French–German contract dedicated to a SDR Demonstrator realisation. He joined Motorola in January 2001 and was the Technical Manager of the European research project IST/TRUST for SDR. He is coordinating SDR activities within the Wireless World Research Forum (WWRF) and in 2002 edited the WWRF SDR White Paper 'Reconfigurable SDR Equipment and Supporting Networks – Reference Models and Architectures'.

Soodesh Buljore

Dr Soodesh Buljore is currently employed as Research Staff Engineer Senior, since 1998 at the European Communications Research Labs Paris, MOTOROLA LABS. He received his PhD degree, in Electrical Engineering, in 1996, from Ecole Doctorale Sciences pour l'Ingénieur, Ecole Centrale de Nantes. Between 1996 and 1998 he held a Post Doctoral fellowship at the Electrical and Computer Engineering at the University of California San Diego. He has been involved in the design and specifications of UMTS W-CDMA FDD, namely in new modulation and transmit diversity schemes. He is currently the Technical Manager of the IST project SCOUT and is also involved in the operations of other internal research programs. His research interests include, mobile radio communications systems, Software Definable Radio enabling technologies, 3rd Generation Air Interfaces and End to End Reconfigurable mobile radio systems for beyond 3G.

Genevieve Conaty

Genevieve Conaty is a human factors researcher at Motorola Labs in Basingstoke, England. Genevieve's work includes gathering information about user needs for future wireless applications, using a variety of methods such as consettual inquiry and group interviewing. Her work also involves the design, prototyping, and evaluation of new multimedia user interfaces. She completed her first degree in mathematical and computational science with a special interest in the history and philosophy of science at Stanford University. Her masters degree in human–computer interaction at Carnegie Mellon University included a team project about the design of wearable devices for teenagers. In addition, she worked on speech interfaces for information systems and games as part of the ICIE/RHINO Labs.

Lucas Elicegui

Lucas Elicegui joined Motorola Labs, Paris, in 1999 after receiving the Engineering Degree in Telecommunications from Ecole Nationale Supérieure d'Ingénieurs, Limoges. He initially carried out research in the field of radio systems performance analysis with an emphasis on radio resource management for UMTS. Then, he worked on flexible spectrum-sharing techniques and was responsible for the advanced spectrum-sharing activity within the TRUST project on reconfigurable radio. He is currently working in the Spectrum Engineering Centre (Motorola Labs) and is in charge of projects addressing spectrum engineering issues and radio systems coexistence analysis involving systems such as 2G, 3G and WLAN.

Michael Fahrmair

Michael Fahrmair received his Diplom-Inf. degree in Computer Science in 1998 from the University of Technology, Germany. Currently he is completing his PhD in Computer Science at the Munich University of Technology, Germany, while working as a researcher at the chair for Software and Systems Engineering of Prof. Dr Manfred Broy. His main research field is mobile and ubiquitous computing.

Jafar Faroughi-Esfahani

Jafar Faroughi-Esfahani (Motorola Ltd) works for Motorola's Wireless and Broadband Systems Group within the Semiconductor Products Sector. He obtained his BEng in Electronic Engineering from Southampton University and a Masters Degree in Software Engineering from Bournemouth University. He joined Motorola as a graduate in March 2000, doing research on software download and security issues for reconfigurable terminals for the IST–TRUST project. At present he is involved with the security team, researching the overall security architecture for SDR systems within the IST–SCOUT project. Furthermore, he is part of Motorola's 3G Systems Architecture team, focusing on subscriber platform software and security architecture.

Vangelis Gazis

Vangelis Gazis received his BSc in Computer Science from the Department of Informatics at the University of Athens, Athens, Greece, in 1995, his MSc in Computer Science

(Communication Systems and Networks) from the same department in 1998 and his MBA from the Athens University of Economics and Business in 2001. From 1995 until now he has been a research associate of the Communication Networks Laboratory (CNL) at the Department of Informatics in the field of mobile *ad hoc* networks and cellular systems (DECT, GSM/GPRS, UMTS). In parallel, he worked with a number of established companies in the Greek IT Services Sector (IBM Greece S.A., BYTE Computer, BP Oil International, etc.) as an IT networking consultant. He is currently pursuing a PhD in the Department of Informatics. His research interests include intelligent and reconfigurable service provision schemes and quality of service issues in 3G/4G mobile telecommunication networks.

David Grandblaise

David Grandblaise (*david.grandblaise@motorola.com*) received an MSc degree in Electrical Engineering with an emphasis in Telecommunications from ESME (Ecole Spéciale de Mécanique et d'Electricté, Paris) in 1998 and a Mastère degree with specialisation in mobile radio communications from ENST (Ecole Nationale Supérieure des Télécommunications, Paris) in 1999. In 2000 he joined Motorola Labs (European Communications Research Lab, Paris). After carrying out initial research activities on the physical and system aspects of UMTS transmit diversity techniques, his research activities have focused on system aspects of SDR since September 2000. David also contributes to SDR and reconfigurable radio research projects partly funded by the European Commission R&D program. He has contributed to the advanced spectrum-sharing techniques activity in the IST TRUST project. Currently he is responsible for the dynamic spectrum allocation and reconfigurability research activity in the IST OverDRiVE project. David's research interests cover SDR and reconfigurable radio with special emphasis on flexible spectrum management, spectrum and SDR regulation, radio resource management and wireless system performance analysis for dynamic multi-radio networks inter-operability.

Stoytcho Gultchev

Stoytcho Gultchev received a diploma degree in Electronics from EBTC Bulgaria in 1996. In 2000 he graduated as BEng in Information Systems Engineering from the University of Surrey, UK. In the same year, he started his PhD with the Mobile Communications Research Group in the Centre for Communication Systems Research (CCSR), the University of Surrey. His research is in the field of Software Reconfigurable Radio Technology looking into management and control architectures for reconfigurable terminals and networks. He works on the definition of Soft-Terminal Design Aspects contributing to the Mobile VCE work area on Software Based Systems.

Nikos Houssos

Nikos Houssos obtained his BSc degree in Informatics from the University of Athens in 1998 and his MSc (with distinction) in Telematics (Communications and Software) from the Department of Electronic and Electrical Engineering, University of Surrey, UK, in

1999. He is a staff member at the Communication Networks Laboratory of the University of Athens, working in the area of mobile service provision. He is involved in the projects MOBIVAS (including workpackage leadership) and PoLoS of the European Union IST framework. He is also currently pursuing a PhD at the Department of Informatics and Telecommunications, University of Athens. His current research interests relate to flexible value-added services provision in 3G/4G mobile communication networks and in particular to the design and implementation of service management and reconfiguration control platforms, intelligent service adaptation mechanisms, network reconfigurability and advanced business models. He has more than 15 publications in the above areas.

Péter Kacsuk

Péter Kacsuk is the Head of the Laboratory of the Parallel and Distributed Systems in the Computer and Automation Research Institute of the Hungarian Academy of Sciences. He received his MSc and doctorate degrees from the Technical University of Budapest in 1976 and 1984, respectively. He received the kandidat degree from the Hungarian Academy in 1989. He habilitated at the University of Vienna in 1997 where he is a private professor. He is a part-time, full professor at the University of Westminster, the Eotvos Lorand Science University of Budapest and the University of Miskolc. He served as visiting scientist or professor several times at various universities of Austria, England, Germany, Spain, Australia and Japan. He has published two books, two lecture notes and more than 140 scientific papers on cluster and grid tools, mobile computing, parallel computer architectures, parallel software engineering and parallel logic programming.

Tereska Karran

Tereska Karran is a Senior Lecturer in Information Systems at the University of Westminster where she teaches On Line Analytical processing and does research in complex information systems. She has contributed to several European Union projects.

Maria Koutsopoulou

Maria Koutsopoulou has received her diploma with distinction (ranked first in her class) in Electrical Engineering from the Demokritos University of Thrace, Greece, in 1996. In 1997 she received a fellowship for post-graduate studies from the Institute of Informatics and Telecommunications of the National Centre for Scientific Research 'Demokritos'. Since 1999 she has been a PhD candidate and member of the Communication Networks Laboratory at the Department of Informatics, University of Athens, working in the area of the software reconfigurable radio systems and networks and 3G mobile systems and data networks as well. Currently she works as a researcher in the IST project 'MOBIVAS' (Downloadable MOBIle Value-Added Services through Software Radio and Switching Integrated Platforms). Her research interests include design of open and flexible architectures and platforms, charging, billing and accounting issues for GPRS and UMTS systems and general IP networks.

David Lund

David Lund has worked as Senior R&D engineer with HW Communications Ltd for 5 years where his major responsibility guides the technical direction of their reconfigurable digital signal processing (RDSP) group. With a substantial track record in digital signal processing for communication applications, the group currently provides a major technical contribution to the 'Configurable Radio with Advanced Software Technologies' (CAST) project and the 'Two Dimensional Optical Storage' (TWODOS) project. CAST and TWODOS are both European Commission funded projects. CAST investigates the intelligent reconfiguration of SDR based mobile networks. TWODOS provides advanced optical storage research using SDR physical layer techniques for algorithm evaluation. David's recent PhD thesis entitled 'Reconfigurable Channel Coding for Future Communication Systems' tackles the problems associated with the practical implementation of information theoretical aspects of SDR.

Jijun Luo

Jijun Luo received his MEng degree form Shandong University, China, in 1999 and an MSc degree from Technique University Munich, Germany, in 2000. He joined Siemens in 2000 heading for his PhD affiliated to RWTH Aachen University, Germany. Until now he has published numerous technical papers mainly in the fields of information theory, mobile communication systems, SDR and radio resource management.

Mehul Mehta

Mehul Mehta graduated with a BEng (honours) degree in Electronic Engineering in 1994 and a PhD in 1998, both from the University of Southampton. His doctorate research focused on power control, air-interface and multiple access design considerations for non-geostationary satellite communication systems. In 1997 he joined Roke Manor Research Ltd (a Siemens Company) as a senior member of the technical staff. From 1997 to 2001 he worked on several research and development projects, including MAC layer design, wireless local loop link air-interface, 3G propagation modelling and planning tool development, UMTS link-level demonstrator and SDR. Since December 2001 he has been with Synad Technologies Ltd as a Principal DSP Engineer, where he is responsible for WLAN physical layer algorithm design and implementation, together with overall system performance analysis and modelling. Mehul's technical expertise includes algorithm design, DSP software development and system level analytical studies, all within the scope of wireless communications. He has authored several conference and journal papers. His current research interests include OFDM, CDMA and space–time signal processing. Mehul is a member of the IEEE and the IEE.

Stefano Micocci

Stefano Micocci was born in Pordenone in 1968. In 1993 he obtained his degree in Telecommunication Engineering from the Italian State University (Bologna). In 1995 he joined Italtel Central R&D. He has been active on multimedia and internet platforms. He participated in the MEDEA A109-2GMS project, where he has been active in the definition and integration of video pumping processes on real-time environments in a distributed architecture video server. In 1999 he joined Siemens Mobile Communication R&D and was leader of the integration activity in the IST-1999-10684 STARLITE project which focused on interworking between intelligent networks and the Internet domain. He is now taking part in the IST-2001-34091 SCOUT project where he is active in the definition of a novel network architecture to support terminal reconfiguration. He is co-author of several technical papers.

Klaus Moessner

Klaus Moessner works as a Senior Research Fellow in the Mobile Communications Research Group at the Centre for Communication Systems Research (University of Surrey) and leads a pan-university research team within Mobile VCE's software-based systems work area. Klaus is actively involved in the Reconfigurability Interest Group of the Wireless World Research Forum (WWRF) and is part of the editorial group issuing a 'living' white paper on research efforts and trends for reconfigurable systems. Klaus received his Dipl-Ing (FH), MSc and PhD degrees in Electrical Engineering from University of Applied Science in Offenburg (Germany), Brunel University and University of Surrey (both UK) in 1996, 1997 and 2001, respectively.

Eiman Mohyeldin

Eiman Mohyeldin received a MSc degree from the Technical University, Munich, Germany, in 2000. In 2001 she joined the Mobile Network Division at Siemens and was developing software for layer 1 of the EDGE signal processing unit. Later she joined the System Engineering Division where she was responsible for SDR design of network management and driving concepts for reconfigurable systems. She has published several technical papers mainly in the field of mobile communication systems, SDR and radio resource management.

Nicolas Motte

Nicolas Motte became in 2000 Engineer of Ecole Centrale de Lille (France) and became Diplome-Ingenieur in telecommunications of the Munich University of Technology (Germany). He joined Motorola in April 2001. He currently holds the position of Research Engineer inside the European Communications Research Lab (ECRL) of Motorola Labs in Paris. He is working on internal projects in the domain of SDR, more specifically on the system aspects of SDR. He is focusing on the design of Advanced Radio Resource Management rules and on the development of Flexible and Dynamic Spectrum Allocation schemes. He is also involved in the European IST-Project SCOUT (follow-up of TRUST) and holds the position of work package leader on Joint Radio Resource Management schemes. He is an active contributor to WWRF.

Raquel Navarro-Prieto

Raquel Navarro-Prieto is a human factors researcher at Motorola Labs in Basingstoke, England. Her work includes multiple aspects of the user-centred research process, from in-field observational studies to lab-based prototype development and evaluation. Raquel has been involved in the working and coordinating work at several European projects. She has done research in diverse areas addressing the impact of cognitive processes on the use of information from multimedia interfaces. Before joining Motorola, she investigated the cognitive processes involved in searching for information on the Web, and the role of external representations in these processes at Sussex University. She completed her PhD in Cognitive Psychology, specialised in Human Computer Interaction, at the University of Granada in the Department of Experimental Psychology. Prior to her PhD she earned a BA in Psychology at University of Granada.

Nikolas Olaziregi

Nikolas Olaziregi received his BEng degree in Industrial Electronics Engineering from Mondragon Unibertsitatea, Basque Country, Spain, in 1998, and his MSc from the Department of Electronic Systems and Centre for Microelectronic Systems at the University of Westminster, London, UK, in 1999. He entered the Centre for Telecommunications Research, King's College London, in late 1999 where he is currently pursuing his PhD degree under the supervision of Professor A. H. Aghvami. Nikolas worked on the IST programme TRUST project for two years (2000–2001), joining the staff of CTR as a Research Associate in early 2001. He is currently involved in the TRUST follow up project SCOUT, which is also part of the IST programme.

Spyros Panagiotakis

Spyros Panagiotakis was born in 1973 and received his BSc in Physics and his MSc in Electronic Automation from the Department of Physics of the University of Athens, Greece, in 1997 and 1999, respectively. In 1998 he received a four-year fellowship for postgraduate studies from National Center for Scientific Research 'Demokritos'. Since 2000 he has been a PhD candidate at the Department of Informatics of the University of Athens and member of the Communication Networks Laboratory at the aforementioned Department, working in the area of 3G mobile systems and data networks. Currently he works as a researcher in the IST project 'MOBIVAS'. His research interests include mobile systems, software reconfigurable radio systems and networks, adaptability, service management and provision, location-awareness and design of open architectures and platforms.

Christian Salzmann

Christian Salzmann is a post-doctoral researcher in the Department of Computer Science at Munich University of Technology. In the group of Manfred Broy he leads several projects in the area of service-based systems and SDR. His main research interest is in software development for *ad hoc* systems and model-based design. Christian graduated in computer science at the University of Karlsruhe and at the University of Massachusetts at Dartmouth. He obtained his PhD in 2002 from Munich University of Technology.

Egon Schulz

Egon Schulz was born on April 6, 1955. He received his diploma degree in physics from the University of Siegen, Germany, in 1982 and his PhD from the Department of Electrical Engineering, Technical University of Darmstadt, Germany, in 1988. From 1982 to 1988, he was employed as a Research Assistant working on the design and analysis of channel coding for communication systems at the Technical University of Darmstadt. In 1988 he joined the Mobile Network Division at Siemens AG, Munich, and investigated and developed radio link control strategies for GSM and other standards. This has given him a wide experience of protocols in mobile radio systems. In addition to system modelling, he was a member of the ETSI standardisation group for the GSM Half Rate Speech Codec. From 1992 to 1993 he was a Professor of Electrical and Telecommunication Engineering at Fachhochschule of Darmstadt. In 1993 he returned to Siemens as Director for System Engineering for radio in the local and for wireless private branch exchange systems based on DECT and WB-CDMA. Since 1998 he has been serving as Director for radio access network simulation for third generation mobile systems.

Georgios Vardoulias

Georgios Vardoulias obtained his diploma in electronic and computer engineering from the National Technical University of Athens, Greece, in 1997 and his PhD degree from the University of Edinburgh, Scotland, in 2000. In 2000 he joined Motorola Ltd, working on the TRUST software radio IST research project and on system modelling for 3G wireless platform architecture. His research interests include the network aspects of SDR technology, dynamic spectrum allocation techniques, and CDMA synchronisation issues.

Acknowledgement

Work presented in this chapter has been a part of Core 2 Research Programme of the Virtual Centre of Excellence in Mobile & Personal Communications, Mobile VCE, *www.mobilevce.com*, whose funding support is gratefully acknowledged.

More detailed technical reports on this research are available to Industrial Members of Mobile VCE.

Introduction

For a Software Defined Radio (SDR) system to be useful as an adaptable future-proof solution, and to cover both existing and emerging standards, it is required to have elements of reconfigurability, intelligence and software programmable hardware. In addition, the emerging user requirements on reconfigurable mobile systems and networks are paving the way for the introduction of reconfigurability in future mobile systems.

This book deals with reconfigurability aspects in various layers on the network and terminal, as well as the introduction of reconfigurability management middleware and protocol adaptability mechanisms, which have become key factors for the support of flexible and adaptable service provision to mobile users.

Chapter 1 describes the reconfigurable systems in a heterogeneous environment. The traditional SDR concept introduces flexible terminal reconfiguration by replacing radios completely implemented in hardware by ones that are configurable or even programmable in software to a large extent. These concepts include reconfiguration of the antenna, the radio transceiver and the baseband. Recently, the concept of terminal reconfiguration has been extended and now includes reconfiguration of applications and services, as well as network-based reconfiguration support, provided by a dedicated network infrastructure. The reason for this development is that applications and services are likely to be affected by changing transmission quality and changing Quality of Service (QoS) resulting from vertical handover from one radio mode to another and, therefore, service aspects have to be taken into account in handover decision-making. A network infrastructure that provides support for terminal reconfiguration (e.g. for mode detection, mode monitoring, handover decision-making, software download) can increase the performance of vertical handovers or other bearer service adaptations and, at the same time, release the terminal from tasks that are computationally demanding and therefore may negatively affect the performance of active services.

Chapter 2 deals with the user requirements for SDR terminals. This chapter describes user studies to identify the initial user requirements for reconfigurable systems and networks. User requirements are of high importance in terms of identification of the functionalities to be supported by future reconfigurable systems and networks. The potential benefits of SDR can only be realised if the tasks and expectations of end-users are taken into considerations throughout the development of the technology.

Chapter 3 explains the need for network reconfigurability management. Mobile communications are evolving from a world of monolithic and inflexible systems and service models to a highly dynamic environment, contributed to by a wide variety of market players. Here, a plethora of ubiquitous services will provide valuable help to end-users with every aspect of their everyday activities. The realisation of this vision requires denouncing the proverbial rigidity of telecommunication networks with the introduction of programmability and reconfigurability at various levels of network functionality. Nevertheless, several challenges need to be addressed before the successful application of these concepts in real-life environments. Intelligent, mediating service management frameworks can be an important factor for successfully facing those challenges. Moreover, standardisation of open, third-party accessible reconfigurability interfaces is another crucial enabler in the demanding but hopefully rewarding series of efforts in academia and industry towards reconfigurable networks.

Chapter 4 deals with adaptive protocols. Despite the strong trend to support an evolutionary dimension of mobile communication architectures and to reuse the existing infrastructure investments as much as possible, a number of wireless access technologies have evolved over the last decade. Targeted at different application domains and aimed at satisfying different QoS levels, wireless system architectures tend to exhibit a varying number of similarities and dissimilarities, both in structure and function. However, reconfigurability as an emerging enabler for system integration and flexible service provision support brings about the impending support for advanced software development practices in mobile communication system software. Numerous research efforts have been undertaken or are currently under way in the fields of adaptive, parameterisable, or extendable protocols. The adaptive protocols are characterised by their ability to be configured and initialised on session initiation based on specific communication conditions. They can be reconfigured and re-adapted properly, even at run-time, following the changes in their communication environment. This chapter introduces a survey on protocol adaptability, and presents the origins, drivers and general requirements behind protocol adaptability and protocols adaptation management. Furthermore, special attention is given to the applicability of adaptable protocol infrastructures in future communication systems.

Chapter 5 describes network architectures and functions. Analyses of system aspects and network functions supporting terminal reconfiguration are very important for SDR implementation. Other important issues are related to mobility management in order to achieve seamless service provision, mode switching, radio resource management and software download over the air. These issues can be addressed within a coherent framework encompassing distributed intelligence for the management and triggering of the required functionalities. For this purpose, a generic software radio system solution is proposed, combining reconfigurability with intelligence. The reconfigurable elements are not limited to the physical layer only, and reconfigurability management is distributed in all layers. The Intelligent Reconfiguration Controller interfaces with and collects data from three different environments: the user environment, the radio environment and the network environment. The reconfiguration of the radio network is then based upon the results of the processing of these data sets by the intelligent support system. Within this universal consett, reconfigurability entails the pervasive use of 'software re-definition', empowering upgrades and patching of network elements, and of all services and applications running on it.

Chapter 6 describes a self-learning and adaptive system called Complex Organic Distributed Architecture (CODA). CODA represents a new generation of decision-making systems that includes a means of monitoring and controlling objectives to allow the enterprise to evolve dynamically with a certain degree of autonomy. It applies a theoretical foundation for modelling evolutionary enterprises provided by organic principles, defined by the Viable System Model. CODA refines and adapts Beer's Viable Systems Model with reference to distributed decision supporting systems. Although originally applied to help define management principles, the model is further adapted to the control of data flow in a complex decision support system. The result is a flexible, organic and adaptable business intelligence architecture. CODA intelligent structures are based on business intelligence concepts. Business intelligence systems are complex distributed systems that consolidate data from different data sources, using different formats and performing complex analytical processing on the data. Such complex systems require solid reference architectures in order to simplify their development and management and to provide the best decision-making support.

Chapter 7 studies open Application Programming Interfaces (APIs) for flexible service provision and reconfiguration management. In order to materialise the vision of fully reconfigurable systems and networks that will be able to support flexible service provision, an important enabler is the introduction of open APIs that will allow reconfigurability management entities to trigger appropriate reconfiguration actions towards various applications and layers on the underlying network and terminal. The introduced APIs will enhance and complement current standardisation efforts with functionality related to reconfiguration triggering and management based on various policies identified by respective reconfigurability and service provision management entities.

Chapter 8 presents a framework for charging and billing for reconfigurable service. In dynamic, reconfigurable mobile environments, the requirements related to the provision of a number of services by independent application and service providers to users through a variety of network providers are now becoming feasible. Among the important issues that need to be resolved in order to facilitate the establishment of these advanced models is the support of innovative, flexible and efficient charging, accounting and billing mechanisms. An important aspect is the introduction of a generic framework that will consider advanced, reconfigurable charging accounting and billing mechanisms as a discrete sophisticated service that copes with the ascending requirements for flexible service provision in reconfigurable mobile environments.

Chapter 9 describes communication profiles for reconfigurable systems. In this chapter the schematic architecture for communications profiles is established. All the entities that are part of the architecture and all the interaction and information exchanged between different entities are reflected. Communication profiles are stored in a software module, called a registry, that takes care of the management of the profiles, such as accessing and efficient updating. An Extensible Markup Language (XML) structure is proposed for the profiles. To efficiently reduce the signalling overhead, profile data are clustered in classmarks by meaningful criteria. By efficient coding of temporary clusters of profile data, dynamic classmarks are introduced. Dynamic classmarks capture all the features of the information entities in an abstract and succinct manner. Strategies for efficient coding and compression of the proposed profile structure are discussed that will have an impact on both the amount of resources needed for storage, the processing power to access, and the bandwidth necessary to transmit, profile information within a distributed environment.

Chapter 10 discusses radio resource management in heterogeneous networks. This chapter introduces the general concepts of radio resource management in conventional radio networks based on different access protocols. Problems, modelling and analysis techniques are explained. As a part of reconfigurable technologies, joint radio resource management (RRM) techniques are pointed out with capacity gain analysis. A general functional architecture based on heterogeneous networks and the importance of interworking between different radio resource manage (RRM) layers, between multimedia service types and between coupled sub-networks are emphasised.

Chapter 11 is a study of spectrum-sharing methods and joint RRM. The purpose of this chapter is to present a vision for the evolution of spectrum and RRM in a SDR consett, focusing on different spectrum-sharing strategies that can be envisaged with the introduction of reconfigurable equipments and the potential gain associated with a more flexible management of the spectrum. Attention is first given to current spectrum allocation and regulation. Flexible and dynamic spectrum allocations are then studied in detail, from the perspective of research in some European projects to the complete formalisation of flexible spectrum management. Following this, the chapter focuses on one possible application of RRM in heterogeneous networks. Finally, specific emphasis is placed on future research on the evolution of RRM in a hybrid *ad hoc* environment.

Chapter 12 describes the mode identification and monitoring of air interfaces. This chapter concentrates on the following areas within SDR:

- characterisation of the energy distribution in the channel and in adjacent channels, and
- recognition of the mode of the incoming transmission.

The chapter describes the problems and requirements either in terms of the Initial Mode Identification (IMI), the search for all possible modes following the power of the terminal; or in terms of the Alternative Mode Monitoring (AMM), monitoring other modes when the User Equipment (UE) is already using a certain mode. It presents the relevant Global System for Mobile Communication (GSM) and Universal Mobile Telecommunication System (UMTS) procedures as an example of present-day operations, and introduces a generic framework for the calculation of the worst-case mean cell selection time in a dynamic spectrum allocation environment.

Chapter 13 introduces the reconfiguration of the network elements. A reconfigurable network element, such as a base station or a mobile terminal, in a mobile communications network has the potential to revolutionise the field of wireless communication by providing customised and adaptable services to end-users. This chapter provides essential technical information for the reconfiguration of such network elements in SDR systems and Reconfigurable Radio Networks (RRNs). The comprehensive overview in this chapter describes the reconfiguration technology used in base stations and mobile terminals, the role of local intelligence, the abstract modelling of the hardware, and optimisation algorithms that can be used for the purpose of resource management. Classification of network elements including processing and connection elements is described. For heterogeneous systems, some standard interconnect architectures are described including: global interconnect networks, hierarchical interconnect networks, and crossbar interconnect architecture.

Chapter 14 deals with management, control and data interfaces. SDR requires the definition and design of communication equipment with sufficient flexibility to adopt virtually any existing radio standard as well as any non-standardised radio–air interface. This prerequisite generates high complexity security implications, system and reconfigurability management issues, and software compatibility problems. The reconfiguration of terminals requires a somewhat complex logistic to support that terminal, and network information can be collected and provided to the entity that makes the decision for a reconfiguration. In addition, the software necessary to perform a reconfiguration has to be obtained.

Furthermore, it has to be ensured that regardless of where the software originated, it is downloaded securely and properly installed. This must be done in exactly that partition of the software definable radio hardware platform for which this particular software module has been designed and intended. This chapter deals with key issues regarding management and control in reconfigurable networks.

Chapter 15 discusses the reconfiguration principles for adaptive baseband. In the consett of SDR, the objective for baseband reconfiguration is to redefine the functionality, behaviour and performance of the baseband transceiver chains. This means that the digital signal processing algorithms make up these chains should be redefined so that their post-reconfiguration functionality is different. Reconfigurable networks and systems require many new support mechanisms to allow dynamic air interface operation to be a reality. Amongst Information Society Technologies (IST) Framework V projects sponsored by the European Commission, the Configurable Radio with Advanced Software Technologies (CAST) project and the Transparently Reconfigurable Ubiquitous Terminal (TRUST) project consider many such mechanisms.

In this chapter three software primitives are described which are required in order to define a baseband chain:

- Baseband Processing Cell (BPC)
- Process Leaf class
- ConfigMap

In addition, a framework is presented for the control of configurable processing resources in order to facilitate these three primitives. This is achieved by careful abstraction of hardware capability using Object Oriented methods. Finally, an example is detailed to help demonstrate how the Application Programming Interface (API) is used to configure a resource in a baseband subsystem.

Markus Dillinger
Kambiz Madani
Nancy Alonistioti

Part I

Reconfigurability in Heterogeneous Networks

1

Reconfigurable Systems in a Heterogeneous Environment

Markus Dillinger
Siemens AG

Soodesh Buljore
Motorola SA

1.1 Reconfigurable Systems in Future Networks

As we look beyond the Third Generation of Mobile Communications, we will experience the integration and intercommunication of existing and future networks together with radio access technologies. The convergence towards an Internal Protocol (IP)-based core network and ubiquitous, seamless access between 2G, 3G, broadband and broadcast wireless access schemes, augmented by self-organising network schemes and short-range connectivity between intelligent communicating appliances, enforces common terminal and network entity platforms (Figure 1.1).

In this 'composite radio environment' where several highly standardised legacy radio transport schemes exist, the medium-term goal would be to develop reconfigurable network and terminal techniques to enable interworking and so deliver diverse and exciting applications using the most appropriate radio access scheme(s). Appropriate in this sense refers to the dynamic choice of access scheme(s) to achieve seamless, uninterrupted delivery to the user, customised to the user needs in terms of content, quality of service (QoS) and cost. In such an environment, vertical handover may take place between different access systems (the cellular layer down to the personal network layer, e.g. Bluetooth), combined with real-time service and resource negotiations to seamlessly achieve the desired QoS. The interworking, mobility management and roaming would be handled via the medium access systems and the IP-based core network.

Thus, one of the key elements in the exploitation of the potential benefits, for example ubiquitous connectivity with improved QoS from a heterogeneous radio network, is the

Software Defined Radio: Architectures, Systems and Functions. Edited by M. Dillinger, K. Madani and N. Alonistioti
 ISBN: 0-470-85164-3

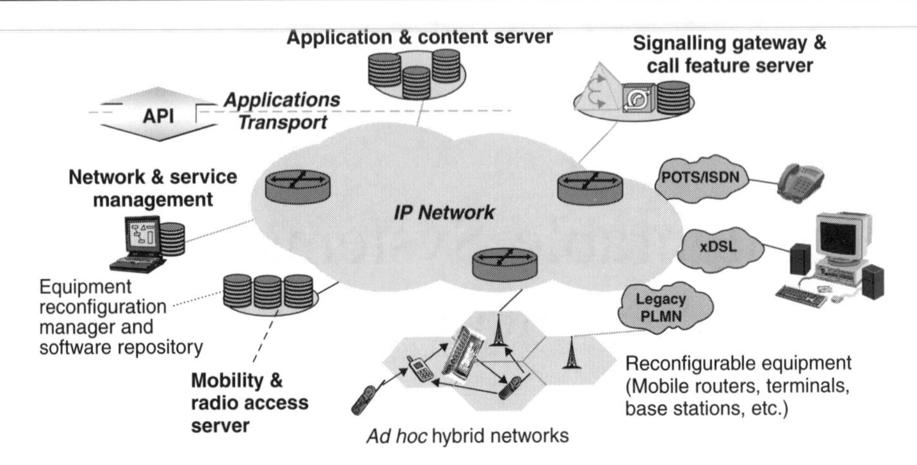

Figure 1.1 IP-based networks supporting reconfigurable terminals

availability of Software Defined Radio (SDR) terminals with reconfigurable protocol stacks (and its equivalent on the network side).

SDR usually is compared to a radio PC, which can host different air interface applications and the major focus is on the access system. However, it is necessary to broaden the scope and to include all layers for optimising network resources and improving user satisfaction.

However, a challenge for mobile reconfigurable devices originates from the intended flexibility of an open mobile terminal, offering an open software platform. Experiences from the PC platform have shown that the user welcomes the opportunity to install and use third-party software on his system for personalisation issues. This particular feature and the equally reconfigurable protocol software in combination with the traditionally high security demands in communication systems, lead to very complex software systems that need a demanding software security concept. First steps have been undertaken (Java 2 Mobile Edition, MExE) but the increasing complexity and the possibility for complete system reconfiguration are demanding new security and validation concepts. Moreover, currently the proliferation of Java-enabled terminals and the provisioning of multi-mode 2.5 and 3G terminal platforms are also becoming key turning points to more flexible terminals enabled by SDR terminal platforms.

The traditional SDR concept introduces flexible terminal reconfiguration by replacing radios completely implemented in hardware by those that are configurable or even programmable in software to a large extent. These concepts include reconfiguration of the antenna, the radio transceiver and the baseband. Recently, the concept of terminal reconfiguration has been extended, and now includes reconfiguration of applications/services as well as network-based reconfiguration support provided by a dedicated network infrastructure. The reason for this development is that applications/services are likely to be affected by changing transmission quality and changing QoS resulting from vertical handover from one radio mode to another and, therefore, service aspects have to be taken into account in handover decision-making. A network infrastructure providing support for terminal reconfiguration (e.g. for mode detection, mode monitoring, handover decision-making, software download) can increase the performance of vertical handovers or other bearer service adaptations and, at the same time, release the terminal from tasks that are computationally demanding and therefore may negatively affect the performance of active services.

Systems supporting that extended notion of reconfiguration require a clear software architecture that integrates the many layers and functions and offers a reasonable function split between terminal and network infrastructure. The architecture has to be both open and flexible, allowing the exchange of existing functionality as well as integration of new functionality in order to keep pace with the rapid development of terminals and radio technology. User and operator demands as well as regulatory requirements have to be taken into account in order to achieve an optimal solution.

Hence we will refer to reconfigurable systems as a concept to provide reconfigurable mobile communications systems which aim to provide a common platform for multiple air interfaces, multiple protocols and multiple applications thereby increasing network and terminal capability and versatility by software modifications (downloads). With the proliferation of open Application Programming Interfaces (APIs), software from different vendors can run on proprietary hardware platforms. On such platforms, the air interface protocols and applications are executed under the control of a common software environment.

Reconfigurability therefore affects virtually all communication layers (from the physical layer to the application layer) of the radio interface and impacts both the mobile terminal and the radio access networks.

Figure 1.2 shows an example of an all-IP heterogeneous mobile access communications network architecture supporting reconfigurable terminals. The reconfiguration of the layers of the protocol stack may be supported by Home (in the IP backbone) and Local Reconfiguration Managers (in the radio access networks). Note that several

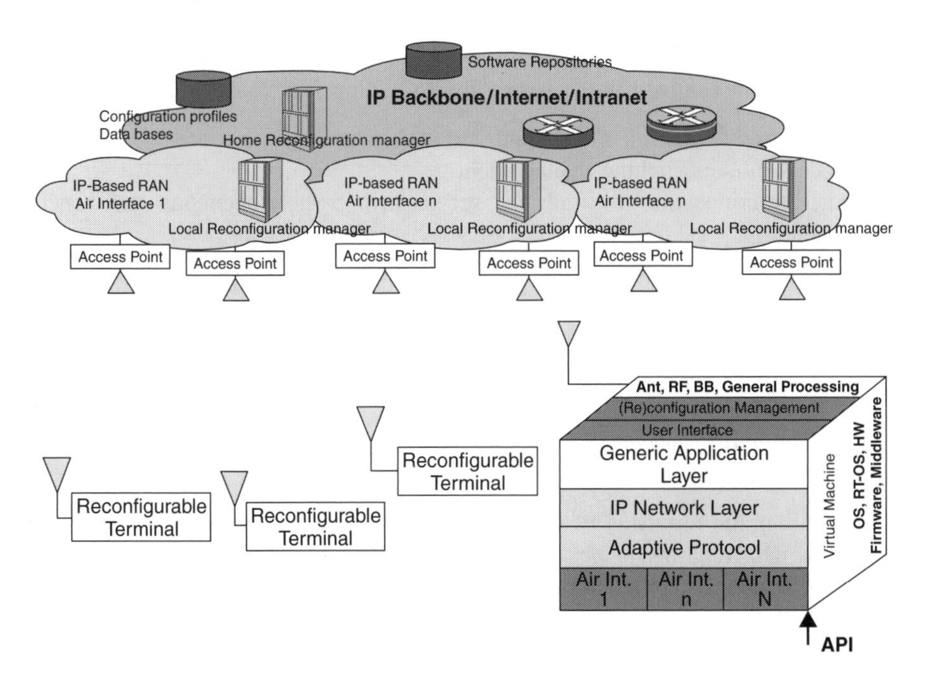

Figure 1.2 Communications layers subject to reconfiguration in the terminal

functions of these reconfiguration managers may be further defined, as is done below in Section 1.1.10. The layered functional architecture in Section 1.1.10 is mainly derived from the TRUST/SCOUT architecture [1].

1.1.1 Key Objectives and General Requirements for Reconfigurable Systems

As key objectives, reconfigurability should provide the means for:

- Adaptation of the radio interface to varying deployment environments/radio interface standards.
- Provision of possibly new applications and services.
- Software updates.
- Enabling full exploitation of flexible heterogeneous radio networks services.

The provision of SDR poses requirements on the mobile communication system, which fall into three distinct groups:

- Radio reconfiguration control.
- Creation and provision of services to reconfigurable terminals over converging networks and different radio access modes.
- User environment (user profile, terminal, access network(s), and location) management.

Moreover SDR should consider and take into account appropriate security functions that allow reliable operation and avoid any potential abuse despite the high flexibility provided by SDR.

1.1.2 Radio Reconfiguration Control

The reconfiguration control has to provide the following functions:

- Radio access mode identification, blind and assisted.
- Radio access mode switching management.
- Simultaneous connection to multiple services in a heterogeneous radio network environment.
- Secure software download including authentication, hierarchical capability exchange and integrity assurance.
- Efficient algorithms to realise flexible, robust radio access schemes.
- Flexible, reconfigurable terminal/base station software and hardware architectures.

1.1.3 Creation and Provision of Services over Converging Networks and Different Radio Access Modes

The fast creation and provision of (scalable) services is a prerequisite for reconfigurable radio systems. This requires the possibility to activate/download temporarily or over longer time periods protocols, applications or even complete services. These applications and services should

- be independent from the underlying network technologies, environment and traffic conditions; and

- adapt to different underlying network technologies, environments and/or traffic conditions.

The adaptation to different underlying network technologies may require the possibilities to download and activate corresponding protocol stacks to network entities and into the terminal.

1.1.4 User Environment Management

User environment management with respect to SDR is the functionality to match a user profile to the present capabilities of the terminal, network and/or radio interface. The user profile may be stored in a database in the network and/or in a user identification module of a terminal. The user environment management tries to provide adaptively to the user always the same or variable service environment (services, applications, QoS, etc.) depending on the user preferences, the time and/or the location, as far as it can be supported by the capabilities of the terminal, network and radio interface. If these capabilities exceed the user profile, the provided services will be limited to the user profile.

1.1.5 Logical Architecture for Reconfigurable Systems

The logical architecture has to support the following functions:

- Management of the terminal, user and service profiles in the network entities and the terminal.
- Efficient download control and reconfiguration management for terminals and network entities.
- Negotiation and adaptation functionalities for services and Radio Access Technologies (RATs) (e.g. vertical handover); see, for example, ref. [2].
- Assurance of standard compliance.

These functions are logical functions, i.e. they can be implemented in different places in the network. Moreover, they can be distributed within the network and between the network and terminal.

1.1.6 Constraining Considerations

Owing to their great flexibility and to the possibility to change nearly all parameters of the radio interface or higher layer parameters (e.g. parameters in the transport layer), reconfigurable systems are potentially subject to standardisation if mixed operations (a mix of different hardware and software vendors) and open API's between modules are required.

Related topics to be considered are, for example:

- Security functions for reliable and trusted software download (e.g. software download limited to manufacturer approved builds available only from a manufacturer's secure server to protect the manufacturer's regulatory liability for system integrity).
- For the terminal: separation of functionalities used for applications and for radio-specific software.
- For the terminal concerning new applications and services: request user confirmation before software update to avoid incompatibility with other already installed software.

1.1.7 Definition of Reconfigurable Systems

Reconfiguration means supporting different system properties and providing multiple views of functionality. In this sense reconfigurable systems are not only an implementation method but have many consequences for system functions to be addressed in standardisation or manifested in industry agreements. We consider the following characteristics for reconfigurable terminals and network entities:

- *Multi-band.* Multi-band systems support more than one specific frequency band, which is used by a harmonised standard such as the Global System for Mobile Communication (GSM), Digital Enhanced Cordless Telecommunications (DECT) and the Universal Mobile Telecommunication System (UMTS). The Radio Frequency (RF) front-end must be capable of fulfilling the necessary RF specifications and, moreover, must be tunable over a wide range of frequency bands. When considering variable duplex distances in Frequency Division Duplex (FDD) systems, additional complexity is added.
- *Multi-homing.* Conceptually, independent radio standards may be processed simultaneously by the platform and, when using it for a user equipment, a flexible partition of the traffic streams are possible thereby providing better QoS and in general a higher connectivity. In particular, multi-standard base stations must be addressed in this context.
- *Multi-function.* The reconfigurable platforms should be application independent and provide a multi-tasking environment for air interface processes, higher layer protocols and user services and applications.
- *Reconfiguration levels and scenarios.* These can be differentiated according to time and usage and the following reconfiguration types are important to outline:
  - Partial Reconfiguration – this refers to the situation when one or more modules are reconfigured without changing the operating standard, i.e. intra-standard reconfiguration. For example, certain modules may be reconfigured in order to improve QoS whilst remaining at the current operating standard. This is also applicable when only parts of the terminal are being reconfigured (e.g. digital baseband) whilst applications or user interfaces are still active.
  - Total Reconfiguration – this refers to the situation when the signal chain is reconfigured from one standard to another, i.e. inter-standard reconfiguration. For example, from GSM to UMTS. This reconfiguration corresponds to exhaustive changes in the functionality, behaviour and interfaces of the constituent modules.
  - Static Reconfiguration – this refers to when, at the time of manufacture or in off-line mode, new capabilities are programmed by, for example, a smart card or by other means.
  - Background Reconfiguration – this refers to the situation when software is downloaded, installed and initialised at a certain event or time stamp. Usually, this is realised by providing shadow and active modules, or even a complete chain of modules, which are then swapped when user services are terminated.
  - Transparent Reconfiguration – this refers to the situation when software is downloaded, installed and initialised at a certain event or time stamp without impacting current user activities and services.

Its is evident, that transparent reconfiguration in conjunction with adaptive radio multi-homing is the most sophisticated design approach for terminals, and provides the highest QoS, connectivity and user satisfaction.

1.1.8 *Reconfigurable Terminals in Heterogeneous Radio Network Environments*

Today, it is clear that multimode (multi-RAT) terminals are becoming more and more a *de facto* feature in mobile radio terminals [3]. Typically, at least two or more radio access technologies [e.g. short range such as blue tooth and wide-range such as General Packet Radio Service (GPRS)] are integrated on the same device. In parallel, the proliferation of Java enabled terminals and the provision of multi-mode 2.5 and 3G terminal platforms are also becoming key turning points to any future flexible terminals enabled by reconfigurable terminal platforms, where all the layers will be subject to reconfiguration.

Such a reconfigurable terminal platform is represented in Figure 1.2 on three planes: the hardware and associated operation system plane, the software plane and the management plane. The gluing technologies for the three planes are well defined APIs and Middleware. An API can then be seen as an abstract interface definition, which is not code, a program or application, but a description of the relationships between related software and/or hardware modules, such as bi-directional flow of data and control information. It describes the relationship between modules, not the implementation of those relationships. The interfaces should be independent of the implementation. Middleware technology, very popular in the Information Technology (IT) domain, can be seen as an enabling layer of software that resides between the application and the networking layer of heterogeneous platforms and protocols. It decouples applications from any dependencies on the underlying layers, which consist of heterogeneous operating systems, hardware platforms and communication protocols.

1.1.9 *Functional Abstraction Layers between the Mobile Radio Heterogeneous Entities and the Reconfigurable Terminal*

Given that all the layers of the protocol stack are subject to reconfiguration, the functional interface between the terminal (which could also be assumed to be distributed, functionally

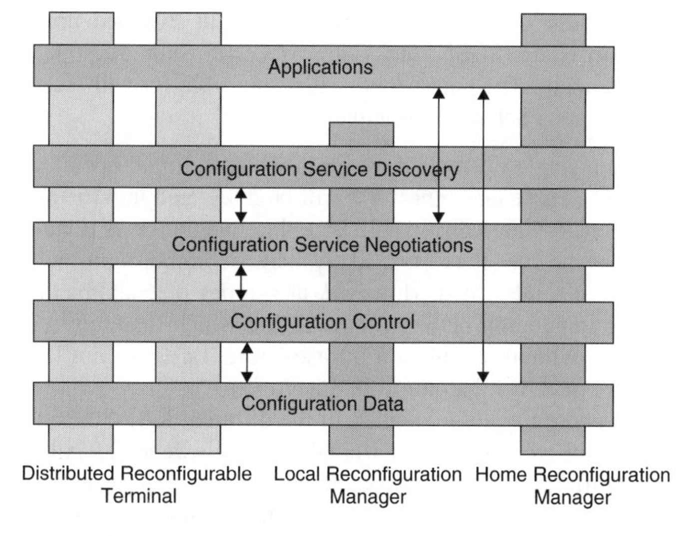

Figure 1.3 Example of functional interfaces between the different open platform entities in a mobile access network

or physically), the radio access networks and the core network entities need to be devised adequately in order to fully benefit and also minimise the burden, in the context of open platforms (terminals and radio access network entities).

Several of these abstraction layers – configuration service discovery, negotiations, configuration control and data – are shown in Figure 1.3. Moreover, one can imagine interactions between the abstraction layers in order to further optimise the configuration and usage of the open platforms. Note also that given the potential burden introduced by the flexibility provided by open platforms, a management policy based on user, terminal, radio access networks profiles and location is a means to minimise the complexity of such reconfigurable radio terminals and networks. These profiles and policies are stored in corresponding databases in the terminal, radio access networks or core network.

1.2 System Functions for Reconfigurability in Mobile Networks

The discussion of a specific reconfiguration support functions for reconfigurable terminals in this chapter is based on the general overall structure of (cellular) mobile networks. It is assumed that the cellular network serving as a reference network in which other RATs are integrated has a clear separation between the Radio Access Network (RAN) and the Core Network (CN), as shown in Figure 1.4. These may be regarded as separate layers with a defined interface. This separation is realised by the current UMTS architecture, as demonstrated in [4], where an idealised architecture for the UMTS network is displayed. It clearly shows the separation between the RAN and the CN:

- The RAN contains radio-technology-specific entities such as the Radio Network Controller (RNC) and the Node Bs, and takes care of tasks such as radio resource management and other access stratum functions.
- The CN hosts entities such as the Mobile Switching Centre (MSC) and the Gateway Mobile Switching Centre (GMSC) for the circuit-switched domain or the Service GPRS Support Node (SGSN) and Gateway GPRS Support Node (GGSN) for the packet-switched domain. The core network is responsible for call management and for establishing connections between participants.

We now focus on a single operator scenario in which a single operator offers different RATs to its customers. These different RATs will be integrated into the operator's cellular network. Several scenarios are conceivable how this integration is done. For the present investigation, we assume that it is most likely that a common core network is used in which several RANs are integrated. However, also types of couplings are possible that require only a loose integration between these RATs, which therefore remain essentially independent networks without a common infrastructure. Last, but not least, the coupling depends on the types of RATs that are to be integrated.

Regardless of the chosen type of coupling of the different RATs, the additional requirements for the support of reconfigurable terminals and the extended flexibility provided by handovers between different RATs, in particular consideration of the additional parameters in decisions concerning handovers, requires a specialised infrastructure that provides the necessary support in the network that is beyond the capabilities of the currently deployed network entities.

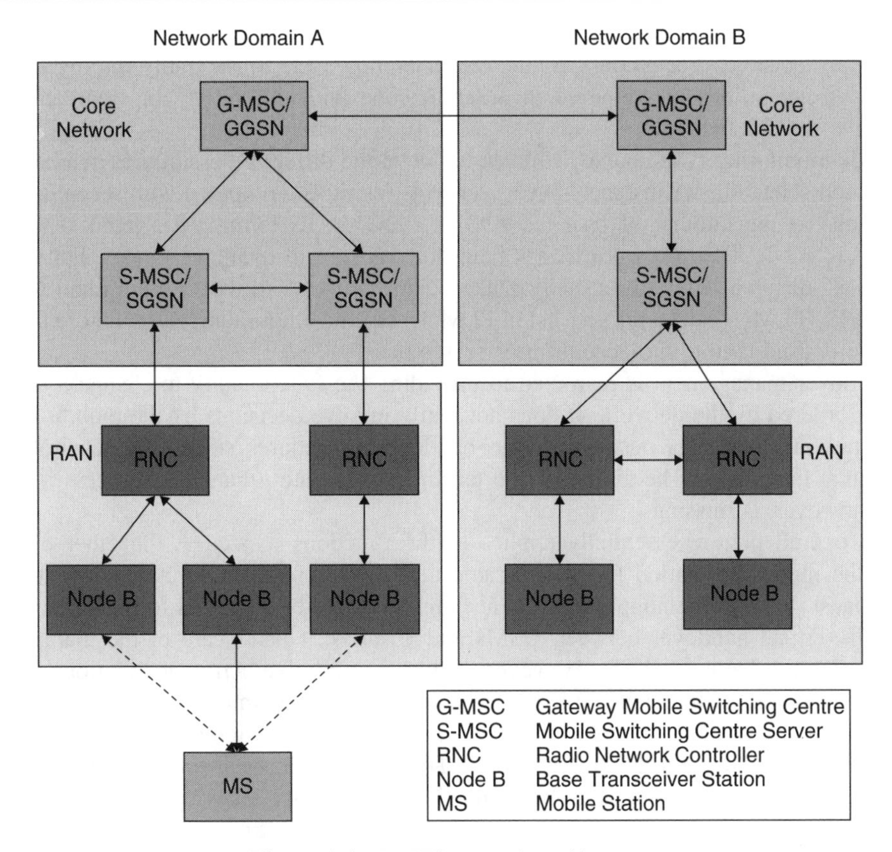

Figure 1.4 UMTS network architecture

For the inclusion of additional functions related to reconfigurable terminals, however, some functions have to be added.

- Handover management is already available at all entities, as shown in [5]. At the G-MSC it is required only if handover to external networks is to be performed (see, for example, in [7]). MSC/SGSN and RNC are regularly involved (depending on the scenario) in horizontal handover management in UMTS.
- Mode filtering and mode monitoring are done by the mobile station and the RNC (see, for example, in Figure 1.16 [6]).
- The handover decision function is located at the RNC.
- The access stratum function which provides radio layer parameters such as, for example, measured link layer parameters to assess the quality of the connection.

Compared with the functionality required for the full support of reconfigurable terminals, these functions are, however, very restricted and only comprise a small part of the functionality required for reconfigurable terminals.

Starting with the mobile station, the functions in the UMTS and the required extensions are discussed. The access stratum functionality is realised to a great extent in UMTS terminals. It provides the interface to all access stratum parameters; in particular it takes care of signal strength measurements and thereby provides the input parameters for the handover decisions. For reconfigurable terminals, however, considerably improved measurement capabilities are required in order to cope with the variety of simultaneously available radio networks.

Mode monitoring is of course available to detect the different possibilities available for initial cell selection, for instance, but its scope is restricted compared with reconfigurable terminals, as the number of possible RATs for 2G or 3G terminals is fixed. This also applies to mode filtering, a prime task being the ranking of available modes. For 2G/3G terminals, different RATs are usually ranked using a list of allowed Public Land Mobile Networks (PLMNs) and a priority list of PLMNs. For reconfigurable radios a more flexible scheme is used taking into account user preferences.

Handover management is restricted to providing the functionality to change a serving cell, if ordered by the network. It does not really involve decisions by the mobile station itself but just performs network driven handover procedures since it concentrates on horizontal handovers. The ability of the terminal to initiate a handover processes on its own, however, is missing.

The overall picture essentially applies to the functions located at the other entities. Only the mode negotiation function located at the RNC requires additional attention. It is the least developed among the different functions already presented for UMTSs. In the case of vertical handover between GSMs and UMTSs, it takes care of the mapping of bearer services between RATs. However, it does not involve active negotiations, as the properties of the different bearers are fixed and known. Of importance, however, may be the availability of resources in the target network, as some multimedia calls may not be handed over due to a lack of resources. The present approach for UMTSs, however, does not explicitly negotiate this beforehand, but implicitly takes care of this problem if a handover request to the target network fails by attempting a handover to another target cell.

1.2.1 Reconfiguration Support Architecture

Reconfiguration management for reconfigurable systems, however, requires in particular extended support from the network architecture. It is suggested that the radio network architecture is supplemented by Reconfiguration Support Functions (RSFs), which provide the additional functionality necessary to take into account additional parameters. For the sake of a clear presentation, the RSFs are treated as separate entities. Deployment within a given network architecture, however, would in general integrate the support functions into existing network entities.

In order to provide the required support from the network architecture for reconfiguration management and vertical handovers in reconfigurable systems, a complementary architecture is considered. It consists of RSFs that provide the additional functionality required for improved handover support. This architecture mirrors the layer structure of the cellular network model architecture shown in the preceding section.

Figure 1.5 shows the network located entities for an architecture supporting reconfiguration. According to the separation of layers, the responsibilities of the RSFs are divided:

- At the terminal, the Terminal Reconfiguration Support Function (TRSF) is responsible for issues concerning the capabilities, resources and configuration of the terminal,

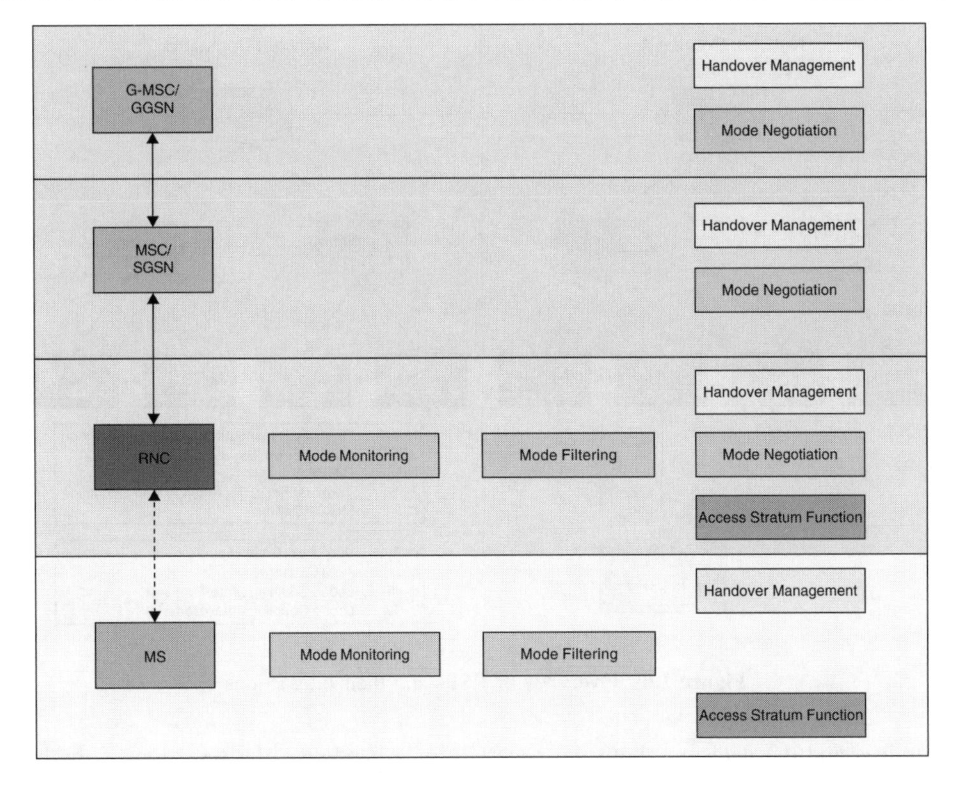

Figure 1.5 Function distribution in network architecture

terminal-based mode monitoring and, in particular, QoS monitoring and the reconfiguration of applications.

- In the RAN the RAN Reconfiguration Support Function (RRSF) is located at the entity responsible for radio resource management. For instance, in a UMTS, the RNC is the deployment site.
- In the core network, the Serving Reconfiguration Support Function (SRSF) would be situated at the Mobile Switching Centre (MSC).
- Communication with other networks (inter-operator scenario) is taken care of by the introduction of a Gateway Reconfiguration Support Function (GRSF).

1.2.2 Functional Distribution

The functions of the different types of RSFs are shown in Figure 1.6. It shows the different RSFs and their associated functionalities. According to their different positions in the negotiation scheme and their association with the RAN or core network, they provide different functionalities.

- TRSF is responsible for all tasks concerning the handover management on the reconfigurable terminal. This includes the handling of air interface modes, also called mode management, the generation and processing of triggers, QoS management for active

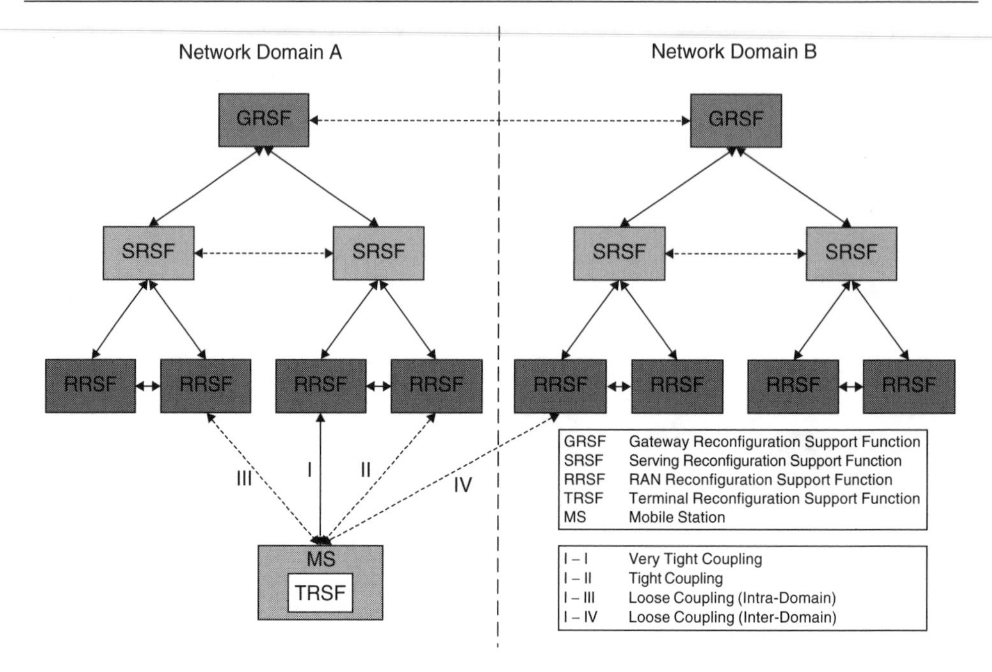

Figure 1.6 Overview of RSFs and their interactions

services and the decision about the necessity of a handover, the selection of the best suited air interface modes, and the handover execution if required.

- RRSF is the only entity located at the RAN; it provides access stratum functions. Moreover, mode filtering and mode monitoring functions are present. In order to speed up decisions, certain parts of the profile management functions may also be present in the RRSF. Finally, handover management itself is an important function of the RRSF.
- SRSF it contains the general handover management and associated mode filtering and mode negotiation functions. Moreover, context management, profile management and policy management are provided by the SRSF.
- GRSF serves as a gateway to other domains; its functions are restricted to handover management and mode negotiation as far as they are required to perform inter-operator negotiations.

As can be seen from Figure 1.7, some of the functions, e.g. handover management or profile management, are distributed between different reconfiguration support functions. How the involvement of the different functions looks is, however, dependent on the coupling scenario. It determines which entities are involved and which functions on the reconfiguration support functions are used.

1.2.3 Network Coupling and Reconfiguration Support

The integration scheme used for the coupling of two networks determines to a great extent the processes and the RSFs needed for vertical handover. To allow for service continuity and predictable handover behaviour, vertical handovers between different networks require

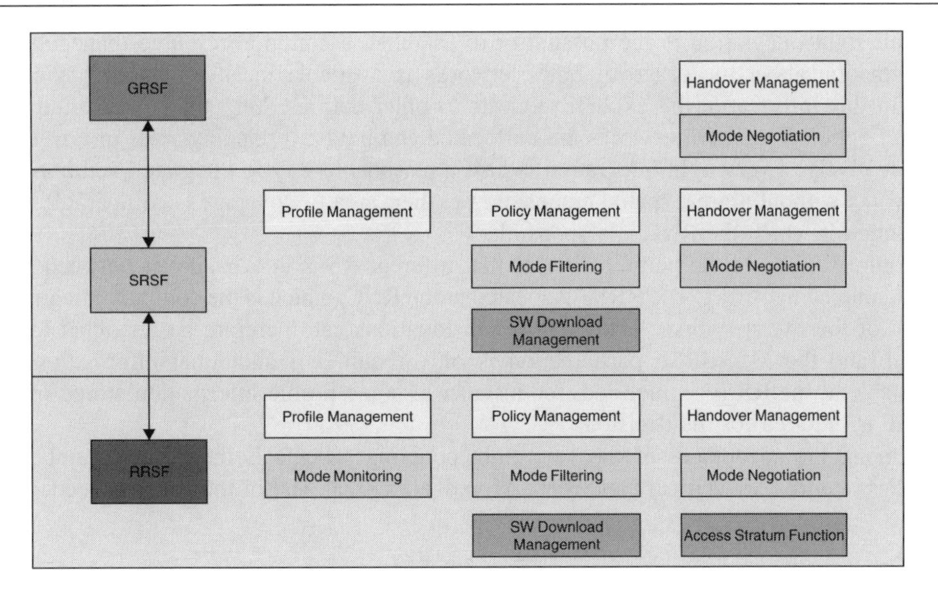

Figure 1.7 Functions of the RSFs

a connection between the participating networks for handover management. According to the different types of networks that are coupled and depending on the type of coupling, there are different integration schemes [8, 9]. In what follows we assume a UMTS-type cellular network as standard into which other possible networks may be integrated and adopt the UMTS terminology for the entities. In general, four different types of integration/coupling can be distinguished:

- Very tight coupling
- Tight coupling
- Loose coupling
- No coupling

The simplest scheme (no-coupling) requires no integration of the two networks at all. It does not even provide a mechanism for the interchange of information between the networks. All actions have to be coordinated by the mobile station which acts as the only relay for the handover between the two networks. This option is possible for all types of networks; however, information about the new network and the possibility to perform a seamless and reliable handover is very restricted, unless a 'make before break' strategy is employed for handover, which usually requires two radio transceivers.

The other schemes, however, deserve further discussion, as vertical handovers between networks integrated according to the different schemes involve the reconfiguration support architecture in rather different ways.

1.2.3.1 Very Tight Coupling

In the very tight coupling case both networks are integrated at the RAN layer. The entities responsible for both radio access technologies are attached to the same RNC. An advantage of this tight integration is the possibility to establish a common resource management. Information about the usage of both networks is available locally, without having to obtain this information by explicit requests to other entities. Very tight integration is a useful option if the two networks are co-located and have overlapping areas of coverage. Alternatively, a very tight integration is also a reasonable way to integrate local hot-spot networks without a great spatial extension, i.e. their area of coverage is comparable with the area for which the RNC is responsible.

Figure 1.8 shows the participation of the different RSFs in a handover between very tight coupled networks. The RRSF is located at the RNC, which is the common integration point of the two networks. The handover negotiations can therefore be restricted to the RRSF and the MS. SRSF participation is only required if additional information not available at the RRSF is required, for instance if some profile information stored in the SRSF is required for the decision.

Among the advantages of the very tight coupling scheme is the limited number of entities required for vertical handovers. Moreover, a great deal of information needed for

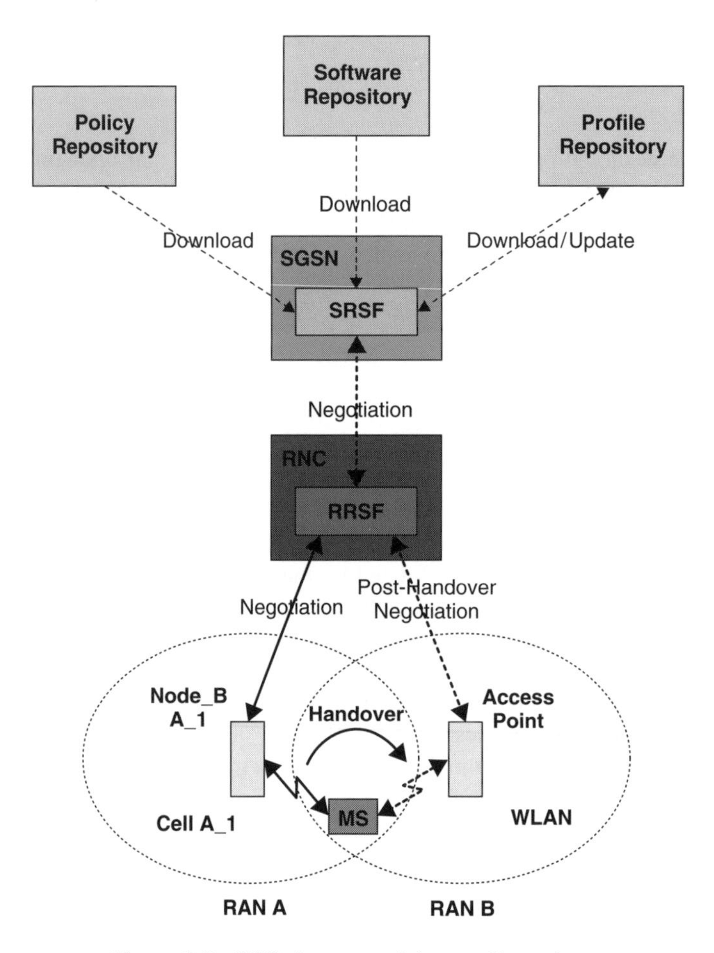

Figure 1.8 RSFs in a very tight coupling scheme

the decision process is at hand. Therefore delays are reduced to a minimum with this coupling variant.

An additional task of the RSFs is to provide an infrastructure for the provision of software downloads required for the configuration. The figure shows the most extended download path leading via the RRSF and the SRSF to a software download server. By local caching on the RSFs this path may be shortened. In particular it is conceivable that the software for a common terminal is stored at the RRSF, thereby again reducing the reconfiguration time.

1.2.3.2 Tight Coupling

The tight coupling scheme integrates two networks at the RAN layer, as is done in the very tight coupling scheme. However, it uses two different RATs working together with a single core network. The coupling point is the MSC [or the SGSN for the Packet Switched (PS) domain], to which both RANs are attached. In contrast to the loose coupling scheme integration between the two RANs is achieved by the existence of a coupling between the RNCs.

This coupling scheme may be applied to the case of co-located networks. A prominent example is the integration of a GSM and a UMTS. In this case both RANs are connected to the same MSC (or SGSN) taking care of the connection management for both networks.

Although this type of integration does not provide the same degree of cooperation as the very tight coupling scenario (e.g. common resource management), the integration is nevertheless sufficient to provide a fast and reliable interworking for vertical handovers.

However, the handover negotiation process, shown in Figure 1.9, is more complex than in the very tight coupling scheme. For serving and target network there are different responsible RRSFs. The Mobile Station (MS) therefore has to negotiate with both of them during the handover decision-making. The only contact point for the MS is the serving RRSF of the current RAN. This RRSF is responsible for contacting the target RRSF and performing the negotiations required between the two entities. Negotiations between the RRSFs are done directly without involving the SRSF making use of the direct connection between the RNCs. The dashed lines between the RRSFs and the SRSF denote negotiations with the SRSF which are only required for retrieving policies, software or profile information.

A more detailed view of the process is shown in Figure 1.10 which shows the parts of the RRSFs that are participating in a handover negotiation and the handover execution. It shows the functions that are needed at the respective locations and the interactions between the functions at different locations. Whereas all functions of the source RRSF are needed, the target RRSF involves only profile management, handover management, mode negotiation and the access stratum function. The other functions are not involved in the negotiation process at the target RRSF, only at the source RRSF.

1.2.3.3 Loose Coupling (Intra-Domain)

Loosely coupled mobile networks in the intra-domain case do not have a connection at the RAN layer. Instead the networks are connected via the core network, i.e. they share

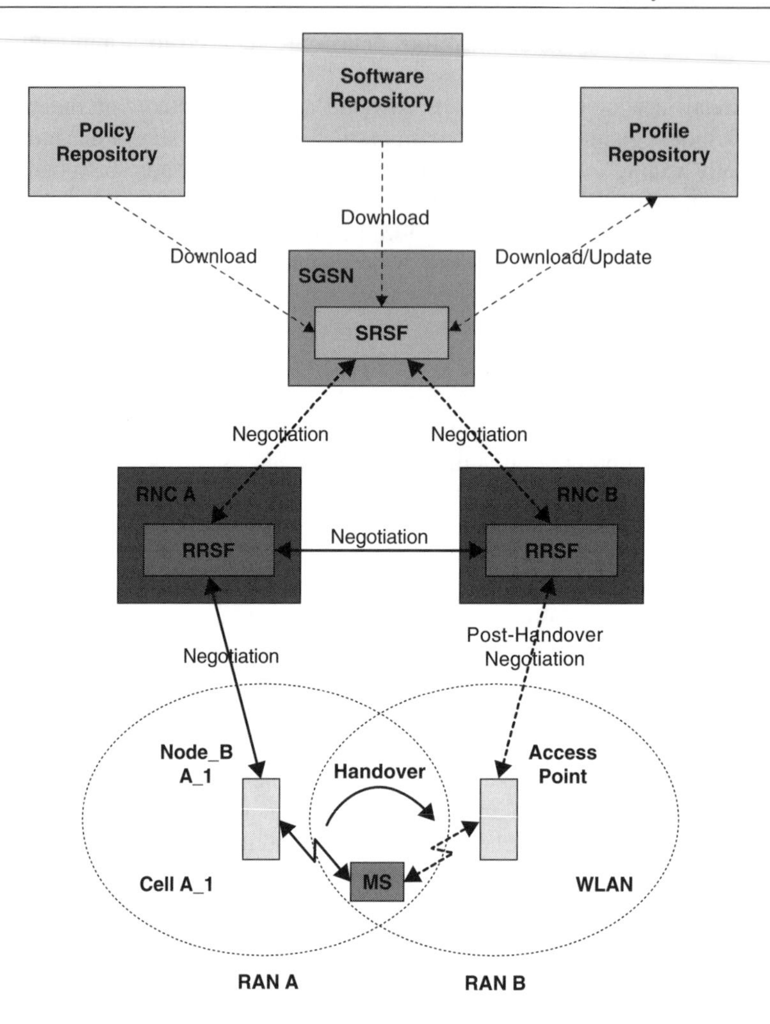

Figure 1.9 RSFs in a tight coupling scheme

the same core network infrastructure. Figure 1.11 and Figure 1.13 show two different scenarios for intra-domain loose coupling.

Figure 1.11 shows the case when only a single SGSN is involved in the handover. This is the standard case assuming that the responsibilities of the SGSN are divided along geographical lines. As no direct connection between the RNCs is available, the negotiation is done via the SGSN.

A more special case is shown in Figure 1.13 involving the SGSNs. This may occur only rarely, as it is required only if the SGSN area also changes with the handover, but it is nevertheless discussed for the sake of completeness.

The detailed required functionality is shown in Figure 1.12 and in Figure 1.14 for the case of two participating SRSFs. As can be seen from Figure 1.14 the second SRSF's responsibility is mainly to act as a relay between the two networks.

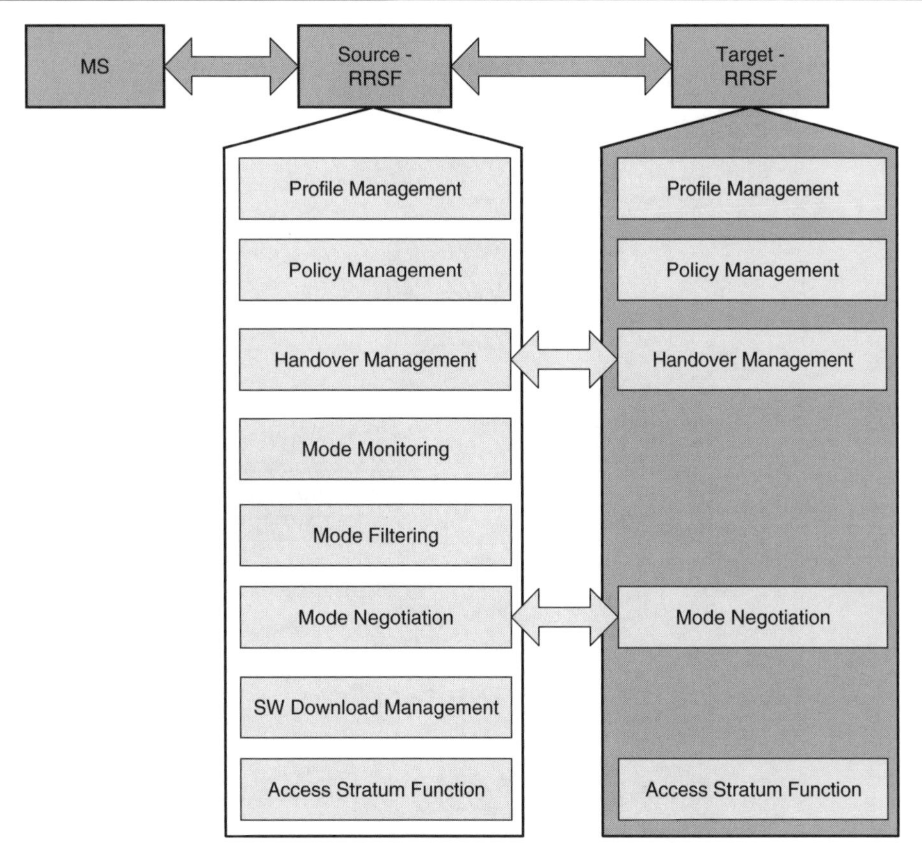

Figure 1.10 RSFs (functional view) in a tight coupling scheme

1.2.3.4 Loose Coupling (Inter-Domain)

Loosely coupled networks offer a common interface for the exchange of information between the networks as required to guarantee service continuity. This may be implemented by a GMSC that takes care of the required interaction. Loosely coupled networks may be either geographically co-located or separated by a geographical boundary. The extent of information available for assessing the necessity and possibility of a handover, however, is restricted. An exchange of information is possible, but requires considerably more time than in the tighter integrated scenarios.

Accordingly, the handover negotiations are considerably more complex compared with the tight and very tight coupling schemes. In this case the complete spectrum of RSFs is involved. The MS starts the negotiation with the RRSF, which performs negotiations with the responsible SRSF, which contacts the GRSF. The GRSF contacts a peer GRSF at the second operator's network, which acts as a negotiation partner for the serving network. The target network's GRSF in turn is responsible for triggering the required reconfiguration using the target network's SRSF and RRSF.

Figure 1.15 shows the participation of the RSFs in this scheme in detail. The main difference compared with the intra-domain scenario (Figure 1.13) is the participation of the

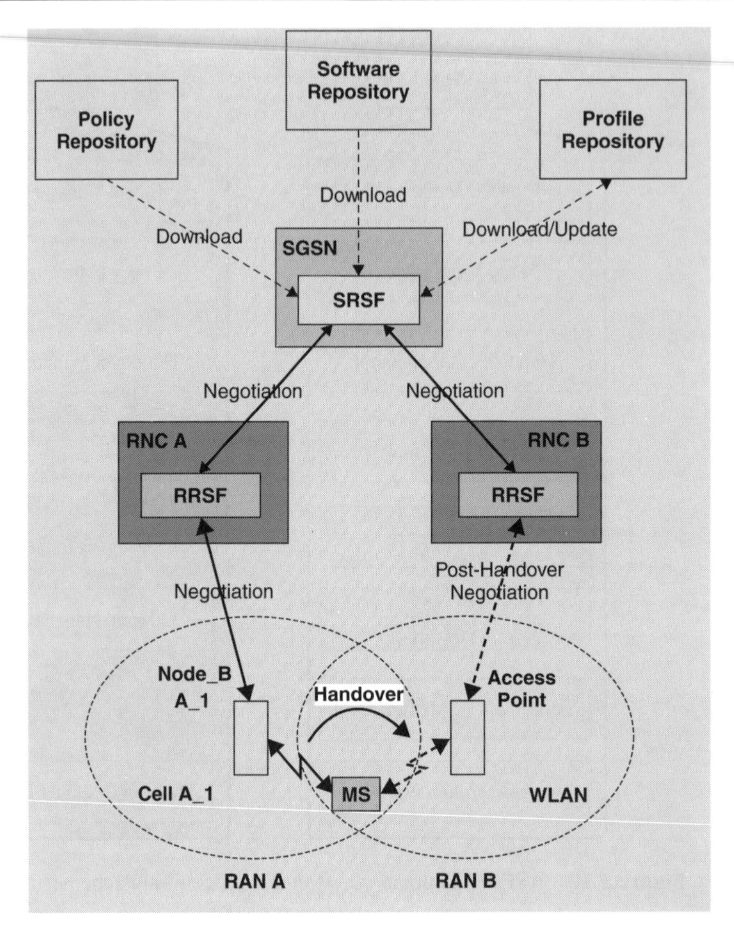

Figure 1.11 Reconfiguration in an intra-domain loose coupling scheme

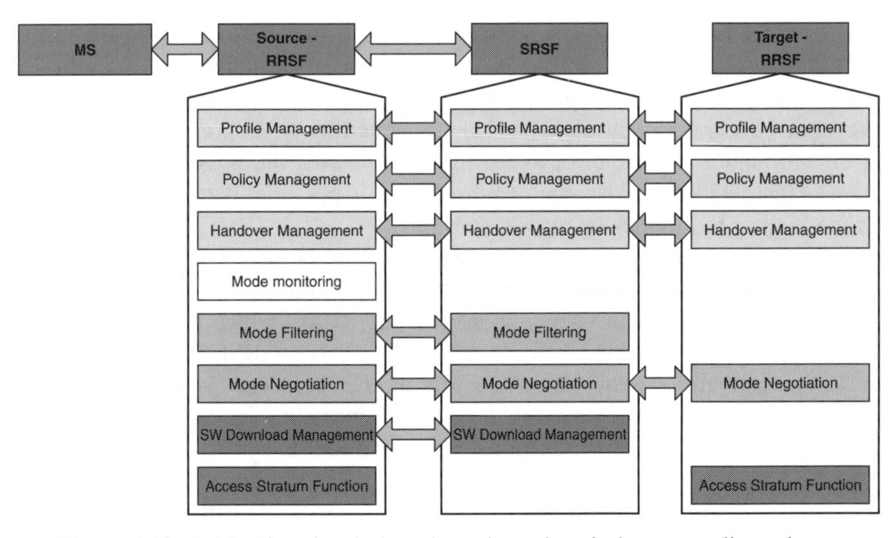

Figure 1.12 RSFs (functional view) in an intra-domain loose coupling scheme

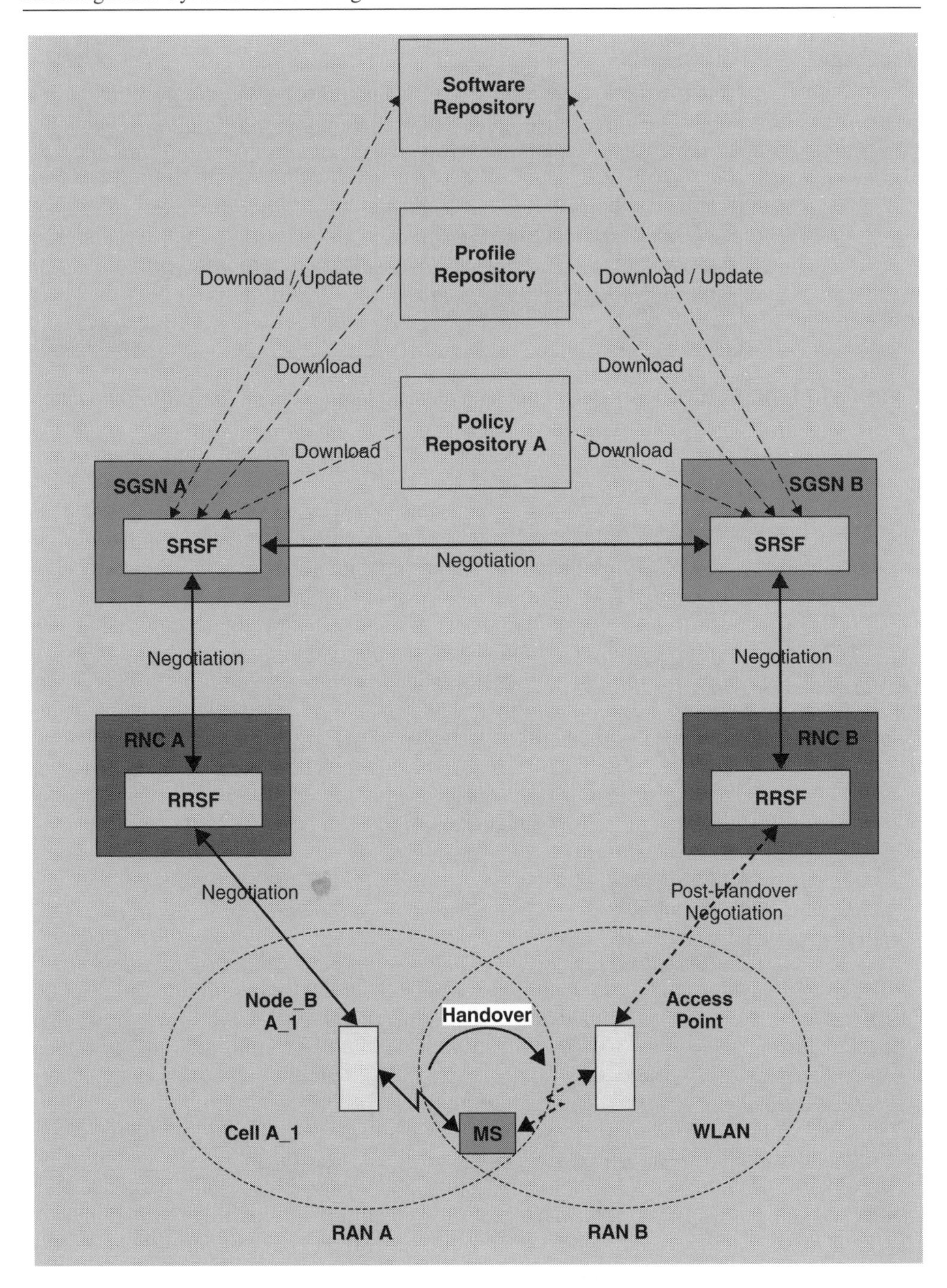

Figure 1.13 Reconfiguration in an intra-domain loose coupling scheme

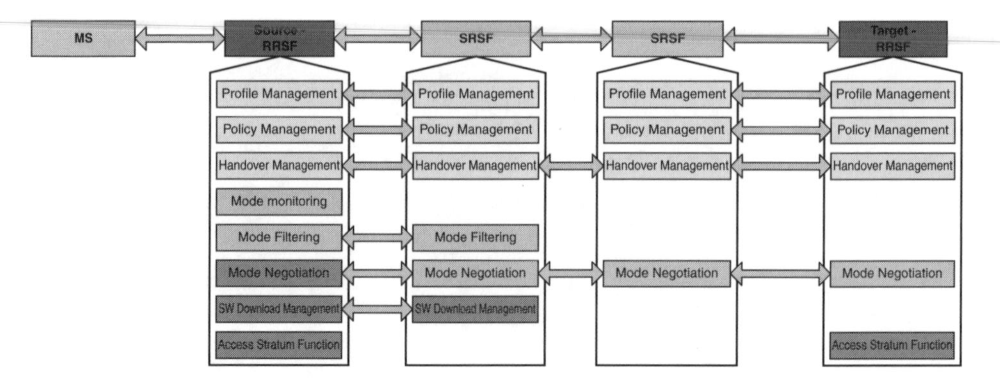

Figure 1.14 RSFs (functional view) in an intra-domain loose coupling (inter-SRSF) scheme

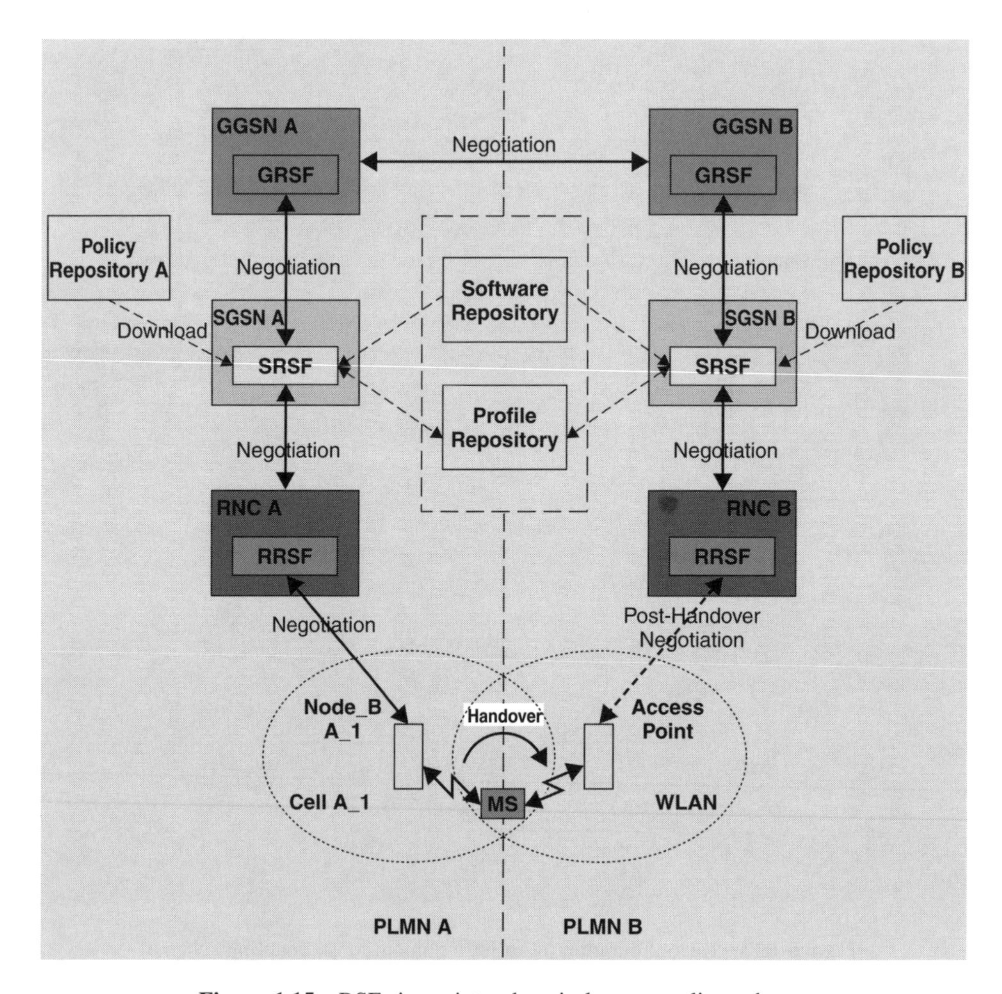

Figure 1.15 RSFs in an inter-domain loose coupling scheme

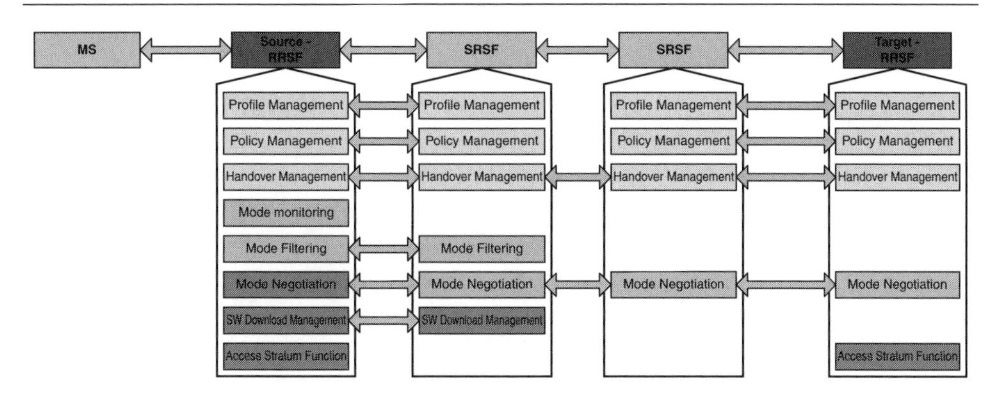

Figure 1.16 RSFs (functional view) in an inter-domain loose coupling scheme

GRSF, which is necessary for dealing with different networks. As shown in Figure 1.16, the additional entities compared with the intra-domain case deal only with mode negotiation, however only by relaying information or by taking decision on a high layer of abstraction, e.g. whether a mode may actually be used by customers of the particular requesting operator, and handover management.

Acknowledgements

The authors would like to acknowledge the contributions of colleagues from the SCOUT project, and in particular from Christoph Niedermeier and Reiner Schmid from Siemens AG.

References

[1] *http://www.ist-scout.org*
[2] 3GPP TS 22.934, 'Feasibility study on 3GPP system to Wireless Local Area Network (WLAN) interworking'.
[3] 3GPP TS 23.934, '3GPP system to Wireless Local Area Network (WLAN) interworking: Functional and architectural definition'.
[4] 3GPP TS 25.501, 'UTRAN overall description'.
[5] 3GPP TS 22.129, 'Handover requirements between GERAN and UTRAN'.
[6] 3GPP TS 25.215, 'Physical layer – measurements (FDD)'.
[7] 3GPP TS 23.009, 'Handover procedures'.
[8] 3GPP TS 25.992, 'Radio resource management strategies'.
[9] 3GPP TS 25.881, 'Improvement of RRM across RNS and RNS/BSS (Proposed Technical Report)'.

Part II

Requirements for Reconfigurable Terminals

2

User Requirements for SDR Terminals

Raquel Navarro-Prieto and Genevieve Conaty
Motorola

2.1 Introduction

This chapter describes our user study to identify initial user requirements for Software Defined Radio (SDR) systems. Our work was completed for the Transparently Reconfigurable UbiquitouS Terminals (TRUST) project, conducted as part of the Information Society Technologies (IST) Programme of the European Union's Fifth Framework 2.4 [10] from 1999 to 2001.

SDR is a developing technology and although its consequences for end-users are not yet known, we expect it to offer them many advantages over current wireless systems, such as allowing them to move seamlessly between different types of networks to optimise their quality of service or minimise the cost of using the device. Cook studied the potential of SDR for the user, and concluded, among other benefits, that SDR will enhance the user's experience with mobile devices by enhancing roaming capabilities without changing the terminal and by allowing over-the-air download of applications software as and when required. However, these potential benefits of SDR can only be realised if the tasks and expectations of the end-users are taken into account throughout the development of the technology. Without the involvement of users, we run the risk of implementing systems that satisfy the technological goals but will not be accepted by users because they do not match the users' expectations, constraints, and desired functionality.

The user-centred design process [4, 7] is a well-known way to bring user expectations and needs into technology development, shifting the emphasis in a project from the development of technology 'for technology's sake' to the development of systems that support particular user needs in an accessible and usable way. For example, in the Wireless World Research Forum's *Book of Visions* (2000) [14, 15], researchers seeking to understand the future of wireless technology observed, 'The principal issue is user satisfaction. A major shift can be observed from a device-driven world to a service- and experience-centred

Software Defined Radio: Architectures, Systems and Functions. Edited by M. Dillinger, K. Madani and N. Alonistioti
 ISBN: 0-470-85164-3

world. It will become more and more important how the users perceive the service and the emotional impact and pleasure that the service creates and maintains.'

In recognition of the importance of understanding user requirements for interactions with SDR technology, the TRUST project sought to apply a user-centred approach to technology development. The user study described in this chapter was part of our effort to understand the information that technology developers required from users in order to develop the most useful and usable systems possible, and to gather this information directly from potential users of SDR.

One challenge of studying user requirements for a developing technology such as SDR lies in the fact that there are few tangible representations of the technology (such as working prototypes) that can be shared with users. In addition, SDR systems do not fit into the expectations that users have built up about mobile systems based on their experience with current systems [9]. Finally, the technology is so new that even the technology developers themselves do not know exactly how it will operate or what ramifications it may have for users. We used a scenario-based approach [2] throughout the TRUST project to help us overcome these difficulties and to portray SDR capabilities for users while the technology is still in an immature state.

We based our research on user scenarios generated at an earlier stage of the TRUST project. We designed a user study to answer the questions about SDR introduced by those scenarios, and used interactive presentations of the scenarios to communicate the concept of SDR to the participants in the study. A total of 23 people participated in our interviews. The focus of these interviews was the eight areas that have provoked the most pressing questions about how users will expect SDR technology to operate: mode switching, air interface download, user profiles, video, group communication, security, battery life, and system resources. In addition, we gathered more general information about the perceived advantages and disadvantages of reconfigurable radio and catalogued recurring concerns that participants mentioned about the concept. The results from this study will be described in this chapter. We consider these data to be very important because, to the best of our knowledge, they constitute the first detailed qualitative data gathered directly from users about SDR. Nevertheless, owing to the limitations of this study, the data should be considered as an initial step in the technology development cycle that must be validated by further research.

This chapter is organised as follows. Section 2.2 describes in detail the methodology of the user study, including the selection of the topics to focus upon, the structure of the research sessions, and the characteristics of the participants. Section 2.3 presents the results of the study, organised by topic, and includes the participants' observations about the SDR concept in general. Finally, Section 2.4 summarises the results and conclusions of the study and presents some directions for future work that will deepen our understanding of how users will interact with future reconfigurable systems.

2.2 Methodology

In this section we briefly describe the methodology used in the user requirements work for the TRUST project. We then describe in more detail the methodological approach as well as the design and materials used in our particular user study.

2.2.1 Overview of the TRUST User Requirements Methodology

Figure 2.1 shows the overall user requirements process for TRUST. This work was conducted in three phases. In the first phase, a user group was defined and preliminary scenarios of use for SDR were generated using information provided by this group through interviews, questionnaires and focus groups. In the second phase, four of these preliminary scenarios were further developed and refined, and the issues for SDR introduced by each user interaction shown in the scenarios were identified. Finally, in the third phase, the issues raised by the scenarios were grouped together as 'research topics' and we conducted a user study to gather feedback, particularly about these research topics, using the scenarios as the locus for discussion. This chapter focuses upon the third phase of this process.

The four scenarios specified in the second phase of the research highlighted particular capabilities of SDR technologies. Each scenario depicted a particular user activity and related interactions with an SDR terminal:

Scenario 1: 'Download'. The protagonist accesses confidential multimedia information from his corporate intranet whilst mobile.

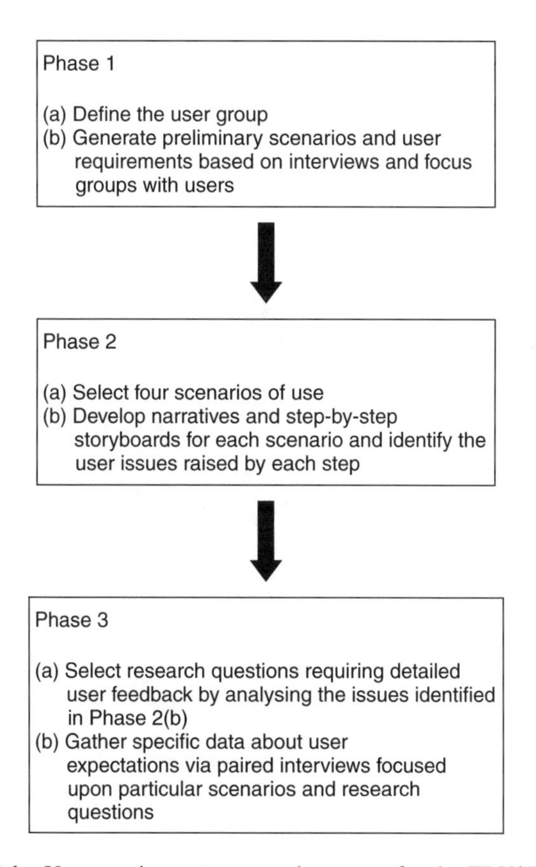

Figure 2.1 User requirements research process for the TRUST project

Scenario 2: 'Always available'. The protagonist sets up his device to allow him to receive an important call whilst driving through an area of patchy coverage.

Scenario 3: 'Watching high-quality video'. The protagonist, a film director, views film footage whilst travelling.

Scenario 4: 'Mobile videoconference'. The protagonist participates in a group multimedia call with three friends, all of whom have a different level of service (i.e. audio-only, video and audio).

For each scenario, our technical partners in the TRUST project helped identify the end-user issues that would be encountered during the technical implementation of that scenario. In addition, interactive storyboards were developed for each scenario using Macromedia Director. A screenshot from the interactive storyboard of Scenario 1, 'Download', is shown in Figure 2.2.

On the basis of the analysis of the user issues for the scenarios, eight particular research topics were selected for investigation in the detailed user requirements phase described in this chapter. The methods used in this phase are described in detail in the next section.

2.2.2 *Overview of Detailed Requirements Gathering Methodology*

In order to plan the user study for gathering detailed user feedback about SDR, it was necessary to understand which research topics appeared most consistently across the scenarios. These topics were identified as the following:

- *Mode Switching.* Mode switching refers to the process by which the device reconfigures to a new network mode. Should the terminal complete this process automatically or

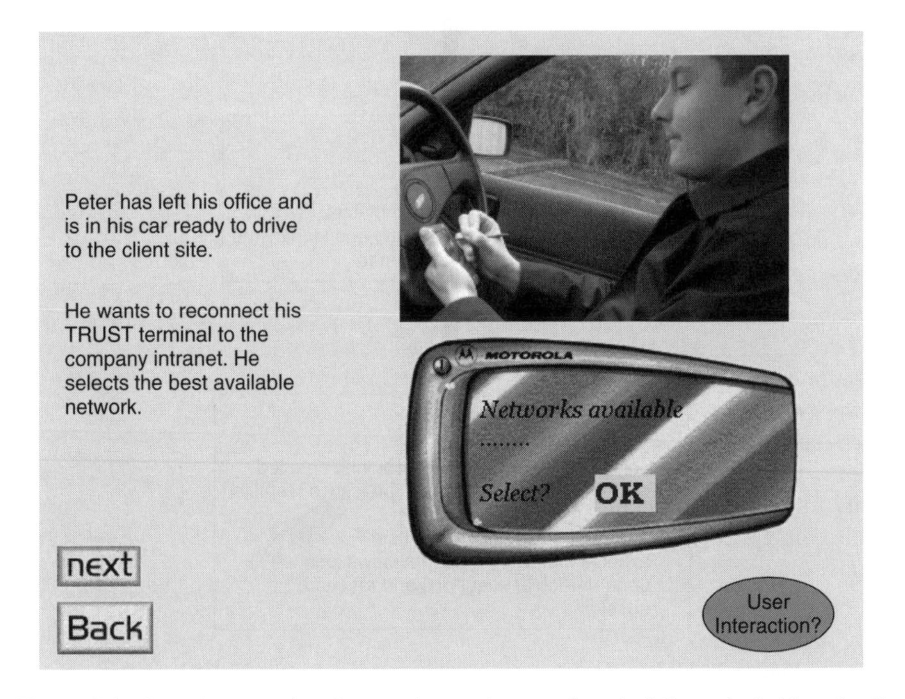

Figure 2.2 Sample screenshot from an interactive storyboard of Scenario 1, 'download'

should the user do it manually? What would the user need to know in order to decide whether to switch to a new mode or not?

- *Air Interface Download.* When the user needs to download a new air interface in order to enable his/her device to operate on a new type of network, should this download be done automatically or manually? What would the user expect to happen to the interface when a new air interface is downloaded?
- *User Profile.* The user profile refers to the information that the device will 'know' about the user in order to customise the behaviour of the device for that user. Research questions include the user interface for the user profile (how to set it up, how to change it, use of a user interface wizard), what should the content of the user profile be and how the user profile should operate.
- *Video.* This refers to the streaming or downloading of a video from the network to the device. A user interface needs to be defined that includes the information required before, during and after the video download.
- *Group Communication.* This refers to situations in which a distributed group of users is communicating with each other, perhaps using multimedia. What information will users need about the other callers before, during and after a group call?
- *Security.* This refers to the security aspects of both the information and the software that is downloaded to the device. The TRUST research explored how to handle secure login and logout for a corporate intranet and suitable interaction dialogue to enable users to log off as well as to receive feedback to confirm the log-off.
- *Battery Life.* This refers to the information that the user needs about the battery of his/her device. For example, how should the system provide the user with feedback on the power 'costs' for different functions (e.g. listening to the radio, sending email, having a voice call, downloading files)? How should the system provide recommendations for conserving the device battery life? How can we improve the visualisation of the impact of different applications on the available power?
- *System Resources.* System resources are the diverse system components that affect the performance of the device, such as the processing power and storage space. What does the user need to know about these resources? How should the system warn the user that the device's resources are almost fully used (or are insufficient for desired tasks)? How can we help the user in the steps that need to be taken either by him/her or by the device?

We then specified which questions needed to be answered for each research topic, and filtered these lists of questions so that all the questions could be addressed by the users without functional implementations of SDR (for instance, we could gain no information about the system in actual use). We then planned a user study in which we discussed each research topic with the users to understand their expectations and needs for that particular research topic.

In the following subsections we specify in more detail the methodological approach used for data collection and data analysis, as well as the design and materials used in our study.

2.2.2.1 Data Collection Methods

As discussed previously, four scenarios were developed in the second phase of the TRUST user requirements research and were used as a basis for the detailed user requirements phase that is described in this chapter. The aim of these scenarios is to illustrate how users will use future SDR terminals and the issues that they may encounter during their interaction with the device. We used these scenarios to present the concept of SDR to the users, as well as to illustrate the different research topics about which we wanted to gather user feedback.

Therefore, the approach taken in this study has been the use of scenario-based requirements for gathering techniques. Scenario-based design techniques have been proposed in human–computer interaction research as effective ways to capture, analyse and communicate user's needs [1, 3, 6]. The basic principle of this approach is that presenting scenarios of use of the technology (or the foreseen technology) to users and developers can bridge the gap between the description of the users' tasks and the design of new technology to accomplish these tasks. Several methods have been proposed to carry out systematic development and utilisation of each scenario [2]. For instance, user scenarios can be analysed during a workshop with a group of users and developers, proposing system features and creating 'claims' (a set of possible positive and negative consequences) for each feature. In our case, we were interested in gathering specific information about a wide range of topics, and did not require the direct involvement of developers in the gathering of feedback because of their involvement in previous stages of the research, such as their help in identifying the user issues raised by the scenarios.

Because of the goals of our research, we decided to use 'paired interviews' to gather information about the research topics. In paired interviews, two users are encouraged to discuss and collaborate with each other during the interview session. The use of paired interviews has been claimed to minimise some of the problems of think-aloud methodologies with single users. In a typical think-aloud study, the researcher captures the user's train of thought while performing a task by asking him/her to verbalise what he/she are doing, why, and what he/she expects to happen next. One problem related to these methods is the workload and attention cost of switching between performing the task and verbalising the actions. The principle underlying paired interviews is that 'the buddy system of learning... allows two people to capitalise on each other's skills in confronting a common learning or problem-solving task' [12].

In addition, in the scenario-based discussion section of the interview we also designed a second task for the participants to complete. For the objectives of this research, it was important to understand how the information about several research topics (such as mode switching or group communication) should be presented to the users in the user interface. The first step toward designing a user interface is to know which information has to be presented. The second step is to decide upon the organisation of the information in the interface. Finally, the third step is to take all the information specified and understand what the best representations are for this information. Because of the scope of this research, we wanted to gather data about the first two steps, namely which information is required and how it should be organised in the interface. Further research will need to investigate which representations to use for each of the types of information identified as important in this study.

The first question, regarding which information needs to be presented to the users, was addressed during the discussion of the topics in the interview. The second question, how that information should be organised, was addressed using a paper-based grouping task. Grouping tasks are exercises in which the participants are asked to sort types of information according to a particular criterion; for instance, the information they need to perform one step in a problem-solving task [5]. This type of task has been used in Cognitive Psychology as a Knowledge Elicitation technique to assess users' mental models of an artefact. In principle, these tasks record how people structure a particular topic and the relationships between concepts related to this topic in their memory. The assumption underlying Knowledge Elicitation techniques is that if the way the information is presented in the interface matches the way the information is organised in the user's memory, then the interface will be intuitive and easy to use. In our case, we used the information gathered during the first part of the interview, when we asked the participants which information they would want to have in order to download or stream a video, change modes, use their user profile or participate in a group call. Participants were asked to use this information to walk through the task (i.e. downloading video or changing mode), and to record on a blank sheet of paper, using Post-its and pens, when they would want to have each piece of information. Because the data gathered in this task is paper-based rather than verbal, during the rest of this chapter we refer to this part of the interview as the 'paper-based task'. The data gathered during this task complemented our data from the interviews, giving us a better understanding of how and when to present the required information to the users in the interface.

2.2.2.2 Number of Participants and Participants' Profile

Following the approach taken in the first stage of the TRUST user research [13], we decided to continue with the 'lead user' approach in which we identified users currently performing tasks that will be relevant for future tasks [11]. The selection criteria for the participants were based on the activities presented in the scenarios to make sure that these activities would make sense to the users. The criteria are listed in Table 2.1.

Twelve paired interviews were scheduled to cover all the topics that were described in Section 2.2.2. Since one scheduled interviewee did not participate in the study, a total of 23 participants took part in our research: 17 male and six female. Their ages were

Table 2.1 Selection criteria for the participants in the study

Selection criteria
Full-time employment for at least the past two years
Owns a mobile phone and, if possible, a PDA
Is an experienced Internet user (daily/weekly use) Uses email for business and leisure purposes Downloads video and/or software over the Internet Watches streaming video and/or listens to streaming audio over the web
Needs to use/access confidential information (personal or for work)
Travels for work (at least once a month) and needs to be contactable while travelling (email and/or phone). Needs to organise or conduct group activities while travelling (conference calls, plan social events, etc.)
Uses PC or laptop as part of his/her work

between 24 and 51. Their professions were diverse, from managers to designers to IT professionals.

2.2.2.3 Structure of the Sessions

In order to cover the eight research topics in the 12 interview sessions, two topics were discussed in each session and each pair of topics was discussed in three separate interview sessions (for a total of six users per research topic). Thus, there were four different 'styles' of sessions, each style focusing on two topics and one of the scenarios and presenting another scenario as an introduction. The order of presentation of the scenarios was balanced across the four sessions styles. Therefore, each scenario was shown in one session style as illustration and in another session style as a basis for discussion. All the scenarios were presented to the same number of users.

Each session lasted between one hour and one hour and a half. Two participants and two researchers attended each interview. At the beginning of the session the participants were briefly introduced to the topic of the research and asked to sign a consent form. Then the researchers spent several minutes explaining the concept of SDR, especially the concept of switching between different types of networks. The researchers then answered any questions that the participants had.

The structure of a generic session was as follows:

- Introduction and consent form
- Illustration scenario
- Discussion scenario
- Question-based discussion on the first topic
- Paper-based task about the first topic
- Question-based discussion on the second topic
- Wrap-up questions

In particular, the researchers stopped at particular points in the discussion scenario to raise particular questions. For example, at a scenario step showing the protagonist running low on battery life, the researchers would ask questions about the management of battery life.

2.2.2.4 Materials

As explained previously, four types of sessions were designed to address the eight research topics listed in Section 2.2.2. Each session covered two of these topics, plus the same set of 'wrap-up' questions. For example, Table 2.2 presents the questions used for one session style, covering the topics of Video and System Resources.

The data were recorded using handwritten notes and audio recording, along with the raw materials of the paper-based tasks.

2.2.2.5 Data Analysis

In order to analyse the user data, we began by transcribing the handwritten notes for each session and organising these according to the questions that were asked. We then categorised the user responses to each question to form a consolidated view of the answers

Table 2.2 Example of a session plan: session style 1, showing interview questions about video and system resources

Question reference no.	Interview questions
[Warm-up]	When and for what do you expect playing or watching video to be useful?
Video 1	Would you expect any differences between downloading and streaming video?
Video 2	Which information would you expect to have in order to download video? (Would there be any differences between the information needed for download vs. streaming video?) Would you expect to have this information before/while downloading a video? Which information do you expect to have after the download or at the end of the streaming video?
Video 3	Do you know what a codec is? Do you expect that a required codec would be downloaded automatically or do you expect to ask the device to download a particular codec? Do you expect to pay for downloading a codec?
Video 2	Paper-based task: user interaction for video download/streaming. What information is required at which stage and how should it be organised?
System Resources 1	Assuming that your terminal will have limited resources, what do you expect to be the factors influencing what you can do with your terminal? Would you expect to manage these resources or do you expect the terminal to do it? Would that depend on your circumstances or activities?
System Resources 2	How and when do you expect the system to inform you that your memory is almost full? And what if it is insufficient for the desired task? Which information do you expect to need in order to make the decision about how to continue?
Wrap-up	Do you have any questions about the way it [SDR] might be used?
Wrap-up	Do you see advantages or disadvantages compared with your current ways of communicating? Is there anything that concerns you about this new idea?
Wrap-up	Do you have any problems with your mobile phone or PDA that you don't think this system will solve?

to that particular question. In some cases users disagreed with one another or made fine distinctions about their own preferences; these differences of opinion have been included in the results presented in this chapter. The materials from the paper-based tasks were also analysed by making consolidated representations of the results based on the contributions of each pair of users. Because of the nature of the information that we aimed to gather in the study (that is, qualitative, not quantitative, data), we have not performed a statistical analysis of the results. Rather, we present the qualitative data that we gathered for each of the research topics.

2.3 Results

This section presents the information that we gathered about each of the research topics during our interviews, along with the more general information gathered throughout the interviews and especially during the 'wrap-up' questions. Sections 2.3.1 through 2.3.7 discuss the individual research topics such as Mode Switching and Security. Since most of the issues related to Mode Switching and Air Interface Download refer to similar situations, we decided to report both of them together in one section. Section 2.3.8 contains the users' other observations, including concerns and perceived possible implications of reconfigurable radio systems. Finally, Section 2.3.9 discusses the effectiveness of the methods chosen for the study.

Although we use the terms 'users' and 'participants' interchangeably in this section, all our data refer to the small sample of potential SDR users who participated in our interviews.

2.3.1 Mode Switching and Air Interface Download

As mentioned earlier, in this section we summarise the main findings for the two issues that we consider intrinsic to reconfigurability: Mode Switching and Air Interfaces. Each of these activities supports a type of 'reconfiguration'; mode switching refers to the transition between network modes, and air interfaces refers to the download of new air interface software to enable the terminal to operate on a new type of network.

One of the most important questions for mode switching is whether the terminal should switch to a new network mode automatically or whether the user should initiate the reconfiguration process. The participants indicated that depending on the circumstances, they might want either of these approaches to apply. They wanted to be able to set up the terminal so that sometimes they could control the mode switching manually. They were comfortable with automatic mode switching when they were not personally paying for the new network (for example, when their employers were paying for their use).

In particular, the factors that would affect the choice between automatic and manual mode switching were:

- whether they were using the terminal to complete a task at the moment and did not want to be interrupted by a mode switch;
- what they were doing in the larger context of use (i.e. the terminal should perform an automatic mode switch while they were driving);
- whether the user likes getting involved in the inner workings of the terminal (some people may prefer to take control, other people might not) [8].

In general, users did not want to be interrupted by an automatic mode switch, particularly if they were in the middle of a voice call. They wanted the option of a 'do not disturb' setting that would postpone the mode switch until they had finished the current task. The same factors applied to the downloading of a new air interface, if required.

We were interested in what users are going to need to know in order to decide whether to reconfigure, what they need to know while the terminal is reconfiguring, and what they need to know afterwards. Table 2.3 presents a summary of our findings. This list does not

Table 2.3 Information required before, during and after a group multimedia call

Stage of the reconfiguration process	Information needed by users
Before reconfiguration	• Which other networks are available • The compatibility between the network and the terminal • The speed of the transition (how long is it going to take to switch?) • The cost of the new network, including the cost of actual network use and the cost to download the software to switch to this network • Comparative cost and bit rate of two (or more) networks • The image/video/speech quality available on the network (and the trade-off with cost) • The coverage of the new network (where can I use this network?) • The applications/services/functions available on a network (for a quick decision, services are thought to be the critical factor) • Storage space needed for any new software • Security information about network • Quality of any new software • If reconfiguration is not automatic, instructions for how to reconfigure the terminal manually
During reconfiguration	• 'What is happening' (e.g. a progress bar, a 'signal bar' or a 'clock ticking away' in the interface)? • Option to abort or pause the reconfiguration • Options to perform other tasks
After reconfiguration	• Notification that the process has been completed • Instructions about how to use the new network (especially if there are changes in security, functionality, cost, etc.) • Services available at any given time

imply that all users will need to have all this information; some of it will only be needed at certain times, and different users may have different needs for this information.

Because of its importance for technology development, we would like to highlight the participants' requirement to be able to perform other tasks that did not 'need connectivity' during the reconfiguration. For example, one participant said she would expect to be able to 'write email' while the new air interface was being downloaded, but not to 'send email'. Also important for the development of the technology was that participants emphasised that downloading a new air interface should not alter the 'core functions' in the terminal. After the reconfiguration they would want their ability to use these core functions to remain the same.

2.3.2 User Profiles

In general, the participants liked the idea of being able to set up their preferences (for instance, about cost, incoming calls and networks) and then not have to worry about those issues. On the other hand, they were aware that the need to set up a user profile might add an additional layer of complexity to their interactions with the device.

Most of the participants wanted to have several profiles or sub-profiles, including at least a work and a personal profile. (These profiles may be distinguished from each other by the costs to allow, the services required, which information to share with other users, and the kind of call filtering to perform.) In addition, users would like the possibility of having specific sets of user preferences or sub-profiles (e.g. for different locations, activities and times of the day), as well as generic scenarios that would be applied across many different sub-profiles. These profiles or sub-profiles would need to be overridden to allow the reception of particular calls; however, to avoid problems with these 'overrides' they must expire after a certain amount of time.

Most of the participants thought that call filtering and cost control would be the main functions of the profile. Among other types of information to be included in the user profile were their preferences for networks in terms of quality, coverage and services needed, and whether to download air interfaces automatically or not. Users were concerned about the complexity of setting up the user profile. Therefore, the user profile set-up needs to be easy, for instance utilising a 'wizard', as well as linked to the phone book for a quick selection of characteristics associated with contact people. One way to make the input of the preferences easy is to allow the device to learn from the user when it should present particular information, filter a particular call, or download in advance a particular set of information (that is, have the system use adaptive user profiles). Nevertheless, such automatic changes must be incremental to prevent the device from behaving in unpredictable ways, and should be reversible by the user. Finally, the user profile should not be overwritten when the device switches to a new network or air interface.

2.3.3 Video

As a general question, we wanted to know whether participants would want to download or stream video to their terminals while mobile. The participants reported that both video files and the use of live video (video communication) would be useful for them. They divided their responses into work and personal uses. Among the uses that they suggested for the work situation were leaving 'video messages' (for instance, using video to express an opinion if they could not be in a meeting), and recording a meeting if they could not attend. Expected personal uses included viewing entertainment (i.e. watching a video while travelling, international TV broadcasts, news), and receiving video from loved ones (e.g. for grandparents to see their grandchildren).

In addition, we investigated the participants' knowledge about the differences between downloading and streaming video files and the trade-offs between these activities when using a mobile device. Although some participants found the concepts of streaming and downloading to be difficult to understand fully, in general participants were able to report several trade-offs between downloading and streaming. For example, they reported that if they downloaded the file they would have it stored on their device and therefore could send it to other people. They also predicted differences in the time required to stream a file versus downloading it. With streaming video, they could see the video more quickly than if they had to download the video file first; therefore streaming is seen as saving time. On the other hand, because streaming is more immediate they thought that it might be more expensive than downloading. Also, streaming video is perceived as taking less (that is, no) storage space and having lower quality than video files that have been downloaded. On the basis of the previous differences, participants predicted that they would choose between streaming and downloading a video based on an 'urgency versus quality' trade-off – that is, if they needed to see the video urgently, they might stream it, but if they needed the best quality, they would download it.

The information and options that the participants wanted to have in order to download or stream a video are summarised in Table 2.4.

After opting to download or stream a video, users said that they would like to be asked for confirmation that the process should start. During the download, users would like to know what the system is doing (e.g. an icon representing the progress through the task), and also would like to be able to use the device to do other tasks while downloading. After the download they want a 'summary' of the download that includes the cost and total time of the download, as well as where the video has been stored in their system.

Participants preferred that if they needed a particular video codec to be able to view a clip, this codec should be downloaded automatically along with the clip. They expected that the codec would be free, or at least included in the price of getting the video.

2.3.4 *Group Communication*

We first wanted to understand whether participants expected this type of communication among groups to be useful. The participants liked the idea of 'mobile videoconferencing', saying that it would be an improvement over the two options available at the moment (i.e.

Table 2.4 Information required before streaming or downloading a video clip

1. Information about accessing video over the network (How fast will it be? How much will it cost? How good will the quality be? How much battery will it use up? What is the security of the connection to the network?) • Time and cost trade-offs before downloading/streaming (with a option to upgrade their software to make downloading faster) • Speed/quality trade-off • Network being used and comparative information among networks (i.e. which network is the cheapest to download this video?) • Any disruption that might happen (i.e. going through a tunnel) • Express button to get the video as quickly as possible • How much battery life is going be used up to watch/download the video. Power/quality trade-offs • Security of the system (i.e. 'is it safe for me to download this, or could it be eavesdropped?')
2. Information about this particular video clip • Preview of video content/video title, etc. • Preview of quality of the video (size, etc.) • Codec required to play the video
3. Information about other available clips • Index of video clips already stored in your system • Ways to look for video clips outside your system: search engine to find videos 'in the world'; on which network/s is it available?

mobile 'voice only' conferencing available in GSM networks, or fixed videoconferencing). But they warned that in order for group communication to be useful, none of the callers should have intolerably poor media quality or frequent network drop-outs.

We also asked users about their perception of 'mixed media calls', that is, the possibility that some of the participants in the call might have different levels of media available in their terminals (i.e. some participants might have video available and some might only be able to use voice). The participants' preference was that the system should allow this asymmetry in the media capabilities of different callers, and leave it up to individual callers to decide whether it was appropriate for that particular communication situation.

We discussed the possibility that someone wanting to conduct a video call with someone who is on a voice-only network might reconfigure that person's terminal to a network that supports video calls, for example. The participants agreed that they would not want anybody to be able to change their communication mode without asking their permission; they wanted the option to accept or reject the change. However, participants said that if they were in a business context and the change in mode were required for business use (i.e. a group of colleagues all participating in a video call), then it would be acceptable for someone else to change their communication mode.

We also wanted to understand the information that users would need before, during and after a group multimedia call. We asked our participants to detail exactly what information they would want to have at each point during a group multimedia call. A summary of their answers is presented in Table 2.5.

We would like to highlight two aspects. First, the participants detailed a large amount of information, and therefore these types of information should be taken into consideration in further research that implements a user interface for a group multimedia call. However, not all of this information will be needed for every group call; the information that is required will depend on the goals of the call, where the callers are located, previous interactions among the callers, etc. Second, participants suggested that they will require a preparation phase for their business calls, to take place well before the call itself. For instance, as part of the preparation process, they would like to plan the functionality and the visual information that might be required for the call (i.e. documents to be discussed) and some information about the other callers. That way, they could download some of the required information before the call, perhaps at a moment when the network service is cheaper, and also could be sure that their device was configured to support the communication functionality planned for the call. However, this type of preparation was not expected to be necessary for personal (not work-related) calls.

2.3.5 Security

Security was a very important issue for users, who were concerned about the risk of exposing corporate or other secret information and about the reliability of the new software to be downloaded. On the other hand, users accepted that their employers might access certain information about them (i.e. diaries, file storage, or location under specific circumstances), depending on the depth and content of the information that was being sought. In any case, they want to know who is trying to get the information about them. When asked whether they expect to re-authenticate to secure networks after reconfiguration, the participants stated that they would not like that unless it was very easy.

Table 2.5 Information required before, during and after a group multimedia call

Stage of the group multimedia call	Information and functions needed by the users
Preparation for the call (business calls only)	• Which functionality (video, voice-only) will be used in the call • History of interactions among these callers • Personal profiles of the callers
Immediately before the call	• Projected cost of the call • Agenda for the call • For all callers: time zone, location, availability, task status (i.e. have they finished the task that needs to be discussed in the call?), communication functionality available, signal strength, battery life
During the call	• Ongoing cost of the call • Agenda for the call • For all callers: time zone, location, availability, communication functionality available, signal strength, battery life • Data coming from the other callers • Information about what other callers are receiving • Notification if another person is trying to call them during the group call • List of which other applications can be used during the call • Other system functions: option to record the call, automatic assignment of action items
After the call	• Summary of the call (time elapsed, total cost, data sent and whether it was received) • Automatic transcription of meeting minutes

2.3.6 Battery Life

We learned that limited battery life is one of the participants' biggest complaints about their current devices. Thus, special care needs to be taken to avoid these complaints with SDR technology. In order to reach this goal, users will need to be advised when any critical service will be affected by low battery life, but the control of the device should be left to the user. Under some conditions, participants thought they would be willing to have the device shut down particular services or applications, but they must have previously prioritised which services or applications should keep running. In addition, they wanted to know how much battery they have left, and how a particular activity, especially if it is an unfamiliar activity, would impact the battery life. One solution proposed by participants to any possible battery shortage would be to have a back-up battery in reserve.

2.3.7 System Resources

Most of the users interviewed stated that they wanted to have the option to manage their system resources sometimes, while at other times it may be done automatically by the device. Specifically, they prefer to make the decisions about system resources unless they are busy at work (e.g. in a meeting), in which case the device would use the prioritisation of services that they had entered previously to make decisions about how to manage their resources. One of the reasons they wanted to make the decisions is that sometimes they would like to go ahead with a task even if they know that they will not be able to finish it completely.

2.3.8 General Requirements

In general, the participants advised us that the notifications and warning signals delivered by the device should be situation-sensitive. For instance, if the user were driving, the device should use primarily audio information. This context awareness extends to the input mechanisms of the device, as well; for example, while driving, users would like to use voice input.

We also gathered important data about the concerns that users have about reconfigurable radio as portrayed in the four scenarios. The first concern among the participants was the system complexity and possible information overload due to the amount of information that would need to be handled by the users. The participants proposed several ways to avoid this information overload, such as having the user profile make automatic decisions and integrating principles of good interface design. A few participants also mentioned the need for a very easy way to get started using the reconfigurable terminal, avoiding a complex set-up process. Another serious concern expressed by the users was the size of the future device. They required it to be at least as portable as their current Personal Digital Assistant (PDA) or mobile phone. On the other hand, they were also worried about the difficulty of inputting and displaying information on a small device. Other concerns included the physical design of a multifunctional device, integration with other devices, overall cost and cost control, and network issues such as coverage. Related to their concern about incompatibility between data formats, they suggested that they would like to have their information stored 'somewhere else' in the network, not in their device, so that access to their own information would be device-independent. Some possible negative implications have been reported which may impact on the individual (i.e. the user becoming too reliant on the device, as well as on their social interaction.

Several possible positive implications were also identified. Users were particularly enthusiastic and impressed by the possibility of having one device that would provide many of the functions currently provided by their mobile phones, PDAs, desktop PCs, and laptops. They also liked the flexibility afforded by the system, and the fact that it would 'save time' and 'make their lives easier'. Finally, they were very positive about the idea of being more 'contactable' and less dependent on any one network.

2.3.9 Methodological Critique

In the execution of the study, scenario-based design and feedback gathering proved to be an effective mechanism for exploring future activities. Users were able to make detailed, thoughtful comments about things they had never done using technology they had never seen before, because the scenarios walked them step-by-step through the ways that they might interact with this technology. The activities portrayed in the scenarios were ones that they could imagine doing and thus they were able to offer constructive criticism about how to make these activities easier for future users of reconfigurable systems. In conjunction with interview questions based on the scenarios, we also used the paper-based grouping tasks to discuss how to organise the information in the interfaces. Consolidation of the data from different pairs of participants also proved possible and useful because of a high degree of overlap in the ways that the users envisioned the tasks. The paired-interview technique was a good choice for this study, and we observed that having two participants prevented the interviewers from having to do much 'prompting'.

2.4 Conclusions

This chapter has presented the results a user study about SDR, conducted as part of the TRUST project in the autumn of 2001. Our study focused upon eight issues about user interactions with SDR: mode switching, air interface download, user profiles, video, group communication, security, battery life, and system resources. The plethora of end-user questions for each issue were identified and then distilled into a comprehensive set of questions to explore with the users. We used interactive presentations of four user scenarios as the main way of communicating the features of SDR to the users, in accord with the scenario-based design approach.

The user study consisted of 12 interviews with pairs of users (for a total of 23 participants). Each research session used one of the scenarios as an illustration of the concept of SDR, and one as the main focus for discussion in the session. In each session, one of the eight topics was addressed in depth and with an accompanying paper-based task about the required information and organisation for that topic, while another topic was addressed only with interview questions. Thus, we obtained detailed feedback from six users on each of the eight issues, and comprehensive feedback from all 23 users on general questions and issues about 'reconfigurability'.

The main results of the study were as follows.

- *Mode Switching and Air Interface Download.* We discussed two types of 'reconfiguration' with participants. The first was switching from one network to another of the same type ('mode switching'), while the other was downloading and installing an entirely new air interface ('air interface download'). In both cases the participants wanted to be able to choose between having this done manually and having the device carry the task out automatically when the user is busy. If the process were manual rather than automatic, there was a range of information required in order to select a new network or air interface, thus requiring settings in the user profile and good information design to reduce complexity. In general, users were reluctant to be interrupted by what they perceived as technical tasks, and wanted some functionality to be available even during reconfiguration. They did not want the core functionality of the device to be altered by every reconfiguration.
- *User Profiles.* The participants recognised the potential of user profiles to relieve them of some of the complexity of using a reconfigurable device. An adaptive profile that learned from their behaviour would be an easy way of tailoring the device to their needs. The participants anticipated a need for multiple profiles and sub-profiles based on their activities, environment, etc. They wanted profiles to handle such functions as call filtering and cost control.
- *Group Communication.* The participants saw many uses for group multimedia calls, as depicted in scenario 4, in both business and personal contexts. They required a

great deal of information not only about their own system capabilities but also about the other callers (i.e. personal profile information) and the other callers' systems (i.e. battery, available services, what they are receiving).
- *Video.* The participants envisioned a wide range of uses for video information on a reconfigurable device. They identified various trade-offs between streaming and downloading, such as cost versus quality and quality versus memory required to store a clip. In order to make the decision between streaming and downloading for a particular video clip, they would require many different pieces of information (such as the time required to download a clip).
- *Security.* Participants were concerned about keeping control of personal and corporate information (including ensuring secure connection to their corporate intranet). Under some circumstances and for some information, they were willing to share personal information, such as location, with their employers or with their friends. They also wanted to know prior to downloading new software that this software was safe, reliable and compatible with the device.
- *Battery Life and System Resources.* The participants wanted to control their system's use of battery life and system resources (processing power and memory), although they wanted the device to handle some aspects automatically. In particular, they agreed that power and system resource management could be done automatically if they had previously set up a prioritisation of applications and services, perhaps as part of their user profile. They wanted continuous information about the resources they had available in the device, as well as notifications and warnings in advance if a particular action were going to consume a lot of power or other resources.
- *General Concerns about Reconfigurable Radio Systems.* The participants' biggest concerns about the general concept presented in the scenarios were the complexity and potential for information overload, the portability of the device, the input and display constraints on a small device, and the general ease of use, especially for novice users. Other concerns were integration with other devices, centralisation of information, cost control and network issues.
- *Positive Implications of Reconfigurable Radio Systems.* The idea of one device that could perform many of the functions that now require a phone, PDA and PC was very pleasing to the participants. They thought that such devices would add flexibility to their lives, 'saving time' and 'making life easier'. They enjoyed the fact that they would not be dependent on one network and that they could obtain new functionality without having to obtain new devices.

Developers and interface designers for future reconfigurable radio systems will be able to use these results to guide their work. In addition, further user research will validate, refine and expand upon the results of this study. The next step in this research is to build low-fidelity interface prototypes to present the information identified as important by the users in this study. These prototypes could then be evaluated to ascertain whether the information described in this chapter was required to complete the given tasks and whether the user interface mechanisms were appropriate. The iterative development of these low-fidelity prototypes could lead to higher fidelity prototypes, such as working 'reconfigurable' handheld devices, which could then be evaluated in actual contexts of use. These next steps will be carried out as part of a new project called SCOUT (Smart user-Centric cOmmUnication Environment), also conducted in the IST Programme of the European Union's Fifth Framework.

Acknowledgements

This work was performed in the framework of the IST project IST-1999-12070 TRUST, which is partly funded by the European Union. The authors would like to acknowledge the contributions of their colleagues from Siemens AG, France Télécom R and D, Centre Suisse d'Electronique et de Microtechnique S.A., King's College London, Motorola Ltd, Panasonic European Laboratories GmbH, Robert Bosch GmbH, Telefonica Investigacion Y Desarrollo S.A. Unipersonal, Toshiba Research Europe Ltd., TTI Norte S.L., University of Bristol and University of Southampton. Special thanks to David Williams and Kate Cook for their contributions to the process described in this chapter.

References

[1] J.M. Carroll and M.B. Rosson, 'Getting around the task–artefact cycle: How to make claims and design by scenario', *ACM Transactions on Information Systems*, **10**(2), 181–212, 1992.

[2] J.M. Carroll, *Making Use: Scenario-Based Design of Human–Computer Interactions*, Massachusetts Institute of Technology Press, Cambridge, MA, 2000.

[3] G. Chin Jr and M.B. Rosson, 'Progressive design: Staged evolution of scenarios in the design of a collaborative science learning environment', *Proceedings of CHI'98*, pp. 611–618, 1998.

[4] A. Dix, J. Finlay, G. Abowd and R. Beale, *Human–Computer Interaction*, Prentice-Hall, Hillsdale, NJ, 1993.

[5] D.J. Gillan, S.D. Breedin and N.J. Cooke, 'Network and multidimensional representations of the declarative knowledge of human–computer interface design experts', *International Journal of Man–Machine Studies*, **36**, 587–615, 1992.

[6] H. Karasti, 'Bridging the analysis of work practice and system redesign in cooperative environments', *Proceedings of DIS'97*, pp. 185–195, 1997.

[7] D.J. Mayhew, *The Usability Engineering Lifecycle: A Practitioner's Handbook for User Interface Design*, Morgan Kaufmann, San Francisco, 1999.

[8] R. Navarro-Prieto and G. Conaty, 'What do end users want from reconfigurable terminals?', *Proceedings of IST Mobile and Wireless Telecommunications Summit*, Thessaloniki, Greece, 2002.

[9] W. Orlikowski, 'Learning from notes: Organizational issues in groupware implementation', in: *Conference Proceedings on Computer-Supported Cooperative Work* (Toronto, Canada, November 1992), ACM Press, pp. 362–369, 1992.

[10] TRUST Homepage: *http://www.IST-TRUST.org/*

[11] G.L. Urban and E. von Hippel, 'Lead user analyses for the development of new industrial products', *Management Science*, **34**(5), 569–582, 1988.

[12] D. Wildman, 'Getting the most from paired user testing', *Interactions*, **2**(3), 21–27, 1995.

[13] D. Williams, E. Ballesteros, C. Martínez and E. Morata, User assessment: Reconfiguration scenarios and requirements. IST-1999-12070 TRUST Internal Report D2.1/1, 2000.

[14] Wireless Strategic Initiative, IST-WSI, *Book of Visions*, p. 9, 2000.

[15] *http://www.wireless-world-research.org/*

3

The Need for Network Reconfigurability Management

Nancy Alonistioti and Nikos Houssos

Communication Networks Laboratory, University of Athens

Mobile communications are evolving from a world of monolithic and inflexible systems and service models to a highly dynamic environment, contributed by a wide variety of market players, where a plethora of ubiquitous services will provide end-users with valuable help with every aspect of their everyday activities. The realisation of this vision requires denouncing the proverbial rigidity of telecommunication networks with the introduction of programmability and reconfigurability at various levels of network functionality. Nevertheless, several challenges need to be addressed before the successful application of these concepts in real-life environments. Intelligent, mediating service management frameworks can be an important factor for successfully facing those challenges. Moreover, standardisation of open, third-party accessible reconfigurability interfaces is another crucial enabler in the demanding but hopefully rewarding series of efforts in academia and industry towards reconfigurable networks.

3.1 Introduction – The Evolution of Mobile Service Provision Towards 3G and Beyond

Supported by rapid technological advances in various research areas, the increasing trend towards open, deregulated markets is already beginning to bring about substantial changes in the telecommunications industry worldwide. The traditional model of telecommunication service provision, characterised by the rigid incorporation of service functionality into 'black-box' network equipment, proved to be successful for a number of years, in both wired and wireless networks. Nowadays this paradigm seems to require a major reconsideration, which is expected to lead to an irreversible shift in the way we develop, deploy and manage communication systems and services, by prioritising flexibility and openness over reliability and performance. This section introduces the chapter by examining the

Software Defined Radio: Architectures, Systems and Functions. Edited by M. Dillinger, K. Madani and N. Alonistioti
 ISBN: 0-470-85164-3

evolution of mobile communications towards future, more dynamic environments and designating the emerging significant role of network reconfigurability in this context.

3.1.1 Service Provision in the Pre-3G Era

Second generation (2G) cellular systems experienced phenomenal success and managed to introduce mobile communications to the everyday life of people worldwide, although they only offered a very limited range of services, based on plain voice telephony and later on short text messaging. The development and delivery model for those services, depicted in Figure 3.1, was inherited from traditional wired telephony networks. The main technical characteristics of the current generation of cellular mobile systems and services can be summarised as follows.

The service intelligence is placed on a limited number of core network nodes [e.g. Customised Applications for Mobile Network Enhanced Logic (CAMEL) Service Environment, Short Message Service (SMS) Gateway nodes]. The rest of the network equipment (e.g. switches, base stations) are responsible only for delivering basic transport

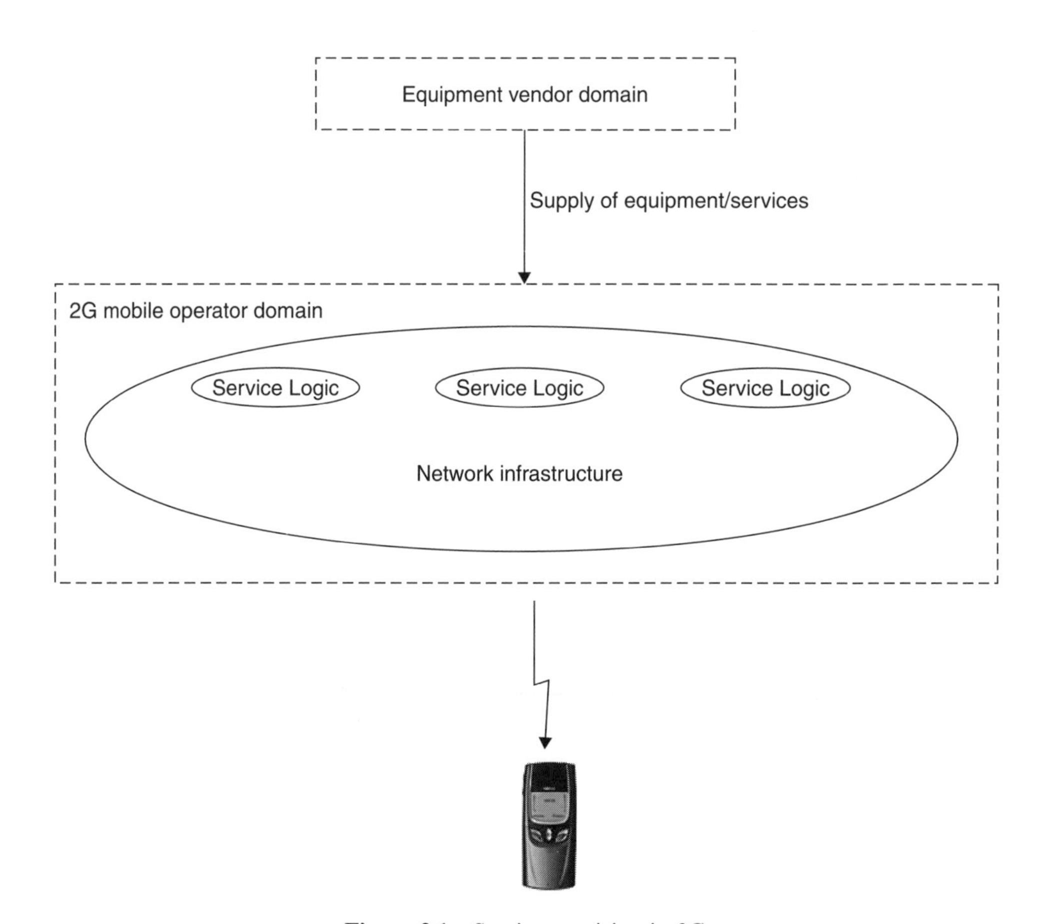

Figure 3.1 Service provision in 2G

services, while end-user terminals have very limited capabilities and thus the application-related logic they incorporate is restricted to basic user interface and communication functionality.

Services are developed and managed using telecommunication-specific and mostly proprietary tools and technologies. They are vertically integrated into network equipment, which typically exposes to third parties proprietary and, with limited functionalities, interfaces for programmability and management.

The aforementioned paradigm leads to important restrictions in telecommunication service provision:

- Services typically are created and deployed solely by operators and/or equipment vendors. Third-party contributions to service development and delivery do not normally involve service programming; typically they include solely the provision of application-specific content.
- The range of available 2G value-added services is restricted to no more than a few dozen for a typical mobile operator. Therefore, these applications do not at present constitute a significant revenue source for operators.
- Service time-to-market is of the order of months or even years, since the coupling of services with the underlying network makes service development, testing and deployment a tedious and time-consuming task.
- Most services are bundled together with network access in various subscription-based offerings by network operators and are accessible exclusively to the subscribers of a single operator.

3.1.2 Mobile Services in 3G and Beyond – the Vision

Third generation (3G) cellular networks aim to pave the way towards a more flexible environment, where the focus will be shifted from simple standardised services such as voice and SMS to non-standardised multimedia value-added applications that substantially enhance the end-user mobile service experience [1–3]. These applications differ from their 2G predecessors in a number of ways:

- Their functionality can encompass almost everything imaginable (e.g. advanced messaging, entertainment, content access, e-commerce, e-government), departing from plain extensions and enhancements of basic telecommunication mechanisms (e.g. call control).
- Their logic can be distributed to a number of distinct, accessible over Internet Protocol (IP), physical locations that reside in different administrative domains, including enterprise applications servers as well as mobile terminals, which in 3G can incorporate significant hardware and software capabilities. Moreover, they are portable to different networks and can in general be seamlessly and consistently delivered over a variety of contexts, characterised by differing technological infrastructure (terminals, radio access technologies) or other parameters such as user preferences and status.
- They are developed using general-purpose, popular in the Information Technology (IT) world programming languages and technologies [e.g. Java, Common Object Request Broker Architecture (CORBA), Single Object Access Protocol (SOAP)/Extensible Markup Language (XML)].

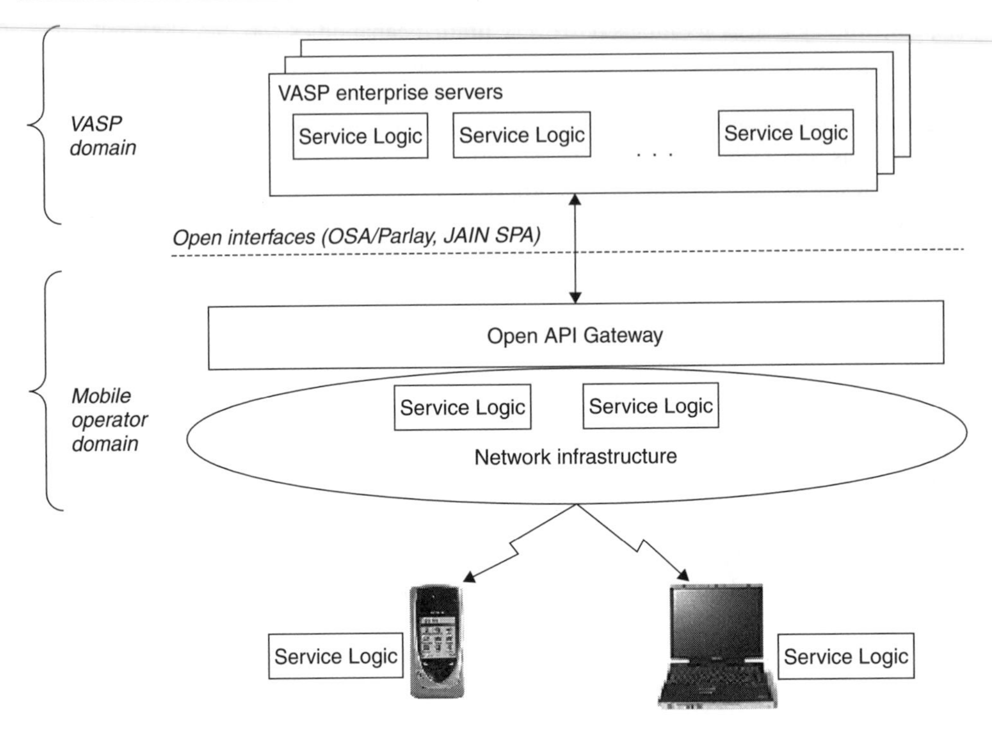

Figure 3.2 Service provision in 3G

The service provision marketplace should therefore bear new characteristics (as shown in Figure 3.2):

- Services will offer a new level of end-user experience by combining personalised access to rich content and functionality with full exploitation of the potential of the underlying communication technologies (e.g. high data rates, user location and presence information).
- Service development and management models will enable (and in practice, demand) the contribution of many different business players (e.g. mobile operators, equipment vendors, application developers, content providers, financial institutions), since the required expertise typically will not be available within a single organisation.

These objectives are to a large extent reflected in the design principles of 3G networks:

- Packet switching, and in particular IP-based networks, have been adopted for the transport of data [4] (and in the future also voice and multimedia [5]) traffic. Besides the advantages in potential capacity gains and ease of deployment, this choice leads also to greater flexibility in network and service design and deployment, as well as the introduction of sophisticated billing schemes that can be tailored to specific applications and thus be more attractive to the user.
- There is a clear distinction between the network and the service/application domain. End-user services in themselves are no more subject to standardisation (very few exceptions to this rule have survived the 3G specifications); merely service capabilities are

being standardised. Moreover, open Application Programming Interfaces (APIs), such as OSA/Parlay [6, 7], have been specified, enabling third parties to create advanced applications by making use of selected network functionality.
- The Virtual Home Environment (VHE) concept has been introduced for consistent anytime, anywhere user access to personalised services [6, 8].
- Standardised Mobile Execution Environments (MExEs) [9] have been defined, turning the terminal into a potential host for sophisticated application functionality.
- The access and core networks are to a significant extent decoupled [10], so that service provision over diverse access technologies is possible.

The potential of mobile services, however, given recent technological developments in areas such as wireless networking and software engineering, goes far beyond 3G systems and application provision. The ultimate goal of future systems, commonly referred to as fourth generation (4G) [11, 12], can be envisioned as a pervasive computing environment where a multiplicity of valuable, affordable services will impact almost every aspect of users' daily activities and significantly increase the quality of their lives. With this ambitious objective in mind, the following basic characteristics of next generation mobile systems and services could be identified:

- *Focus on the user:* Ideally, end-users instead of technologies should be the dominating factor in the design and delivery of value-added services. These services should be provided over converged networks, incorporating many different connectivity types [e.g. cellular networks, wired Internet, Digital Video Broadcasting (DVB)]. The access technology employed to deliver a particular application at a specific time should be totally transparent to the user.
- *Ubiquity:* Future communication networks are expected to comprise a vast space of devices of disparate sizes that incorporate computing and wireless communication capabilities and are densely distributed in the physical environment (e.g. embedded in all kinds of tangible objects). This could lay the ground for the emergence of a multitude of pervasive services of unprecedented variety and functionality.
- *Multiple access technologies:* One of the commonly identified features of 4G systems is the seamless co-existence of competing and/or complementary connectivity technologies [e.g. cellular networks, wireless Local Area Networks (LANs), satellite systems, digital broadcasting] in a single, integrated system, as perceived by the end-user [13, 14]. A dynamic switch between different access technologies (i.e. vertical handovers) should be possible. The optimal way of connectivity at a particular time instance will be determined on the basis of a variety of context parameters.

It is apparent from the above that mobile services in future systems beyond 3G will be delivered in a variety of circumstances, as depicted in Figure 3.3, that will not be a priori predictable. Context parameters typically will include information concerning the human user (e.g. identity, location, personal preferences, current activity, emotional status), the physical environment (e.g. temperature, light or noise level) as well as the network environment (nearby devices and/or network infrastructure, capacity and quality of communication links). The constantly changing context could influence many aspects of service delivery, such as the services that can be used as well as the optimal access technology, network configuration or service composition. Thus, reconfigurability and adaptability to rapidly varying conditions of the various layers of functional components

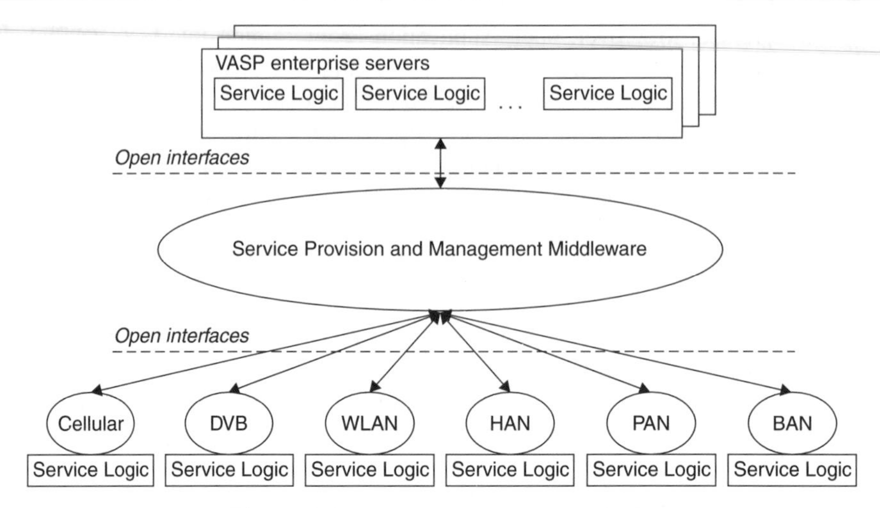

Figure 3.3 Service provision beyond 3G

involved in service access and provision, ranging from end-user applications and mobile middleware down to terminal and network elements, becomes an important prerequisite for the development of next generation networks.

3.2 The Need for Network Reconfigurability

3.2.1 Network Reconfigurability – Enablers for Next Generation Mobile Service Provision

The lack of dynamic reconfiguration capabilities in current mobile networks is a result of the inflexible service provision model employed in those systems. However, as described in the previous section, in 3G and beyond, mobile service provision is expected to change substantially, leading to the creation of an open market, where entities such as independent Value Added Service Providers (VASPs) and content providers will be able to contribute to the development of a plethora of diverse services, which will be targeted to a variety of environments besides cellular mobile networks (e.g. computers in an Ethernet LAN accessing the Internet over wired connections or utility devices connected to a remotely manageable home area network).

Adequate support for such an environment cannot be provided by the current service development paradigm, according to which services are incorporated in specialised network nodes and are thus tightly coupled with the underlying communication technology and network configuration, as well as with the hardware and software of the specific network equipment they are executing on. In the new era, applications should be portable to a large range of environments, whose characteristics cannot be fully known a priori (at the time they are being developed). Consequently, to a large extent, services should be independent of the technological platforms they run on, since it would not be feasible to develop separate versions for all possible execution contexts.

Bearing in mind the above-described situation, the need for more flexible networks that can adapt dynamically to the requirements of the disparity of services that are provided over them, can be clearly identified [15]. Thus, network reconfigurability becomes an issue critical to the successful development of the next generation telecommunications marketplace according to the expectations of end-users as well as market players that have substantially invested in technology.

In particular, network reconfigurability mechanisms are crucial enablers for a large variety of functionality pertinent to mobile service provision:

- Automated service deployment
- Application of high-level policies based on terminal, service, user and network profiles
- Handling the convergence of various access systems
- Protocol and service adaptability
- Support for flexible service discovery, provision and charging
- Protocol/software downloading

Moreover, network reconfigurability management can act as a significant complementary and supporting factor for terminal reconfiguration management.

3.2.2 Reconfigurability Functionality in Mobile Networks

Reconfigurability should be applicable to the whole of the layered functionality of mobile networks, from the firmware of network elements to management applications at the network level. To this end, since the reconfigurability aspects of various layers should be adequately addressed, the provision of holistic solutions towards the support of reconfigurable mobile environments implies the introduction of advanced reconfigurability management and control. Reconfigurability management and control is essential for the support of network-wide reconfiguration actions (including also support for terminal reconfigurability). Reconfigurability management and control depends on various aspects. Some basic aspects that affect the reconfiguration management span across low-level as well as high-level functionalities and features. On the basis of some of these features, reconfigurability management can be identified in:

- Service provision dependent reconfiguration management
- Policy dependent reconfiguration management
- Network, terminal and user context dependent reconfiguration management

An example architecture for service provision dependent reconfiguration management based on a distributed architecture is shown in Figure 3.4.

To apply this approach, intelligent mechanisms should exist for identifying the particular high-level requirements of an application and mapping them to appropriate reconfiguration operations on the underlying hardware and software infrastructure. These operations could be triggered during service deployment to a network or even in the course of service activation and execution. The application of such operations requires the availability of appropriate lower-level mechanisms for reconfiguration of the network elements that comprise the underlying infrastructure (see Chapter 13 in this volume).

Some examples of the types of reconfiguration actions that would be useful in a mobile network are the following:

- *Quality of Service (QoS) provisioning.* Network equipment (e.g. routers) could require changes in their configuration so that they can identify the transport flows corresponding to the usage of a particular service and provide them with the desired QoS. The desired

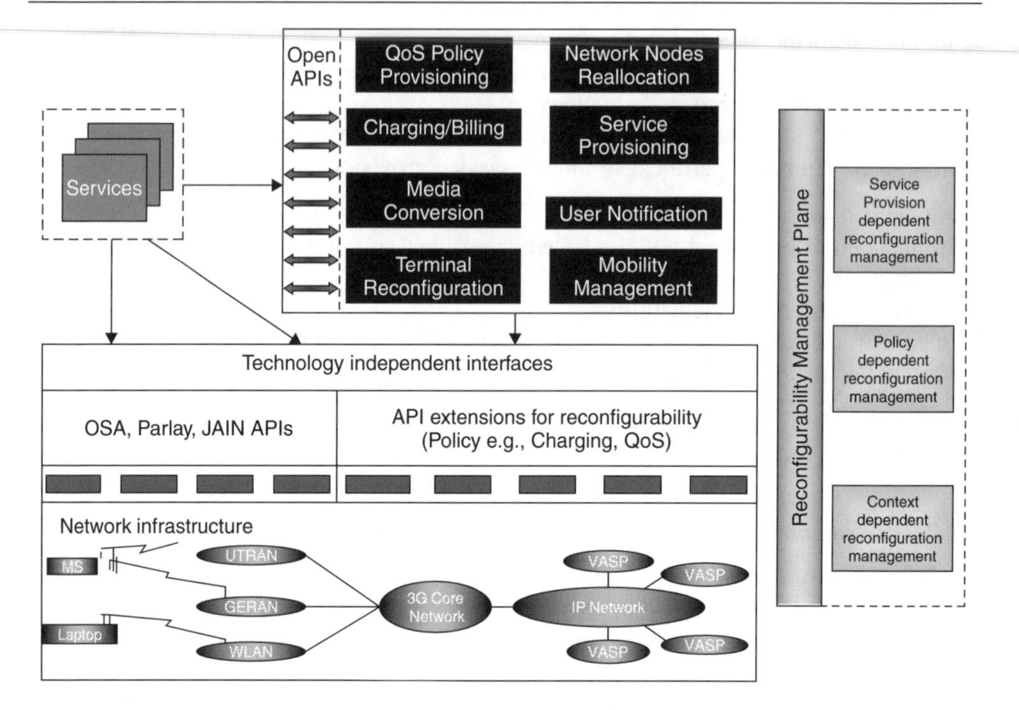

Figure 3.4 Profile-based reconfigurability management in mobile networks

QoS for a service access session by a particular user may be specific to the service (e.g. certain services may require a minimum QoS level to be accessed) and may also depend on the identity and preferences of the user. QoS adaptivity is highly important, since it can lead to more efficient network operation, as well as contribute to service provision personalisation.

- *Charging and billing.* The system employed to gather service usage data, process it and calculate the corresponding charges for the end-user should be dynamically reconfigurable. This is a powerful way to take into consideration the various charging-related events occurring in the network (tariff updates, tariffing policy changes) and subsequently produce an accurate user bill.
- *Dynamic software downloading.* The optimal provision of a service may necessitate certain software elements to be installed dynamically to the terminal or some place in the network during service deployment or activation. For example, owing to limited bandwidth available at the radio communications link, certain service content (e.g. images, audio) should be drastically compressed, so that it can be transmitted in real time to the user terminal. To do that, an appropriate codec could be downloaded to a node at the edge of the mobile operator's network as well as the terminal. Furthermore, more radical downloading approaches, closely related with the SDR vision, could be applied in 4G networks for dynamic downloading and installation of complete protocol stacks to network nodes and terminals.
- *Monitoring functionality in the network.* The collection of network traffic and usage data, which typically is performed by network devices such as switches and routers, is

of particular importance in a network, since the information gathered supports a range of vital operations like QoS provision, security mechanisms, billing, resource management as well as fault restoration. Different policies for monitoring may apply at different times during network operation as well as at different network locations (e.g. detailed monitoring can be either disabled to improve performance or enforced for selected network equipment that is prone to error or seems to demonstrate strange/unexpected behaviour).
- *Radio Access Technology (RAT).* RAT can be considered as a service that can be downloaded to flexible platforms for terminals and Radio Access Networks (RANs). An important aspect related to reconfigurability support and management is the introduction of intelligent service and RAT management platforms that act as mediators between service providers, end-users and mobile network operators. This simplifies the extremely complex task of service management and provisioning, as well as reconfiguration actions in various layers in the protocol stack, terminal and network that employ various policies.

To this end the need for network reconfiguration functionality to be part of open, standardised interfaces that provide access to mobile networks, such as the OSA/Parlay APIs is clearly identified. However, OSA/Parlay at present does not include explicit support for reconfigurability to the extent desired by next generation services. Future OSA/Parlay extensions with reconfigurability interfaces will be an important enabler for the development of advanced portable services and management applications in mobile systems from 3G and beyond. Moreover, a quite promising initiative towards the introduction of modular radio base stations featuring open internal interfaces appears to be the recent formation of the Open Base Station Architecture Initiative (OBSAI), which aims to complement current activities in existing standardization organizations (for example, 3GPP). The Open Base Station Architecture intends to revolutionise radio base station development and to allow next-generation radio base stations to be built using best-of-breed, shared platforms and modules, available on an open market, whilst allowing network suppliers to differentiate on system and network element levels.

3.3 Towards the Realisation of Network Reconfigurability Management

Reconfigurable mobile networks have become a realisable vision, due to the continuous, considerable progress achieved by communications, networking and software technologies. Putting network reconfigurability into everyday practice, however, is definitely far from trivial since it requires addressing a variety of technological as well as non-technical issues. This section identifies and discusses a number of these issues and proposes the introduction of mediating software platforms for advanced mobile service management as a step towards their resolution.

3.3.1 Challenges in Realising Network Reconfigurability Management

A non-exhaustive list of critical aspects for the incorporation of reconfigurability in mobile networks is presented in the following:

- *Identification of the desirable reconfiguration actions and management of the reconfiguration procedure.* Reconfiguration of the network infrastructure should reflect the requirements of optimal service provision to end-users and take into account the policies and constraints imposed by various factors such as security, performance and reliability. The entities coordinating the overall procedure should be aware of this information (e.g. service and user requirements, network status data, operator policies), and should thus operate at least at the network level and ideally at the service level.

- *Mechanisms for access to the application of reconfiguration actions.* The dominating way to automatically reconfigure a network/network element is by using programming interfaces exposed by the network infrastructure to appropriately modify the behaviour/operation of the network at run-time. This requires the availability of programmability at various granularities, such as network elements (through management interfaces supported by devices) as well as entire networks (through specialised servers such as OSA/Parlay gateways) or at an even higher level, with service requirements expressed at an abstract level and being mapped to actions on the underlying networks. Programmability requires the definition of suitable standardised interfaces [16], an undoubtedly very demanding task. The development of relevant activities is well under way by various international organisations [6, 7, 17, 18]. Another, more radical way of achieving network reconfigurability has been proposed by the active networks research community: network operation could be affected by mobile code that could be carried by individual packets [19]. This approach enables a higher degree of flexibility, at the cost, however, of significantly increased complexity of development and deployment, leading to difficulties in the application of these ideas in practice [20]. Thus, in the near to short term, standardised, open APIs should be the prevailing paradigm for performing automated network reconfiguration at all levels.
- *Network security.* Opening up a network to third parties raises significant security concerns for an operator. Rigorous security procedures that guarantee the safety of the infrastructure against unauthorised/malicious clients are thus a major requirement. An important step towards resolving this problem is the introduction of generic, unified secure access mechanisms (e.g. such as those employed by the OSA/Parlay framework) that apply to all applications that interact with the network. Similar approaches may be adopted for securing the reconfiguration of individual network devices or components.
- *Feasibility of third-party access to reconfigurability functionality.* Although the capability of a network to be dynamically reconfigured could by itself be a very powerful tool for service adaptation and delivery, its full potential cannot be realised if such features are not accessible by third parties. Employing reconfiguration for a restricted number of operator- or equipment-vendor-provided services limits its impact, while offering reconfigurability capabilities to service provision platforms and applications could enable the creation of a dynamic environment where flexible, personalised, revenue-generating services will be within the anytime, any place, any terminal reach of the end-user. It is worth noting that third-party access should not only be technically possible [e.g. through the support of standardised, secure accessible APIs such as OSA/Parlay and Java APIs for Integrated Networks (JAIN)], but also be feasible in business terms for any entity that is able to conceive and develop useful value-added services. For example, the management overhead and cost of having a contract directly with a network operator providing reconfigurability APIs to third parties can be unaffordable, or at least discouraging, for software development Small and Medium Enterprises (SMEs) without particular telecommunications expertise. A business entity in the role of mediator

between network providers and Value-Added Service Providers (VASPs) could be beneficial for these situations.

An approach that seems to bear significant potential in addressing the above issues is the introduction of intelligent service management platforms that act as mediators between VASPs, end-users and mobile network operators. Such platforms could incorporate a certain functionality, which is crucial for mobile service provision in 3G and beyond and can be more efficiently handled in a unified way by a mediating framework residing between the network operator and the VASPs than by entities such as individual applications and mobile networks [21, 22]. The main aspect of this functionality is managing the reconfiguration of the network infrastructure based on the profiles of the services, users, terminals and networks involved in the service provision context, as depicted in Figure 3.5. According to this approach, service deployment as well as lower-level operations that typically require network reconfiguration (e.g. specification of service flow monitoring policies) are performed by VASPs via open, secure interfaces exposed by the platform (see also Chapter 7 in this volume). Moreover, the platform is able to undertake, on behalf of the VASP, the required management interactions with the network operator so that an application is authorised to access standardised APIs to network the functionality. This could lead to drastically decreasing the overall cost of mobile service provision for the involved market players and lowering entry barriers for new third-party application providers.

3.3.2 A Reconfiguration Management and Service Provision Platform (RMSPP)

This section introduces a distributed software platform for the flexible provisioning and management of advanced services in next generation mobile networks. The platform aims to fulfil the requirements described in the previous section through the incorporation

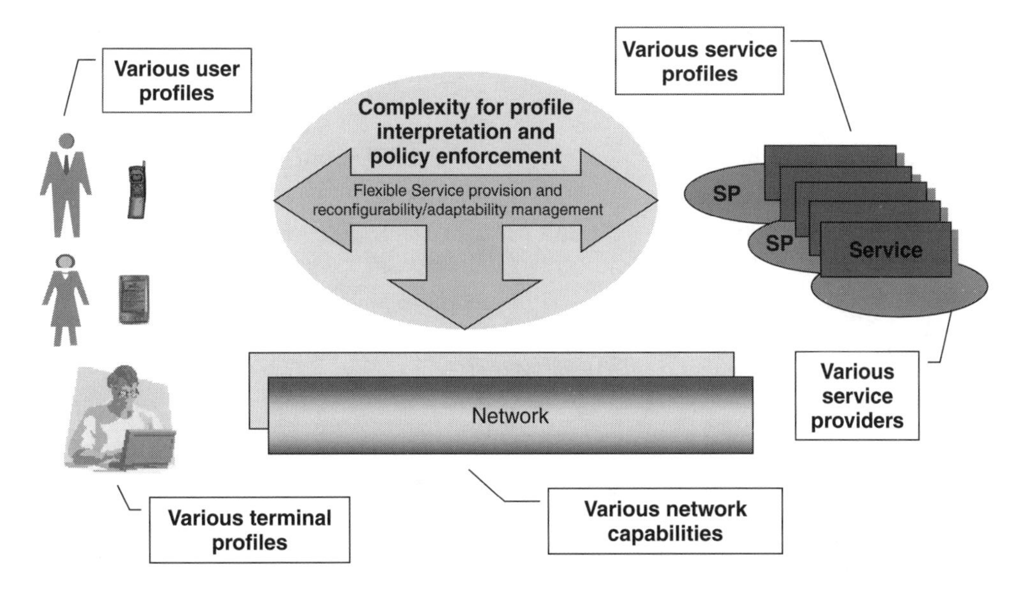

Figure 3.5 Intelligent context-aware service management through mediating platforms

of intelligent adaptation mechanisms and the support of reconfigurability extensions to standardised open APIs.

3.3.2.1 Business Model

The introduction of such platforms enables the application of advanced business models that depart from monolithic, operator-centred approaches and encourage among market players the establishment of strategic partnerships, which ultimately benefit end-users by significantly enhancing service provision and thus paving the way for the evolution of mobile systems and services towards 4G. Before elaborating on the platform architecture and functionality, we briefly describe these models with a particular focus on the role of a new actor, namely the *service platform operator.*

The proposed framework is designed to accommodate flexible, advanced business models that enable the contribution of business actors other than mobile operators to service delivery. The core roles and relationships in such models are depicted in the Unified Markup Language (UML) class diagram of Figure 3.6.

In the proposed paradigm the following roles are defined:

- *(Mobile) User:* The actual consumer of the available services. The user requests the provision of value-added services and applications from a service platform operator. To formulate and manifest these requests, the user employs a communication and computing infrastructure contributed by another entity in the business model, the

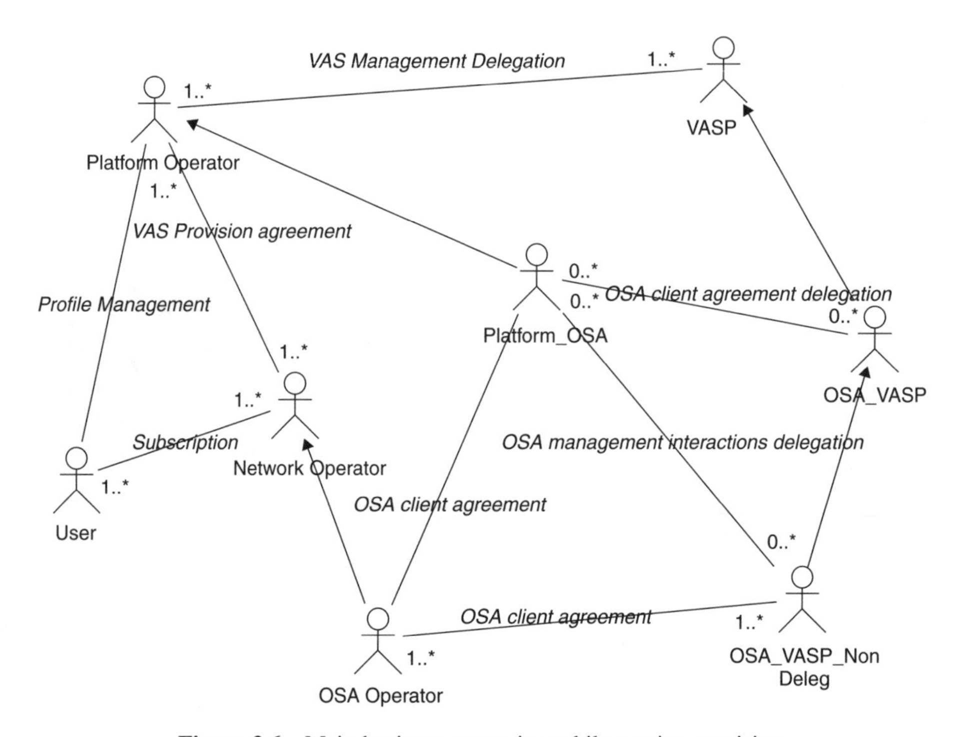

Figure 3.6 Main business actors in mobile service provision

network operator, with whom the user maintains a business relationship in the form of a subscription.

- *Mobile network operator:* The mobile network operator provides the network infrastructure and transport medium for authenticated and authorised mobile subscriber access to standardised circuit- and packet-switched services (e.g. voice telephony, Internet connection), as well as value-added services developed by third parties. It maintains a customer relationship with the user via a subscription arrangement. The network operator would also typically provide independent software vendors with access to network functionality through open, standardised APIs (e.g. OSA/Parlay), thus undertaking the role of the *OSA/Parlay operator*. In our model, we assume that the OSA/Parlay operator is a specialisation of the network operator (in other words, the OSA/Parlay operator is itself a network operator). Although this is not strictly imposed by the Third Generation Partnership Project (3GPP) and OSA/Parlay specifications, we believe that it is going to be the massively prevailing paradigm in 3G networks. In Figure 3.6 there is a clear grouping of the actors that are introduced to deploy and provide services (of the Mobile Specialised Services category, as described in Section 2.3) that make use of the OSA/Parlay interface.
- *Service platform operator/provider:* A business entity that mediates between service developers, network providers and end-users by operating a software platform for service management and provision. The platform operator enters into business-level agreements with network operators and VASPs. These agreements concern the provision of services and applications owned by the latter to subscribers of the network operator. A specialisation of the platform operator is an entity (*Platform_OSA_Operator*) that is able to undertake the deployment and provision of services that make use of the OSA/Parlay APIs. This is considered to be situation that deserves to be modelled by a separate role, since the management of OSA/Parlay applications introduces significant additional administrative effort as well as new business relationships compared with the case of plain (non-OSA/Parlay) services.
- *Value-Added Service Provider (VASP):* A business entity that controls the computational infrastructure directly employed in the process of developing and realising applications and services. Such services can range from Voice over IP (VoIP) and teleconferencing to mobile banking and electronic commerce. Service providers establish business-level agreements with platform operators, outsourcing to them the deployment and provisioning of their services to various 3G networks. Business-level agreements between VASPs and users are not required; however, they are not precluded. A specialised class of VASPs encompasses the providers of services that make use of the OSA/Parlay API (*OSA_VASP*). These providers may choose to undertake the required business interactions with the OSA/Parlay operator (this corresponds to the OSA_VASP_Non_Delegator role) themselves or delegate them to the Platform_OSA_Operator. The latter possibility enables small software vendors in the OSA_VASP role to avoid the management overhead and cost of having a contract directly with the OSA/Parlay operator. Thus, it is a step towards making entry into the telecommunication services market easier for independent software enterprises.

It is worth noting that it is possible for one single entity to undertake several of the roles described above and vice versa. For instance, a mobile network operator may also operate a service provision platform, as well as develop value-added services.

3.3.2.2 Platform Functionality and Architecture

This section provides an overview of the main platform functionality and architecture. The platform functionality would include:

- A service provision and reconfigurability management middleware enabling reconfigurability and adaptability actions in various layers in mobile networks, systems and terminals. These actions could include service-level operations, such as service deployment and modification of service tariffing policies, as well as lower-layer actions such as monitoring of traffic flows.
- Generic, re-usable mechanisms for adapting applications as well as service provision procedures in the current context (identified, among others, by terminal capabilities, network characteristics, user locations and status).
- A one-stop shop offering mobile users highly personalised, location and context-aware service discovery and access through a unified, customisable interface.

The proposed framework, depicted in Figure 3.7, comprises several components, namely the Reconfiguration Control and Service Provision Manager (RCSPM), the Charging, Accounting and Billing (CAB) system and the Mobile Terminal Software (MTS). The following paragraphs elaborate on the functionality of these components:

- The *RCSPM* is the central component of the platform. It co-ordinates the required procedures for dynamic application deployment (including management of underlying network reconfiguration) and personalised, consistent and reconfigurable discovery, downloading, execution and management of services to mobile users. For that reason, it maintains information about the services offered by the platform, as well as user

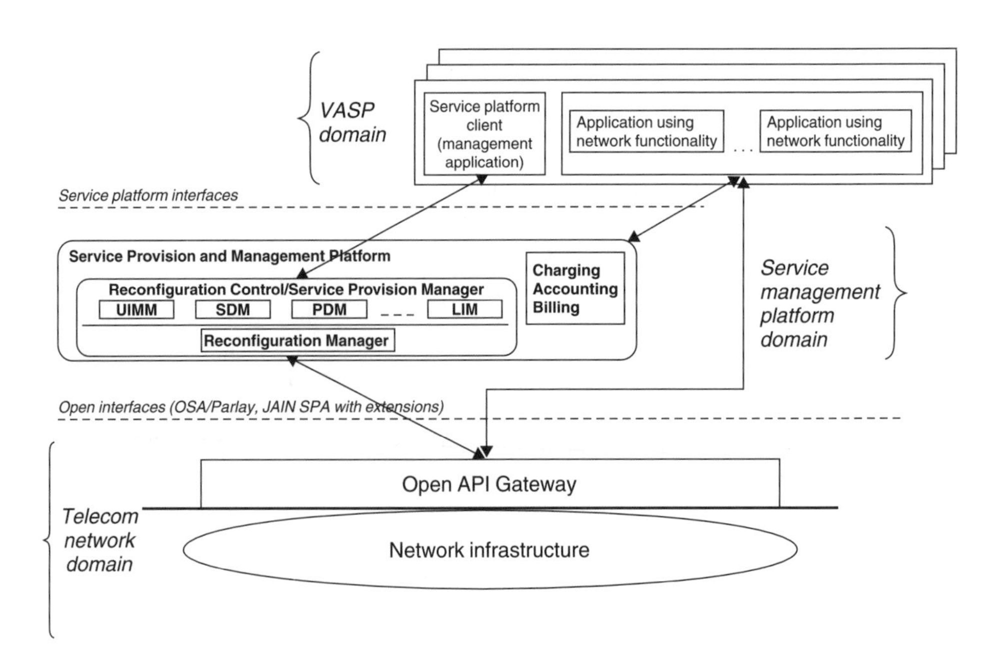

Figure 3.7 Service management platform – overview and placement

profile data and undertakes profile and policy management tasks based on the various user/service/network/terminal/charging profiles (see also Chapter 1 in this volume). The RCSPM also hosts the *Reconfiguration* and the *Location Information Manager* modules described later in this chapter. Detailed analysis of the functionality provided by the RCSPM and its internal structure can be found in a subsequent section.
- The *CAB* system is responsible for the overall control of the charging, accounting and billing processes. It collects charging information from both mobile networks as well as IP networks, applies the appropriate pricing models, calculates the portions that are due to each business entity and produces a single itemised bill to each subscriber [23, 24]. Additionally, the CAB provides advanced charging services through open APIs in order to enable the RCSPM and independent VASPs to configure the applicable pricing policies dynamically, to retrieve statistical information concerning VAS usage (e.g. the list of users that currently execute a specific VAS, or of the VASs that are currently being executed by a specific user) and to be informed about the current status of user and VASP accounts.
- The *MTS* includes functionality such as service downloading management, Graphical User Interface (GUI) clients for service discovery and selection, capturing of event notifications as well as service execution management. The MTS is able to identify and communicate to the RCSPM context information required for customised service provision (e.g. terminal capabilities).

3.3.2.3 RCSPM Functionality and Internal Architecture

In this section we elaborate on the functionality and architecture of the RCSPM introduced above. The RCSPM is the main component of the proposed framework and comprises several internal components. These components are the following (as depicted in Figure 3.8).

1. *Service Deployment Manager (SDM).* Through this module the VASPs are able to register their services with the RCSPM framework and thus make them available to mobile users. Service deployment includes storing service information in the service database maintained by the platform operator, as well as performing certain reconfiguration actions in the network (e.g. configuring network devices to produce service usage monitoring information or allocate resources to provide the appropriate QoS to users of the service). During the deployment of the application, the VASP provides a high-level description of service attributes and requirements. This service information is encoded in XML and should conform to a universally common XML Document Type Definition (DTD). The SDM incorporates intelligent procedures that are used to map this compact service description to specific, network-wide reconfiguration actions that are necessary for the optimal delivery of the service. This approach has the advantage that the burden of service deployment is, to a large extent, moved from the VASP to the platform operator, thus facilitating entry to the application provision market for independent software vendors.

2. *Reconfiguration Manager (RCM).* This module undertakes network, terminal, RAT, platform and service reconfigurability. Based on SDM input, it triggers the appropriate reconfiguration actions on the underlying network during service deployment, on behalf of the VASPs. Communication with the network infrastructure can be accomplished through standardised (e.g. OSA/Parlay/JAIN, COPS) or proprietary APIs and protocols. The RCM therefore includes clients of the corresponding network gateways. The design of the

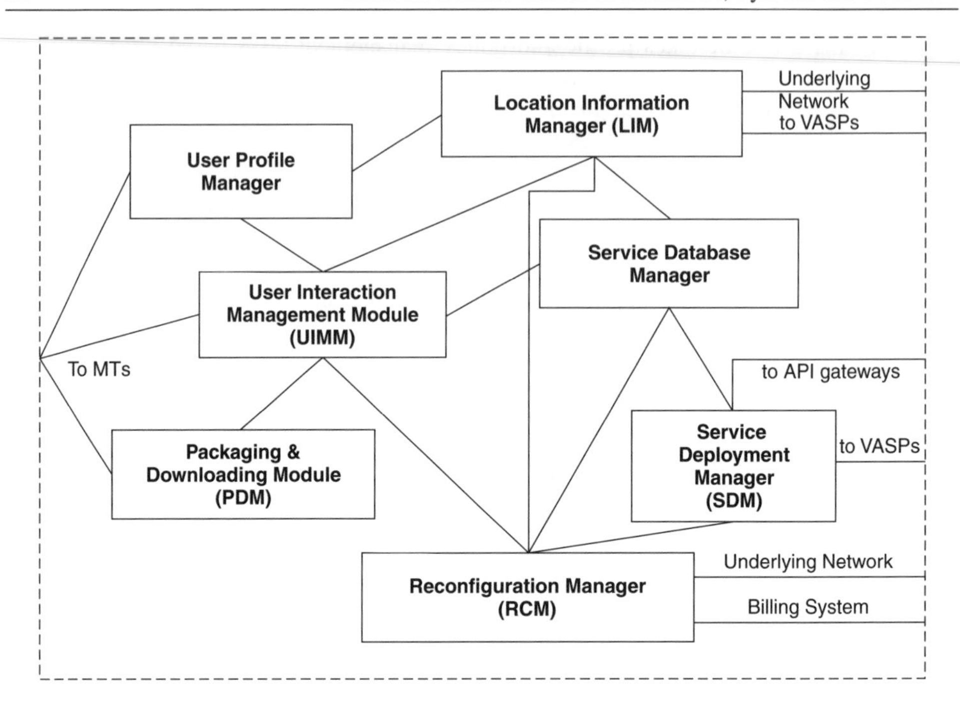

Figure 3.8 RCSPM internal architecture

RCM is modular and dynamically extensible, so that new reconfiguration management logic as well as additional network clients (e.g. supporting communication over new protocols/interfaces) can be plugged into the platform at run-time. The RCM also provides interfaces to end-user applications, in essence extending APIs such as OSA/Parlay, thereby offering authorised third parties the capability to perform certain reconfiguration operations on the underlying networks and systems.

The RCM also comprises a generic adaptation module [25] that is used for intelligent profile matching. An important characteristic of the adaptation module is that it is independent of the types of profiles that need to be matched and is thus able to accommodate profiles of any kind without modifications in its code. In the context of the proposed architecture the adaptation module is used for the customisation of the service discovery procedure and for service adaptation. These functions are accomplished by employing the adaptation module to perform appropriate matching between user, terminal, network and service profiles. Different types of profiles are retrieved from appropriate locations/repositories/network servers (e.g. service database, user database, MTS, OSA/Parlay gateway) and may come in various formats (e.g. Resource Description Framework (RDF), XML, relational database records). Before being processed by the adaptation module, all profiles are subsequently converted to an internal RCM format, which represents arbitrarily complex hierarchical structures. This conversion is performed through appropriate adapters that can be dynamically added to the system during execution.

3. *User Interaction Management Module (UIMM).* This module is responsible for all interactions between user and platform. During the lifetime of a user session with the platform, it maintains and manages session state information. It carries out service discovery;

that is, it presents to the user listings of the available services, tailored to the terminal capabilities and user preferences and location. Moreover, it handles service selection, namely detection of the event when the user chooses an application to access and initiate procedures necessary for service activation (e.g. packaging and downloading of the application client). To achieve this, the UIMM collects information from appropriate repositories, such as service and user profile databases, as well as the terminal. The data are cached and re-used, if required within a user session. This information is provided as an input to the RCM that performs the required intelligent matching operations and returns the corresponding results to the UIMM, which further exploits them to co-ordinate the user experience of customised service provision.

4. *Location Information Manager (LIM).* A very important feature of the RCSPM is the management of location information and the support of location-based reconfiguration policy provision. As depicted in Figure 3.7 and Figure 3.8 the LIM [27] can be considered as the internal functional entity of the RCSPM responsible for retrieving, managing and exploiting the information related to the location and mobility of subscribers. This is performed in a generic way, enabling the LIM to interact with the location information sources of the underlying network infrastructure (e.g. the Location Services (LCS) or the Presence Server) in order to track the location and mobility of the subscribers. Then, location and mobility data along with the preferences of the corresponding subscriber, taken from his user profile, are processed properly to form new advanced location aware services and policies. The location-sensitive policies generated by the LIM can concern the reconfiguration of various resources including network nodes, user equipments, services and applications. The applicable policies are propagated to the appropriate elements through the Reconfiguration Manager of the RCSPM. Figure 3.9 illustrates the interactions between the LIM, the RCM, location and user profile sources and reconfigurable elements to accomplish the provision of location sensitive and reconfigurable service.

The LIM retrieves the required location information by accessing the Location Server [26] of the underlying network infrastructure through the associated open API provided by the underlying network infrastructure. Furthermore, for service transparency the functionality provided by the LIM is accessed by internal modules of the RCSPM as well as independent VASPs through an open API provided by the RMSPP to authorised entities. The open API providing for location information management enables:

- Retrieval of the user location with the specified accuracy. Location retrieval can be either immediate (in the case that the current location of the user is requested) or deferred (in the case that the location of the user is requested when a specific event takes place). Making use of appropriate spatial data bases the LIM is able to map the current user location taken by the Location Server of the underlying network infrastructure (expressed in geographical coordinates or network areas) in the requested format (e.g. street address or predefined geographical zones). The RCSPM, for example, uses those methods to achieve location-based service provisioning in terms of location-sensitive service discovery and execution.
- The creation/deletion/modification of specific location-based events, relevant to the current location or mobility of end-users.
- The registration of end-users and other third parties to receive location-sensitive event notifications.

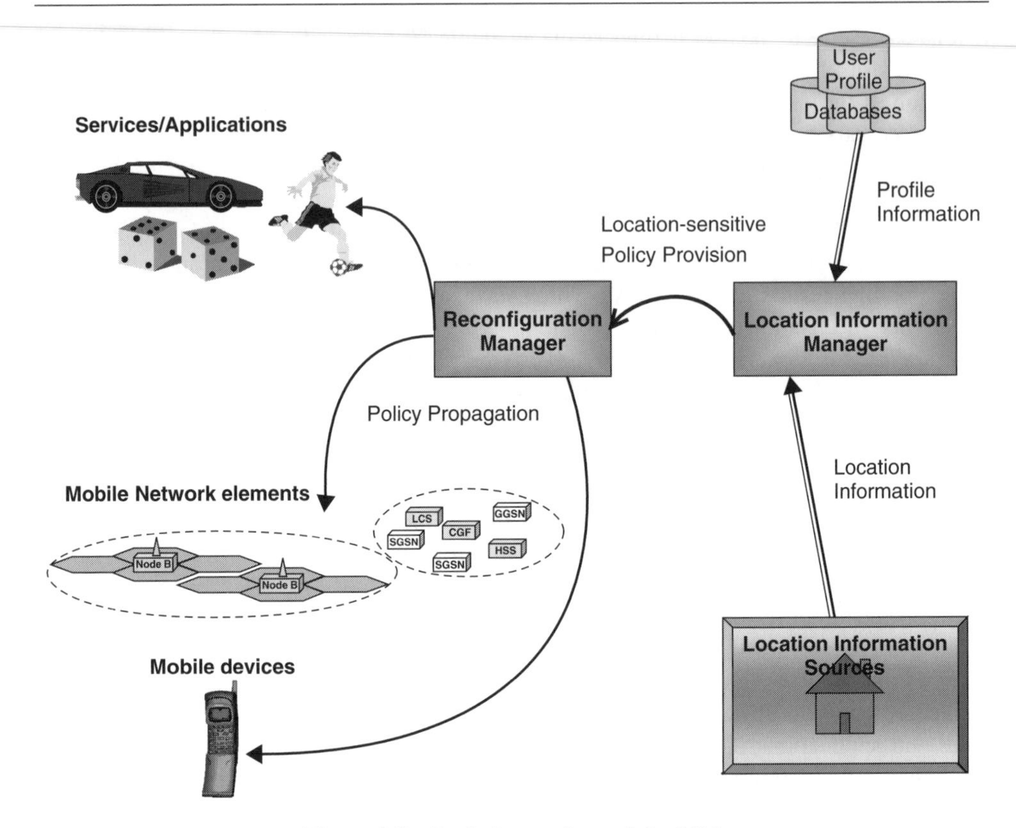

Figure 3.9 Basic interactions of the LIM

- Notification to end-users and third parties of available policies, restrictions, updates, tariffs, reconfigurations and other events that are associated with the current location of end-users or induced due to location updates that occur.
- The activation/de-activation of various special location-based services offered by the LIM (e.g. location-based charging/billing [28]).

The LIM enables the location and mobility based reconfiguration of network nodes assisting the RCSPM to provide users and VASPs with context-aware and customised service provision. Furthermore, it enriches even non-location aware VASs with location features enabling better personalisation and customisation. Location tracking takes place with respect to user privacy settings included in the user profile. Hence, the LIM prompts the involved user for authorisation prior to any location retrieval action, in the case that his profile requests so.

By further deploying the operations provided by the LIM the functionality of the RCSPM could be enhanced to include novel open, advanced location aware services such as: (i) Advanced mobility management providing optimised network planning, location registration, paging and handover management. (ii) User equipment reconfiguration with software and protocol updates/upgrades. (iii) The selection and reallocation of network nodes based on user mobility, for network performance optimisation and better service offering to the end-user. (iv) Conversion of multimedia services for content adaptation (e.g. voice to text, video to voice) based on user location and associated user preferences.

5. *Packaging and Downloading Module (PDM).* The PDM is responsible for packaging all the software components and other supporting resources (e.g. images) required to execute a service in a single archive and provide it for download to the mobile client. The single archive produced is dynamically tailored to the context of the particular service selection request (e.g. terminal and network characteristics, user preferences).

An important design issue, specific to service adaptation, is where the application packaging and bundling function is performed. In the wired Internet world, where terminal connectivity constantly becomes cheaper, higher data rate and more reliable, current approaches [29, 30] propose that individual components constituting or supporting an application are downloaded separately by the client (Figure 3.10 (a)). In a wireless network, however, this leads to significant overhead as far as the use of the scarce resource of bandwidth is concerned. Thus, we have chosen to place this function at the server side (Figure 3.10 (b)), where all executables, libraries and supporting software that are required to run an application are packaged in a single file and downloaded by the terminal over a single network connection. The major advantage of producing a single file containing all the data necessary for service execution is that a client is able to download the archive over a single network connection, thus avoiding superfluous network traffic and excessive signalling.

6. *User Profile Manager.* The RCSPM includes user-profiling logic to enable service discovery, adaptation and provision according to user preferences. The user profile contains the following types of information:

- User identification and security data. The primary key for user identification is the email address, so that it is independent of the access network employed for terminal connectivity in each user session.

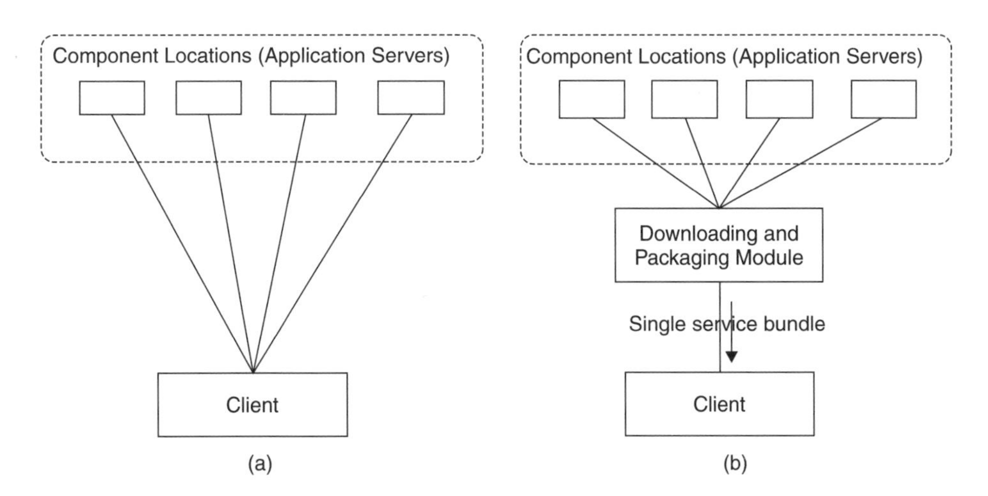

Figure 3.10 Different alternatives for the location of the component downloading and application packaging functions: (a) client-based, as in existing approaches on the internet; (b) the proposed, mediator-based approach

- Generic, service independent user preferences (e.g. language, default tariff class, security preferences).
- User interface preferences (e.g. font size, preferred media type).
- A list of user-specific favourite services.

The master copy of the user profile is stored in an appropriate database within the platform. Caching the profile, for efficiency reasons, is possible in alternative locations [e.g. the terminal, the User Interaction Management Module (UIMM)]. The system gives the user the ability to view and update user profile information at any time.

7. *Service Database Manager.* This module is an interface to a database of services that is maintained by the platform operator. This information is necessary to various other modules of the RCSPM that are responsible for service discovery, adaptation, management and provisioning. The service database may be dynamically accessed and updated by the VASPs.

The structure and content of the service profile are based on a service model that has been defined a priori. A major design goal of the proposed scheme is to be generic, to be able to represent a very broad range of services, as well as to be independent of the service implementation; that is, application programmers need not be aware of this scheme at the development stage for the application to be compatible with it. This an important step towards service abundance and diversity, features that typically benefit all market players involved in service provision by increasing revenues as well as end-user satisfaction.

A generic, high-level model of a service is depicted in Figure 3.11 as a UML class diagram. For simplicity, only the data attributes of the various classes that are relevant to service adaptation are shown.

Essentially, an application (ServiceSoftware in Figure 3.11) offered to mobile users comprises two parts: a *stationary* part (ServiceServerSoftware), which encapsulates the main service logic and the required interactions with other entities in the fixed network (e.g. back-end systems, other enterprise servers), and a *downloadable* part (ServiceClientSoftware), which typically is downloaded to the end-user's terminal on demand (when a user chooses to access a service) and implements functionality such as a graphical user interface, communication with the stationary part and potentially other application logic. It is worth noting that certain services (e.g. a stand-alone gaming application) may not include a stationary part.

The movable part of an application (ServiceClientSoftware) normally comes in a variety of versions (ServiceClientVersion). Each version is targeted to certain classes of terminals, networks and users. This fact is reflected in the TermReq, NetworkCaps and UserPref data member attributes of a ServiceClientVersion. Service versions abstract an aggregate service (made of components) and present different alternative deployment configurations with regard to the structure and/or content of a specific application. A service version consists of a mandatory part (CorePart) and zero or more optional parts (OptionalPart). Each part comprises one or more components (ServiceComponent) that may have additional terminal or network requirements and support further user preferences than the corresponding ServiceClientVersion. These service components can be single, autonomous, re-usable among different services, functional components. The decision whether or not to include optional components in the application package that is downloaded on demand by a particular user is based on these additional parameters. Optional elements can themselves contain further optional components (OptionalComponents), whose suitability for

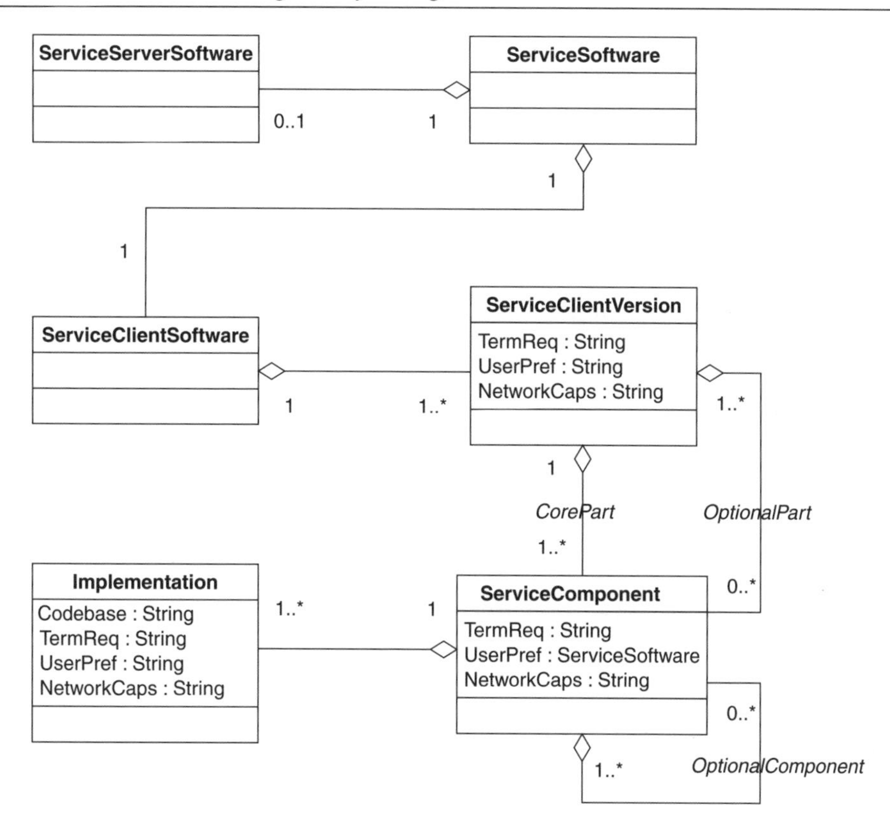

Figure 3.11 A high-level service model

a particular service client installation should be recursively determined by the service adaptation function. Each component has one or more different implementations (Implementation), each identified by a unique URL, where the corresponding code can be found. The selection of the single implementation that is packaged together with a downloadable application client is again customised to the terminal, network and user involved in the specific service access request.

The above-described model is characterised by generality; it supports a vast range of services. It is worth noting that the model is not restricted to end-user applications (i.e. applications that the user is aware of and interacts with); in future systems it can also represent what nowadays are considered as system services (e.g. dynamic selection and switching between radio access technologies) that need to be customised to the parameters of the current context, since the implementation and deployment configuration of those services typically would follow the stationary–movable part structure.

In the platform prototype implementation, while service profiles are stored in a relational database service data that is exchanged between the Service Database Manager and its client is encoded in XML, according to a specific Document Type Definition (DTD). This approach facilitates interoperability of service information representation. A service version fragment of the service XML descriptor DTD is depicted in Figure 3.12.

```
<!ELEMENT ServiceClients (Version+)>
<!ELEMENT Version (VersionName, VersionDescription, ClientSoftware, TermReq, UserPref,
        NetworkCaps, Communication?, Components, IPAddr, IPPortLow, IPPortHigh,
        PricingModelNumber, TariffClassNumber, CostDescription, L4Timeout, MexeClassmark,
        OSAInfo)>
<!ELEMENT Communication (CommunicationService?, TransportProtocol, QoSIndicator)>
<!ELEMENT VersionName (#PCDATA)>
<!ELEMENT VersionDescription (#PCDATA)>
<!ELEMENT ClientSoftware (CorePart, OptionalPart?)>
<!ELEMENT CorePart (Component+)>
<!ELEMENT OptionalPart (Component*)>
<!ELEMENT Component (Description, Implementation+, OptionalPart?)>
<!ELEMENT Implementation (Codebase, TermReq, UserPref, NetworkCaps)>
<!ELEMENT Codebase (#PCDATA)>
<!ELEMENT IPAddr (#PCDATA)>
<!ELEMENT IPPortLow (#PCDATA)>
<!ELEMENT IPPortHigh  (#PCDATA)>
<!ELEMENT PricingModelNumber (#PCDATA)>
<!ELEMENT TariffClassNumber (#PCDATA)>
<!ELEMENT CostDescription (#PCDATA)>
<!ELEMENT L4Timeout (#PCDATA)>
<!ELEMENT MexeClassmark (#PCDATA)>
<!ELEMENT TermReq (#PCDATA)>
<!ELEMENT UserPref (#PCDATA)>
<!ELEMENT NetworkCaps (#PCDATA)>
<!ELEMENT CommuicationService (#PCDATA)>
<!ELEMENT TransportProtocol (#PCDATA)>
<!ELEMENT QoSIndicator (#PCDATA)>
<!ELEMENT OSAInfo (RequiresOSA, FwOpData, OSAService+)
```

Figure 3.12 XML DTD fragment for service client data

3.4 Reconfigurability Management as Adaptability Enabler

The introduction of reconfigurability management in mobile networks is expected to have positive effects on various others aspects of mobile service provision and in particular it will promote the adaptability of services and protocols. This contribution of reconfigurability can be considered as twofold: first, by enabling the applicability of adaptive protocols and applications, and secondly, by the re-use of network reconfigurability management mechanisms in application and protocol adaptation procedures.

Adaptive protocols can be applied and fully exploited provided that the underlying infrastructure can be dynamically reconfigured to employ on-the-fly the protocol version and configuration that is optimal given the current context of communication. This behaviour is mostly desirable in 4G systems, where the technologies/protocols used even in the lower layers of the network communications will need to be modified at run-time in order to provide end-users with the best possible quality of service in a given context.

The overall procedure of reconfiguring/adapting an entity comprises certain generic phases that are independent of the type of entity and the type of action that is performed in each particular case. A high-level identification of these phases can be the following:

1. Context identification, which corresponds essentially to collecting/monitoring the appropriate information from the current environment.
2. Decision on the most suitable (given the current context) transition to another state/behavior of the entity.
3. Application/activation of the selected transition.

In many systems there is no clear demarcation between the implementations of these phases, which results in a tight coupling among them and hampers the re-usability of certain functions that can be useful in other contexts. In highly dynamic environments like those of the forthcoming next generation mobile communication systems, functions 1 and 2 above can be re-used in a variety of adaptation and reconfiguration actions, provided that they are developed in a generic manner, so that they are fully independent of the type of target entity and desired action in a reconfiguration operation. The design and implementation of such a generic mechanism that takes advantage of the decoupling between distinct reconfiguration phases can be found in ref. [31].

3.5 Platform Operation Scenarios

3.5.1 Service Deployment

A major operation of the platform is service deployment, which is triggered by the VASP via an appropriate (e.g. web-based) interface. The platform determines how the service should be deployed and provided making use of the service definition, which is provided by the VASP, either by the above-mentioned interface or in the form of a precompiled XML document, conforming to a universally common DTD. The service profile includes a large number of parameters such as available service versions, location of downloadable service clients, terminal and network requirements, networks over which the service should be deployed, security information, etc. The Service Deployment Manager first checks the request parameters and, in the case that they are acceptable, stores the appropriate service data in the service database. Subsequently, the RCM, based on the profiles of the new service and the target networks, determines the appropriate reconfiguration actions and

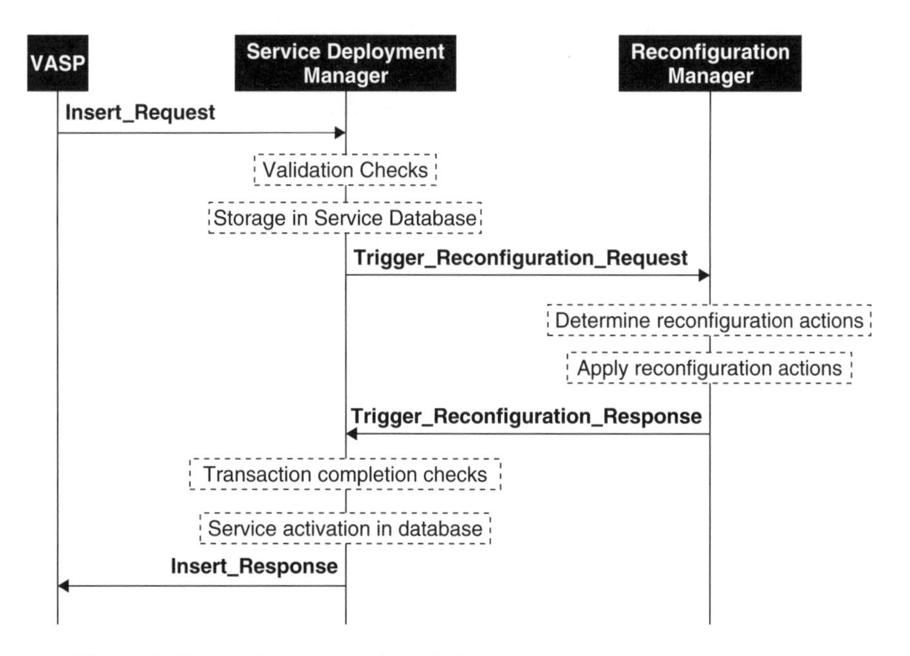

Figure 3.13 Platform interactions during the service deployment operation

triggers their application (e.g. via open APIs or proprietary protocols). Once all necessary actions have been performed, the transaction is considered complete, the service is made available to mobile users (included in the user portal for discovery) and the VASP receives a response message indicating the outcome of the entire operation.

The VASP is then able subsequently to remove a service from the platform or perform complex modifications of its definition (e.g. change service tariffs, insert, delete or modify individual service versions). It is worth noting that all the above operations are dynamic in the sense that they are performed on-line and that overall service provision and system operation are not affected or interrupted during them. A diagram depicting the high-level interactions and procedures taking place during the deployment of a new service is shown in Figure 3.13.

3.5.2 *Service Discovery, Adaptation, Downloading and Execution*

In this section we describe in detail the proposed mechanism for service adaptation. This is illustrated in Figure 3.14, where a sequence diagram is presented that includes all the interactions between the user terminal, the service provision platform and the VASP required for user login/logoff, service discovery, selection, adaptation and downloading.

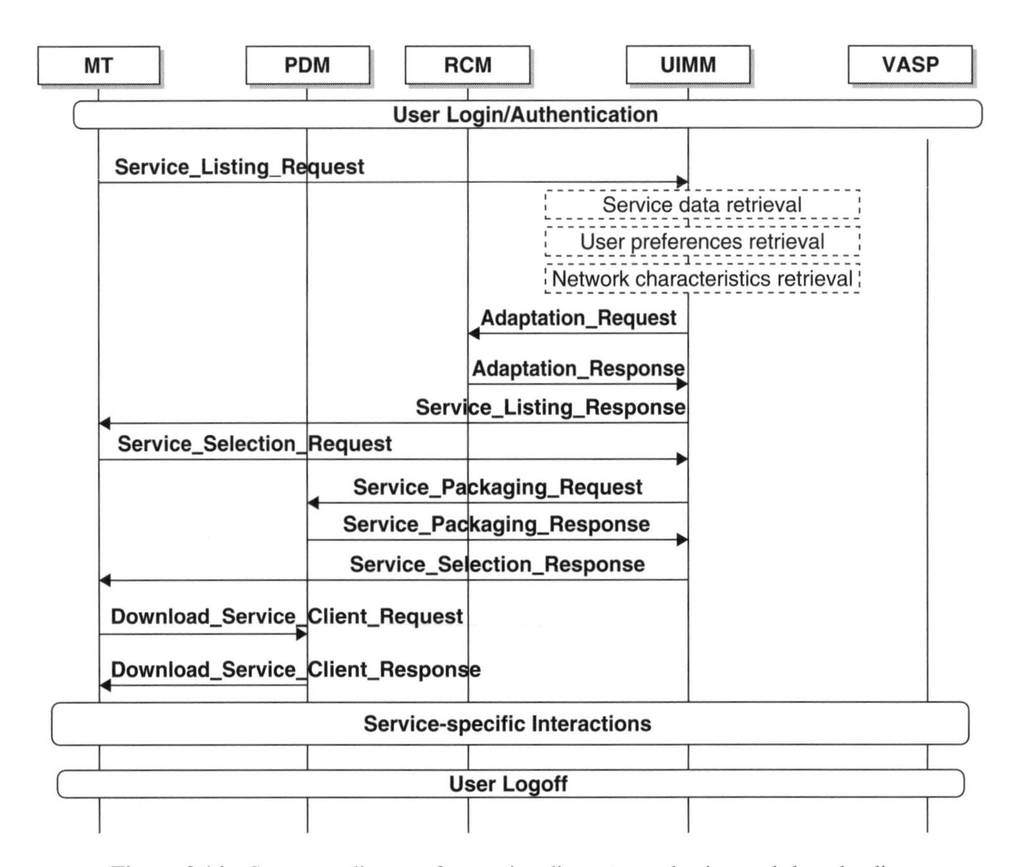

Figure 3.14 Sequence diagram for service discovery, selection and downloading

The above-mentioned functions are performed through the following steps:

1. The user logs in the platform, an operation that includes user authentication.
2. The user issues a request for a service listing, with criteria such as service category and keywords. Terminal capabilities and possibly also network characteristics are sent to the UIMM as request parameters.
3. The UIMM retrieves from the service database the profiles of applications that match the user query and formulates the service listing according to terminal capabilities, network characteristics and user preferences. To achieve that, it collects all the required profile data and forwards it to the adaptation module in the RCM, which performs the necessary matching, by identifying the services/service versions that are suitable for the specific user/terminal/network and also determining the appropriate configuration (e.g. inclusion of optional components or not) of each service version. The customised listing is returned to the user.
4. The user selects a service from the list.
5. The UIMM triggers the PDM to perform packaging of the adapted service, and after completion of this task returns the URL of the corresponding customised package to the user.
6. The user downloads and executes the customised service. Service execution typically results in interactions between the downloadable and stationary service parts.
7. Steps 2–6 can be repeated as many times as the user desires to discover and access other services. At the point where the user does not wish to use any other applications, he logs off the platform.

References

[1] UMTS Forum Report No. 9, 'The UMTS third generation market – Structuring the service revenues opportunities', available from *http://www.umts-forum.org/*

[2] UMTS Forum Report No. 10, 'Shaping the mobile multimedia future – An extended vision from the UMTS Forum', available from *http://www.umts-forum.org/*

[3] UMTS Forum Report No. 11, 'Enabling UMTS third generation services and applications', available from *http://www.umts-forum.org/*

[4] 3G TS 23.060: 'General Packet Radio Service (GPRS); Service description'.

[5] 3G TS 23.228: 'IP Multimedia Subsystem (IMS)'.

[6] 3GPP TS 23.127: 'Virtual Home Environment'.

[7] Parlay specifications, available from *http://www.parlay.org/specs/index.asp*

[8] N. Houssos, M. Koutsopoulou and S. Schaller, 'A VHE architecture for advanced value-added service provision in 3rd generation mobile communication networks', *Proceedings of the Globecom 2000 Workshop on Service Portability and Customer Services Environments, San Francisco, USA*, **November**, 69–79, 2000.

[9] 3G TS 23.057: 'Mobile Station Application Execution Environment (MExE)'.

[10] 3G TS 23.002: 'Network architecture'.

[11] J. Pereira, 'Beyond third generation', Wireless Personal Mobile Communications (WPMC) 1999, 22 September, Amsterdam, The Netherlands, 1999.

[12] V. Gazis, N. Houssos, A. Alonistioti and L. Merakos, 'Evolving perspectives of 4th generation mobile communication systems', 13th International Symposium on Personal, Indoor and Mobile Communications (PIMRC 2002), 15–18 September, Coimbra, Portugal, 2002.

[13] W. Mohr, 'Access network evolution beyond third generation mobile communications', *IEEE Communications Magazine*, **December**, 122–133, 2000.

[14] M. Annoni, et al., 'Radio access networks beyond 3G: A first comparison of architectures', IST Mobile Communications Summit 2001, 4 September, Barcelona, Spain, 2001.

[15] N. Alonistioti, N. Houssos and S. Panagiotakis, 'A framework for reconfigurable provisioning of services in mobile networks', Invited paper at the International Symposium on Communications Theory & Applications (ISCTA '01), July, Ambleside, Cumbria, UK, 2001.

[16] A. Lazar, 'Programming telecommunication networks', *IEEE Network Magazine*, **September/October**, 1997.

[17] J. Keijzer, D. Tait and R. Goedman, 'JAIN: A new approach to services in communication networks', *IEEE Communications Magazine*, **January**, 2000.

[18] IEEE P1520 working group, *http://www.ieee-pin.org*

[19] D. Wetherall, J. Guttag and D. Tennenhouse, 'ANTS: Network services without the red tape', *IEEE Computer*, **32**(4), 42–49, 1999.

[20] D. Wetherall, 'Active network vision and reality: Lessons from a capsule-based system', *Proceedings of the 17th ACM Symposium on Operating System Principles (SOSP '99)*, **December**, *Kiawah Island, SC*, 1999.

[21] N. Houssos, et al., 'Value-added service management in 3G networks', IEEE/IFIP Networks Operations and Management Symposium (NOMS 2002), 15–19 April, Florence, Italy, pp. 529–545, 2002.

[22] N. Houssos, V. Gazis and A. Alonistioti, 'A flexible management architecture for the support of advanced business models in 3G mobile service provision', 1st International Conference on Mobile Business, 8–9 July, Athens, Greece, 2002.

[23] M. Koutsopoulou, C. Farmakis and E. Gazis, 'Subscription management and charging for value added services in UMTS networks', IEEE Semiannual Vehicular Technology Conference VTC2001, May, Rhodes, Greece, 2001.

[24] M. Koutsopoulou, A. Kaloxylos and A. Alonistioti, 'Charging, accounting and billing as a sophisticated and reconfigurable discrete service for next generation mobile networks', Proceedings of the IEEE Semiannual Vehicular Technology Conference (VTC2002), 24–28 September, Vancouver, Canada, 2002, accepted for publication.

[25] N. Houssos, S. Pantazis and A. Alonistioti, 'Generic adaptation mechanism for the support of context-aware service provision in 3G networks', 4th IEEE International Conference on Mobile Wireless Communication Networks (MWCN 2002), 9–11 September, Stockholm, Sweden, 2002, accepted for publication.

[26] 3GPP TS 23.271: 'Functional stage 2 description of LCS'.

[27] S. Panagiotakis and A. Alonistioti, 'Intelligent service mediation for supporting advanced location and mobility aware service provisioning in reconfigurable mobile networks', *IEEE Wireless Communications Magazine*, **October**, 2002.

[28] S. Panagiotakis, M. Koutsopoulou, A. Alonistioti and A. Kaloxylos, 'Generic framework for the provision of efficient location-based charging over future mobile communication networks', PIMRC, September, Lisbon, Portugal, 2002.

[29] Java Web Start technology, Java Network Launching Protocol (JNLP), *http://java.sun.com/products/javawebstart/*

[30] The Open Software Description Format Specification, W3C Note, available at *http://www.w3.org/TR/NOTE-OSD.html*

[31] N. Houssos, S. Pantazis and A. Alonistioti, 'Generic adaptation mechanism for the support of context-aware service provision in 3G networks', 4th IEEE International Conference on Mobile Wireless Communication Networks (MWCN 2002), 9–11 September, Stockholm, Sweden, 2002, accepted for publication.

4

Adaptive Protocols

Spyros Panagiotakis and Vangelis Gazis
Communication Networks Laboratory, University of Athens

Despite the strong trend to support an evolutionary dimension of mobile communication architectures and to re-use existing infrastructure investments as much as possible, a number of wireless access technologies have emerged over the last decade. Targeted to different application domains and aimed to satisfy different quality of service levels, wireless system architectures tend to exhibit a varying number of similarities and dissimilarities – both in structure and function. However, reconfigurability as an emerging enabler for system integration and flexible service provision support brings about the impeding support for advanced software development practices in mobile communication system software. Numerous research efforts have been undertaken or are currently under way in the fields of adaptive, parameterisable or extendable protocols. The adaptive protocols are characterised by their ability to be configured and initialised at session initiation based on specific communication conditions (e.g. user-, terminal-, network- or service/application-dependent) and then to be reconfigured and re-adapted properly, even at run-time, following the changes in their communication environment. This chapter, which can be considered as an introductory survey to protocol adaptability, presents the origins, drivers and general requirements behind protocol adaptability and protocol adaptation management. Furthermore, special focus is given to the applicability of adaptable protocol infrastructures in the new era of communications.

4.1 Introduction

Over the last decade, the remarkable evolution of mobile communication technologies has spawned numerous wireless access technologies, each tailored to a particular application. Despite the strong trend to support an evolutionary dimension of mobile communication architectures and to re-use existing infrastructure investments as much as possible, a number of revolutionary wireless access technologies matured to commercial production levels while several more are expected to emerge in the near future.

This continuing proliferation of wireless access technologies and the respective system architectures tends to make their manageability of a difficult task. As a result of

Software Defined Radio: Architectures, Systems and Functions. Edited by M. Dillinger, K. Madani and N. Alonistioti
 ISBN: 0-470-85164-3

being designed with varying performance objectives and interworking requirements, the architectural blueprints of wireless systems bear numerous similarities and dissimilarities – both in structure and function [1]. For instance, the majority of cellular system architectures specify some form of mobility management (MM) functionality. However, since the development instruments used to implement such functionality are typically proprietary and lack support for modern software development practices, implementation re-use has remained an elusive goal in mobile communication system software. That has led to higher development costs and an increased time-to-market for new mobile communication products, an undesirable effect that could potentially hamper the long-term economic prosperity of the mobile communication industry.

Consequently, it is no surprise that the research community has been gradually shifting its focus to make up for that particular handicap. Numerous research efforts have been undertaken or are currently under way in the fields of adaptable, parameterisable, extendable and, generally, reconfigurable protocols [2]. This chapter[1] addresses various issues related to the protocol adaptability and the research activities in this area. Section 4.2 presents the definition and the extent of applicability of adaptive protocols. Section 4.3 discusses the requirements induced on the SDR architecture by the insertion of adaptive protocol architectures in a communication environment and Section 4.4 introduces some technical issues related to the provision of adaptive protocols. Section 4.5 discusses the requirements for generic management architectures providing for protocol adaptability and reconfiguration.

4.2 Scope and Applicability

4.2.1 Definition

According to a broad definition a system can be considered a product, process or service that converts a set of inputs into a set of outputs. From this point of view an adaptive protocol can be considered to be a reconfigurable system that alters its functionality or form so that it meets the communication conditions (the 'inputs') in its current communication environment. In order for an adaptive protocol to be more than simply 'configurable', it must be possible to change either whilst it is in use or by taking it out of use for a short time [3]. Hence, adaptive protocols are characterised by their ability to be configured and initialised on session initiation based on specific communication conditions (e.g. user-, terminal-, network- or service/application-dependent) and then to be reconfigured and re-adapted properly, even at run-time, following changes in their communication environment.

Different models, studying how protocols can be reconfigured in terms of granularity, intelligence and the control method required, have been proposed. However, most models conclude that adaptive protocols can be divided into:

[1] This chapter constitutes an introductory survey on protocol adaptability focusing especially on the applicability of adaptable protocol infrastructures in the new era of communications and presenting the general requirements, origins and drivers behind protocol adaptability. This chapter complements Klaus Moessner's work 'Protocols and Network Aspects of SDR' in the companion volume *Software Defined Radio: Enabling Technologies*, pp. 339–364, John Wiley and Sons, Ltd, 2002, which explores protocol adaptability from a different point of view. The authors encourage readers of this chapter to read also Moessner's work in order to gain a global view on this issue.

- *Extendable protocols:* modular protocols that extend, improve or enhance their functionality by incorporating into their architecture the appropriate updates or complementary modules (e.g. new codecs, classes or boosters). Such architectures include the appropriate environments for downloading and installing dynamically the new software [8] as the networking environment changes (e.g. due to the end-user's mobility).
- *Parameterisable protocols:* protocols that modify at run-time their functionality and behaviour according to the parameters applied in their input, despite their initial parameterisation. By supplying them with the appropriate monitoring and management Application Programming Interfaces (APIs), users of these protocols (e.g. the services/applications on the terminal and network sides) are allowed to control the protocol behaviour in a structured manner, whilst protocols remain re-usable, modular and extendable [4]. The on-demand provision of service data flows with Quality of Service (QoS) is the main example of parameterisable protocols.

An adaptive protocol can be simply extendable or parameterisable or both (parameterisable and extendable).

4.2.2 Protocol Adaptability in a Generic Reconfigurable Environment

The concept of Software Defined Radio (SDR) has conflicting definitions, but a detailed discussion in ref. [7] is summed up as

> we envision Reconfiguring on demand not only the terminal but also the serving network(s) and the services they provide... Upon this open framework, we envision truly 'platform'-independent applications, no longer exclusively developed by or for operators, capable of adjusting themselves to the serving network capabilities ... and the terminal characteristics, negotiating with the network to obtain the best possible service taking into account the user profile.

Hence, the basic premiss is that everything is going to be reconfigurable. This ranges from the protocols at all levels [e.g. using Global System for Mobile Communications (GSM) or Universal Mobile Telecommunications System (UMTS)], through network management (e.g. reconfiguring a network to cope with different terminal/user/network/service loading profiles), to applications (e.g. an application that normally provides streaming video switching to still monochrome images if the QoS is insufficient) [3]. The general message is that nothing is predefined and that much of this is, as its name implies, software defined. Adaptive protocols belong to this raising reconfigurable world.

One of the most critical challenges for future multiple access communications is very likely to be the enormous range of operational parameters and QoS metrics that should be met in such an heterogeneous and dynamic environment. Consequently, the prevailing design assumptions for mobile networks, devices and protocols must be re-evaluated in light of the volatility and dynamism that dominate the wireless world. Design priorities may shift from vertical goals, such as optimum performance, toward more horizontal concepts, such as sustainable operation and service ubiquity.

It is, of course, possible to conceive of a heterogeneous and reconfigurable system comprising blocks of different sizes and with different levels of intelligence. The function of the overall system is affected by the arrangement and internal configuration of its constituent blocks. In the reconfiguration chain the adaptive protocols constitute intelligent, multi-function building blocks which can be reconfigured to alter the function they perform. By re-adapting their functionality they affect the total system function, while the changes in their adjacent environment may affect, equally, their configuration. Since adaptive protocols are involved in the communication session establishment, naturally their reconfiguration is reflected end-to-end on the involved session. This probably affects the adaptability of the protocol infrastructure on an end-to-end basis including end-user terminals, interconnecting networks and service architectures. Thus, reconfiguration of adaptive protocols does not affect solely the service experience on the one side of a session. Rather it is extended to both ends and intermediate nodes, affecting vertically and horizontally the communication among involved parties. The vertical adaptability concerns the adaptability of the protocol infrastructure in a single resource, including not only the individual protocol layers' adaptability (e.g. Data Link Layer, Network Layer, Application Layer), but also the adaptability of the entire protocol stack. The horizontal adaptability concerns the protocol infrastructure on an end-to-end basis. It includes the adaptability of protocols on both the terminating and intermediate nodes (e.g. terminals, networks and services) in order to ensure sustainable service operation and maintenance.

In European research adaptability at the protocol level is examined in various projects [e.g. Configurable Radio with Advanced Software Technology (CAST) [5], Downloadable Mobile Value Added Services through Software Radio and Switching Integrated Platforms (MOBIVAS) [16, 17], and Transparent Ubiquitous User Terminal (TRUST) [35] and Smart user-Centric cOmmUnication environmenT (SCOUT) [36]], and from various points of view. Research efforts cover reconfiguration of protocol stacks at various layers, from lower to higher. Attention, for example, is given to:

- Demonstrating architectures for intelligent reconfiguration of the physical layer in wireless communication networks. This covers interoperability between GSM and UMTS at the bottom level of a communications system, together with adaptation to prevailing conditions and user requirements. Hence, a mixture of hardware and software reconfiguration is demonstrated.
- Reconfiguration actions mainly on the higher layers of the protocol infrastructure. This encompasses an intelligent and integrated service platform that mediates between end-users and service providers to orchestrate reconfigurability and provide end-users with adaptable, personalised and context-aware service provision. From the protocol's point of view adaptability includes the dynamic reconfiguration at run-time of network protocols to provide end-users with the desired QoS.
- Dynamic adaptation of the available services to the terminal capabilities, supplementing services and terminals with the appropriate JAVA classes, where is possible.
- Demonstration of how the Data Link Layer can be dynamically enhanced with link-layer boosters that improve the perceived QoS over wireless links.

The economic necessity for rapid time-to-market in an environment of complex, evolving standards is one of the reasons for the interest in reconfigurability in this area. Furthermore, the constraints on the size and power consumption of mobile equipment, along with the enormous versatility requirements in terms of applications, networking and physical environments, constitute really challenging issues that instigate the research activities [3].

4.2.3 *Applicability*

4.2.3.1 Multi-Access and Heterogeneous Systems

As mentioned at the beginning of this chapter, the ongoing proliferation of wireless access technologies suggests that future mobile communication environments will be technologically heterogeneous, pulling together numerous different wireless access technologies in a complementary manner to form an integrated system that consistently behaves as a coherent whole [25]. This suggests a system whose architectural blueprints differ significantly from the design assumptions embedded in current communication protocol software. Typical mobile communication devices, such as commercial cellular phones, are not capable of using multiple wireless access technologies simultaneously; most restrict user registration to a single cellular network at any time instance, even when radio coverage from multiple mobile network operators is available. Evidently, in multi-access and multi-operator wireless environments, the aforementioned practices fail to maximise utilisation of available system resources. For instance, for a nomadic worker whose mobile device is under UMTS and wireless Local Area Network (LAN) radio coverage, one could imagine a scenario where the end-to-end signalling between application endpoints is routed via UMTS because of its predictable performance and QoS guarantees, while the media stream is routed via a nearby wireless LAN to take advantage of its significantly greater bandwidth capacity. Ideally, as the radio coverage varies from multiple wireless access systems to a single wireless access system, the mobile device should be able to automatically detect the change and adapt its behaviour to employ the most efficient combination of available system services at any given time instance.

4.2.3.2 System Performance Enhancement

Communication protocols are designed, validated and implemented under a specific set of assumptions about their operating context. In the world of mobile communications, where new, unexpected technological developments emerge almost daily, the design assumptions of communication protocols may not remain valid over prolonged periods. As a result, deployed implementations of mobile communication protocol may fail to reach their expected performance goals.

To overcome such problems, a new approach in designing protocol software is necessary. Rather than committing to a rigidly integrated design in order to squeeze out every possible bit of performance, protocol architects should opt for a more loosely coupled design that allows a certain degree of run-time control and post-deployment evolution. By leveraging object-oriented techniques such as abstraction and encapsulation to reap system modularity and favouring composition over inheritance to enable delegation, it is possible to design protocol software that can evolve at run-time [26], both in terms of structure as well as behaviour. Abstracting the behavioural aspects of a protocol's algorithm as programmatic constructs (i.e. classes) and establishing a contract (i.e. an interface) for interaction with the rest of the protocol's software entities, facilitates the exchange of behavioural protocol implementations at run-time, thus making behavioural adaptability possible.

4.2.3.3 Global Roaming

Reconfigurability is a particularly valuable capability in technologically heterogeneous environments, especially if the exact set of possible technologies cannot be known a priori. Global roaming is one such case, primarily because spectrum regulation has not been uniformly applied on a global basis or under a common policy. Despite substantial harmonisation efforts, existing wireless access systems differ significantly across major regions of the world, both in terms of their operational parameters (e.g. allowable spectrum band, transmit and receive power levels) as well as in terms of certified mobile equipment for each region and system [27]. Without a common, globally applicable regulation policy, wireless access systems will continue to exhibit such differences, making the design and implementation of mobile devices with global circulation capabilities a difficult task.

With regard to global roaming, adaptable protocols can prove of great assistance. One can easily imagine a scenario where a mobile terminal bearing an reconfigurable protocol stack and an adaptable location/mobility management functionality enjoys global roaming by automatically detecting the set of available wireless access systems, selecting the most appropriate ones and downloading the required software and protocol implementations to use them. In this particular scenario there is no need for the mobile device to have been pre-configured for any particular wireless access system, since it can detect and dynamically download all required software modules over the common signalling channel.

4.2.3.4 *Ad hoc* Networking

Ad hoc wireless networks consist of mobile nodes interconnected by multi-hop communication paths. Unlike conventional, i.e. cellular, wireless networks, *ad hoc* networks have no fixed network infrastructure or administrative support. The topology of the network changes dynamically as mobile nodes join or depart the network or radio links between nodes become unusable.

In such a topologically dynamic environment, the mobile nodes may need to exhibit disparately varying behaviour for the *ad hoc* network to remain operational, e.g. a mobile node may need to relay its neighbours' traffic in addition to its own traffic. Realising such an adaptable behaviour is a challenging task that should be performed in a resource-optimal manner, given the inherent volatility of the quality of the wireless link and the finite energy budget of mobile devices.

For example, the continuously evolving topology of mobile *ad hoc* networks is not particularly friendly to the classic Internet Protocol (IP) routing algorithm. However, the widespread deployment of the IP protocol, the availability of mobility-specific extensions for it and its hop-by-hop operation, make it a strong candidate for *ad hoc* network protocols. Ideally, an IP implementation should be equally applicable in wireless, *ad hoc* and wireline networks, with a minimal performance penalty. To achieve that goal, a design approach that carefully identifies operating assumptions and isolates the varying aspects of functionality into programmatic constructs is the best one to follow. For instance, one could encapsulate the IP routing algorithm as a class and declare an interface that exposes the routing algorithm's decisions. Provided appropriate system support functionality is available, that would facilitate changing the protocol routing algorithm at run-time. For a mobile node that roams from a totally infrastructure-based (e.g. cellular) network to an infrastructure-less (e.g. *ad hoc*) network, that would allow significant performance enhancement (e.g. by enabling the use of multiple interfaces to multiple networks and traffic relaying across network interfaces). Furthermore, such improvements do not necessarily come at the cost of an implementation bloated with rarely used or under-utilised functionality, particularly if over-the-air download of hot-plug software module is possible.

4.3 Requirements for the SDR Architecture

In principle, protocol adaptability may manifest itself by a change in structure, behaviour, or both. Structural change relates to protocol layering and, from an architectural perspective, corresponds to either the addition or removal of a protocol layer. Behavioural change relates to protocol specification and corresponds to the manifestation of varying behaviour without violating the correctness of the protocol operation.

For structural change to be possible, candidate protocols must conform to a specific abstract contract, henceforth termed an *interface specification*. Being abstract, the interface specification does not impose any kind of commitment on a particular implementation. Many modern programming languages support such interface specifications or semantically equivalent programming structures (e.g. IDL, Java, Smalltalk, C++).

However, for behavioral change to be possible, the logical architecture or the protocol implementation itself must provide an appropriate separation between those parts of the protocol's algorithm that may exhibit varying behaviour and the ones that should not, i.e. the parts that should abide by the protocol specification in all cases. That suggests that the aforementioned principles and requirements pertaining to structural – or inter-protocol – change must also hold within the internals of the protocol implementation itself, i.e. intra-protocol.

The reconfiguration process deals primarily with the identification of the most suitable choice among a set of available solution implementations, thus revealing the notions of specification and implementation that were mentioned as the cornerstones of protocol adaptability:

- The notion of specification concerns an abstract contract for a particular well-understood functionality. For instance, specification of a transport protocol such as a Transmission Control Protocol (TCP) would include the service primitives provided by the transport protocol, its finite state machine, its prerequisite implementations, as well as other protocol metadata.
- The notion of implementation is associates with a particular specification and concerns a real-life instance of the functionality described by the specification. Many modern development environments and programming languages provide support for managing the association between specification and implementation – or between the abstract and the concrete, if you like.

When examining the multitude of wireless access systems available today, one can readily identify numerous commonalities in various parts of the system architecture. Location management, mobility management, paging, call control, session management and numerous other functional entities of existing wireless access systems exhibit significant commonalities, yet are typically designed, implemented and deployed vertically in independent implementations, even in the case where these reside in the same physical device. As the set of wireless access technologies that must be realised in a single device increases, the vertical model imposed by this legacy practice leads to significant resource penalties that lengthen time-to-market and hinder the development of innovative products and services.

To achieve re-usability – and reconfigurability – of protocol software, protocol architectures must not only build upon the principles of abstraction and implementation, but also on hierarchical aggregation of common functionality. Whereas abstraction and implementation are fundamental reconfigurability enablers, hierarchical aggregation of functionality enables software re-use and simplifies reconfigurability management. By exploiting inheritance among protocol classes, protocol implementations become more robust and flexible. For instance, instead of incurring the cost of designing, implementing and validating distinct location management protocols for GSM, General Packet Radio Service (GPRS) and UMTS, it would be possible to agree on the common functional denominator of the aforementioned location management functionality and define an specification for GSM, GPRS and UMTS by inheriting from such a base specification [2]. That would greatly simplify protocol development by facilitating the re-use of a reference implementation validated and certified for correctness of operation across multiple different systems, thus allowing the realisation of adaptable and extendable protocol stacks.

Besides being adaptable and extendable, protocol stacks for future mobile communication systems should also be readily parameterisable. The term 'parameterisation' refers to accessing and managing the operational state information that resides within each particular protocol. Currently, protocol parameterisation capabilities are virtually non-existent, primarily because the majority of existing protocols exchange signalling with their peers in an end-to-end fashion (e.g. TCP), providing little or no support for the out-of-band management of protocol state from a higher-layer intelligence.

Over the last decade the field of policy-based management and policy provisioning has emerged to address the dynamic parameterisation of protocol operation in a framework supporting higher-layer protocol management. More specifically, policy provisioning prescribes the specification of a virtual database, termed *policy information base*, for each protocol that records the minimal set of a protocol's operational parameters in an organised structure [28, 29]. Using the Common Open Policy Protocol for Policy Provisioning (COPS-PR) [30], Policy Decision Points (PDPs) may autonomously provide policy information to Policy Enforcement Points (PEPs). Each PEP concerns a particular functional protocol and different functional protocols are supported by multiple independent PEPs. Likewise, each PDP concerns a particular functional protocol and may provide policy information to multiple PEPs. Recent proposals have highlighted the applicability of policy-based management for third generation mobile communication networks [31].

4.4 Issues on Adaptive Protocol Provision

4.4.1 Protocol Downloading

As a property, run-time protocol adaptability is derived from the architectural decisions made during protocol design and implementation [26]. The aforementioned software development practices (i.e. object-oriented design, abstraction, encapsulation, use of interface contracts) facilitate the interoperable specification and implementation of protocol components. Complementary to these desirable architectural properties, protocol downloading facilitates the evolution of available protocol components in a communication device after deployment and throughout its operational lifetime. By downloading and installing protocol software, mobile devices can be empowered with new or improved functionality on demand, e.g. to deal with use cases that were not anticipated during the protocol design stages. By choosing to define and implement a protocol architecture in terms of simpler functional components [32] (e.g. an error-checking component), protocol extensibility by over-the-air downloading of protocol modules is greatly simplified. No longer is it mandatory to download an entire – monolithic – protocol or protocol stack over the precious wireless interface; only the necessary protocol components for the desired function's extensions need to be downloaded; already installed protocol components are preserved and reused. Figure 4.1 illustrates the reconfiguration required on a typical protocol stack to accommodate a newly downloaded software/protocol module.

However, a number of technical as well as administrative and commercial issues must be addressed before protocol downloading becomes an everyday reality:

- *Regulatory regime.* Ideally, protocol downloading should not be restricted to the device manufacturer or the wireless network operator domains, but rather should encompass all players in the software development market. That would increase competition which, in turn, would lead to differentiated – and affordable – offerings of improved quality. However, numerous complex regulatory issues (e.g. certification of correct operation, distribution of liability) need to be worked out first.
- *Protocol run-time system.* For the interoperable use of protocol implementations from different manufacturers to be possible, the specification of internal interfaces between

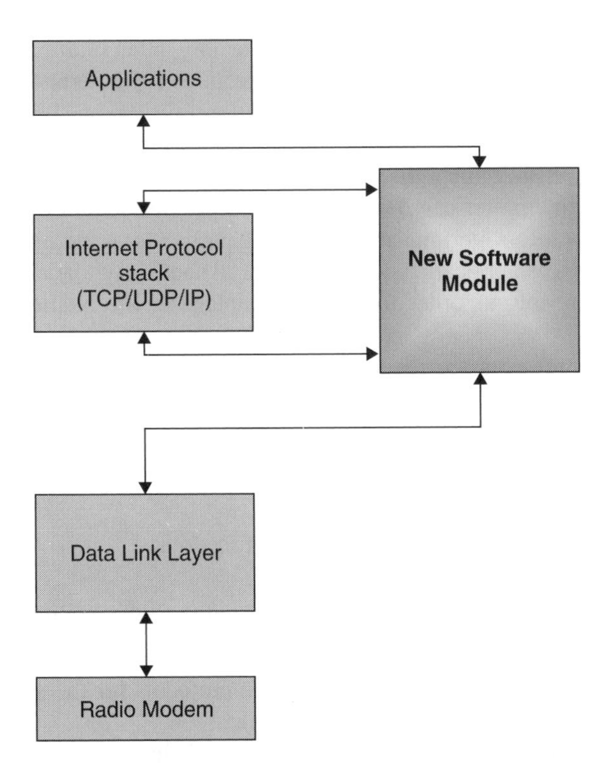

Figure 4.1 Protocol stack reconfiguration following protocol/software downloading

protocol components is not enough. It is also necessary to fully define all required support services from the operational environment (e.g. required API instances, compliance to standards such as POSIX or ISO).

- *Capability exchange.* Whether appropriate certification exists for a protocol implementation, other factors can also determine whether a particular protocol implementation can be successfully downloaded and installed on a particular mobile device. In addition to issues such as required API instances and system libraries, resource load levels play an important role, since resource unavailability or insufficiency may hinder the download of a protocol implementation. Important resources concern both the mobile device itself (e.g. availability of sufficient persistent storage or cache for the download, energy supply levels, visualisation capabilities, available content transcoding codecs) and the currently available network service (e.g. available bandwidth, error resilience of the wireless link). It is therefore necessary to conduct some form of capability exchange between the mobile device, the download server hosting the protocol implementations and the currently employed wireless access networks. Important enough, the capability exchange process should be pessimistically biased towards a negative outcome to account for a degree of uncertainty (e.g. the mobile device energy supply will drain at a faster rate than before if the user engages into more actions).

4.4.2 Adaptive Protocol Stacks

With the rapid growth in the bandwidth of communication systems, it is anticipated that the type of application and content will change unpredictably in the future. In such an evolving environment the idea of dynamic protocol architectures [4] asserts that no fixed set of protocols *can* satisfy the needs of all future applications. On the one hand, it means that no set of currently defined protocols will be suitable for all future communication demands. On the other hand, it underlines the requirement for future protocol architectures to be modular so that an interpreter can demand and load protocols from a library. Modularity in the design of protocols permits the update of protocols as required and allows the communication services/applications to access the dynamically loadable, extendable and parameterisable protocols in order to request maintaining the quality of service at the desired level [3].

Communication protocols are complex because they typically involve several activities that may happen concurrently at different places. As a result, several compositional techniques, which follow a 'separation of concerns' paradigm, have been proposed for designing and analysing protocols. These techniques typically involve the following three steps:

1. Divide the functionality of a protocol into subfunctions
2. Develop protocols for the subfunctions
3. Combine the protocols to obtain a protocol for the original problem

A major advantage of such a technique [12] is that the properties of the composite protocol can be inferred from the properties of the protocols for the subfunctions (which are smaller in size and therefore easier to analyse).

In ref. [13] a new design technique for enhancing existing and developing future modular protocol stacks for mobile communication systems is introduced. It is proposed to identify commonalties of signalling, user data transfer and management protocols of, for example, ISDN BRI, GSM, DECT, 3G, Bluetooth and HiperLan/2, in order to develop a generic protocol stack. This protocol stack is supplemented by standard specific parts to constitute a dedicated air-interface protocol stack. This approach offers adaptive and re-usable protocol software for SDRs and is regarded as facilitating the evolution of mobile communication systems from second to third to fourth generation by extracting and re-using the approved features of existing protocol stacks for future systems.

In ref. [9] an architecture supporting the dynamic deployment of link layer boosters for flow selective link-level packet treatment in wireless Internet access in order to optimise the link-layer transmission and improve the perceived QoS over the wireless link is introduced. To provide per-flow treatment of QoS over wireless links, the authors propose extending the standard logical protocol architecture of both mobile terminals and access points with four additional building blocks providing:

- Identification of the flows of multimedia applications.
- Mapping of individual flows to specific link-layer protocols with respect to their particular structure and QoS requirements.
- Scheduling the access of flows to the shared transmission channel.
- Access to the special features of the radio channel, such as the transmission power or the usage/non-usage of multiple antennas, through a specific common device driver API.

This reconfigurable terminal is dynamically updated by software. The appropriate protocol modules (link-layer boosters) are downloaded to update or extend the existing code on the terminal side. The proposed architecture supports a three-stage approach for the management of the life-cycle of the link-layer boosters, including automatic discovery of the code, code transfer and code installation in the terminal protocol stack. Implementation of this architecture has partially taken place within the IST project MOBIVAS.

Another research effort in the area of adaptive protocol stacks, targeting to improve the wireless Internet access, is the Remote Sockets Architecture (ReSoA). The basic idea behind the ReSoA proxy architecture [6] is to split the legacy TCP/IP (or UDP/IP) protocol stack into two pieces, moving the TCP/IP processing from the terminal to the access point and leaving on the terminal only the processing of the layers above the TCP. Communication between the two aforementioned parts of the same split protocol stack is achieved through establishment of a remote socket. To make the socket calls available in the terminal over the wireless link, a two-layer protocol structure consisting of the EXport Protocol (EXP) and the Last Hop Protocol (LHP) is used. The design of EXP is wireless-technology independent. On the one hand, the EXP ensures the remote execution of socket calls, and on the other hand, it is coupled to the TCP/IP protocol engine ensuring that the socket call semantics remains unmodified. The LHP covers the functionality of the link layer and is in any case wireless-technology dependent. Figure 4.2 illustrates the ReSoA architecture.

4.4.3 Execution Environment for Adaptive Protocols

The execution environment within which an adaptive protocol works should be flexible enough to support the easy and secure reconfiguration of hosting protocols. Furthermore, it should be structured by re-usable building blocks that will not bound or limit to any

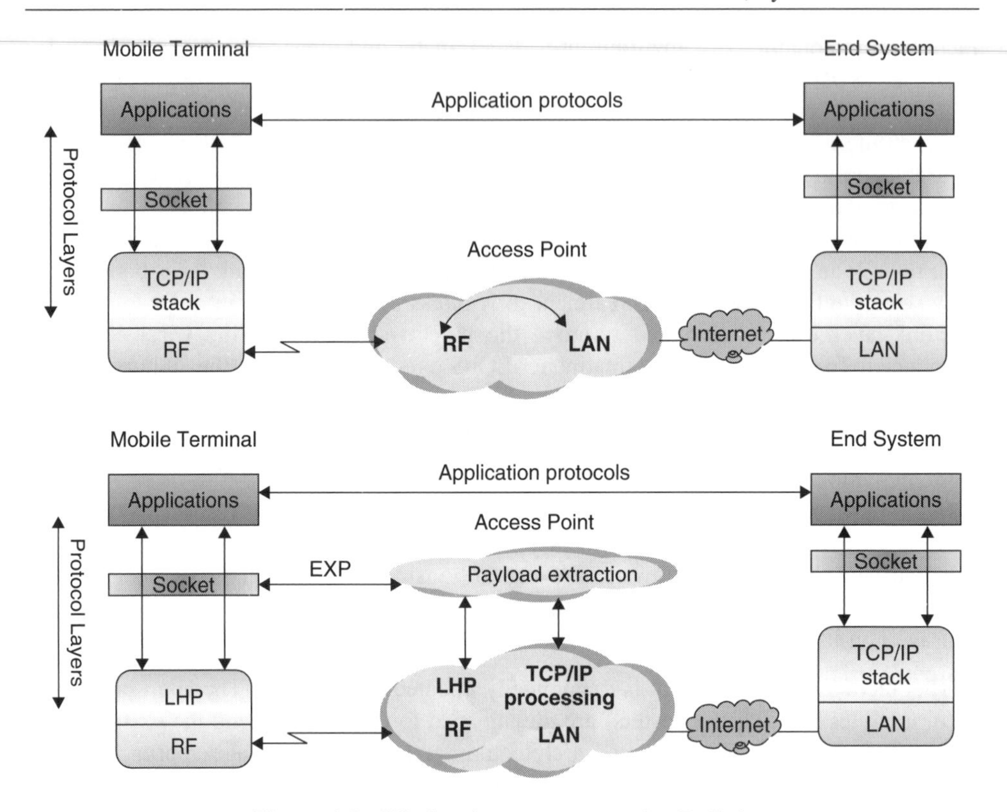

Figure 4.2 Wireless internet access using ReSoA

specific protocol. Figure 4.3 illustrates the structure of such an execution environment on the terminal side. In general an execution run-time environment for adaptive protocols should provide:

- A security infrastructure that will enable adaptive protocols to be securely reconfigured by service/applications. The features that such a security infrastructure should provide include:
  - fine-grained access control;
  - easily configurable security policy;
  - easily extensible access control structure;
  - security checks to all programs, including applications as well as applets.
- A service environment that should include APIs that enable services/application to access the system resources and dynamically reconfigure the underlying protocol stacks or individual protocol elements, changing the parameters at their input so that they meet the required functionality. It should also include monitoring and control APIs that will allow services/applications to control the protocol behaviour in a structured manner without affecting the benefits of re-use and dynamic composition. Through these APIs the clients will be also able to embed monitoring objects in the underlying protocol infrastructure [4].

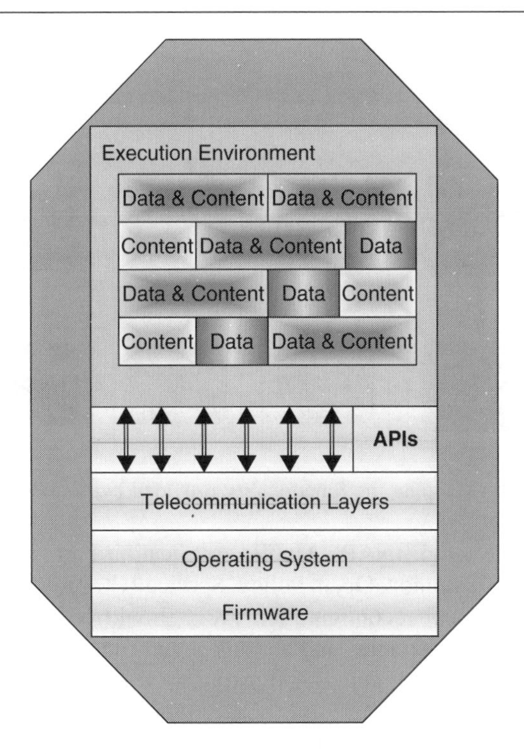

Figure 4.3 Example execution environment on the terminal side

- Core software update that would provide trusted parties (e.g. manufacturers and operators) with the means to download, install and upgrade protocol stacks and individual protocols in terminal devices in the field with new modules, updates, supplements and technologies, so that they can update their architecture and improve their performance as the communication environment changes.
- Multiple protocol support, which is required by the adaptive protocols' execution environment to provide services/applications with a consistent environment, allowing services and protocols to run predictably regardless of whether other protocol software, beyond that required by the run-time environment, has been installed on the device.

The key elements in such infrastructures are the protocol adapters [11], which recognise both human and binary forms of protocol descriptors interpreting reconfiguration policies from one form to another. Every run-time execution environment should contain a list of adapters, each corresponding to an individual protocol, which are activated on demand at run-time. The protocol adapters hide the implementation details of the underlying infrastructure and allow services to be seamlessly delivered to multiple networks and end-user device types. The API provided for communication between application/services and underlying infrastructure receives, maps and redirects the requests made by various applications/services (and the reconfiguration policies that might be included) to the appropriate protocol adapters. In turn, the protocol adapters communicate with the underlying services and components using the appropriate industry standard protocols [e.g. Intelligent

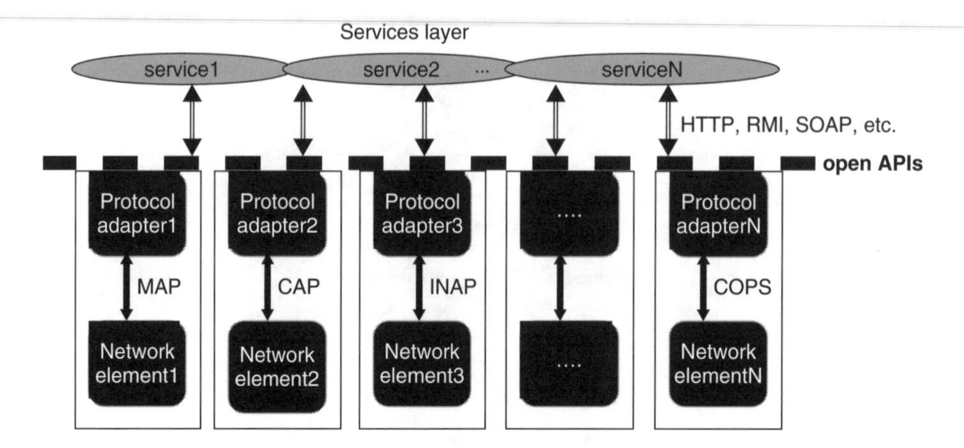

Figure 4.4 Mapping the functionality provided by protocol adapters

Network Application Protocol (INAP), Mobile Application Part (MAP), CAMEL Application Part (CAP) and Common Open Policy Service (COPS)] to dictate the required reconfiguration. The protocol reconfiguration APIs should be error sensitive, detecting the errors on reconfiguration policies and actions at run-time, without breaking the operation of each individual layer. Figure 4.4 illustrates the mapping functionality provided by such protocol adapters.

Two very common execution environments are the Java sandbox [23] and the Linux kernel [24]. The former provides a secure execution environment that is used widely in Internet browsers. One of the most important recent developments for Linux is the dynamically loaded kernel modules. The Linux kernel design is similar to that of classic UNIX systems: it uses a monolithic architecture with file systems, device drivers and other pieces statically linked into the kernel image to be used at boot time. The use of dynamic kernel modules allows the writing of portions of the kernel as separate objects that can be loaded and unloaded at run-time on a running system [10]. The Linux kernel modules can be used for dynamically loading or updating protocols.

4.4.4 Ad hoc Networking

Ad hoc networking is recognised as one of the most challenging research areas in mobile communications. The volatile topology of *ad hoc* networks and the varying traffic requirements of user applications form a highly dynamic communication environment. Designing a protocol architecture to fit such a dynamic environment is a truly challenging task, particularly because of the relaxed assumptions one can make about the operating context of the protocol. Rather than committing to a specific set of assumptions and designing for efficiency or optimal performance, protocol engineers should opt for a more prudent approach, namely designing for run-time-evolving behaviour [33].

Instead of limiting the set of possible communication partners by fixing the set of supported protocols during deployment time, a mobile node in an *ad hoc* network may find it easier to continuously monitor, discover and retrieve applicable protocol implementations from its neighbouring mobile nodes [34]. Obviously, a number of issues need to be addressed before this vision materialises into reality:

- *Common control channel:* global commonly applicable signalling for resource discovery and retrieval, where the resources in question may be protocol implementations, protocol algorithms, etc.
- *Security control channel:* security policy to prevent masquerading of rogue mobile nodes and injection of malicious protocol code into their unwary neighbouring nodes.

4.5 Generic Management Architecture Providing for Protocol Adaptability and Reconfiguration

4.5.1 Introduction

In order to reconfigure a system, there must either be an overall controller, initiating and executing the change, or else the parts of the system must respond to external stimuli by changing their own organisation [3]. The management mechanisms can be categorised as follows:

- *Central, external intelligent manager:* the system or its modules are 'downloaded' with a new arrangement from outside as required.
- *Central, internal intelligent manager:* part of the system can calculate and execute rearrangement of the remainder.
- *Distributed, intelligent:* each part of the system can decide on the need for rearrangement and negotiate changing itself, or other parts.
- *Distributed, unintelligent:* each part of the system is modified according to predefined rules in response to external events.

Although reconfigurability research at its first steps focused primarily on the radio domain (RF processing, A/D conversion, etc.), currently a more innovative and forward-looking view is increasingly drawing interest. According to that view reconfigurability encompasses the entire service provision domain, extending from the mobile terminal through the network infrastructure to application services [7]. On the other hand, the intelligence required to seamlessly integrate networks, terminal devices and applications with different capabilities moves from applications and terminals to the network side in order to orchestrate reconfigurability and allow reconfigurability actions to take place transparently to applications and terminals.

Most of the proposed platforms dealing with protocol adaptability incorporate into their architecture a central and overall, network-sided, manager, responsible for managing reconfiguration and software downloading on various parts of the architecture. The placement of this manager varies according to the platform under study, from the radio access network [e.g. the Access Network Added Intelligence (ANAI) in MOBIVAS or the Proxy Reconfiguration Manager (PRM) in the TRUST and SCOUT approaches], to the core network [e.g. the Global Intelligent Reconfiguration Component (GIRC) in CAST], or the trusted-third-party domain [e.g. the Value Added Service Manager (VASM) in MOBIVAS]. Figure 4.5 depicts an example placement of the reconfiguration management controller (as it is adopted in MOBIVAS) where the manager communicates with the registered Value Added Services (VAS) and the underlying network infrastructure through open APIs. Although in MOBIVAS the reconfiguration manager is situated in the third-party domain, different placements are not precluded.

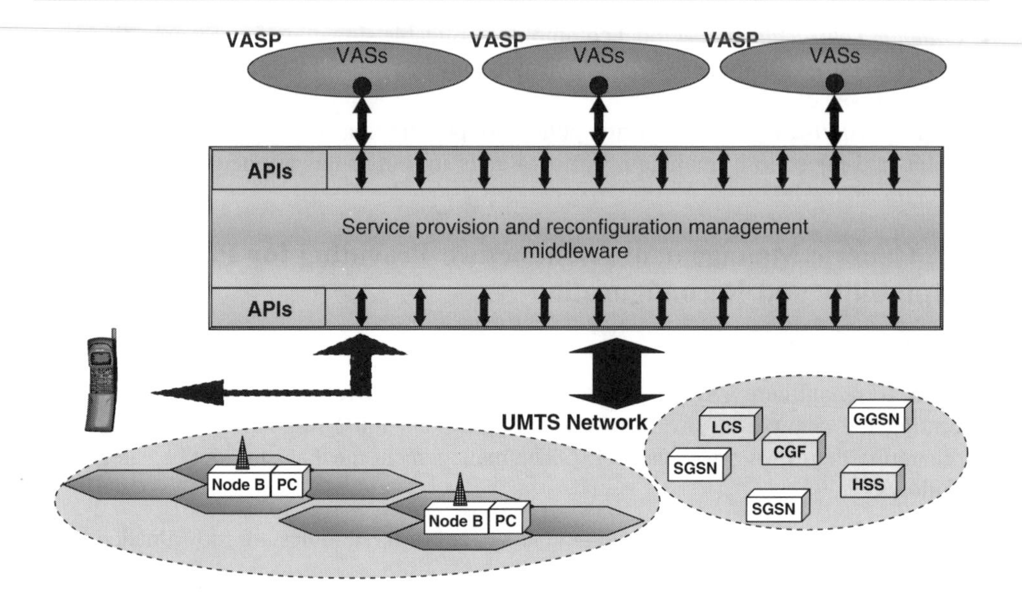

Figure 4.5 Example placement of the reconfigurability management controller

The existence of a central controller in the reconfigurability infrastructure does not imply that intelligence is concentrated solely on one component. Almost every part of a reconfigurable system should host the required intelligence to decide whether it needs rearrangement and hence to negotiate changing itself, or other parts. From this point of view the overall manager is responsible for initialising and performing some certain reconfiguration actions on other parts (e.g. the extendibility of various components with protocol modules and the initialisation of software downloading based on certain events), while the peripheral components are responsible for initialising, triggering or performing their own reconfiguration (e.g. the re-parameterisation of protocols through APIs). The rest of this section focuses on the functionality required by a reconfiguration manager to configure and readapt protocols.

4.5.2 Issues on Protocol Adaptability and Reconfiguration Management

The optimal provision of services to mobile end-users very often necessitates certain software elements to be installed dynamically to the terminal or some place in the network during service deployment or activation. For example, due to limited bandwidth available at the radio communications link, certain service content (e.g. images, audio) should be drastically compressed, so that it can be transmitted in real time to the user terminal. To do that, an appropriate codec could be downloaded to a node at the edge of the mobile operator's network as well as the terminal. According to the dominant aspects on protocol adaptability the intelligence required to implement the adaptation functionality that enables service portability is congregated in the network infrastructure, thus lowering requirements on mobile terminals and extending the service provision domain.

To provide adaptability and reconfiguration for protocols the mediating platform is required to face the following challenging issues:

- *The deployment/storage/testing of protocols.* To facilitate the discovery and downloading of protocols (and other software in general) a common repository of protocol profiles, with information related to the protocol provision, should be provided. Each protocol profile in such a repository describes the attributes of a single protocol (e.g. its name, its version, its size, its provider, the URI from where it can be downloaded and the minimal requirements from the targeting execution environment). To achieve communication of protocol profiles across heterogeneous platforms and devices the profiles should preferably be XML based. Each protocol, prior to its deployment and general availability, should be tested thoroughly to ensure its conformity to the corresponding standards as well as its interoperability within the targeting environment.
- *The discovery and selection of protocols by authorised parties.* Whenever an authorised client (e.g. a service/application running on the terminal or the network) requests a specific protocol or the discovery of available protocols, the manager filters the records in the protocol repository in order to provide the client with a list of the available software. Filtering of records is based on various parameters such as the capabilities profile of the requesting client (i.e. the run-time platforms and the installed software and hardware profiles in the device), the current needs of the client at the specific instance, its network location and the traffic conditions). Through this listing the client is able to select the desired software for downloading.
- *The management of the procedures required for protocol downloading/installation/activation.* Since many different versions of a software may exist, before a protocol or software is downloaded and installed in a device it should be tested again to ensure its interoperability within the targeting execution environment. The secure packaging and downloading of the software and the associated supplemented modules (e.g. drivers, classes, certificates) to the targeting device is followed by the authorisation, validation and verification checks *in situ* in the device to ensure proper transfer. The final step includes the initial configuration, parameterisation and activation of the downloaded protocol in the serving execution environment. Figure 4.6 illustrates the steps required to achieve software downloading and/or update.
- *The reconfiguration of applicable protocol stacks.* Reconfiguration, to a great extent, is event driven. Events can be user specific (e.g. an update on the location of a user that results in a change in the associated user preferences for QoS [22], expressed in

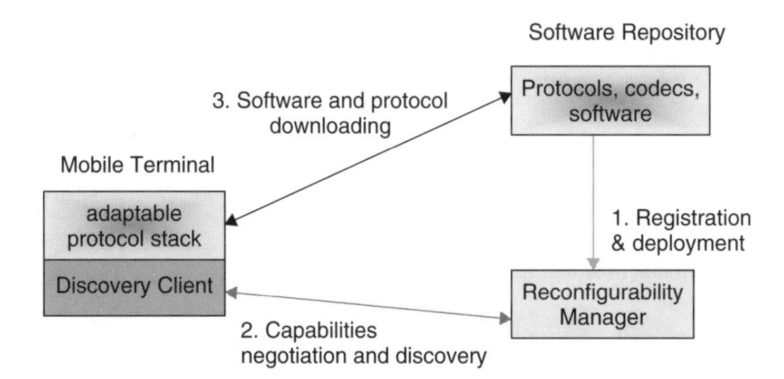

Figure 4.6 Steps required for software downloading and/or update

the user profile), network specific (e.g. as users move from one place to another the serving network and the traffic conditions change accordingly), terminal specific (e.g. failures in the terminal devices or degrades on the battery level), or service specific (e.g. the initialisation of a video data flow from a multimedia service to a terminal device that may require downloading the appropriate codec or some additional classes to the terminal). The reconfiguration manager should be able to decode all these events and translate them into the appropriate reconfiguration actions for adapting service provision to the current conditions. Reconfiguration actions to various entities might are expressed in terms of policies [14, 15]. This requires the targeting entity or the mediating component to be aware of the applied policy rules, so that it translates them properly into the required reconfiguration actions. The provision of open APIs [19–21] to authorised entities by various components for the application of policies and the propagation of events to the registered users is of great importance in reconfigurability [14, 15]. Figure 4.7 depicts an overview of the software/protocol downloading under the supervision of the reconfiguration manager.

- *Profile management.* It has been obvious hitherto that maintaining profiles of various kinds offers great dynamism, context awareness and opportunities for adaptability at all levels of service provisioning, including software adaptability. The profiles, which can be user, terminal, network and service specific, should describe univocally the features, attributes and requirements of the corresponding element. However, to exploit the awareness and the dynamism provided by profiling, innovative mechanisms for

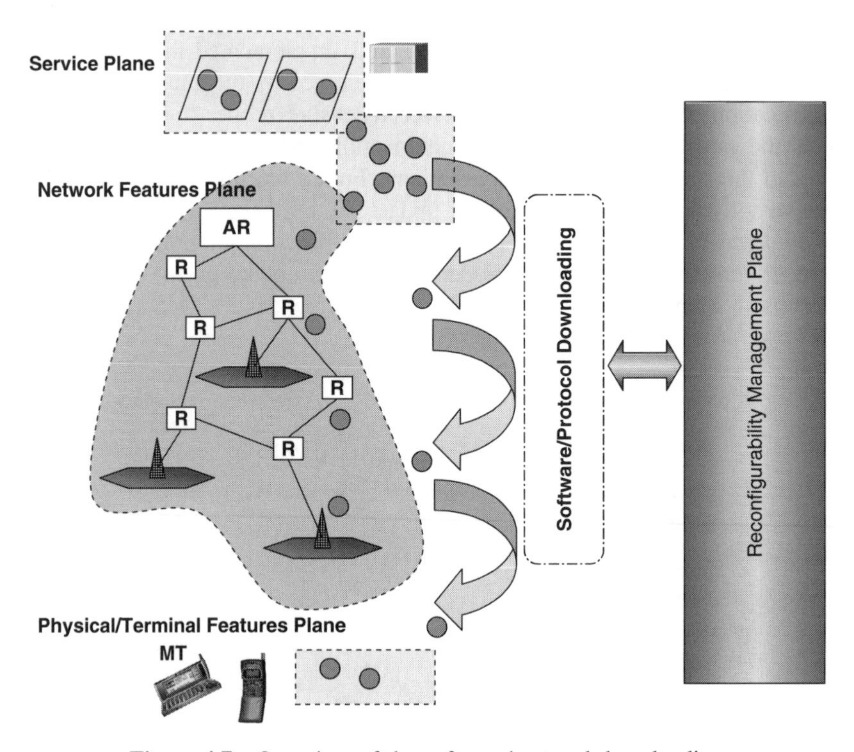

Figure 4.7 Overview of the software/protocol downloading

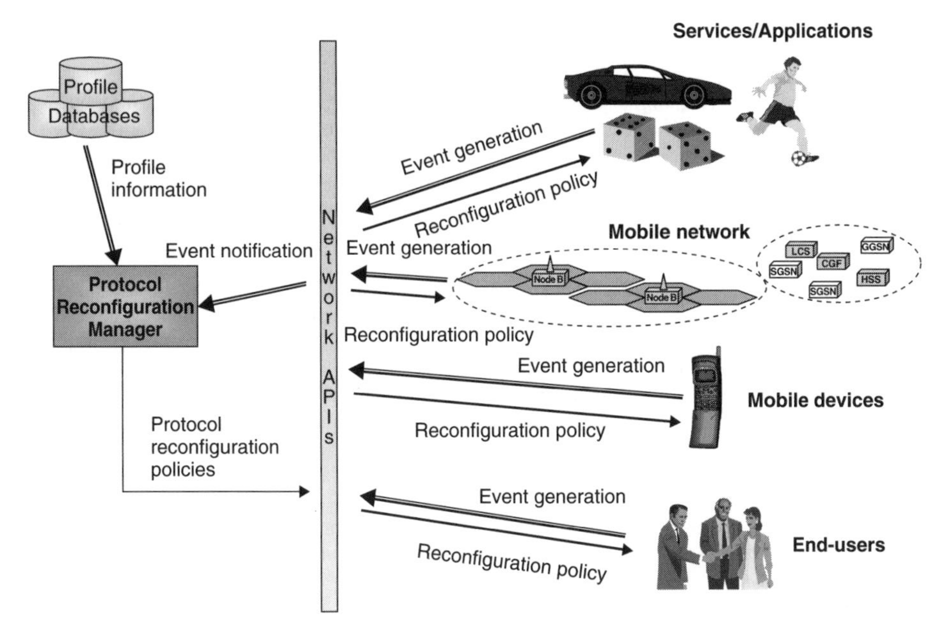

Figure 4.8 The event-driven and profile-assisted nature of protocol reconfiguration

intelligent matching between profiles and respective filtering of content should be developed [18].

Figure 4.8 illustrates the event-driven and profile-assisted nature of protocol reconfiguration. Events from various components (users, terminals, networks and services) are combined properly by the reconfiguration manager with information taken by various profiles to decide the required reconfiguration actions and produce the corresponding policies for instructing the reconfiguration of protocols and infrastructures.

References

[1] M. Annoni, R. Hancock, T. Paila, E. Scarrone, R. Toenjes, L. Dell'Uomo and D. Wisely, 'Radio access networks beyond 3rd generation: A first comparison of architectures of 4 IST projects', Mobile Communications Summit 2001, 10–12 September, Barcelona, Spain, 2001.

[2] M. Siebert and M. Steppler, 'Software engineering in the face of 3G/4G mobile communication systems', 10th Aachen Symposium on Signal Theory, available from *http://www.comnets-rwth-aachen.de/*

[3] A. Carter, 'Using dynamically reconfigurable hardware in real-time communications systems; a literature survey', University of York, Department of Computer Science, Real Time Systems Group, November 2001.

[4] J.S. Crane, N.G. Pryce and J.N. Magee, 'A dynamic protocol architecture for multimedia communications', Technical Report, City University, 1998.

[5] K. Madani et al., 'Configurable radio with advanced software technology (CAST) – initial concepts', IST Mobile Summit, 2000.

[6] IST-1999-10206 MOBIVAS, Vol. 4: Architecture and Design of the Access Network Added Intelligence and Downloadable Communication Protocols, December 2000.

[7] J.M. Pereira, 'Re-defining software (defined) radio: Reconfigurable radio systems and networks', IEICE Transactions on Communications, **E83-B**(6), June 2000.

[8] C. Hoene, I. Carreras and A. Wolisz, 'Voice over IP: Improving the quality over wireless LAN by adopting a booster mechanism – an experimental approach', SPIE ITCOM 2001, Denver, August 2001.

[9] C. Hoene, I. Carreras, T. Chen and A. Wolisz, 'Design and deployment of link-layer boosters for per-flow improvement of QoS in wireless internet access', European Wireless 2002 (EW2002), Florence, Italy, 26–28 February 2002.
[10] M. Welsh et al., 'Implementing loadable kernel modules for linux', Dr Dobbs Journal, **20**(5), 1995.
[11] IST Projects Cluster, 'Reconfigurability: Vision on reconfiguration in mobile networks', Draft *5.2, 23.10.01*.
[12] G. Singh and M. Sammcta, 'On the construction of multiphase communicating protocols', Department of Computing and Information Sciences, Kansas State University, October 1995.
[13] M. Siebert and B. Walke, 'Design of generic and adaptive protocol software (DGAPS)', 3G Wireless 2001.
[14] N. Alonistioti, N. Houssos, S. Panagiotakis, M. Koutsopoulou and V. Gazis, 'Intelligent architectures enabling flexible service provision and adaptability', Wireless Design Conference (WDC 2002), London, 15–17 May 2002.
[15] S. Panagiotakis, N. Houssos and A. Alonistioti, 'Integrated generic architecture for flexible service provision to mobile users', IEEE 12th International Symposium on Personal, Indoor and Mobile Radio Communications (PIMRC 2001), San Diego, California, 30 September–3 October 2001.
[16] N. Houssos, V. Gazis, S. Panagiotakis, S. Gessler, A. Schuelke and S. Quesnel, 'Value added service management in 3G networks', 8th IEEE/IFIP Network Operations and Management Symposium (NOMS 2002), Florence, Italy, 15–19 April 2002.
[17] S. Panagiotakis, N. Houssos and N. Alonistioti, 'Generic architecture and functionality to support downloadable service provision to mobile users', Third Generation Infrastructure and Services Conference (3GIS), Athens, Greece, July 2001.
[18] N. Houssos, S. Pantazis and A. Alonistioti, 'Towards adaptability in 3G service provision', IST Mobile Communication Summit, Thessaloniki, Greece, 16–19 June 2002.
[19] 3GPP TS 29.198: 'Open Service Access (OSA); Application Programming Interface (API); Part 1–12 (version 4.3.0)'.
[20] Parlay Group, 'Parlay API Spec. 2.1', July 2000, available from URL *http://www.parlay.org/specs/index.asp*
[21] J. Keijzer, D. Tait and R. Goedman, 'JAIN: A new approach to services in communication networks', IEEE Communications Magazine, **January**, 2000.
[22] S. Panagiotakis, M. Koutsopoulou and A. Alonistioti, 'Advanced location information management scheme for supporting flexible service provisioning in reconfigurable mobile networks', IST Mobile Communication Summit, Thessaloniki, Greece, June 2002.
[23] *http://java.sun.com*
[24] *http://www.linux.org/*
[25] V. Gazis, N. Houssos, A. Alonistioti and L. Merakos, 'Evolving perspectives of fourth generation mobile communication systems', PIMRC 2002, Lisboa, Portugal, 15–18 September 2002.
[26] E. Gamma, R. Helm, R. Johnson and J. Vlissides, Design Patterns – Elements of Reusable Object-Oriented Software, Addison-Wesley.
[27] Main results of WRC-2000, Istanbul, Turkey, 8 May–2 June 2000, available from *http://www.itu.int/ITU-R/conferences/wrc/wrc-00/results/index.html*
[28] RFC 3198, 'Terminology for policy-based management', available from *http://www.ietf.org/*
[29] RFC 2748, 'The common open policy service (COPS) protocol', available from *http://www.ietf.org/*
[30] RFC 3084, 'COPS usage for policy provisioning', available from *http://www.ietf.org/*
[31] Internet Draft, 'COPS-PR for outsourcing in UMTS: UMTS go PIB', available as draft-hamer-rap-cops-umts-go-00 from *http://www.ietf.org/*
[32] P. Mahonen, T. Saarinen, G. Orphanos and N. Passas, 'Platform-independent IP transmission over wireless networks: The WINE approach', IEEE Personal Communications Magazine, **8**(6), 32–40, (2001).
[33] C. Shirky, 'In praise of evolvable systems', available from *http://www.shirky.com/writings/evolve.html*
[34] H. Chen, A. Joshi and T.W. Finin, 'Dynamic service discovery for mobile computing: Intelligent agents meet Jini in the aether', Technical Report, CS Department, UMBC, 2000, available from *http://www.cs.umbc.edu/dna/papers.shtml*
[35] IST project TRUST (Transparently Reconfigurable Ubiquitous Terminal): *http://www.ist-trust.org/trust_frameset.html*
[36] IST project SCOUT (Smart user-Centric cOmmUnication environmenT): *http://www.ist-scout.org/*

Part III

Networks Supporting Reconfigurable Terminals

5

Network Architectures and Functions

Nikolas Olaziregi
King's College London

Stefano Micocci
Siemens AG

Kambiz Madani, Tereska Karran, George R. Ribeiro-Justo and Mahboubeh Lohi
University of Westminster, London, UK

David Lund, Ian Martin and Bahram Honary
HW Communication Ltd, Lancaster, UK

Sándor Imre, Gyula Rábai, József Kovács, Péter Kacsuk and Árpád Lányi
MTA, Budapest, Hungary

Thomas Gritzner and Matthias Forster
Netage Solutions GMBH, Munich, Germany

5.1 Requirements for the Reconfiguration Process

5.1.1 System Aspects

TRUST concluded that a distributed processing environment and configuration management are necessary to support seamless service provision and scalable Quality of Service (QoS), involving the usage of smart devices.

Analysis of the needs and regulatory issues reveals a mapping of requirements and constraints onto a set of key system support functions [1], namely the creation and provision of services over converging networks and different radio access modes, user

Software Defined Radio: Architectures, Systems and Functions. Edited by M. Dillinger, K. Madani and N. Alonistioti
 ISBN: 0-470-85164-3

environment management and distributed processing framework supported by appropriate middleware(s) and radio reconfiguration control. These functions constitute a core of operative procedures grouped in the following system aspects.

5.1.1.1 Advanced Spectrum-Sharing Techniques

Current spectrum allocations do not exploit the flexibility of reconfigurable radio concepts. This is because they are intended for radio systems with rigidly defined frequency specifications that would not be able to share spectrum resources easily. Of particular importance is the impact that resource-sharing has on system performance and the potential performance benefits of the various techniques for spectrum-sharing. The scenarios comprise either spectrum allocations with one network operator using multiple network technologies, or with different operators sharing the same frequency allocation.

The prime requirement is therefore to incorporate Joint Radio Resource Management (JRRM) schemes, which translate into either combined or cooperative structural and functional architectures.

5.1.1.2 Alternative Mode Identification

In order to enable mode selection or reselection in reconfigurable radio terminals, it is necessary first to identify which modes are available to the terminal. Identification and monitoring techniques can be classified as either 'blind' or 'assisted'. Whereas 'blind' methods impose a severe workload on terminals, for they proceed entirely on their own with no external support or advance knowledge and involved complexity, 'assisted' methods are more relaxed as the terminal is presented with some information about the environment.

The conclusions drawn from the 3GPP/ETSI work on pilot channels show that the likelihood of an international global channel (or set of channels) being provided is very low. This is not as a result of any technical limitations, but as a result of political and commercial considerations.

Considering the points above, it appears that the support methods most likely to gain wide acceptance are the bulletin board approaches and alternative mode advertisement services. These are approaches where a terminal can update information about frequency allocations and system deployment when in service. This in turn has to be supported by network entities responsible first for gathering and then transmitting the information according to terminal needs (i.e. PRM).

5.1.1.3 Mode Switching

Mode switching encompasses the decision criteria needed to switch from one mode to another. As user requirements must be strictly considered, the mode selection must take into account user profiles, user activities and all the available modes and bearer services (found by mode identification), as well as the signal levels measured in the monitoring process.

Bearer service parameters in the User Service Profile should be based on future proof concepts (i.e. UMTS QoS classes). This will provide the means to transport different types of data with different QoS requirements ensuring a proper interworking of all existing network technologies.

The Virtual Home Environment (VHE) is an important concept for mode switching since the ultimate goal of reconfigurable radio is to provide a transparent service to the end-user regardless of the network system the terminal is attached to. End-users may not even be aware that the terminal has been reconfigured to work in a new Air Interface and is using services from a new network. To achieve this goal, the terminals try to present the same 'look and feel' to users of the service regardless of when, where and which network and what service is being used corresponding to the terminal capabilities.

There should also be a checking/reporting mechanism on the capabilities and performance of reconfigured modules involved in the provision of the requested service. Capability and content negotiation proposed in MExE can be extended to cater for the compatibility and capability negotiation between the downloaded software module and its targeted hardware entity.

The specific environment for negotiation needs to be studied, although current trends lean toward delegating such heavy processes onto intelligent entities located in the fixed system side. Thereby, the prerequisites to provide seamless switching capabilities rely on distributing the intelligence of the terminal between the latter and the network; proxy entities (i.e. PRM) being the target architectural additions. Furthermore, they could be enhanced with agent systems to allow for delegated asynchronous and disconnected operations.

5.1.1.4 Radio Resources for Software Download

Download channel radio aspects consider that downloading of software over the air requires mechanisms that can efficiently support this functionality in a generic manner. This will enable the use of the same mechanisms over different Radio Access Network (RAN) technologies (such as GSM or HIPERLAN/2). The two aspects to downloading that have been considered are the implications of the download to normal user traffic and mechanisms to ensure download efficiency. Adaptive Radio Resource Management (RRM) algorithms, compression techniques and wireless-optimised protocols need to be developed to minimise the network performance implications of terminal reconfiguration. The impact of software downloads on other users also needs to be investigated to identify the most appropriate schemes.

Consequently, the applicability of Joint Radio Resource Management (JRRM) units envisages an increase in system capacity and user QoS. Collaborative architectures in this field are complementary to spectrum-sharing, aiming to enhance the usage of radio resources as presented previously. Additionally, a method for augmenting the reduction of the required bandwidth is to prioritise and schedule downloads efficiently along with a number of parameters. This solution utilises a method of extending RRM concepts to bandwidth reduction, wherein interactive processes between Configuration Managers and JRRM entities take into account software already available in the terminal, in order to minimise the upgrades to a small extent.

5.1.1.5 Configuration Management

Configuration management aims to address the issues raised by the co-existence of multiple hardware and software versions. Multiple versions of ostensibly the same device or software component are currently a common occurrence, as hardware and software vendors strive to maintain the marketability of their products. What is certain is that the importance of configuration management will increase greatly as terminals become more reconfigurable, with almost limitless combinations of system software and application.

The SDR Forum has conducted much research into configuration management. Of particular interest are their proposals for a well-defined interface definition and functional description for software components and *in situ* testing. Their work approaches the area of configuration management from several sides. It is suggested that configuration problems can be avoided by properly defining downloadable software components in terms of their Application Programming Interfaces (APIs) and functional description. The SDR Forum additionally proposes to address issues arising from *in situ* testing of downloaded software components.

The use of a common distributed programming environment would enable platform independence in reconfigurable terminals. However, while some problems will be eased, others will arise, such as the use of middleware in a performance, memory and bandwidth constrained environment. More work is needed on middleware optimised for the wireless, real time and embedded systems environments. Although the Object Management Group (OMG) is beginning to address the requirements of the telecommunications community in these areas, the time between initial specification and product availability remains considerable. The distribution of such functionality between terminal and network brings clear benefits for reducing the signalling (i.e. authentication, security), although imposing a compromise on the frequent reports required for updating on terminal configuration.

5.1.2 Reconfiguration Process

The requirements imposed upon the reconfiguration process combine in a generic set of sequential functions. The premises defined provide a high level view of the numerous specific individual scenarios that need to be considered to support the reconfiguration of terminals in all realistic situations.

Figure 5.1 shows the TRUST Reconfiguration Scenario Framework [2]. The reconfiguration process will vary depending on the purpose of the reconfiguration operation, which in turn refers to the scenario that takes place. The scenario is mainly conditioned by the triggering event and the actual state of the terminal. The reconfiguration process can be triggered either by the user or the previously set User Profile, the terminal application, the terminal resource system, the network operator, the service provider or the manufacturer. The state of the terminal is described as idle, starting a session (either user initiated or upon other party's request) and in-session.

In spite of that, some generic tasks have been identified for the process. Even though they might not be applicable to every single reconfigurable scenario, they give a common understanding of the several functionalities the terminal must incorporate in order to satisfy the requirements imposed by each situation. The following list summarises such tasks briefly in order of events:

- Available modes lookup
- Detection of new air interface and monitoring
- Authentication
- Mode negotiation

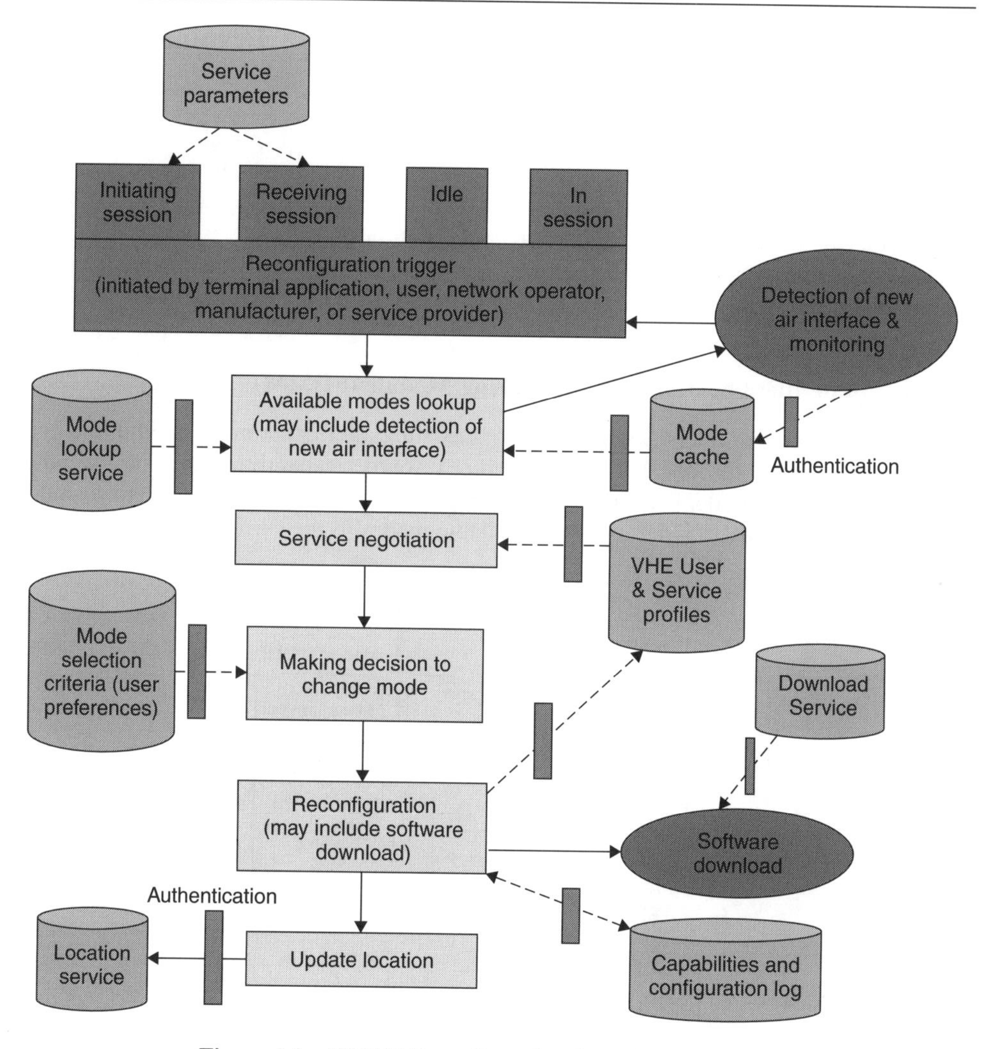

Figure 5.1 TRUST Reconfiguration Scenario Framework

- Making decision to change mode
- Software download Over The Air (OTA)
- Reconfiguration
- Location update

The framework also illustrates the interactions required between these logical tasks, which provide an initial indication of the relationships between these functional (or scenario) groups.

According to the framework, TRUST has developed a system architecture comprising terminal and supporting network entities, which considers distributed processing across the system. In the next section the main basis of such an architecture is shown and explanations concerning the structure and functionalities provided.

5.2 Logical Functions Supporting Reconfigurable Terminals

5.2.1 Terminal Functionalities

The TRUST Functional Architecture, capturing both functional modules and interfaces in Figure 5.2, reflects elements required in the terminal architecture to support dynamic reconfiguration in object-oriented notation [3]. Other components appearing in the architecture refer to actors who may trigger a reconfiguration process.

Concerning the system aspects, the main modules and their associated functionalities are presented following the sequence of events in the Scenarios Framework.

5.2.1.1 Mode Identification and Monitoring Module (MIMM)

This module discovers, identifies and monitors existing alternative modes within the constraints imposed, on the one hand, by terminal resources/capabilities and, on the other hand, by the current mode. During the detection process the terminal will have to scan different frequency bands in order to detect available modes in the surrounding geographical

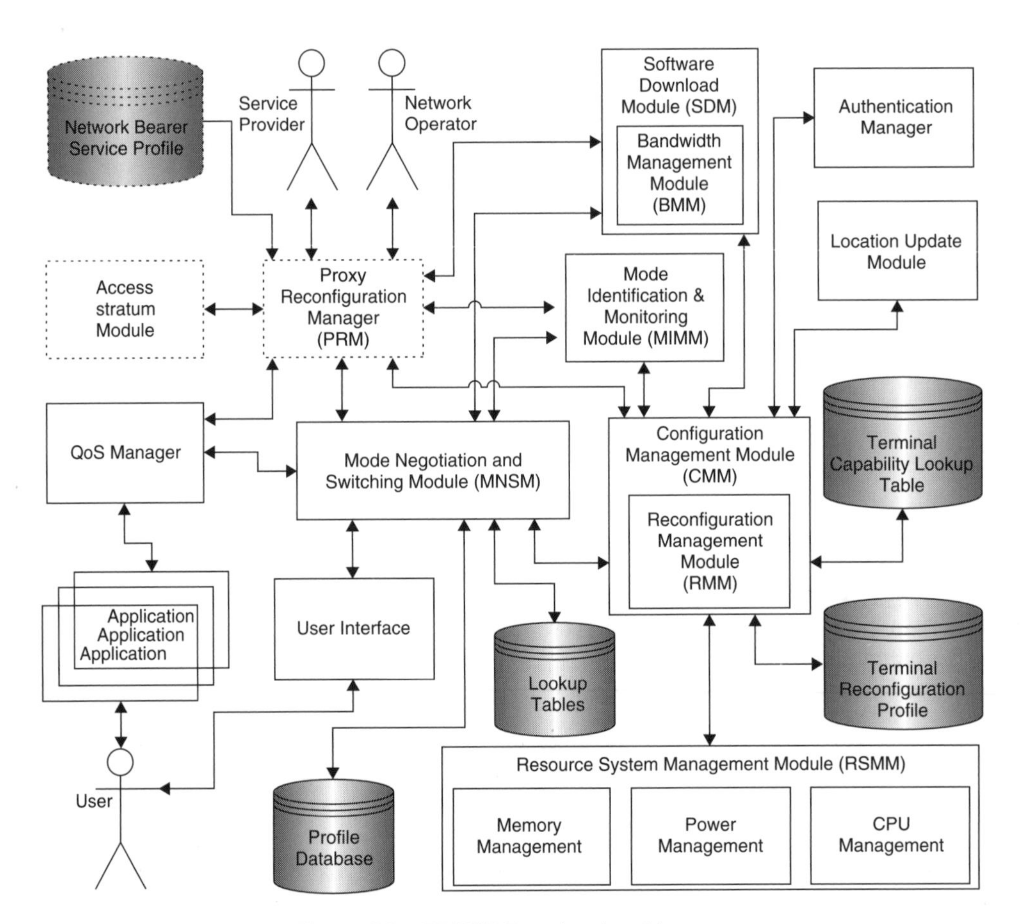

Figure 5.2 TRUST Functional architecture

area of the user. Therefore, the RF front-end must provide accurate dynamic behaviour, i.e. be able to synchronise and retune without causing disallowed spurious emissions in the RF spectrum. Detection of alternative modes is very complex as a result of the constraints imposed by the current mode and the terminal capabilities, in terms of the amount of 'free time' available. Furthermore, if the terminal is in-session, the service must not be affected during the detection process, so the amount of time available is even more constrained.

After this difficult task has been completed and modes have been identified, monitoring of these modes is vital to ensure that a sufficient level of service and link quality is being offered before that mode can be considered as a potential roaming destination. Here again severe time restrictions are imposed by the current mode and by any active services. If the terminal can be assisted by some external entity (i.e. Proxy, other terminals, third party), then the demands placed on the terminal by detection and monitoring activities can be alleviated to a certain degree, although external support cannot always be relied upon.

MIMM has to determine what external support is available (if any), and the constraints of the current mode, from information provided by the Reconfiguration Management Module (RMM). It also has to determine, from information provided by the Resource System Management Module (RSMM), whether there are sufficient resources (battery power, processing power, memory, etc.) available on the terminal to be able to achieve the required mode detection and monitoring without adversely affecting the level of service in the current mode.

5.2.1.2 Mode Negotiation and Switching Module (MNSM)

This module guides the mode negotiation process. It checks service availability in the target RAN, negotiates Bearer Services and makes sure the terminal can provide the desired performance levels in the target mode, regarding the load of the system and link quality. The entire process follows a strategy based upon previous knowledge of the available modes. Thus, a filtering process will generate a ranking derived from both the User Profile, terminal state and triggering event. Later on, keeping in mind outputs from other modules as well, a decision is made to change to the target mode or not. For decision-making it is important to identify the user preferences stored in the Profile Databases, the link level quality provided by the MIMM, the reconfiguration complexity given by Configuration Management Module (CMM), the time expected for software download obtained from the Software Download Module (SDM) and its impact in the active services if any, resource availability and user interaction (agreement) in the case when the User Profile is not set up in detail.

The results of these negotiations are stored in Lookup tables, which should be used for potential future access to the same mode and network. They would save time and resources, as negotiation results would probably be still valid.

5.2.1.3 Configuration Management Module (CMM)

This module controls the current and possible future configurations of the terminal. In order to enable the configuration of different distinct parts of the terminal core software and hardware to be coordinated without restricting the flexibility of the implementation, configuration management architecture with different domains of responsibility is proposed. Two types of entity are introduced to enable interaction between these domains of responsibility, namely domain managers and terminal agents.

The domain managers are responsible for control of the interactions between terminal agents in different domains and provide generic configuration information that is required by all terminal agents. The terminal agents are responsible for performing specific configuration management operations, which include retrieval of detailed information regarding current and possible future terminal configurations and controlling the reconfiguration of distinct parts of the terminal software and hardware.

Terminal agents interact with the RMSM in order to obtain information regarding the resources available before reconfiguration is performed. Each terminal agent is also responsible for requesting (via domain managers) other configurations to be activated or deactivated and for notifying other terminal agents (via the domain managers) of any change in the configuration status controlled by the terminal agent. In addition, the terminal agents perform the necessary reconfiguration management functions in an appropriate way. For example, one terminal agent could be responsible for the baseband section and could create active and shadow transceiver chains and control the switching between the two chains, while another could be responsible for the RF hardware section (both of these examples of terminal agents are likely to be within the manufacturer's domain of responsibility).

Furthermore, terminal agents responsible for protocol stack components may fall within the network operator domain of responsibility and control different parts of the communication protocol software, whereas terminal agents responsible for transport layer protocols and higher layer protocols may fall within the service provider domain of responsibility.

The CMM also provides information regarding the current and possible future modes to the mode-switching module and other terminal modules. Therefore, it provides the mapping between modes [which are abstract representations of the Radio Access Technology (RAT) air interface and higher layer communication protocol combination] and the configurations that provide the support for those modes. This will include the power consumption estimate for a given mode, which can be a very important criterion for making a decision in the mode-switching process.

5.2.1.4 Software Download Module (SDM)

The SDM contains the Bandwidth Management Module (BMM) which, depending on several variable values, calculates the optimum download strategy and passes it to the SDM. The SDM then takes care of either downloading the software from a certain entity in the network, making the most of decentralised download schemes, or retrieving it from the Lookup tables, i.e. libraries, which may contain the required module from a previous reconfiguration. It is also responsible for defining who can initiate, enable and disable software downloading.

5.2.1.5 QoS Manager

The QoS Manager takes care firstly of mapping QoS parameters from user/application requirements to the levels of underlying technology components, which are formed by the Operating System (gathering both the software and the hardware platform) and

Communication components; and secondly obtaining resources from the aforementioned components. Additionally it monitors the status of service, if any, regarding the user/application satisfaction level and dynamically adapts the requirements to potential changes in the resource availability of each component.

The mapping of parameters, subject to the availability of a mapping filter, is preceded by acquiring parameters from several sources. Since each component may incorporate its own QoS control mechanism, the TRUST QoS Manager only addresses matters involved with the current and target RATs. Thus, the sources to be checked are restricted to the capabilities of the target RAT, compared with the current one, and the capabilities of the terminal in terms of its reconfigurable modules' performance in the new mode.

Afterwards, the mapped parameters are compared against the target ones (required by the application and set by the user, either on the User Profile or on demand). The comparison could be preformed at any level of parameter definition, no matter which parameters are mapped.

The QoS Manager uses the outcome information to obtain resources from specific components. With that goal in mind, it negotiates, reserves and adapts the available resources, and thus deploys the requested level of service satisfying the user/application as far as possible. Such procedures are already incorporated within the MNSM and thus the QoS Manager relies on the MNSM for those purposes during the Vertical Handover processes.

5.2.2 Network Functionalities

The network-provided supporting entity, the Proxy Reconfiguration Manager (PRM), serves as a proxy instance for negotiations with other network entities, as illustrated in Figure 5.3, in particular the Serving Reconfiguration Manager (SRM) in the Access Network and the Home Reconfiguration Manager (HRM) in the Home Network.

The SRM is used in each Access Network and is dimensioned according to the number of users that an Access Network can support and what the expected services for the Access network are. The PRM is used to minimise delays, use information on the location and status of a terminal, and negotiate a strategy for reconfiguration between the terminal and SRM. Both the SRM and HRM are key elements for upgrade of a large number of terminals and they relieve currently deployed network entities from several management and security issues that have a critical impact on network performance.

5.2.2.1 Proxy Reconfiguration Manager (PRM) Main Functions

In order to save bandwidth on the wireless link, distribution of the Functional Architecture is considered an important issue. Depending on the user's preferences, subtasks that do not have to be carried out on the terminal can be delegated to the PRM. A benefit resulting from this approach is that the terminal is relieved from expensive computations. Thus, a more economic use of terminal resources (e.g. CPU performance, memory space or battery power) is achieved leading to faster and less disruptive mode switches. In order to make this possible some information concerning user preferences and negotiation results has also to be made available on the PRM.

There are different levels of functionality distribution, ranging from performing the whole process in the terminal to delegating all processes to the proxy side. This approach

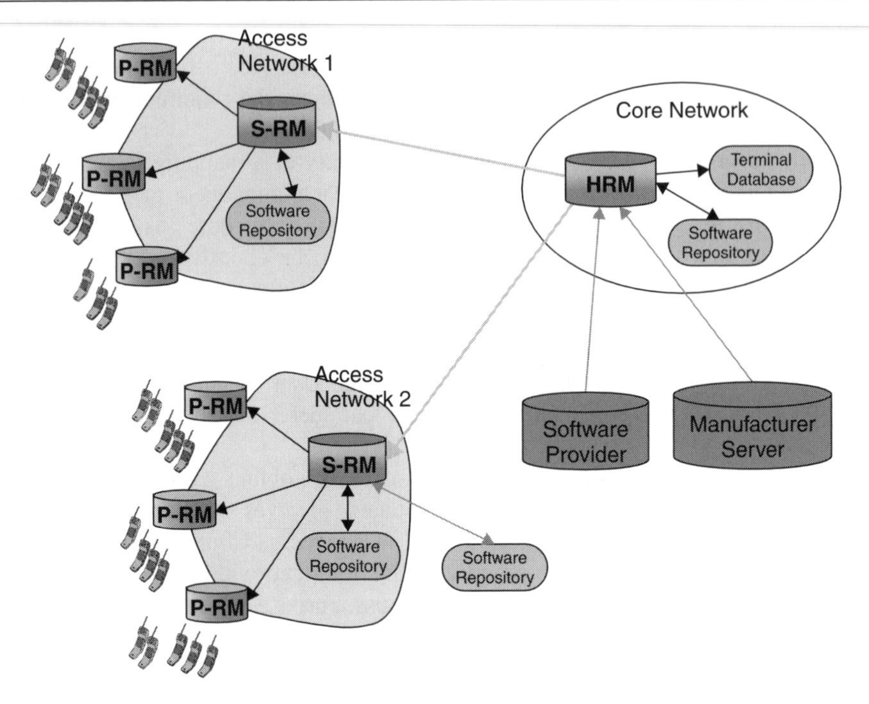

Figure 5.3 TRUST network-centric architecture

proves especially interesting in the case of time constraints and scarce terminal resources scenarios. In what follows a summary of some functionality of PRM is presented.

5.2.2.1.1 Information Broker for the Terminal. Some information required for the *Mode Switching* decision is not available locally on the terminal. Various interactions with entities related to the network or even located in the network are necessary. In order to relieve the terminal from the burden of frequent interactions with network entities, information from the network could generally be obtained via the PRM, which is located in the RAN. Thus, the PRM plays the role of an information broker for the terminal, which distributes data from the terminal to a variety of other entities and vice versa.

5.2.2.1.2 Caches Measurements of Terminals Operating in Specific Mode. This function is required during the *Mode Monitoring* phase. Some terminal measurements have to be transmitted from the terminal to the reporting point (PRM), and information on the available modes needs to be provided. This information is transmitted from the information centre to the terminal through the PRM. The PRM has the necessary knowledge about the terminal capability and can filter the data about the existing modes around the terminal to provide only relevant information. There is no point in informing a terminal about an available mode if the terminal could not be reconfigured to operate in this mode. The PRM is supporting the terminal in both processes: creating the measurements report and monitoring modes. The PRM will cache these measurements for further reconfiguration.

5.2.2.1.3 Caches Negotiations of Terminals Requesting the Same Bearer Services. This function is required during *Mode Negotiation*. Information about the Network Bearer services currently offered and their customisation parameters is contained in the Network Bearer Service Profile that may also be retrieved via PRM where it is cached for negotiations with the terminal.

5.2.2.1.4 Autonomous Service Discovery and Mode Negotiation. This function is required during the *Mode Negotiation* phase. The PRM will act as an Autonomous Service Discovery that provides a service lookup, which is delivering the identity of a set of service providers. Moreover, it can provide contact with the set of intermediates that might finally direct the client to the requested service with an implied negotiation between the parties who offer the services. An interaction with SRM might be required.

5.2.2.1.5 Records Terminal Classmark and Capability Information. Some of the terminal capability parameters play a role in the process of the reconfiguration of the terminal and in evaluating the switching to a new mode. These parameters generally have a two-fold aspect: static and dynamic values. The static features of the terminal are the equipment characteristic, the raw capabilities that the hardware is providing (transceiver RF front, baseband parameters, processing power, etc.). Dynamic capability refers to the instantaneous availability of resources in the terminal. This depends on the state of the terminal, on the current mode as well as on the applications running. This information should then be collected and presented in a suitable format to the different entities that require access to the capability data; the terminal capability with terminal classmark information is usually recorded in the PRM.

5.2.2.1.6 Download Management Support. For the software downloading process, interworking between the PRM and other architecture elements is recommended. This is due to the user/application/terminal requirements for the software download process on the one hand and the available radio and network resources on the other. The latter are further exploited interacting with RRM entities belonging to each RAN, and maximized with coupled system architectures and JRRM schemes (e.g. the terminal could inform the PRM about some time constraints for the download to finish). Related to the QoS requirements of other users and the aim to minimise the impact on the system, the PRM initiates the reservation of resources in the Core and RAN and schedules the download, using either unicast or multicast methods.

Some examples comprise single-user downloads and mass upgrades. In the former the terminal downloads software updates. If other terminals have requested the software previously, then it is already stored in the cache of the PRM and the update can be transferred with less delay and better resource usage. The latter refers to the availability of a software bug fix for a large number of terminals. Simultaneous downloads would easily lead to an overload of the network capacity. Alternatively, the load of the network could be reduced if the software modules are cached in the PRM. Then terminals need not have a connection to the original server. Therefore, the software is downloaded from the manufacturer's server only once per PRM and for upgrade of the terminal only a connection to the PRM is required.

5.2.2.2 SRM and HRM Role and Location

Two additional entities support the PRM in the reconfiguration process, namely SRM and HRM. The main idea is to have a hierarchically distributed architecture that minimises the network load and speeds up the software download, as presented in Figure 5.3.

The HRM is located in the home network of the terminal and is informed by providers about new software upgrades. In this case the HRM notifies the availability of new software to the SRM in the RANs and forwards the software to them in the case of a mass upgrade. If a request for a software download arrives at the HRM, it is also responsible for the authorisation of the terminal in the case of a request to download licensed software. Another point considers accounting of software download. For this the HRM uses a charging repository, which is updated if appropriate software is downloaded.

The SRM is dimensioned according to the number of users that an Access Network can support and what the expected services for the Access Network are. The SRMs are located in the RANs, between the PRMs and the HRM. As mentioned above, they are informed from the HRM about new software and distribute it to the attached PRMs. Because of the possible heterogeneity of the RATs in an IP-based mobile network, the size of the individual RANs may vary. If the software must be transported to a large number of PRMs, then the serving reconfiguration managers have the advantage to minimise the load. In addition, not all software is needed in every Access Network. Therefore the PRMs are trying to reduce the delay and memory needed and are caching only a small number of files. The SRMs, on the other hand, have access to large software repositories and store many more files.

Furthermore, the SRM is involved in the control of the mobility, allocation of resources and security of moving terminals. This includes procedures required for vertical handovers, location update and interworking between different RATs in order to provide the desired QoS.

5.3 Design and Development Considerations for Reconfigurability

5.3.1 Factors Influencing Terminal Reconfiguration

There are many different factors to be considered prior to the reconfiguration process. The state of the terminal influences some of the requirements of the reconfiguration process. For example, the time constraints depend in great part on the state of the terminal. Here time constraints refer to the period required from the point when the need for reconfiguration is observed until the terminal finalises the reconfiguration process.

A terminal in the idle state does not have strong time constraints for monitoring alternative RATs or performing software download. Hence, the mode negotiation and decision-making processes might be relatively relaxed compared with a terminal having a session in progress. On the other hand, the necessity for reconfiguration might be quite limited since there is no active connection; there are no QoS requirements or services to be maintained. Returning to the home/preferred network or getting out of the coverage area of the current one might be some reasons for performing terminal reconfiguration. The great free time available with this state might be used for non-urgent downloads or for retrieving information about available networks and different services offered by them. If the terminal capabilities (remaining power, available memory) allow it, then the software required to reconfigure the terminal to neighbouring networks could be downloaded while the terminal is in the idle mode.

A terminal in a connected state has most probably more reconfiguration triggers compared with the idle mode case; however, it also has many more constraints. Terminal reconfiguration might be a substantial part of the whole handover delay. An acceptable handover should be transparent to the user whenever possible. The meaning of the term 'transparent' might also vary depending on the type of service class. For conversational services only some milliseconds might be allowed, whereas for streaming, the delay might depend on the amount of data buffered prior to the content display. If a download is being performed, a few seconds might still be acceptable to the user. Since the terminal has a connection running there is not much time left in the terminal to perform monitoring of different available networks, so it is important to obtain as much information as possible from outside the terminal to alleviate this problem. The PRM can be used to provide this on, for example, available networks, services offered and QoS guarantees.

If the handover is caused by the establishment of a new session, either by an incoming or an outgoing call, requiring a service not supported by the current network (or supported only with low QoS), then reconfiguration should take place allowing an acceptable delay for the user. Thus, if no other session is in progress, the terminal can be considered as being in the idle mode giving an opportunity for monitoring and negotiation.

The greater the time constraints placed by the current scenario, the more important it is to have assistance from the network side. If the PRM is aware of the requirements placed by the current session(s) running at the terminal and also of some user requirements and preferences and terminal capabilities, it may provide a first selection of RATs fulfilling these requirements. This would partly avoid unnecessary negotiations with modes that do not cover the current terminal needs or do not support the QoS required by the application. Having some indicator that the RAT being considered might be suitable could be a determining factor for terminal reconfiguration. This information might be obtained partly from the result of recent reconfigurations from this and other terminals.

Another factor for a fast decision might be whether the terminal reconfiguration requires some modules to be downloaded or not. The need for a download introduces extra delay in the whole reconfiguration process. If the PRM is aware of the neighbouring cells, it could place the software required for reconfiguration in them. Then the terminal could download these modules in periods when no connection is established, or might perform the download in the background with quite a low priority without affecting the active sessions. This might depend, among other factors, on terminal capabilities and whether a user has to pay for the download or not. A terminal with high capabilities, e.g. a laptop, could use its spare time to download all the possible software required for reconfigurations.

On the contrary, if the download is crucial for reconfiguration, then this should take place with high priority and the active connections might get their priority reduced or even stopped/cancelled to allow the download to complete in time. This scenario highlights the importance of having the modules cached somewhere close to the terminal: the PRM is seen as the most appropriate entity for storing this kind of software. Also, the SRM might collect download requests less frequently.

5.3.2 Centralised vs. Distributed Proxy Architecture

Proxy systems are emerging as an effective approach to improve user experience in networks nowadays. Such improvement is tailored by caching diverse kinds of information in strategic locations, along with providing the servers with extended intelligence so that particular tasks are carried out locally on behalf of a user's terminals.

Distributed systems are based on a collection of (probably heterogeneous) automats whose distribution is transparent to the user so that the system appears as one local machine. This is in contrast to a network, where the user is aware that there are several machines, and their location, storage replication, load balancing and functionality is not transparent. Distributed systems usually use some kind of client–server organisation.

Previous research on the caching subject focused mainly on Distributed Shared Memory (DSM) systems. DSM has been implemented both in hardware and in software, providing the shared memory abstraction on physically distributed memory machines. An extensive introduction on this topic can be found in ref. [4]. Whereas DSM caching has been an active research field since the early 1980s, investigations on caching proxies started a decade later. In ref. [5] a comprehensive list of differences between the two paradigms is discussed, although providing no real comparison. The boundaries between the two approaches are not very clear; thus, it is arguable whether proxy servers can be considered as centralised systems or, by contrast, whether they should be conceptualised as another automat within a given distributed system.

In what follows this issue will be considered only if appropriate; the differing criteria focus on the mechanisms rather than the functionalities. Therefore, as long as the properties are the discussion item, the realisation is left aside open, and it is for implementers to trade off between each of the possible alternatives.

5.3.3 Location of Proxies in Cellular Networks

In this section we study the placement of proxy servers in cellular networks. The approach centres on the various factors that must be considered when placing proxies inside cellular networks [6]. The requirements are imposed on:

- Collection of valuable information
- Network support to reconfiguration functions
- Mobility schemes

Furthermore, coexistence problems between proxies with security features of IP based networks (IP Security) happen to impose requirements as presented in refs [7] and [8].

Traditionally, the general usage of proxies is linked to the enhancement of application and transport level performance capabilities. Common implementations try to adapt content to user preferences and terminal capabilities in the former, with data transfer parameters to the available network bandwidth due to variable congestion status, in the latter.

Additionally, recent research [9] has demonstrated that adaptation can be performed at all layers of the protocol stack to accommodate the dynamics of wireless channels. Accordingly, proxy servers perform channel adaptation, improving the performance of multimedia applications considerably.

5.3.4 *Collection of Valuable Information*

Among the issues to be considered to incorporate proxy services within a cellular network, there is the need to address the physical placement of the servers. In order to perform within acceptable latency ranges, the functions that require fast access to wireless channel conditions need to be located close to the information source; the supply being the RRM entities for each cellular network.

Likewise, as the wireless QoS traffic classes are provided and controlled by these entities, the re-negotiation and re-allocation of resources is also fixed by locality.

5.3.4.1 Network Support to Reconfiguration Functions

Not surprisingly, the amount of signalling required to gather and then process the aforementioned information is also noteworthy. To reduce this, locality relieves the network of undesirable increased congestion.

Local placement, on the other hand, ensures that sensitive information (i.e. user's service profile, network status) remains within secure and reliable boundaries, thus enabling exploitation of the full potential for providing the user with the most suitable service out of those offered by the network, bearing in mind the dynamics of available system resources in a cellular environment.

When analysing the signalling generated, the information period update has to be considered too. Information such as the terminal classmark or User Profile remains the same throughout a connection; on the other hand, channel condition information requires frequent updates in order to make error-sensitive applications, such as real-time multimedia, perform within the QoS ranges stipulated by the user.

5.3.4.2 Mobility Schemes

Provided the advantages of placing proxies close to the RRM entities are being made use of, the drawbacks must be considered in order to trade off between the necessary in-depth grade of the servers within a cellular network, with the maximum depth located in the access point (i.e. base station) and the minimum in the cellular network gateways. Negative grade is measured out of the cellular network and not considered in our research, due to the consequences it brings on latency, traffic overhead and security aspects.

Problems arise when a mobile user moves out of the provision area of the serving proxy. At this point the proxy might need to forward reconfigurability process information to the new proxy. This turns out to be a key feature in particular scenarios (e.g. software downloading), wherein a proxy handover envisages transfer of the session profile (i.e. PDP context) and status to the new server.

Accordingly, the routers involved in data delivery to the user must also be updated so that incoming packets are delivered efficiently to the new location. The trouble increases when considering the differing mobility schemes adopted for micro mobility [10] within IP-based cellular networks (i.e. Cellular IP, HAWAII, BCMP, HMIP); consequently, a generic solution with regard to update of the routing path in an efficient manner is unfeasible in practice. The consequences affect delay and overhead in the handover procedure. As a result the placement of the proxy must take into account the micro-mobility schemes adopted in each cellular network so that handover disruptions are minimised to the maximum extent.

5.4 Network Architecture

The SCOUT project inherits the TRUST proxy concepts and intends to add more functionality to the proxy in order to support the entire reconfiguration process (not only the SW download process) for a more realistic and completed scenario. The availability of different RATs (GSM, GPRS, UMTS, WLAN, etc.) gives the possibility to support a wide range of services, not just voice. This fact is hastening the integration and interworking of different access technologies and also a multi-mode terminal, which is a terminal able to reconfigure and upgrade its communication stack and to support new services. The role of the network, in this evolution, is crucial because it has to implement the capabilities to support the terminal during the reconfiguration process. Moreover, in a future scenario, the terminal should be able to have available a particular service in an independent way from the radio technology access and the mobility of the terminal should be also supported. This means that the network has to carefully manage the network resource in terms of bandwidth according to the service required from the user or the bandwidth available for a particular access technology. Moreover, the service should stay connected during handover procedures. Thus, the SCOUT architecture has to support:

- Vertical handover decision
- Mobility management
- Software download
- Terminal initiated request
- Network initiated request

5.4.1 Interworking Scenario and Vertical Handover Support

The interworking between different RATs is an important topic, especially in the standardisation bodies. Different approaches can be taken, depending on the level of integration that is necessary or possible to reach between RATs. When integration between different technologies is close, the provisioning of the service is more efficient and the choice of the mode in order to find the best radio access is faster, as also is the handover procedure. On the other hand, a higher level of integration requires a bigger effort in the definition of interfaces and mechanisms able to support the necessary exchange of data and signalling between different RANs. Following these considerations, some different coupling scenarios can be identified:

- Open coupling
- Loose coupling
- Tight coupling
- Very tight coupling

5.4.1.1 Open Coupling

The term ‘open coupling’ indicates that there is no real integration effort between two or more access technologies. As reported also in ref. [11] in an open coupling situation, two access networks, for example WLAN and UMTS, are considered independent with only a billing system shared between them.

So, separate authentication procedures are used (i.e. SIM-based authentication for 3G system and simply user name and password for WLAN) and a common database system

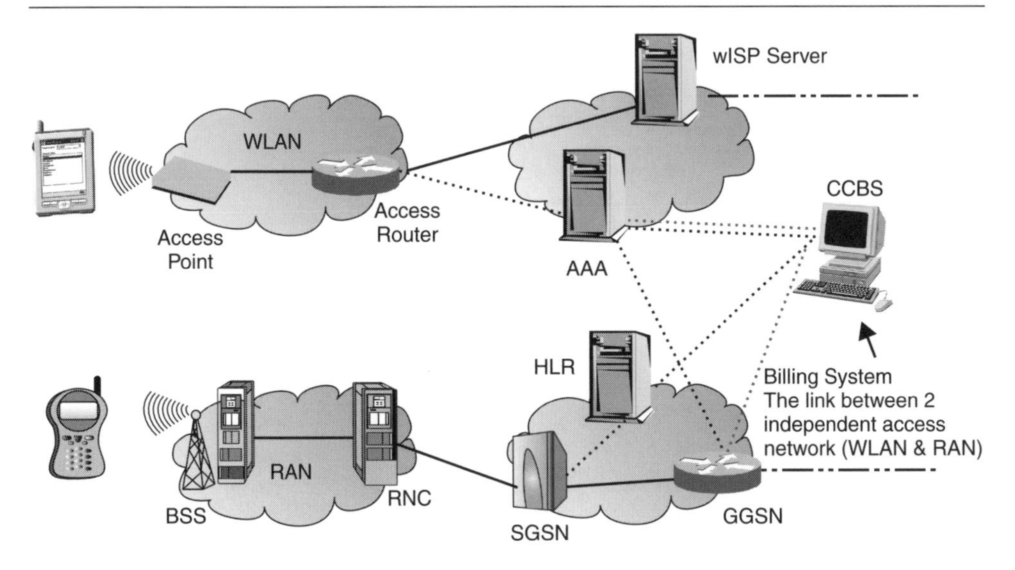

Figure 5.4 Open coupling

is used to handle the billing between the different technologies (see Figure 5.4). In this case there is no integration between different accesses except for the billing.

5.4.1.2 Loose Coupling

Taking ref. [12] as our base, loose coupling is defined as the utilisation of a generic RAT as an access network complementary to current 3G access networks, utilising the subscriber databases but without any user plane IU interface, i.e. avoiding the SGSN and GGSN nodes. The operator will still be able to utilise the same subscriber database for existing 3G clients and new RAT clients, allowing centralised billing and maintenance for different technologies. In this scenario, it is more correct to speak of 'interworking' rather than vertical handover. In fact, one of the consequences of this kind of coupling is that the service in progress is dropped during the switch between the two RATs. For the software download scenario, this means that it is not possible to maintain the download of the software when the terminal changes its type of coverage. Download of the software has to be restarted from the beginning.

For the loose coupling scenario, the core network coordinates sub-networks during the interworking.

For authentication and billing, one customer database and procedure is used, and a new link between the wISP and the 3G core network is provided and needs to be standardised (AAA-HLR link). This means that the user has to perform a unique subscription if the network provider is the same for both network, or alternatively the user has to perform a unique subscription to a certain service that will be available for both access networks.

In Figure 5.5 we show, as an example, the coupling between WLAN and UMTS networks. In this situation, a terminal attaches to a WLAN network, performs the authentication and acquires an IP address. At that stage, it can have access to the services supported from the network, using that IP address, while it remains within the same

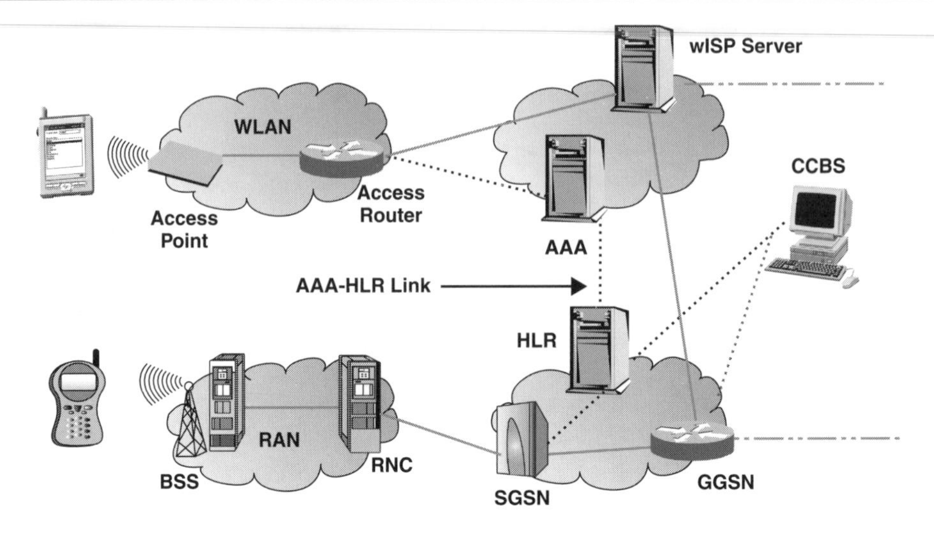

Figure 5.5 Loose coupling

WLAN network. If the terminal moves to a UMTS network, it should re-authenticate and acquire a new IP address, thus creating a new session; it can then continue to use the services subscribed. Any active sessions (i.e. TCP connections or multimedia calls) will be dropped during the switching between the two access networks and a new IP address will be assigned.

5.4.1.3 Tight Coupling

In tight coupling, as shown in Figure 5.6, the generic RAT network is connected to the rest of the UMTS network (the core network) in the same manner as other UMTS RATs (UTRAN, GERAN) using the Iu interfaces by means of an Interworking Unit (IWU).

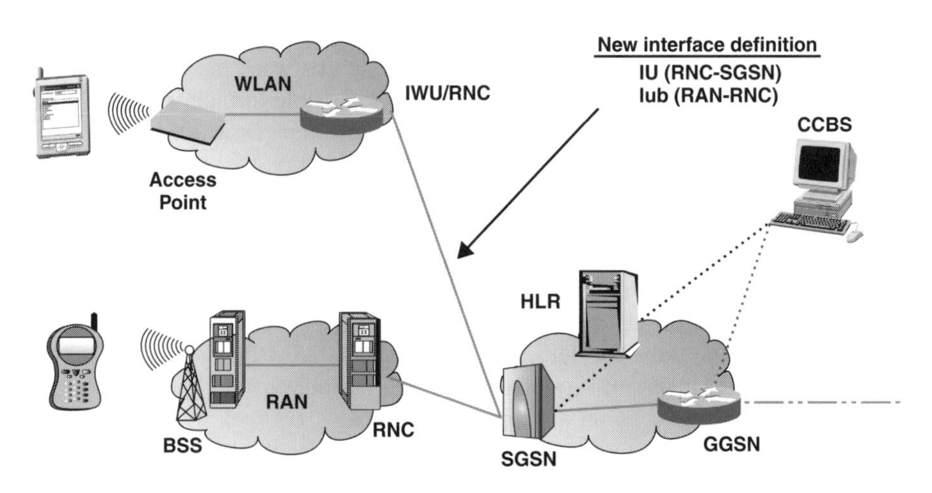

Figure 5.6 Tight coupling

In 3G Systems, as standardised by 3GPP, the Core Network uses two different types of Access Network: the Base Station System (BSS) and the Radio Network System (RNS). The BSS offers a Time Division Multiple Access (TDMA) based technology to access the Mobile Station whereas the RNS offers a Wideband-Code Division Multiple Access (W-CDMA) based technology [13]. Iu is the interface between a Radio Network Controller (RNC) and a Mobile-services Switching Centre (MSC), Serving GPRS Support Node (SGSN) or Cell Broadcast Centre (CBC), providing an interconnection point between the Radio Network Subsystem and the Core Network (CN). It is also considered as a reference point [14]. The Interworking Unit (IWU) provides the functionality necessary to allow interworking between a RAN and the CN. The functions of the IWU depend on the services and the type of CN. The IWU is required to convert the protocols used in the RAN to those used in the CN. The IWU may have no functionality where the service implementation in the RAN is directly compatible with that at the CN.

One of the most relevant aspects is that tight coupling interworking foresees the definition of the Iu interface between different RATs and that vertical handover can be supported. In the tight coupling scenario, the interface between different RANs is located in the core network (i.e. SGSN) and the vertical handover is managed by the core (i.e. SGSN). In the example reported in the figure, a new interface definition is provided between the WLAN IWU and the SGSN, and between the IWU and the 3G RNC. In this way, the WLAN technology is connected to the rest of the UMTS network via the core network (SGSN) in the same manner as other 3G RATs (UTRAN, GERAN).

5.4.1.4 Very Tight Coupling

In very tight coupling, the generic RAT network is connected to the RAN of the UMTS using the Iur interface by means of an IWU. In this new scenario interworking is provided within the RAN, near the RNC and involves the Iur interfaces. The network is able to perform a seamless handover between GSM/UMTS and WLAN networks due to a new interface definition between RNC and WLAN. The example reported in Figure 5.7 shows how the AP of the WLAN connected to the RNC is able to control the radio resources of the area covered by the AP. Iur is the logical interface between two RNCs, whilst logically representing a point-to-point link between RNCs, the physical realisation need not be a point to point link [14].

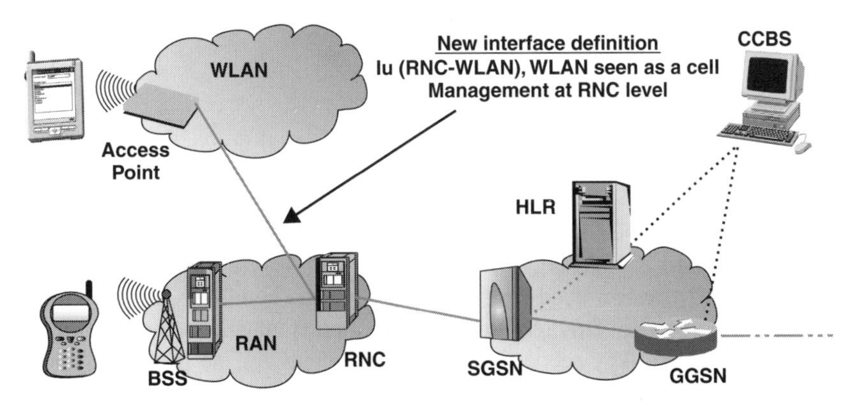

Figure 5.7 Very tight coupling

5.4.1.5 SCOUT Interworking Approach

The SCOUT network architecture has to be able to support a terminal reconfiguration which is particularly relevant when more than one RAT is available. For this reason, the interworking method between different access networks and the choice of the more suitable coupling scenario is important within the SCOUT context. To define the best coupling scenario to apply to SCOUT architecture, some considerations are relevant.

The terminal has to decide quickly if it is possible and necessary to reconfigure the protocol stack. The decision to reconfigure is also led by the resources available in the possible modes, so bandwidth management needs to be implemented in the access radio and it should be able to control every RAT available. The evolution of PRM should include this function.

The services subscribed to by the user should be available in a transparent way from the access technology. This means that high-level integration between different access technologies should be guaranteed.

One important point of the software download is reliability during the handover. If the terminal performs a software download via a PRM and in the meantime (before the end of the process) a change of PRM occurs, then completion of the download should be guaranteed. Therefore the avoidance of packet-losses in a seamless handover can be advantageous. If the previous PRM is informed about the initiation of a handover and about the following PRM, then a connection to this new PRM can be established by the previous one and during the handover process the data packets of the software download can be forwarded to the new PRM. The new PRM buffers these packets and in the case of packet loss the missing packets can be retransmitted.

In conclusion, it is clear that 'open coupling' and 'loose coupling' are not relevant for the SCOUT purpose. In fact, they do not perform a real integration of different RATs and, as a consequence, some of the requirements listed above cannot be addressed. Finally, the SCOUT approach regarding interworking between different RATs is a mixture of tight and very tight coupling, namely a *hybrid coupled heterogeneous network*.

The study will focus on the very tight coupling approach because it allows a very fast handover; to make quick decisions during and before the reconfiguration mode and to have all the information needed for software download in the access network and close to the terminal. This means that the user profile, terminal profile and service profile have to be cached in the access network (i.e. PRM).

5.4.2 Software Download Support

Within the TRUST project, key user scenarios and reconfiguration scenarios were developed. These were used as the basis for the system architecture definition. Some scenarios do not need a seamless handover and therefore mobility could be managed without the need for mobile IP or other seamless inter-mode handover mechanisms (this does not necessarily mean that it should not be, just that it could be). In other user scenarios (e.g. videoconferencing) there is a need for rapid and seamless handover while a videoconference session is in progress. This implies that the mobility management solution could be different for the different key scenarios. The requirements for the SCOUT system architecture should ideally take into account the scenarios and requirements generated within TRUST [2].

Initiating session scenarios, where identified within the user studies activities, is particularly important. They stem from the requirement for a terminal to be highly 'available' and allow sessions to be initiated regardless of the terminal state and 'mode' of operation. This includes interruption of other active sessions and even termination of active sessions to support new sessions (depending on the user preferences). These scenarios present some important implications on mobility management and terminal reconfiguration architecture.

As described in the TRUST project, the configuration and download process can be triggered both from the terminal and from the network. Different functions are involved during respective cases. Hereafter, we present some consideration about the incoming and outgoing session scenarios, and the triggering of the download request from the terminal or from the network.

5.4.2.1 Terminal Initiated Request

When a terminal requests to download software, it interacts directly with the PRM that has cached all the information transmitted from the terminal to the network during the 'Mode Monitoring Phase'. It allows the network to check the real possibility to perform the action required from the terminal.

In this situation the proxy acts as a caching element and as a broker for the terminal, communicating with other network elements in order to accomplish the steps necessary for the software download on behalf of the terminal (i.e. authentication, billing, software repository, etc.).

5.4.2.2 Network Initiated Request

In this scenario the network triggers the download of the software. This can happen when the network detects that the terminal is changing the coverage technology and then a new protocol stack has to be downloaded. This scenario has already been described in TRUST and the necessity to detect the best mechanism to perform the software download to the terminals with the minimum use of resources has been identified. This implies that a new functionality has to be defined in order to analyse the status of the radio link resources, the capability of the terminal and the service requirements, and then decide on the most suitable type of channel to use for the downloading. A factor is the number of terminals that require downloading of the same application within a specific area. In fact, depending on the number of terminals, it could be more indicative to use dedicated channels or broadcast/multicast channels, as reported in Figure 5.8 [15].

This functionality should have interactions with the RRM and fast access to the user profile, terminal profiles and service profiles, as depicted in Figure 5.9. For this reason, this functionality should be located in the RAN and could be an extension of TRUST. In fact, the PRM, located in the RAN and closely related to the RNC and to the RRM modules, can support these decision mechanisms by caching information about system load, terminal measurements, etc.

5.4.2.3 Incoming Session

This scenario is one in which a session is being initiated to a recipient who is using a reconfigurable terminal (or to a recipient terminal connected via a reconfigurable terminal).

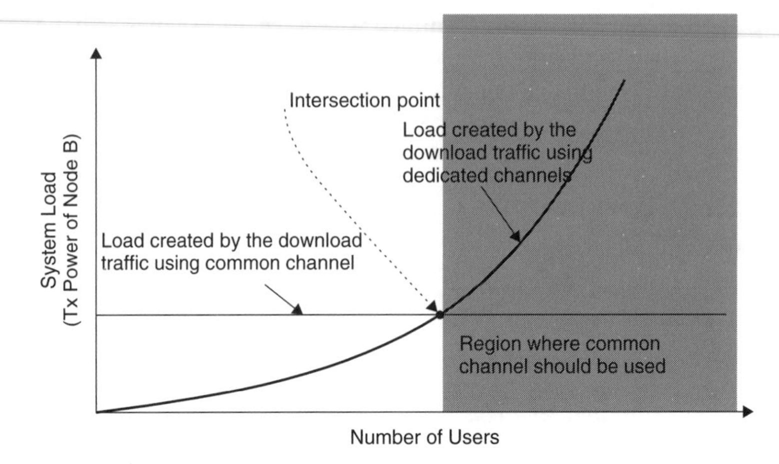

Figure 5.8 Loading profile

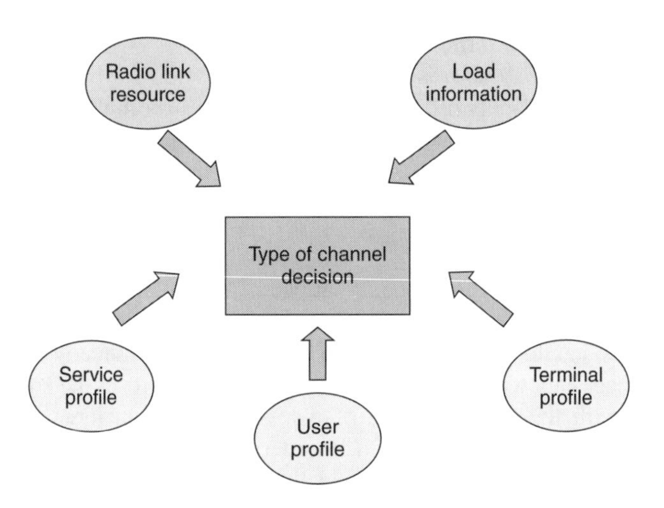

Figure 5.9 Type of channel decision function

The terminal may or may not have existing active sessions. The most suitable 'mode' to support the incoming session may be different from the current 'mode' of the terminal. The most suitable mode depends on the terminal resources (such as battery health), the cost of the reconfiguration, network resources/QoS available for supporting the sessions in the different 'modes', and who is paying for it.

From the mobility management perspective there may be no requirement for handover (and therefore for mobile IP or other micro- or macro-mobility solutions) in this scenario if there are no active session when session initiation takes place. However, reconfiguration of the terminal and routing of the session to the terminal must be performed in a timely manner. For example, this can be handled by a HRM or enhanced SIP server (using the SIP protocol). The security issues may also be very different in this scenario because, if the initiating device is being charged for the session, then there are no requirements to authenticate the recipient prior to reconfiguration or even to set up the session at all (although the initiator may want to check that the recipient is the intended recipient either directly or indirectly).

If the incoming session is to a recipient who is connected via an *ad hoc* connection to a reconfigurable terminal, then there are implications for how the HRM or SIP server and reconfigurable terminal know the address of the recipient (and resources and QoS available in the *ad hoc* network). The reconfigurable terminal could masquerade as the recipient to make it appear as if it were the intended destination. However, there are security implications and service interaction implications of this that need to be considered. If masquerading is used, then the network address translation and other functions need to be implemented within the reconfigurable terminal.

Finally, the issue of interrupting an active session in another 'mode' is another important consideration. There must always be the capability for the active terminal sessions to be interrupted to receive a more 'important' incoming session. This implies a level of pre-emptive priority. Therefore, this must in some way be supported in all possible terminal modes, which may be quite difficult if IP does not extend all the way to the terminal in all the 'modes' supported by the terminals.

5.4.2.4 Outgoing Session

The outgoing session scenario is one in which an outgoing session is being initiated to another entity by a reconfigurable terminal or via a reconfigurable terminal. The session may or may not be to another reconfigurable terminal. The most appropriate 'mode' to support the session depends on the resources of the reconfigurable terminal, cost, etc. as for the initiating session scenario. Since it is likely that it is the initiating terminal that pays for the reconfiguration and cost of the session, security, and in particular authentication of the terminal to the network, is more important. Again mobile IP and other micro- or macro-mobility solutions are not necessary.

If the initiating terminal is connected to a reconfigurable terminal via an *ad hoc* connection, then the issues of the selection of the most appropriate mode to support the session are joint decisions between devices. Also, the authentication of the initiating terminal to the network becomes more difficult. It may not be possible for the reconfigurable terminal to masquerade as the initiating device in this case unless it accepts liability for the costs incurred during sessions initiated via the *ad hoc* connection.

5.4.3 Bandwidth Management Module

As described in the previous section, PRM must closely interwork with RRM algorithms in RNC. A possible approach to use for software download using broadcast/multicast is currently represented by the MBMS strategy (currently standardised in UMTS Rel6 by 3GPP).

5.4.3.1 Multimedia Broadcast/Multicast Service (MBMS)

The previous sections pointed out two difference approaches for the software download process: the use of dedicated channels or broadcast/multicast channels. In the case when

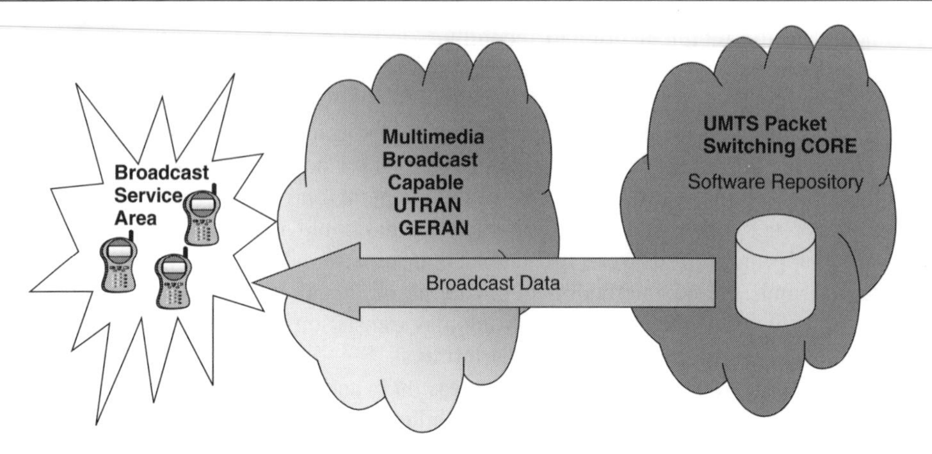

Figure 5.10 Example of the MBMS broadcast mode

the network decides to use broadcast/multicast channels, the MBMS strategy [16] could be a possible candidate as the reference approach. MBMS is a unidirectional point-to-multipoint bearer service in which data are transmitted from a single source entity to multiple recipients.

3GPP has defined two modes of operation:

- The broadcast mode
- The multicast mode

5.4.3.1.1 MBMS Broadcast Mode. The broadcast mode is the unidirectional point-to-multipoint transmission of multimedia data from a single source entity to all users in a broadcast area. In this modality the core network (see Figure 5.10), has the task of defining the 'Broadcast Service Area' used to address the data (i.e. text, audio, picture, video) to all terminals within that area. Efficient use of radio/network resources is achieved by transmitting the data over a common radio channel, for example. The broadcast mode differs from the multicast mode in that there is no specific requirement to activate or subscribe to the MBMS in broadcast mode.

In the SCOUT context, a typical application of this modality is the situation in which the network communicates to all the terminals present in a particular Broadcast Service Area giving information such as advertising or welcome messages.

5.4.3.1.2 MBMS Multicast Mode. The multicast mode allows the unidirectional point-to-multipoint transmission of multimedia data from a single source point to a multicast group in a multicast area. As with the broadcast mode, the radio/network resources are to be used efficiently, for example using a common radio channel, and the core network has the task of defining the 'Multicast Group' and 'Multicast Area'.

In the multicast mode there is the possibility for the network to selectively transmit to cells within the multicast area that contain members of a multicast group. The main difference between the broadcast modes is that the multicast mode generally requires a subscription to a multicast subscription group and then the user can join the corresponding

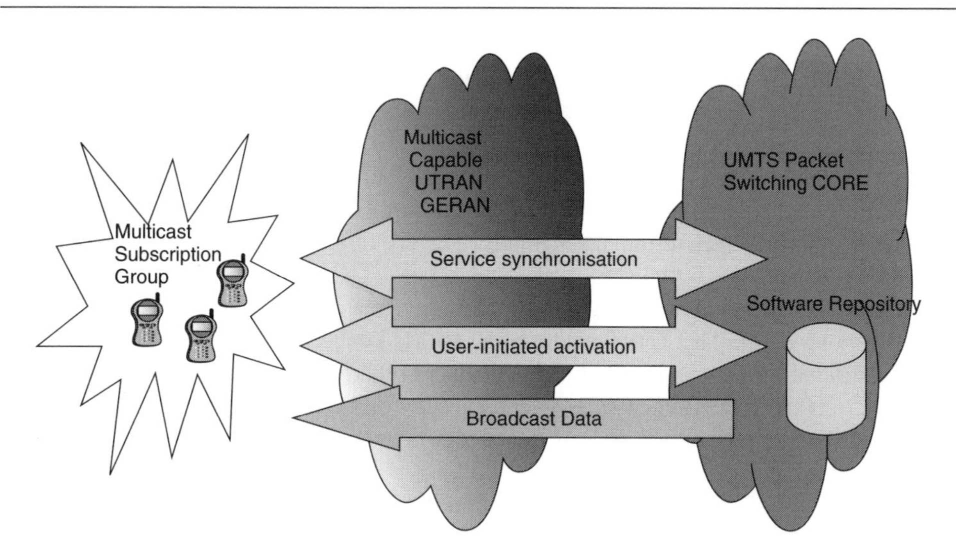

Figure 5.11 Example of the MBMS multicast mode

multicast group, as shown in Figure 5.11. So, in this modality it is expected that charging data for the end user will be generated.

In the SCOUT context, an example of utilisation of this modality is the mass upgrade scenario, when a software upgrade for a class of terminal is available and needs to be sent to a specific class of terminal (multicast group).

5.4.4 System Infrastructure

The objectives of the reconfiguration architecture are to support fast and reliable reconfiguration and to use the available resources in an efficient manner. Furthermore, a common functionality is required to avoid a continuous adaptation of the backbone infrastructure in the case of possible heterogeneity of future RANs. Therefore the main components and functions responsible are located in the particular RAN and, additionally, on the terminal.

The core entities in the reconfiguration process are the proxy reconfiguration managers (PRMs) located in every RAN. The PRMs are the contact points for every terminal attached to the RAN concerning reconfiguration.

In the case of a terminal initiated software download, the terminal signals the need for reconfiguration over the PRM control plane to the current PRM in the RAN and thereafter the PRM is responsible for the delivery of the appropriate software module over the appropriate user plane. For the mode-switching support, the PRM performs additionally different measurements and informs the terminal and the neighbouring PRMs as well.

With reference to the user plane software download, the PRM stores the necessary software modules in its local repository. However, the overall capacity of the storage space in the PRM is not so high. The intention is to have fast access to the most frequently used modules. For less frequent requests for required software there exists an interface between the PRM and an intermediate server database, namely the serving reconfiguration manager (SRM). Thus the request is forwarded and processed by this SRM.

Another reconfiguration supporting functionality is the Inter-PRM Interface. Neighbouring PRMs are connected to each other and are able to exchange some information about the current status of the accompanying radio access or about an ongoing mode change of a terminal.

The reconfiguration of a terminal need not only be initiated by the terminal, but can also be triggered by an external entity. In the case of a new hardware driver version it is inefficient to inform each terminal separately. The use of multicast would help to optimise the content delivery.

We call the Terminal Reconfiguration Serving Area (TRSA) the area of the PRMs served by one SRM. The TRSA encloses all PRMs connected to one SRM. The TRSA is not restricted to one single RAN or RAT and can therefore be larger and cover several RANs (e.g. if the total area covered is achieved by different WLANs) or smaller than a single RAN and cover only a part of the access network (e.g. if a single RAN covers a whole continent).

All the reconfiguration signalling from and to the TRSA leads either across the external connection of the SRM or across the Inter-PRM Interface, if neighbouring PRMs with overlapping cells exchange information.

Figure 5.12 shows an example of a TRSA. In this area different RATs are located. Three hotspot areas with two access points (e.g. IEEE 802.11 or HIPLERLAN2) in each area with lower range but high maximum available bandwidth and one cellular access point belonging to the same TRSA. The neighbouring PRMs with overlapping cell coverage are coupled to each other by the Inter-PRM Interface and every PRM has a connection to the local SRM.

A vertical handover to another access technology might have different time restrictions imposed, among others, by the state of the terminal or by the cause triggering the handover.

The more constraints there are, the more important it is to have assistance from the network side to retrieve as much information as possible. For example, for a faster identification of possible modes the network could inform the terminal about available radio technologies in the vicinity of the terminal's position. The terminal could then limit scanning to the corresponding frequencies. Furthermore, the network could provide the terminal with additional helpful information, for example different measurements from the neighbouring networks, services provided and their general QoS. Also, short-term information (currently occupied resources) could be provided.

Another point considering reconfigurability is the general download of application updates or new driver versions where time constraints are, in general, not so strict. The network could schedule the download and execute the process in a moment of low load and therefore use the resources more efficiently. For this the terminal should provide its reconfiguration requests with a priority indication for the download. The terminal could then compare its own requirements with the information mentioned above and measurements and come to a decision about mode switching.

In the case of an externally initiated reconfiguration, e.g. by the manufacturer's server or HRM, the terminal is notified about the available module. If the terminal is in a busy state and currently not able to receive it, the terminal should inform the PRM about it. Otherwise, if the transmission succeeded completely, the terminal should acknowledge

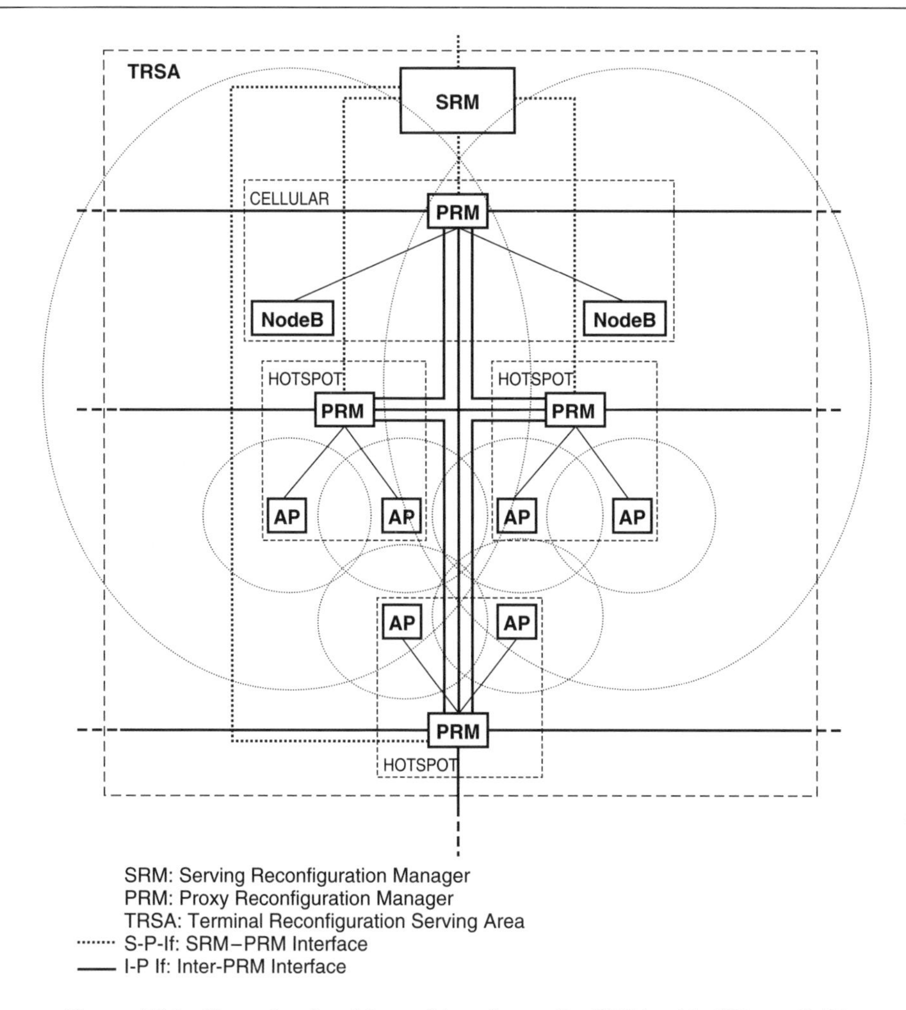

Figure 5.12 Example of entities and interfaces of a TRSA with different RATs

the reception. If errors or packet loss occur during the download process, the terminal could request some retransmissions at the PRM.

If the required software modules for reconfiguration are not available, then the terminal has to request them at the PRM, download the software and reconfigure the hardware before the mode switching can be executed. This might also be a decision parameter if the reconfiguration process is urgent. Modes for which the required software is available at the terminal or at the PRM might be preferred above a different one for which the software must be downloaded from a server since this introduces further delay.

5.4.4.1 Proxy Reconfiguration Manager

As already mentioned, the PRM is the primary contact point for the terminal in the reconfiguration process. Within a RAN more than one PRM could be located. Considering, for example, UMTS, one PRM for the whole UTRAN could be overloaded easily, if a

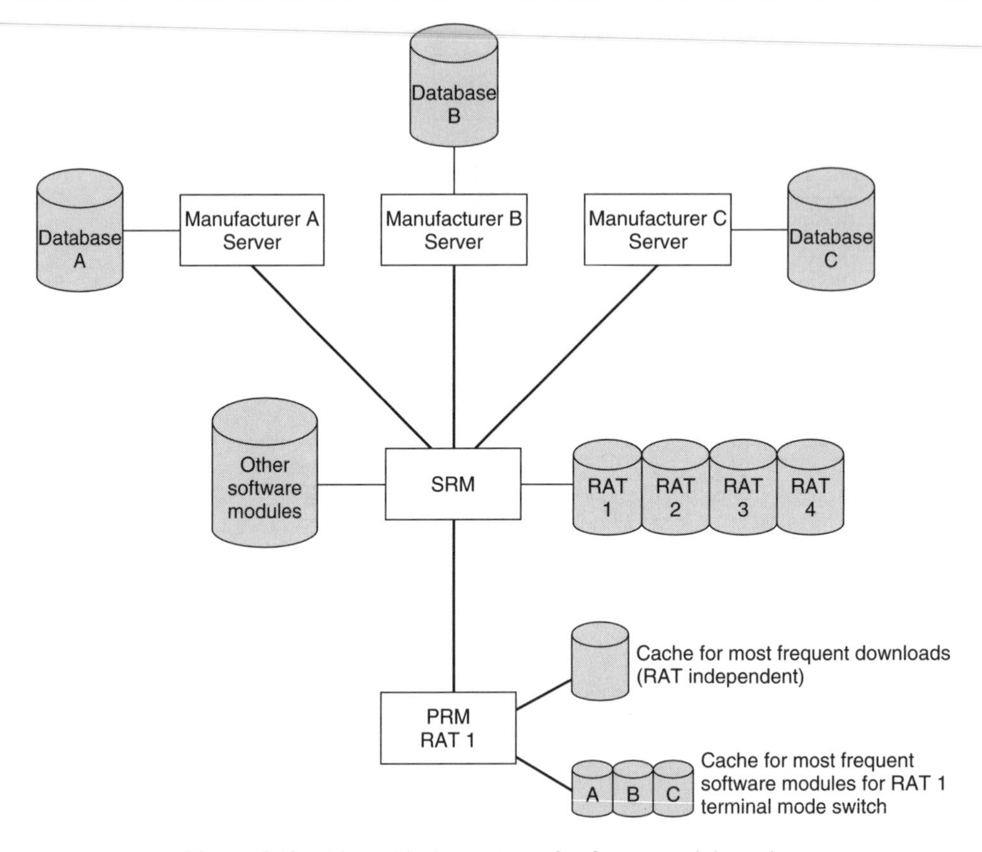

Figure 5.13 Hierarchical structure of software module caches

large number of terminals start a reconfiguration procedure simultaneously. In this case several PRMs distributed in the UTRAN would share and speed up the requests.

The architecture shown in Figure 5.13 includes hierarchical proxy architecture. Every PRM caches the modules, which are more often used in its access network, independent of the RAT belonging to it. Furthermore, the PRM caches a larger number of modules for its controlled RAT. If a terminal were to switch the mode in the foreseeable future and the controlling PRM had not stored the necessary files, the PRM could request them over the Inter-PRM Interface in advance from a neighbouring PRM. The user plane transport would be carried out over the Inter-PRM Interface as well. If the terminal finally decides to do the handover to the other mode, no additional delay occurs because of the software download.

After the mode switch has been initiated, the old PRM transfers useful terminal information to the new PRM. In this way the new PRM can prepare the information for the new terminal concerning the attached neighbouring RANs and provide it to the terminal at an early stage.

As already mentioned, reconfiguration initiated by the terminal can also happen for reasons of minor importance, e.g. an application update. The PRM is notified of the importance of a reconfiguration by a priority indication in the request message. The purpose of this indication is to enable the PRM to schedule the download if possible. If the network load is currently extremely high, but it is foreseeable that this will change in the immediate future, then the PRM could delay the beginning of the download. If the terminal performs a handover whilst a download is queued, the old PRM should notify the new PRM about this schedule and, if necessary, forward the appropriate software module.

5.4.4.1.1 Mass Upgrade. For the mass upgrade of terminals the multicast mechanism is used to avoid network overload. The idea is that each terminal manufacturer, application developer, etc. has its own server. If a terminal is now registered with its profile at the PRM, the PRM knows for which components of the terminal a mass upgrade could happen. After registration of the terminal the PRM joins a multicast session for each possible component. If there is a mass upgrade going on, which a certain server initiated, the software packets are only delivered to those PRMs that have joined the multicast group. In the case where multicast is only employed in the wired link, the benefit of its use is then reduced. For the wireless link either unicast or broadcast need be used depending on the number of users. Moreover, the PRM might take part in the whole decision to perform a unicast or broadcast. In the event that only a few terminals in a cell are involved in the update and the overall load is low, a unicast connection should be established between the PRM and the terminals to ensure fast and reliable download. In the case where a terminal is unable to receive the update at that moment, the PRM caches the software module and delivers it afterwards. If the number of terminals destined to be reconfigured in a cell is higher than a certain threshold, then the PRM can deliver the update over the air interface per broadcast as well. But since the reliability of the transport must be ensured, a re-transmission mechanism must be used. One possibility would be to use unicast NACK and to request the PRM to re-transmit the lost packet per multicast. Another possibility is to use adaptive coding and produce new packets that might be used by several users.

Figure 5.14 shows an example of the delivery of a mass upgrade. The terminals are registered with their profile at the PRMs. The PRMs have joined the appropriate multicast groups and are now able to receive mass upgrades. Now the server of terminal type A manufacturer initiates the mass upgrade, sending the software to the multicast address. The multicast routers in the Internet Backbone forward the packets along the multicast delivery tree. The PRMs, as leaves of the tree, receive the update and transmit it to the terminals of the appropriate type.

5.4.4.1.2 Inter-PRM Interface. The previous sections have already shown that the overall functionality of the PRM is beyond the usual proxy functionality. One extension of the PRM could be a connection to neighbouring PRMs and the exchange of additional information via an Inter-PRM Interface. First of all this could contain long-term information about the general supported QoS in the other RANs (e.g. the priorities of different traffic classes, maximum bit rate, maximum delay, etc.). Thus, the PRM could decide in advance which neighbouring mode is useful and for which mode the terminal should scan. After the detection of an alternative mode, whose general QoS is promising, the PRM could request short-term measurements at the neighbouring PRM. The final decision should be made after consideration of all the measurements.

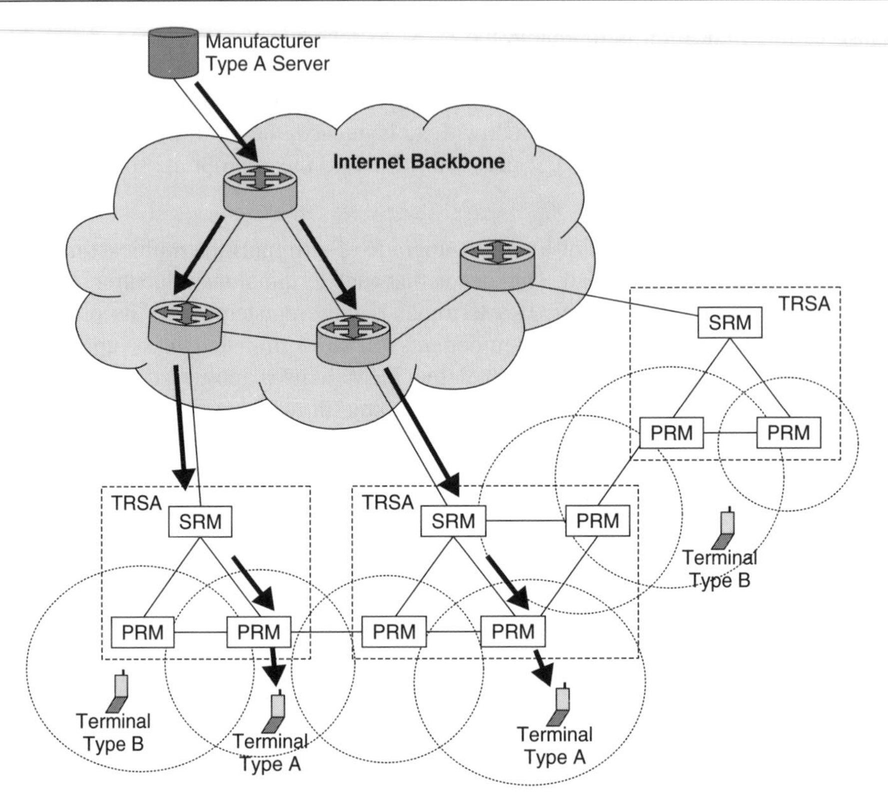

Figure 5.14 Example of a mass reconfiguration process initiated by an external server with the use of IP multicast

After a mode switch has been initiated, the old PRM could transfer useful terminal information to the new PRM over the Inter-PRM Interface. In this way, the new PRM can prepare information concerning the attached neighbouring RANs for the new terminal and provide it to the terminal at an early stage.

As far as measurement and resource information are concerned, each QoS supporting RAN must provide a resource reservation mechanism. If there is a resource manager available, then the PRM can request long-term as well as short-term information from this manager via the Inter-PRM Interface and the neighbouring PRM and report back to the manager in advance if a mode switch is initiated. If there is no resource manager or other resource information entity in a network, then the PRM must measure and store general results on its own.

5.4.4.1.3 Splitting and Location of the PRM. In the previous sections, some aspects have been analysed in order to define the requirements that the network has to satisfy to support the terminal reconfiguration within particular scenarios. Some considerations have been given to the evolution of the cellular network toward an all-IP network, the management of the mobility, the management of the handover between different RATs and the possible coupling scenarios that could be available in the short, medium and long terms. From this analysis, some conclusions can be drawn.

The SCOUT architecture could have benefits following the IP-based RAN approach, described in an earlier section, taking several advantages from its main features, as summarised briefly below:

- Separation from the control plane and the transport plane; one implication is that the control functions (in the c-plane) have the capability to manage and control the radio resource in the u-plane even if they are related to different radio technologies.
- A common transport layer based on IP protocol; this means that the radio base station (i.e. BTS, NodeB, AP, etc.) are connected to the rest of the network via the IP transport protocol.
- Hierarchical distribution of the functions within the CN (functions independent of the radio access) and within the RAN (apart from the radio access). In a medium-term scenario, this approach helps the resolution of problems such as the mobility that is needed to implement different mechanisms for different scenarios (i.e. micro and macro mobility).

So, the availability of different RATs has already led, in the short and medium terms, with different and complementary features, to investigation of the possible coupling scenarios between different RATs. For SCOUT purposes, the tight coupling and very tight coupling have been identified as possible candidates to support the reconfiguration of terminal, seamless handover and efficient mobility management.

One of the most suitable capabilities of the network is the ability to take decisions about the switching mode as fast as possible and the very tight coupling scenario better addresses this feature because the coupling, and consequently the interworking, is performed at the RAN level. On the other hand, the management of the macro mobility is more complicated, so a hierarchical approach could be applied. Following these two scenarios, the PRM should be located in the RAN, which should also have available all the information (user profile, terminal profile, service profile) in order to manage the reconfiguration of the terminal.

An interaction with the RRM is also required because the network could need to decide which are the more appropriate RATs for the service required, when more than one radio access is available in a certain geographic area. Moreover, it should be able to decide which type of radio channel is more suitable to use for safe radio resources (dedicated channel, multicast channel, broadcast channel). Following these indications, a new PRM function, located in the RAN, should be in charge of this task.

After these considerations we can define the position and the structure of the new PRM in SCOUT. As reported above, the PRM is mainly involved in mode negotiation, mode monitoring, mode switching and software download, where information about user, terminal and services and control capability are required. Following the *de-layering* approach the PRM functions can be split into u-plane and c-plane functionalities, as shown in Figure 5.15. The u-plane functions collect all the profiles and the information needed to perform software download. The c-plane functions collects all the functions needed to perform control operations. We define:

- *SPRE:* Software Download and Profile REpository (u-plane) the logical entity that collects:
  - Terminal, user service profile
  - User profile

Figure 5.15 Scout PRM

  - Service profile
  - Software patches
- *SDRC:* Software Download and Reconfiguration Controller (c-plane) the logical entity that hosts the PRM controlling functions, in particular:
  - Software download control functions
  - Reconfiguration control functions

The u-plane PRM (SPRE) should be located in the RAN, near (or included) in the RNC because the following information has to be available and updated as fast as possible:

- Channel and system load conditions
- User service profiles
- Application QoS
- Terminal capabilities
- Network address and protocol

Depending on the type of coupling between different RATs, the time constraints of the service and the need, or not, to perform a seamless handover, the c-plane PRM (SDRC) could be located both in the RAN (near or included in the RNC), shown in Figure 5.17, or in the core network (near or included in the SGSN), as depicted in Figure 5.16.

When both the SPRE and SDRC are located in the access network, the interaction with the RRM is more direct and the micro-mobility management can be managed from a dedicated SDRC function.

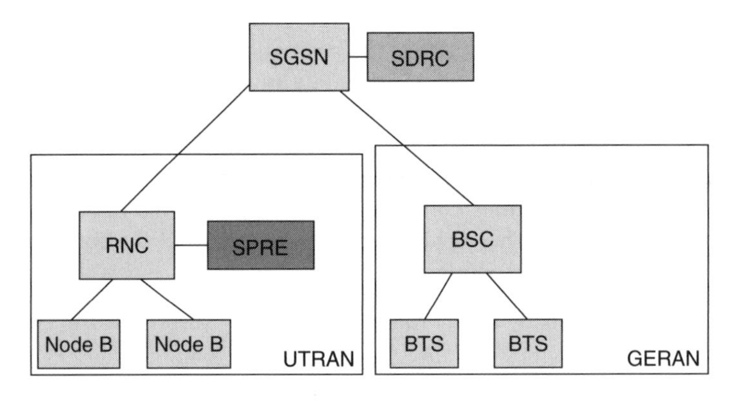

Figure 5.16 Example of proxy control functions in the core network

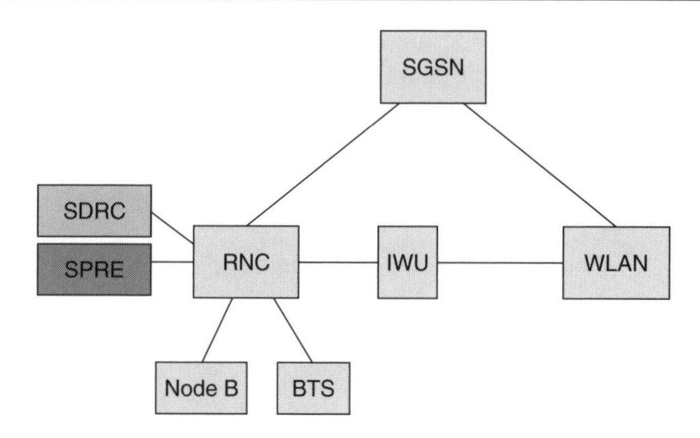

Figure 5.17 Example of proxy control functions in the RAN

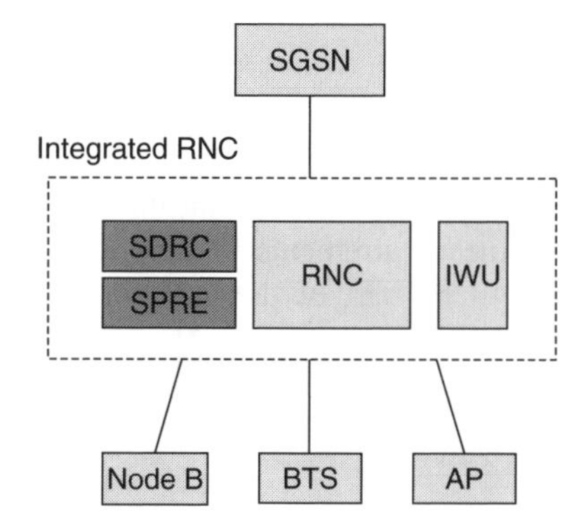

Figure 5.18 RNC with reconfiguration software download functions integrated

Following the 3GPP approach, in the medium term the SDRC and SPRE could be included within the RNC (see Figure 5.18); then, following the All-IP paradigms, they could be distributed in the network.

An advantage to having integrated the PRM functionality into the RNC is that the RNC could directly manage more than one RAT during the reconfiguration and software download process. This means that it could

- Perform seamless handover efficiently
- Efficiently manage radio resources for different radio technologies
- Possibly manage more than one stream by different radio technologies

This possible medium-term evolution can be summarized as in Figure 5.18.

5.5 *Ad Hoc* Network Support

5.5.1 Introduction

TRUST essentially addressed the reconfigurability of terminals roaming between standard modes or radio accesses. During this work interactions between terminals and the networks were identified and some concepts, such as the PRM, were proposed.

Ad hoc and self-organised networks are subject to growing interest, especially when considering wireless networks. SCOUT could extend the work done in TRUST by addressing the reconfigurability of terminals and supporting such networks.

The consideration of *ad hoc* networks will use, as much as possible, the scenarios for reconfigurability already defined in TRUST.

5.5.2 Ad Hoc Networks and Reconfigurable Terminals

In *ad hoc* networks, typically terminals will be able to organise themselves to set up a network. These networks could follow different topologies, such as

- Extension of infrastructure
- Stand-alone network

Moreover, for each of the above usages of *ad hoc* networks, the following two kinds of topology could appear:

- Single-hop network
- Multi-hop network

SCOUT contributes to the study of the decentralisation and self-organisation to build *ad hoc* networks by bringing the requirements to support reconfigurability within those networks. This contribution will be part of the consolidation of the architecture work performed in the project.

5.5.3 Ad Hoc Network Topologies

5.5.3.1 Extension of the Infrastructure

Ad hoc networks could serve as extensions to other existing network infrastructures (with applications, for instance, in airports, railway stations and hotels). A typical application example of *ad hoc* networks extending the coverage of cellular networks would be the following. Imagine someone wants to initiate a mobile phone call but has no network coverage due to the shadowing effect of a large building. Instead, he may use the electronic devices of intermediate pedestrians to relay his data to the base station.

Another type of application is a mobile network, which is the *ad hoc* network itself moving as a group. This is described as the 'train scenario' in the MIND project [15].

The network topology could be either the simple single-hop network, where a terminal acts as a relay to the infrastructure network, or the more complex case of a multi-hop network, as depicted in Figure 5.19.

5.5.3.1.1 Single-Hop Network. Software reconfiguration can be also carried out directly between terminals without any supporting network. One example of such a direct download scheme is presented in ref. [16]. This approach is based on a cellular network architecture with the additional capability of terminals talking directly to each other. This is possible by using a local area radio technology such as Bluetooth or HIPERLAN/2 with direct mode. Even though the network is necessary for any communication, the download of software in this approach is always achieved directly from other terminals. Base stations

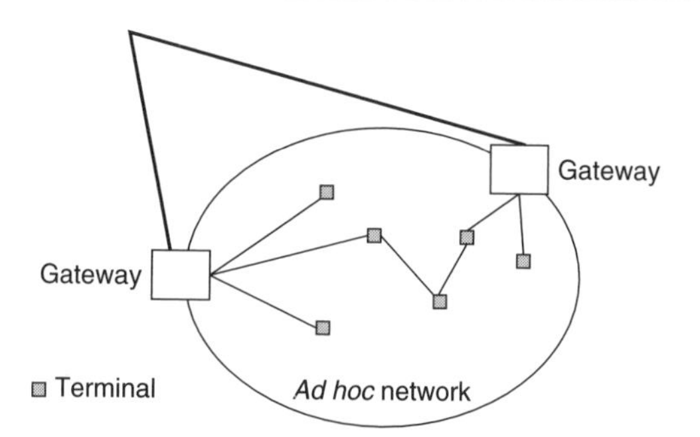

Figure 5.19 Multi-hop network

support this local software download with announcements of current software modules stored in all the terminals within the coverage of a cell. This approach is mainly intended to save download time for software mass upgrade. Although this network architecture is not in the context of *ad hoc* networks, we can still apply the idea to the case of extension of infrastructure, which is a point-to-point scenario.

5.5.3.1.2 Multi-Hop Network. When considering the more complex case of a multi-hop network, the discussion will focus on the connection of *ad hoc* networks to public networks. The reconfigurability aspects that should be considered are both network initiated and terminal initiated.

The case of terminal initiated reconfiguration occurs when the terminal is leaving the *ad hoc* network to join some other (public) network, caused by the user quitting the group of participants of the *ad hoc* network.

A network initiated reconfiguration could arise when a management function of the *ad hoc* network detects that the quality of the network links is degrading (or congestion is arising) and a decision is taken by the network management to move to a better suited mode. Then all the terminals participating in the *ad hoc* network will have to be reconfigured to the new mode in order to resume their sessions.

5.5.3.2 Stand-Alone Network

In this topology, the *ad hoc* network does not have links to other networks, as shown in Figure 5.20. The stand-alone topology might be the one requiring more attention in the

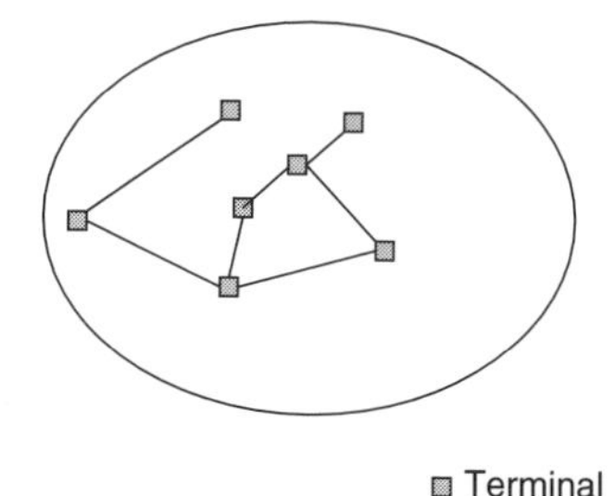

Figure 5.20 Stand-alone network

view of the supporting reconfigurability. The architecture of the *ad hoc* network should be enhanced to provide the functions needed for reconfigurability.

Some of the points to take into consideration are, for instance, the switching decision algorithm (to trigger the reconfiguration), or the provision of the required information to the terminal to perform the reconfigurability (software download).

5.5.4 Some Reconfigurability Aspects

Within *ad hoc* networks, there are some features that might require a reconfiguration of the network, or at least some part of it. One example of this need is the routing algorithm. The choice of routing algorithm is influenced, for instance, by the kind of mobility of the terminals:

- Static terminals
- Slow speed terminals
- High speed terminals

If the nature of the terminals participating in such a network is changing, then there will be a need to change the routing procedures and to reconfigure all terminals in the network.

Another case where reconfiguration could be brought into play is when saturation occurs in the network, and to alleviate it a reconfiguration could take place. This reconfiguration could, for example, transform a single-hop network into a multi-hop network, or change from a Bluetooth-based radio access to an 802.11 based radio access.

Since current approaches only consider a reconfiguration support functionality located in the network, with *ad hoc* networks this functionality has to be carried out additionally inside the terminal.

In the case of a multi-hop extension of the infrastructure, software has to be downloaded over intermediate nodes. Therefore the download process utilises the resources of neighbouring nodes. There are two possibilities:

- The intermediate nodes only have routing functionality for the reconfiguration process.
- The intermediate nodes also have a temporary software repository functionality for caching. In this case the intermediate nodes need more storage capability.

5.5.5 Summary of the Impact on the Architecture

Based on the previous introduction to the aspects combining reconfigurability and *ad hoc* networks, there is a need to refine and enhance the system architecture so that requirements coming from *ad hoc* networking will be taken into account.

At first glance the application of *ad hoc* networks as extensions of infrastructure networks is the least demanding in terms of requirements to modify the TRUST architecture, as it could support:

- The PRM
- Assisted mode monitoring
- Extension of the mobility management

5.6 A Generic Reconfigurable Network Architecture with Distributed Intelligence

5.6.1 Intelligent Reconfigurable Radio Network Concept

For a software radio system to be useful as an adaptable future-proof solution, and to cover both existing and emerging standards, it is required to have elements of reconfigurability, intelligence and software programmable hardware. For this purpose, the CAST project addresses a generic software radio system solution, combining reconfigurability with intelligence [17]. The reconfigurable elements are not limited to the physical layer only, and reconfigurability management is distributed in all layers.

The main concept behind the CAST intelligent reconfigurable mobile radio network is shown in Figure 5.21. The Intelligent Reconfiguration Controller has interfaces and collects data from three different environments: the user environment, the radio environment and the network environment. Reconfiguration of the radio network is then based upon the results of the processing of these data sets by the intelligent support system. Within this universal context, the reconfigurability entails the pervasive use of 'software re-definition', empowering upgrades and/or patching of network elements, and of all services and applications running on it.

The above concept applies to all types of terrestrial and satellite radio systems (including paging, cellular and wireless-LAN), enabling the integration of many systems on the same hardware platform. More importantly, it also interfaces generations of mobile and personal communication systems using the same hardware infrastructure.

Mitola and Maguire have given the following explanation of cognitive radio [18]. Radio etiquette is the set of RF bands, air interfaces, protocols, and spatial and temporal patterns

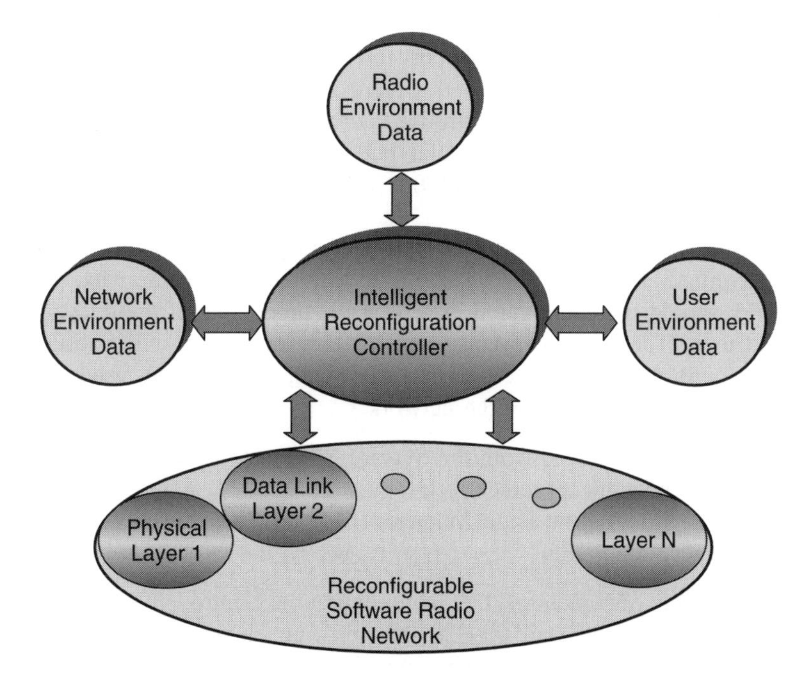

Figure 5.21 CAST intelligent reconfigurable radio network concept

that moderate the use of the radio spectrum. Cognitive radio extends the idea of software radio with model-based reasoning about such etiquettes. Cognitive radio enhances the flexibility of personal services through the use of a Radio Knowledge Representation Language (RKRL). This language represents knowledge of radio etiquette, devices, software modules, propagation, networks, user needs, and application scenarios in a way that supports automated reasoning about the needs of the user.

This empowers software radios to conduct expressive negotiations among peers about the use of the radio spectrum across space, time, and the user context. With RKRL, cognitive radio agents may actively manipulate the protocol stack to adapt known etiquettes to better satisfy user needs. This transforms radio nodes from blind executors of predefined protocols to radio-domain-aware intelligent agents that search out ways to deliver the services the user wants, even if that user does not know how to obtain them. From this explanation it can be understood that the difference between a mobile terminal that uses the software radio concept and one that is based on the cognitive radio concept is that the software radio mobile terminal as presently conceived cannot have such an intelligent exchange of information with a network, since they have no model-based reasoning or planning capability and no standard language to communicate with.

5.6.2 CAST Generic Architecture

The CAST vision of software radio encompasses an intelligent reconfigurable mobile terminal and the serving mobile network [18]. It extends Mitola's cognitive radio concept, and considers low implementation complexity reconfigurability the salient issue and not just defining radio functionality in software. It then further expands this idea by considering the necessary technology to support intelligent reconfiguration, through the use of model-based automated reasoning and a Network Reconfiguration Language (NRL), addressing similar problems to the cognitive radio concept. However, the reconfigurability is not just restricted to the terminal but is applied to the serving mobile network as well. The essential elements of this reconfigurable network are illustrated in Figure 5.22.

To reconfigure any part of the network it is necessary for the network to have intelligence and reconfiguration control. The intelligence is used to decide what part or parts of the network should be reconfigured, based on the processing of the relevant information supplied to it. The Network Manager (NM) then makes reconfiguration decisions, in consultation with the Global Intelligence. In general, there are two ways in which the NM is forced to reconfigure part of the network:

1. A service request is received from the Service Manager (SM), which requires reconfiguration of the network resources.
2. A failure message is via the Fault Manager (FM), which requires remedial action in the form of a reconfiguration.

In either case, the NM requests the Reconfiguration Controller to implement these decisions on the appropriate hardware and/or software elements of the network. Any network element can contain fixed, re-programmable or reconfigurable modules, according to the following definitions:

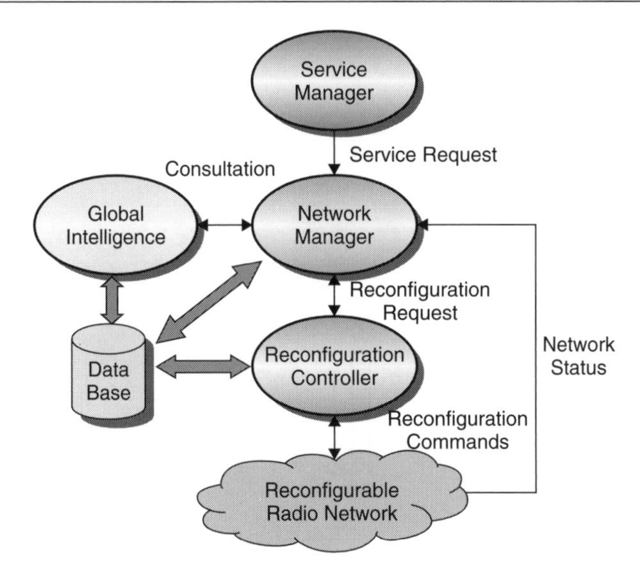

Figure 5.22 The essential elements of the CAST generic architecture

- *Fixed module:* This module performs a fixed function that cannot be changed.
- *Re-programmable module:* This module performs a fixed function, the parameters of which can be changed (i.e. the module has a fixed personality whose character can be modified).
- *Reconfigurable module:* This module performs a function that can be changed by software download (i.e. the module can assume different personalities).

Figure 5.23 shows the CAST Generic Architecture. This design is based on distributed intelligence, which has the following components:

1. Global Intelligent Reconfiguration Controller (GIRC), located in the Mobile Switching Centre (MSC).
2. Local Intelligent Reconfiguration Controller (LIRC), located in each Base Station Controller (BSC) and each Mobile Station (MS).

The GIRC contains the global components of the NM, and the organic based CODA intelligent support subsystem, which share a database. Each LIRC, on the other hand, contains the local components of the Resource Controller (RSC), and the NM, both of which have access to the same database. The reconfiguration decision is made by the NM, in consultation with CODA, via the shared database, and the reconfiguration task is then analysed and passed to the local LIRCs in each corresponding BSC and MS modules. The Layer Controller (LC) in each layer is responsible for the processing of the reconfiguration procedures, normally by forming and executing the required object chains.

5.6.2.1 Distributed Nature of the Intelligence

The GIRC is considered to be the 'heart' of the intelligent reconfiguration technology. It gathers information relevant to reconfiguration via the NM, from the reconfigurable mobile

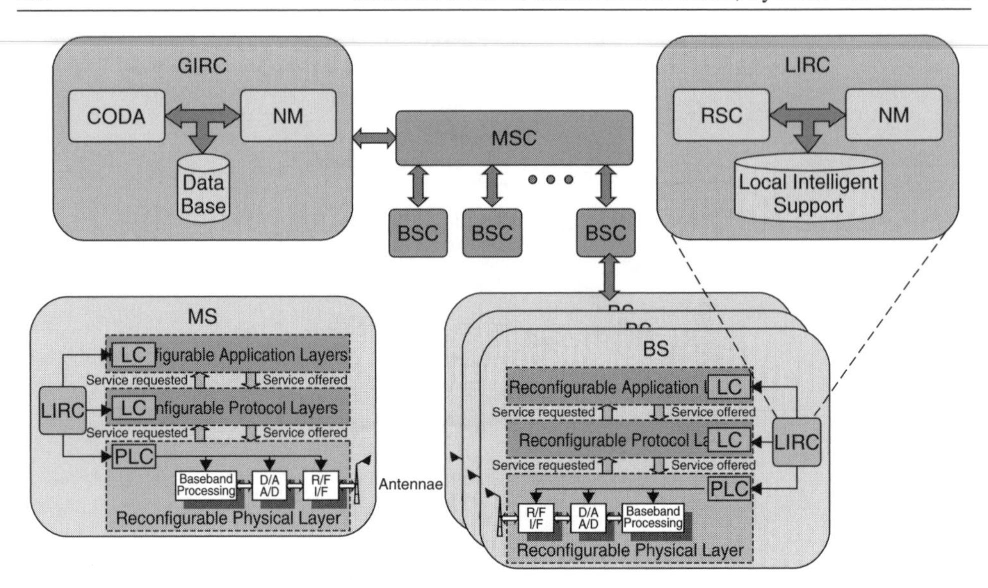

Figure 5.23 The CAST generic architecture

network, about the state of different hardware and software components in the different nodes and protocol layers; and from the end-user applications, about the environment and user profiles. This is stored in the database, were it may be further processed and filtered. The intelligence, working in conjunction with the database, then makes a decision or decisions on what needs to be reconfigured. A model-based automated approach is used to implement the intelligence. The reconfiguration suggestions are then passed to the NM, who then requests the LIRC, in the appropriate network node and protocol layer, to implement it on the relevant reconfigurable hardware and/or software. If the request cannot be implemented, LIRC will inform the GIRC.

Owing to the distributed nature of intelligence, there are generally two types of reconfiguration decisions: globally instigated and locally instigated. In a locally instigated reconfiguration the LIRC will identify the need for reconfiguration in its particular node and protocol layer, and may execute this reconfiguration immediately if it is appropriate to make the decision locally, or ask the GIRC for permission.

Normally, all information gathered from the reconfigurable mobile network is passed to the NM who then passes it to the database or the intelligence as necessary. Also, all globally instigated requests for reconfiguration are passed from the intelligence to the NM who then passes it to the Reconfiguration Controller, and likewise locally instigated requests for reconfiguration are passed from the Reconfigurable Controller to the NM who then passes it to the intelligence. So we can see that all relevant information from and all requests to the mobile network go through the NM to ensure coordination and synchronisation of reconfiguration activities. The only time that the NM is by-passed is when there is an emergency reconfiguration. In this case the intelligence directly informs the Reconfiguration Controller to take immediate actions. We therefore see that intelligence is distributed between the global and local network components.

5.6.3 Reconfigurability Management

The Network Management framework for reconfigurability is based on the idea of using a Local Network Manager (LNM), in each radio network item, including the Mobile Station (MS) and the Base Station (BS). The Global Network Manager (GNM) resides in the MSC.

The Network Management has separate components for single reconfiguration using the Reconfiguration Manager (RM) within either MS or BS or simultaneous reconfiguration of both MS and BS (called 'bi-reconfiguration requests') via the Bireconfiguration Manager (BRM).

The layered Network Management is shown in Figure 5.24. The layered architecture consists of two dimensions: the horizontal dimension is used for Reconfiguration Management and Fault Management, and the vertical dimension consists of five layers as follows:

- User layer (integrated with the MS-LNM)
- MS layer
- BS layer (including the handling of bireconfiguration requests)
- Global layer (MSC)
- Operations layer (addressing the Operator's viewpoint)

The hierarchical Fault Management information propagation path can be described as follows:

1. Faults reported by the MS are forwarded to the LNM-BS FM. This is because the managed entities from the operations point of view are the BSs and not the MSs.
2. Faults reported from any related MSs and from the underlying BS are forwarded from the LNM-BS FM to the GNM layer, which provides the fault information to the Operations Layer as the final receiving destination.

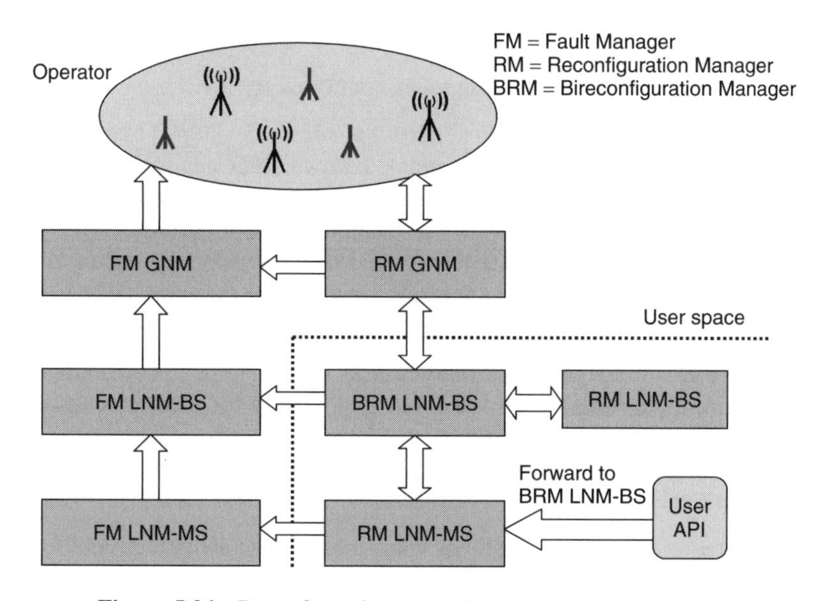

Figure 5.24 Reconfiguration network management architecture

The Reconfiguration Management uses a central layer for the delegation of reconfiguration tasks as follows:

1. The user API allows the user applications to issue the reconfiguration requests, which arrive at the LNM-MS Reconfiguration Manager. Received reconfiguration requests are transferred to the LNM-BS Bireconfiguration Manager except if the MS itself is involved.
2. The operator API is able to initiate global reconfiguration requests, where the global reconfiguration requests are submitted to the GNM Reconfiguration Manager. Globally initiated reconfiguration requests are forwarded from the GNM Reconfiguration Manager to the corresponding LNM-BS Bireconfiguration Manager units.
3. The LNM-BS Bireconfiguration Manager deals with bireconfiguration requests concerning a BS and a MS simultaneously. Bireconfiguration requests are split into their MS and BS parts, submitted to the responsible Reconfiguration Management component of MS or BS, and its execution is synchronised according to the successful completion of both parts, MS and BS.
4. Finally, the success or failure is reported back to the requesting Reconfiguration Manager.

5.6.3.1 Resource Optimisation and Control

The architecture of the Resource Controller is shown in Figure 5.25, and consists of the following components:

- *Queue Handler:* Manages incoming requests.
- *Core Controller Intelligence:* Coordinates inner operation of the RSC, distributes configuration to processing modules.
- *Processing modules:* Find optimal placement on the hardware for a newcomer service, modify settings of an installed service, (re)organises the services on the hardware to utilise the resources more efficiently.
- *Database:* Stores the description of services controlled by RSC.

The Resource Controller (RSC) is an application module developed to intelligently control the configuration of different reconfigurable hardware devices such as DSPs, ASICs, ASSPs and FPGAs in our system. The application uses object-oriented technology, and is designed to demonstrate the operation of the reconfiguration algorithms. The devices we deal with can be configured for different tasks and they can operate simultaneously. One device can often run more than one application (function) at the same time. Constructing a correct configuration for such hardware is complicated for two reasons:

1. Different devices are suitable for different tasks. For example, programmable ASICs are more appropriate for filtering tasks, but DSPs could be used for channel modem and baseband signal processing tasks.
2. In different scenarios different configuration algorithms should be used to satisfy the different requirements

The different requirements that should be taken into account are maximum lifetime, fault tolerance, maximum utilisation and low power consumption. Other important aspects can easily be added to the RSC. The aspects can be handled by various optimisation algorithms

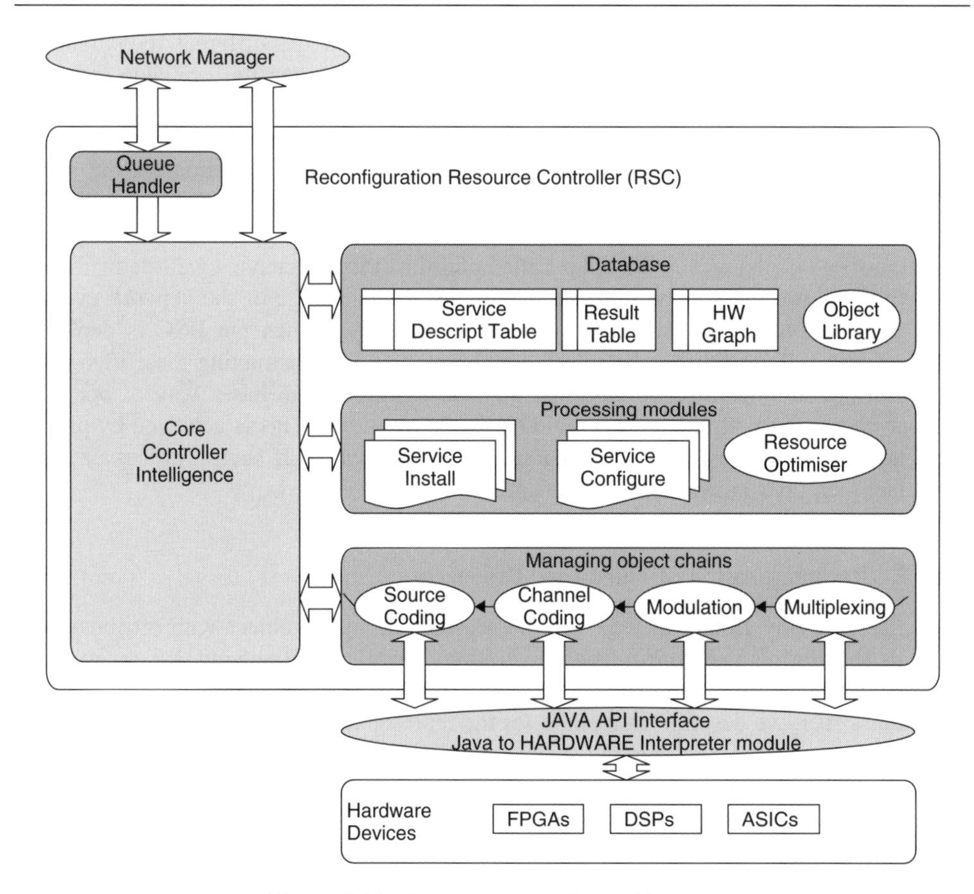

Figure 5.25 Resource controller architecture

such as: the maximum capacity, the hottest first, the coldest first or the blind monkey algorithm. For mobile handsets, the preferred algorithm is the hottest first algorithm. The main goal of this algorithm is that it utilises a minimum amount of hardware. If a resource is not in use it can be powered down, thus saving energy, since a processor can often serve more than one task. When new functionality is needed it can be placed onto resources already powered up. Using this method, considerably lower power consumption can be achieved. For BSs, on the other hand, fault tolerance is a greater issue. In this case even utilisation is the goal, because if we spread our services evenly over a large field of hardware, and part of the hardware fails, only a few services will be affected. For such cases, the two algorithms that accomplish even utilisation are the coldest first and the blind monkey algorithm.

In order to use these algorithms a well-constructed resource controller architecture is needed. This architecture is based on three abstraction layers: The first layer is the Hardware Abstraction Layer (HAL), the second is the Function Abstraction Layer (FAL), and the third is the Chain Abstraction layer (CAL). The role of these abstractions is to hide the irrelevant features of the underlying layers and to provide a clear and well-defined interface to the above layers. Each abstraction layer consists of a set of Java classes and some database tables. The different Java classes represent the different reconfigurable building blocks, and the records in the database tables represent the available resources in the system. When a resource is needed, a new instance of the corresponding Java class is instantiated and is installed as appropriate.

Implementation of these abstraction layers provides a flexible solution for the reconfiguration of different hardware resources and their evolving features. The flexibility can be increased by integrating the intelligent decision-making and resource management algorithms into exchangeable modules called plug-ins. Plug-ins have a well-defined interface which allows them to be replaced as circumstances change or the network evolves. Each plug-in contains a resource management algorithm. When the RSC is deployed, the appropriate algorithm can be selected and installed. In constructing these algorithms and described architecture, we assumed that the RSC software has a way of operating without interfering with the configurable hardware resources. This is achieved by placing the functionality of the RSC on a separate service processor. If such a processor is not available, then distributed resource management should be used.

5.6.3.2 Reconfiguration of Hardware Resources

In our system, any FPGA or DSP can be represented by an object with attributes and methods to describe and control its configuration capabilities. Baseband, RF or any other network functions, often implemented as ASIC, are also described similarly as objects. In a separate paper, we described a method for the representation of these objects [19]. Once all components of the configurable system are described as objects (both hardware and functions) a mechanism is required to organise and manipulate these objects to place real functions onto real processing hardware. The CAST project uses a Java-based mechanism to facilitate the configuration of hardware resources. This provides a bridge whereby one side operates in terms of soft objects and the other manipulates and configures 'hardware', as the methods in the soft objects are called.

The level of native control is highly dependant on the way that the Java Machine (JM) is implemented. The most common implementation of a Java system uses a software emulation of the JM running on an existing processor. This is better known as the Java Virtual Machine (JVM). In this case the native interface to hardware is simply some assembly code that will run on the processor. This code can be made capable of toggling pins or specific ports on the processor device. External configuration mechanisms of configurable devices can be easily controlled.

Another implementation is the byte code processor. In this case the Java code can be written to directly toggle pins or ports of the device to facilitate configurations. Whichever JM implementation is used, the configurable hardware must be accessible to the Java code in an organised way in order to define hardware objects. This requires a partitioning of the hardware into groups of configurable hardware [20]. Each configurable device is defined as the owner of a partition. External peripheral hardware (memory, etc.) which may be attached to the configurable devices must also be considered. All peripheral hardware belongs to a partition which contains the configurable device connected to it. In the case where a particular resource is shared (i.e. PCI bus with slave devices) the resource belongs to the partition containing the configurable device which arbitrates the use of the resource. Each partition is given an address to identify uniquely each partition in the software. Once the hardware is logically partitioned, with regard to configuration, knowledge of this partition must be stored electronically. Database tables are used to store information regarding: each partition's resources (address, owning configurable device, peripherals, and connections between partition devices), and interfaces between each partition (configurable connections). The database information regarding each partition is also linked to a Java class which is capable of controlling its configuration. This class is implemented so as to abstract away, or hide, the functionality of the native mechanism that actually carries out the configuration.

A major problem associated with the configuration of physical hardware is the numerous different types of hardware available. The major difference is the physical method used to configure. For example, different FPGAs will configure in a different way. Some are partially reconfigurable and others can only be wholly configured. DSPs, on the other hand, configure far differently. A configuration mechanism has to liase with the DSP's own operating system in order to send its processing code and activate it.

To make the configuration interface as simple as possible the differences in configuration mechanism must be abstracted away. This should leave a common interface for all types of hardware. Generally, all configurable hardware has a 'space' where the configuration information (or function) can be placed. A particular function will take up a specific 'size' on a device. The function will also be placed at a particular 'location' on a device. These three units of space, size and location will be interpreted differently by each device, therefore a method of understanding this vague interpretation is required. Each partition description has a device-specific map which describes what configurable resources are available at the locations within the space on a configurable partition. This allow resource allocation algorithms to choose what types of resource are available for configuration. Resources are described using the Extensible Markup Language (XML). The XML tags are defined, which indicate location and groups of location, which in turn indicate the specific resource available. For example, a group of locations in an FPGA may represent a Configurable Logic Block (CLB) or a RAM. Each location represents where in the device's configuration space to store the bits that configure that particular resource. Similarly, a DSP memory map can be explained with locations indicating where to place executable code and where to specify the operating system's task list.

With the hardware constructed and its construction classified and stored, all that remains is to deploy functions to the resources. Current compiler technology can limit the system to deploy only pre-compiled functions to their specific device. For this reason it is assumed that pre-compiled functions are stored in a library and deployed where necessary. The function object is defined as being a collection of pre-compiled device codes for different devices. For example, a turbo decoder function object may contain specific implementations targeted to a Xilinx Virtex FPGA, Analogue Devices' TigerSHARC DSP, Texas Instruments' C60 DSP and a PC. Specific resource requirements are stored with each specific implementation. Resource requirements are stored using XML similar to the resource descriptions of the hardware. The whole function object is stored similar to the hardware resource information by using database tables for information and Java classes for operational functionality. A common interface is defined to allow access to operational status during the function's lifetime while configured. With both hardware and functions classified it is possible for resource allocation methods, such as those implemented in the

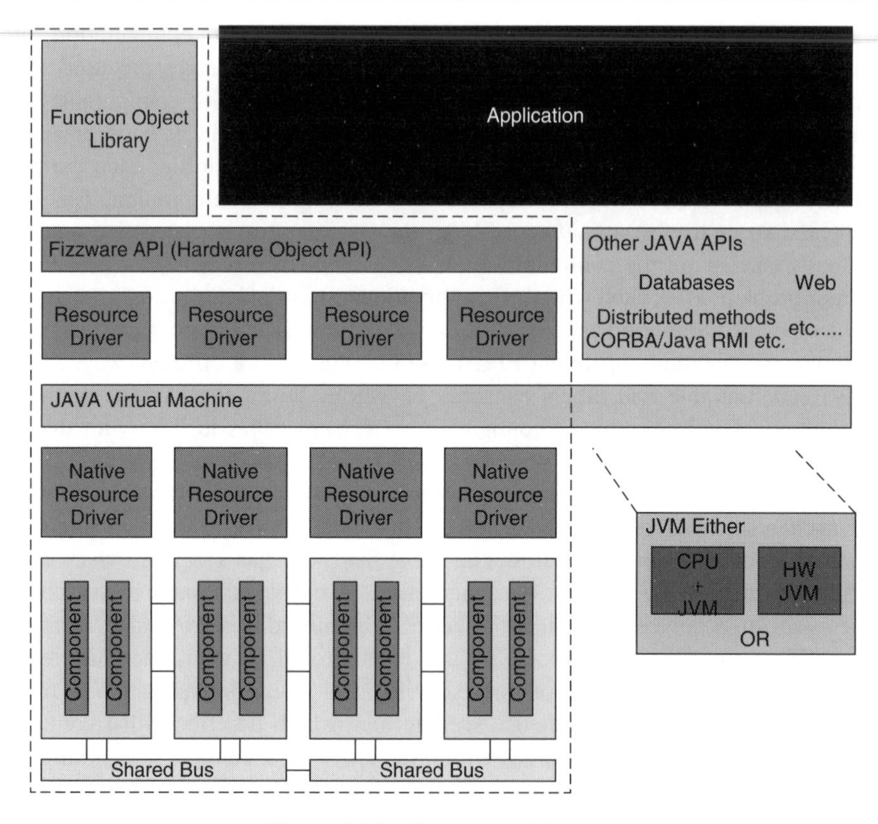

Figure 5.26 Fizzware architecture

CAST RSC, to dynamically allocate functions to the hardware and monitor the status of any configured functions.

Datapath connections between devices are managed using a concept known as 'end-points'. A common communication protocol is defined that manages data transfer between different devices. An endpoint is simply a device-specific function which implements the device communication protocol and distributes the data to and from the functions on the particular device. Function communication with the same device is also specified, but mainly relies on the operating system or other runtime support features of the particular device. The databases that hold all the information regarding hardware and functions are implemented simply via a Java-based Open Database Connectivity (ODBC) driver library. Scalability is ensured by decoupling the implementation from a specific database in this way. The CAST project implements an Oracle 9i database behind an ODBC connection. XML operations, similarly, are carried out using a freely available Java library. The amalgamation of all of these separate mechanisms can be combined to illustrate how hardware configuration can be easily managed once all information is gathered and abstracted. We name this framework 'Fizzware'. Figure 5.26 illustrates the whole system, which gives both hardware and software perspectives. The dotted line in Figure 5.26 encloses the Fizzware framework. From the bottom, the hardware itself is shown. Above the hardware, separate native drivers are shown, each relating to a particular hardware partition. The

JVM forms the centre of the system and defines the barrier between the software and hardware. Resource drivers are formed from database entries and java classes and each implements the common interface for device configuration. All of these resource drivers and several supporting runtime classes form the Fizzware API.

5.7 Conclusion

In conclusion, a generic distributed architecture was described for general reconfiguration of wireless mobile networks. In this design, problems associated with the organic-based intelligent support, network management, resource optimisation and control, and object orientated reconfiguration of resources were investigated and discussed.

Acknowledgements

The work described here is based on the research carried out in the EU-sponsored collaborative IST Framework V projects TRUST and CAST. The authors also gratefully acknowledge the contributions by Tereska Karran, George R. Ribeiro-Justo, Mahboubeh Lohi, Abdolkhalil Lohi, David Lund, Ian Martin, Bahram Honary, Sándor Imre, Gyula Rábai, József Kovács, Péter Kacsuk, Árpád Lányi, Thomas Gritzner.

References

[1] IST-1999-1070 TRUST Project, Deliverable D4.1 'Report on state of the art and requirements on SDR system features', June 2000.
[2] IST-1999-1070 TRUST Project, Deliverable D4.2 'Report on proposed novel solutions on SDR system features', November 2000.
[3] IST-1999-1070 TRUST Project, Deliverable D4.3 'Report on assessment of novel solutions on system aspects of reconfigurable terminals and recommendations for standardisation', October 2001.
[4] V. Milutinovic and P. Stenstrom, 'Special Issue On Distributed Shared Memory Systems', *IEEE Proceedings*, **87**(3), 399–404, 1999.
[5] J.C. Cano et al., 'The differences between DSM caching and Proxy caching', *IEEE Concurrency*, **July–September**, 45–47, 2000.
[6] Z. Jiang et al., 'Incorporating proxy service into wide area cellular networks', *Wireless Communications and Mobile Computing*, **1**, 299–312, 2001.
[7] J. Border et al., 'Performance enhancing proxies intended to mitigate link-related degradations', RFC3135, June 2001.
[8] N. Assaf et al., 'Interworking between IP security and performance enhancing proxies for mobile networks', *IEEE Communications Magazine*, **May**, 2002.
[9] C. Chien, M.B. Srivastava, R. Jain, P. Lettieri, V. Aggarwal and R. Sternowski, 'Adaptive radio for multimedia wireless links', *IEEE Journal on Selected Areas in Communications*, **17**(5), 1999.
[10] A.T. Campbell, J. Gomez, K. Sanghyo, W. Chieh-Yih, Z.R. Turanyi and A.G. Valko, 'Comparison of IP micro-mobility protocols', *IEEE Wireless Communications Magazine*, **9**(1), 2002.
[11] T. Dagiuklas, D. Gatzounas, D. Theofilatos, D. Sisalem, S. Rupp, R. Valentzas, R. Tafazolli, C. Politis, S. Grilli, V. Killias and A. Marinidis 'Seamless multimedia services over all-IP based infrastructure: The EVOLUTE approach', IST Mobile and Wireless Telecommunication Summit, 2002.
[12] ETSI TR 101 957 (v1.1.1 2001-08) Technical Report.
[13] 3GPP TS 23.002 v3.3.0 (2000-03) 'Network architecture', Release 1999.
[14] 3GPP TS 25.401 v5.1.0 (2001-09) 'UTRAN Overall Description', Release 5.
[15] E. Mohyeldin, J. Luo and M. Dillinger, 'Software download management for cell Broadcast Channels in WCDMA', Second Karlsruhe Workshop on Software Radios, 20/21, March, Karlsruhe, Germany, 2002.
[16] 3GPP TS 22.146 v6.0.0 (2002-06) 'Multimedia Broadcast/Multicast Service', Stage 1 Release 6.

[17] *http://www.ist-mind.org/*

[18] M. Dillinger and R. Becher, 'De-centralised software distribution for SDR terminals', *IEEE Wireless Communications*, **9**(2), 20–25, 2002.

[19] K. Madani et al., EC Framework V Project CAST (Configurable Radio with Advanced Software Technology), WP1.2: Architectural Functions, Final Report, September, 2000.

[20] J. Mitola and G. Maguire, 'Cognitive radio: Making software radios more personal', *IEEE Personal Communications*, **August**, 13–18, 1999.

[21] K. Madani et al., 'A distributed approach for intelligent reconfiguration of wireless mobile networks', IST Mobile Summit, Thessalonica, Greece, June, 2002.

[22] D. Lund, B. Honary and K. Madani, 'Characterising software control of the physical reconfigurable radio subsystem', *Proceedings of IST2001*, Barcelona, **September**, 357–362, 2001.

6

Self-Learning and Adaptive Systems: The CODA Approach

Tereska Karran, Kambiz Madani and George R. Ribeiro-Justo

University of Westminster

6.1 What is CODA?

Intelligent agent technology [1] has become important in providing intelligent personal support for users. Agents are however only one possible solution to the problem of intelligent support and in many cases the expectations have been higher than the real results this technology can actually deliver. In addition, an important issue is that in order to offer the best advice in a large and complex system such as mobile networks, a large amount of information may be needed. Intelligent agents are not usually designed to deal with such a large amount of data. A better solution to the problem of analysing large amounts of data is the concept of business intelligence [2]. Business intelligence aims to support business analysis and decision-making by creating consistent, information that is crucial for companies to make timely decisions and respond to changing business conditions. While agent technology focuses on a particular user, business intelligence tries to correlate data from many users in different situations, especially historical data, in order to support global decision-making processes. For example, by combining historical data of network usage with current trends, it is possible to derive future forecasts.

Complex Organic Distributed Architecture (CODA) [3, 4] represents a new generation of decision-making systems that include a means of monitoring and controlling objectives to allow the enterprise to evolve dynamically with a certain degree of autonomy. It applies a theoretical foundation for modelling evolutionary enterprises provided by (organic) principles, defined by the Viable Systems Model. CODA refines and adapts Beer's Viable Systems Model [5] with reference to distributed decision supporting systems. Although originally applied to help define management principles, the model is further adapted to the control of data flow in a complex decision support system. The result is a flexible, organic and adaptable business intelligence architecture. CODA intelligent structures are based on business intelligence concepts. Business intelligence systems are complex distributed systems that consolidate data from different data sources, using different formats and

Software Defined Radio: Architectures, Systems and Functions. Edited by M. Dillinger, K. Madani and N. Alonistioti
 ISBN: 0-470-85164-3

performing complex analytical processing on the data. Such complex systems require solid reference architectures in order to simplify their development and management and to provide the best decision-making support.

Business intelligence systems focus on improving the access and delivery of business information to a wide audience of both information providers and information consumers [6]. They achieve this by providing online analytical processing (OLAP) and information mining technologies, and packaged applications that exploit the power of those technologies. These applications often need to process and analyse large volumes of information using a variety of different tools. A business intelligence system must, therefore, provide scalability, and must be able to support and integrate products from many different vendors.

Building separate data warehouses creates an environment tailored to decision-making and is the foundation for deploying an effective business intelligence solution. The process of collecting, correlating, reconciling, integrating, organising, describing, enhancing and summarising business information is what data warehousing is all about. A data warehouse structures data in a way that makes it easier and more effective to manage, access and analyse.

CODA separates the intelligent information system into five functionally distinct layers, each supported by a data warehouse component and an intelligent component. Each data warehouse component is separated by filter components, which restructure data into formats suitable for the information processing functions to be performed. This approach is based on the way that organic systems manage complex and adaptive behaviour. The advantages of CODA are that it is intuitively easy to manage, and supports complex evolution by using data intelligently. At the same time, CODA minimises information flow between the system components and provides sufficient component autonomy.

6.2 The CODA Architecture

Integrated and distributed information systems have generated new challenges in the development of complex systems. By enforcing the global view of the information system, a structure is offered to manage this complexity. The concept of software architecture aims to provide a description of such structures. There are many definitions of architecture, but at the core of them is the notion that the architecture of a system defines its 'gross structure' [7].

The architecture prescribes the structure of the information systems within the organisation, and the structure must be respected through the remainder of the development cycle. The architecture is an abstraction that helps manage complexity. It has therefore been widely recognised that well-designed architectures are critical to the success of any complex project. The system structure described by the architecture illuminates the top-level decisions as to how the system is composed of interacting parts, where the main pathways of interactions are and what the key properties of the parts are.

The architecture usually describes how the elements fulfil the system requirements, including: how the elements are responsible for which functionality; how they interact with each other; how they interact with the world; and, in the case of computer systems, their dependencies on the execution platform. The purpose of the architecture is not only to describe the key elements of the system but also to enable the architect to decide how the

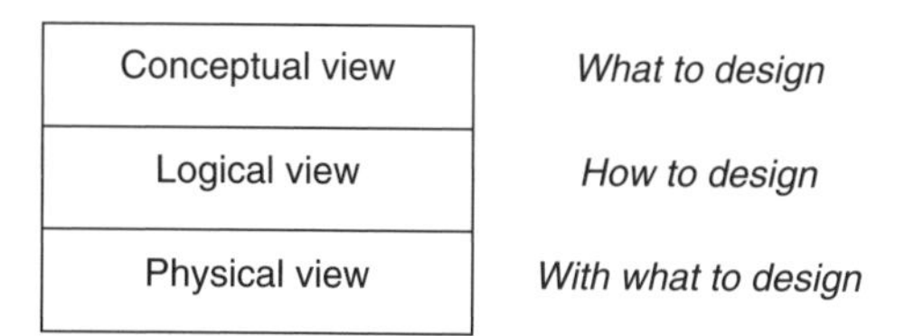

Figure 6.1 Architectural development

system will satisfy its requirements. By exposing key system design concerns, a properly designed architecture helps in guaranteeing that a system will satisfy its requirements. By providing an abstract view of a system, the architecture exposes the main properties whilst hiding non-relevant properties.

Even though some methods have been proposed [8, 9], there is still no universally accepted notational standard or method for architecture description and development [10]. The nature of the documentation for a particular system remains strongly influenced by the needs to which it will be used, in other words the organisational context. A common aspect of most existing methods is the need for multiple views of the architecture. This allows the complexity of the architecture to be managed and usually provides architectural views for different stakeholders. In this chapter we do not follow any particular method but, from our experience, we decompose the architecture into three main views, which clearly correspond to the development steps we will follow, as shown in Figure 6.1.

The conceptual view describes the existing system and its context. It identifies the objectives and principles, which will guide the architecture development. The goal of the conceptual view is to clarify the purpose, fit and limitations of the system with respect with its environment. Following the principles and objectives described in the conceptual view, we can present a high-level design solution for the system in hand. The architecture should define and describe the elements at a relatively coarse granularity. The aim is to focus on the main aspects and leave the details for the design phase. Finally, the physical views describe how the logical design can be implemented. Again, the idea is not to present an implementation but rather the technology options and their possible impact. We will use a standard notation to describe most artefacts used during the architecture development. The notation we have chosen is the Unified Modelling Language (UML) [11]. UML is a standard modelling language for object-oriented software development proposed by the Object Management Group (OMG) responsible for the standardisation and promotion of object-oriented software.

6.3 The Conceptual Architecture

The conceptual model of the business presents a simplified view of the complex reality of the business. It represents abstractions that enable us to eliminate irrelevant details and focus on one or more important aspects at a time. Effective models can facilitate discussion among the various stakeholders in the business. More importantly, the business model can be the basis for other models. In the case of CODA, we use the business model to derive the 'business intelligence information model'.

A business conceptual model is an abstraction of how the business functions [12]. It usually provides a simplified view of the business structure that will act as the basis for communication, improvements or innovations, and define the information requirements that are necessary for a business model to capture an absolute picture of the business or to describe the business in detail. In summary, the objectives for developing a business model are:

- To better understand the mechanisms of an existing business.
- To act as the basis for creating suitable information systems that support the business.
- To act as the basis for improving the current business structure and operation.
- To experiment with a new business concept.

Ideally, the business conceptual model would consist of a single diagram that includes all the important aspects of a business. This is not possible because of the complexity of most businesses. So, a business model is illustrated with a number of different views, each capturing a particular aspect. Each view may contain one or more diagrams.

The diagrams may be represented by different notations. We will follow the method proposed by Eriksson and Penker [12], which is based on the UML. As previously stated, UML is a standard modelling language for object-oriented software development proposed by the OMG responsible for the standardisation and promotion of object-oriented software.

Our main objective in this choice is to create a model of the overall business that can be used to decide which information systems are required, how those information systems can be developed and what functionality the systems should contain. With regard to the initial concepts, one of the most important views of the business model, following the UML approach, is the architecture. The architecture captures the vital parts of the business structure and their relationships in an organised manner [7]. A business is a complex system that has a specific purpose or goal. All the functions of the business interact in order to achieve this goal. The concepts within the conceptual model are defined below:

- *Resources:* the objects that are used and produced in the business. The resources are organised in structures and have relationships with each other. They can be modelled as CODA agents.
- *Processes:* the activities performed within the business during which the state of business resources changes. Processes describe how the work is done within the business. They are governed by rules. They are usually modelled as CODA tasks.
- *Goals/principles:* the purpose of the business or the outcome the business is trying to achieve. Goals can be broken down into sub-goals and allocated to individual parts of the business. Goals express desired states of resources and are achieved by processes. Goals are usually derived from the business strategy (vision). They are assigned to CODA autonomous components.
- *Rules:* statements that define or constrain some aspects of the business and represent business knowledge. Rules govern how the business should be run or how resources may be structured and related to each other. They become CODA critical success factors (CSFs).

6.3.1 A General Conceptual Architecture

Our understanding is that the main goal of a mobile network is to provide the 'best service' for the customer. The 'best service' is defined in terms of quality of the service

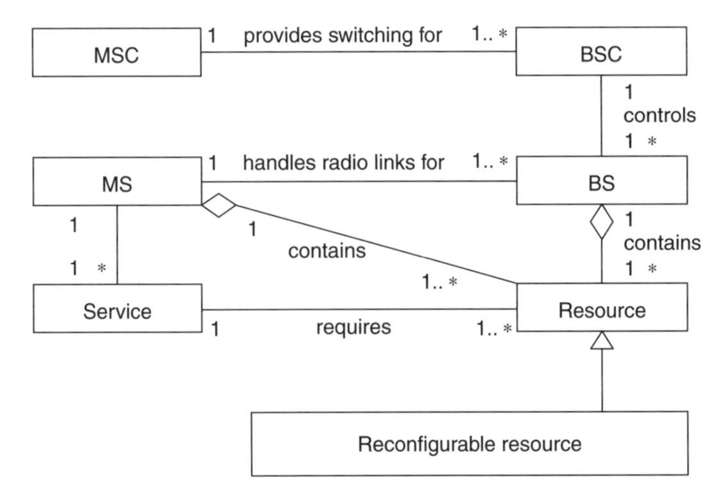

Figure 6.2 Conceptual model of a network used to demonstrate CODA

or the best resources that can be offered to the user in order to support the service. Our business model does not support 'marketing' at this stage. This could be added later. Consequently, no information will be defined about price and profits. The nature of the service is not important in our model either. We will categorise the services according to their resource requirements. For example, we know that a voice service will require fewer bandwidth resources than a black and white view, which also requires fewer resources that a full colour video.

The main resources of the conceptual architecture are defined by the typical components of a 3G network [13, 14], as shown in Figure 6.2. The conceptual model describes a simple but general model, and eliminates several conceptual components. The focus is to demonstrate the intelligent reconfigurability of the whole network in relation to effective service provision. It makes key elements of the CODA architecture easier to identify and test. Observe that we are only interested at the level of detail of the network, which will provide sufficient information for intelligent reconfiguration decisions, as this is the key objective of the architecture we are building. In the CODA test example the network is modelled using a simplified version of the generic model. The main principle is reconfigurability, therefore the primary objective is to provide the best service, and thus the conceptual model should be general enough to deal with general classes of services. We assume, however, that services can be categorised by the amount of resources they require. For instance, a voice service requires fewer resources than a video-on-demand service. CODA should help to show how resources can be best supported by using reconfiguration.

6.4 The Logical Architecture of CODA

An organic model is capable of providing a layered filtered architecture for managing information in a distributed context. CODA achieves this by refining and adapting Beer's Viable Systems Model [5] with reference to distributed information systems. Although originally applied to help define management principles, the model is further adapted to the control of data flow in a complex information system. The Viable Systems Model, as described above, is based on the way a biological organism, such as the human nervous system, processes data in terms of objectives. Incoming data are levelled according to the type of activity performed and filtered so that only the relevant information is presented when decisions are made. According to the cybernetic model of any viable system, there are five 'necessary and sufficient' subsystems involved in any viable organism and organisation [15]. To be viable a system should therefore be organised according to those levels.

Figure 6.3 presents a diagrammatic view of CODA with its principal components and their relationships, as follows:

- *Operations:* This layer deals with simple linear data, which usually correspond to typical transaction processing and business operations. The operational data warehouse usually links together data from the database and from several locations.

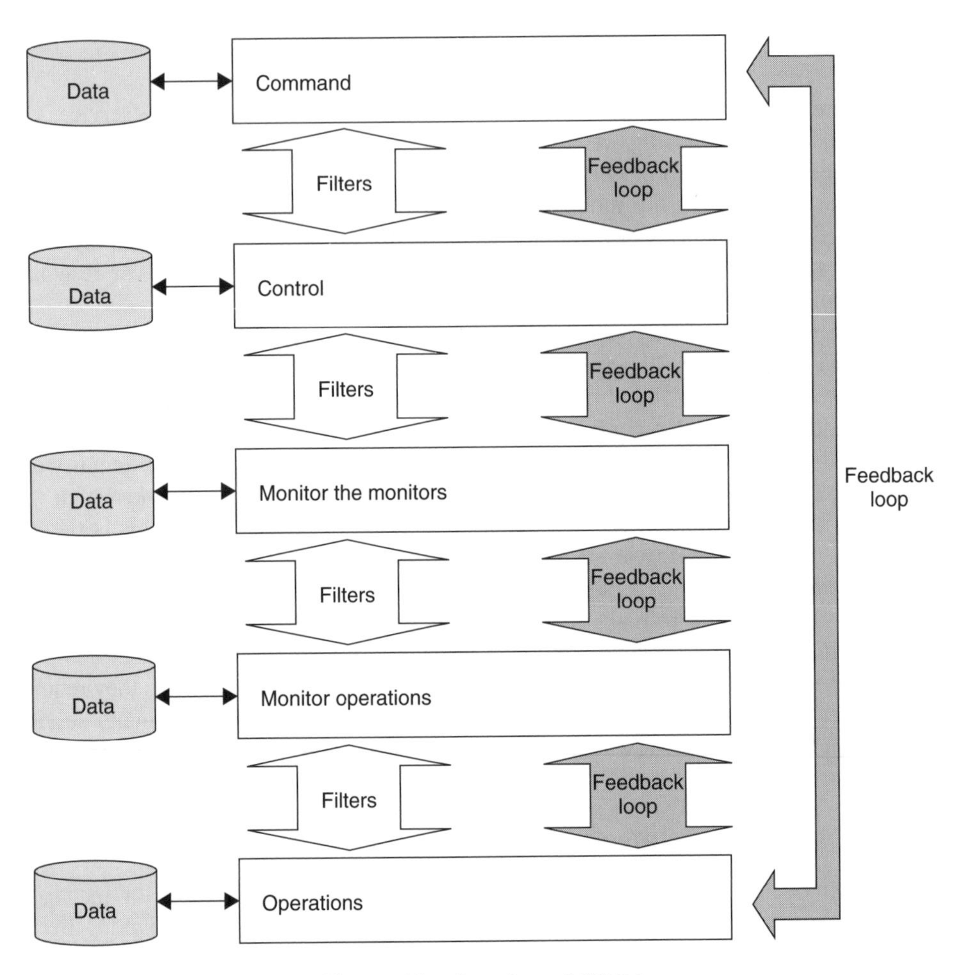

Figure 6.3 Overview of CODA

- *Monitor operations:* In this layer, the data are often dimensional and aggregated. For instance, data are organised by time or group. This layer is responsible for monitoring business operations.
- *Monitor the monitors:* This layer deals with multidimensional data and provides the capability for analysing trend behaviour, and operations are monitored in terms of trends.
- *Control:* This level should be able to 'learn' about simple emergent behaviour, trends and forecasts and be able to run predictions and simulations automatically.
- *Command:* This is the highest level, which should be able to deal with any variety not treated by the lower layers. This means recognising new threats and opportunities.

Table 6.1 illustrates the CODA layers with the respective processing types, which correspond to typical data warehouse and business intelligence components, and also gives an example of a business model for an accounting system. In addition, CODA provides an evolutionary capability for complex information via the notion of a feedback loop [5]. This involves applying Ashby's Law of Requisite Variety [16] using critical success factors (CSFs), in place of coenetic variables [17], as a means of determining goals for each component and dealing with the complexity of the environment The structured breakdown into activities allows systems to evolve new behaviours as follows:

- *Meeting objectives:* The organic model allows system components to meet a variety of objectives by using CSFs within each component.
- *Critical success factors:* CSFs allow the information system to respond to change in data in a dynamic and complex way.
- *Component autonomy:* CSFs operating within a layered architecture allow components to be semi-autonomous. A component only reports unusual failure or success in meeting CSFs within a pre-set range of tolerances. Therefore components act relatively independently, unless an alerting condition is triggered. Each component should only have the required information travelling through it. Any other information produces a huge excess baggage on the successful management of the distributed system. Moreover, the data reaching each component should be of the appropriate type.

A filter separates each level and presents data in the required format for use by that layer. Thus:

Table 6.1 Cybernetic layers of an information system

CODA layers	Component type	A model of an accounting system
Command	AI Systems OLCP type 2	Devise strategies for increasing account activity
Control	Online Complex Processing (OLCP)	Aggregate accounts by profitability and produce trend control mechanisms
Monitor the monitors	Decision systems OLAP type 2	Calculate profits on types and ranges of accounts and assess formulated trends
Monitor operations	Analytical processing OLAP type 1	Aggregate activities by type, location, time and monitor for failure or success
Operations	Online Transaction Processing (OLTP)	Transact accounts

- Information is safely filtered.
- Levels above are not swamped by detail unless an alert condition is triggered.
- A layered architecture is obtained.

Filters present information in a structured way. This is similar to the way biological systems use homoeostats. Homoeostats are the elements within the organism that store variable information about the environment. The information used by a system may not be complete and therefore co-operation between systems is required. An individual operation would be depicted as homeostatically balanced with its own management on one side and its market on the other. These co-operate, filtering information and alerting higher levels only if necessary. The higher levels of the human brain, for example, are not engaged in the everyday activities of the nervous system such as monitoring breathing. The higher brain is only notified of events that cannot be handled by the normal routine. Functions filter information and pass information on to each other only when necessary.

In a business system, the monitoring components monitor the rate of sale of various items. If the predefined rate of sale is not achieved (such as 100 items per day), then the monitoring component can put the item on special offer. Typical offers are: three for the price of two if the rate of sale is almost satisfactory, and two for the price of one if stocks are piling up.

Security is based on a reference model for role-based access control (RBAC) [18]. The RBAC defines user profiles, which grant access to components of a primary layer. However, access to further layers may be given depending on other roles played by the user. This means that access to data is securely controlled based on the organic architecture. In addition, changes to data can be traced back to roles.

6.4.1 Using CODA for Network Reconfiguration

The following conceptual model diagram, which is based on the EC funded IST Framework V CAST Project [19], shows CODA intelligence as a set of interacting semi-autonomous agents. These agents interact as required in order to manage services, resources, traffic load and add new services. They co-operate to execute complex system reconfiguration. Each agent has access to specified data and tasks relevant to their function. There are four levels, in accordance with the four levels of the CODA: operational, monitoring operations, monitoring the monitors and controlling. The fifth command level is not modelled. The authors believe that four layers are sufficient to provide the viability required for most mobile communications applications. The agents shown in Figure 6.4 are able to manage their own tasks and data within specified CSFs. Where CSFs are not met the system will automatically send to other levels for help using the CODA feedback loop mechanism.

As an example, a user may try to obtain a service from a base station transmitter (BST). The BSC monitors the operations between the BST and the user, via simple aggregations and measures to check performance. If the BSC measurements show a possible timeout, or bandwidth failure, or if the BST is unable to service the request, then the Mobile Switching Centre (MSC) agent will be alerted by the change of status in the Base Station Controller (BSC) from green to orange. The MSC agent will try to locate another BSC from the set under its control and divert the call. If this cannot be done from the set of BSCs under its control, then it will notify the next CODA layer for advice. The failure

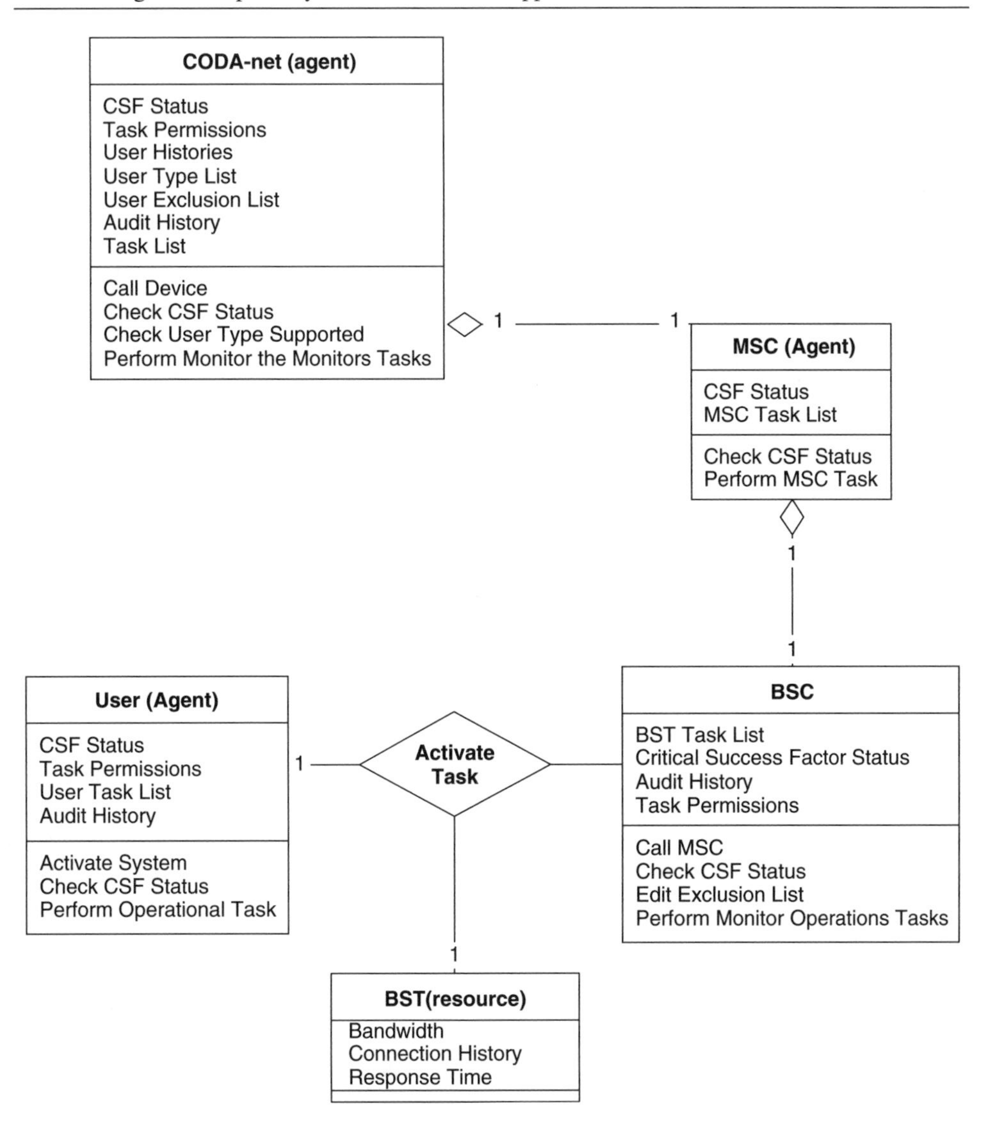

Figure 6.4 CODA agents

can be analysed off line and a solution proposed. Agents at each level are able to draw on stored 'history' information according to pre-set permissions. Each has some control over its own data provided operating tolerances are met. This is usually in a warehouse, although in the case of the user and the BST, the data accessed is a small internal data store. Figure 6.5 shows a static UML use case model of the interactions between the user and the BST via a set of tasks. A use case model is a type of UML model used to describe how a system can be used. Use cases show the functionality of a system.

Figure 6.5 shows the activation of components at the operational and monitoring operations layer. At the operational layer, the user has permission to perform one or more sets

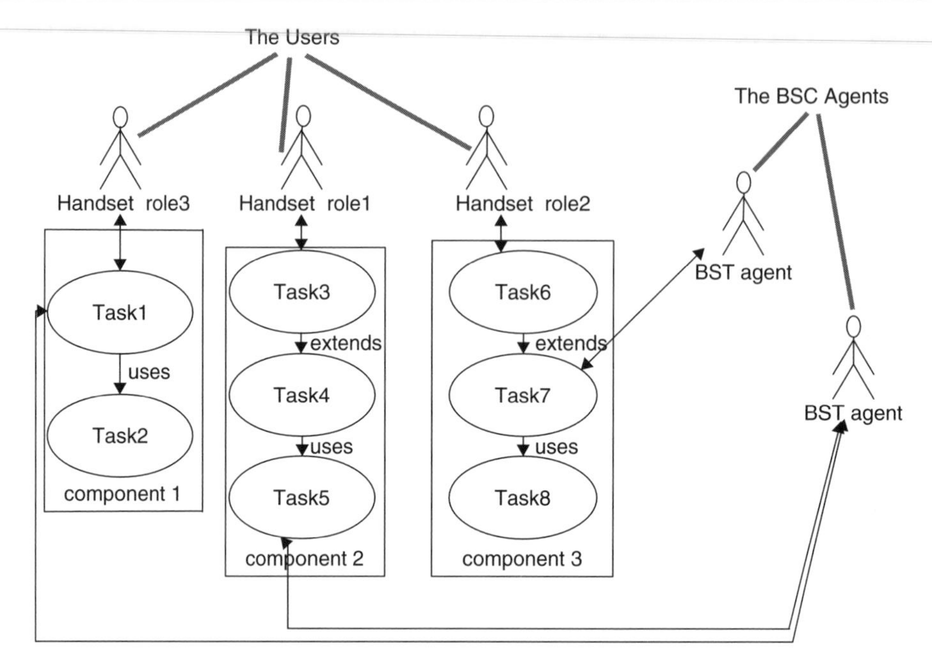

Figure 6.5 CODA use case model

of service request tasks [such as Wireless Access Protocol (WAP), calls, downloading of pictures and text]. Once the user has these permissions, the tasks can be executed via connection to a BST. Adding new permissions will entail reconfiguration of the handset. Note that some tasks contain probes, which allow the BST agent to access a lower layer, or to call out to a higher layer for reconfiguration of resources.

6.4.1.1 CODA Data Warehousing Model

So far we have discussed the task activation processes, which may require intelligent reconfiguration. The systems may also need to be globally reconfigured for performance or other reasons. These reconfiguration requests originate from higher layers of the CODA intelligence. Such requests are based on analysis of information about past task activations which are stored in data warehouses as persistent classes and predictions of future task activations based on present performance and past performance trend analysis. The main objective and test for this part of CODA is to meet the contractual services. This type of information requires a large storage capacity and special software, which must not interfere with operations. In CODA, intelligence of this type is managed by storing information on performance in interconnecting layered warehouses. Each warehouse contains only data pertaining to the type of analysis that is performed. Thus, each layer has its own semi-autonomous agents with specified task and access permissions and a warehouse where data are structured so that tasks can be efficiently executed. For the CAST project we have implemented three layers of the CODA model, namely the operational layer, the monitoring operations layer, and the monitoring the monitors layer.

CODA warehouses at the monitoring operations and monitoring the monitors layers perform two major functions. First, the existing 'legacy' systems can operate with minimal interference. Second, the intelligent reconfiguration of the network can take place transparently, thereby allowing the network to provide extra bandwidth where feasible without a major overhaul of the system. Failure is therefore minimised. The interactions are shown in Figure 6.6.

An innovative feature of the CODA distributed architecture allows the MS, BSC and MSC to filter calls independently at the monitor operations layer, thereby minimising traffic across the network. The call buffer is a CODA wrapper class which acts as a protocol. The MS service requests are initially filtered by the user profile stored on the handset (the first level of filter). The Base Station (BS) will only let this request through if there is sufficient actual and predicted bandwidth (the second level filter). The filter settings are managed by the third level of filter from the MSC, which checks that the BS is operating at sufficient strength to manage the service. This means that at peak times access to a given service for a MS can be restricted by the BS without referring to the mobile switching centre at all.

The setting of filters is managed by the upper layers of CODA. The objective of the system is to meet the contractual services and to discover peaks and troughs in usage Operating information is collected for per timeband. The CODA timeband settings in the demonstrator are set to every three hours, but this can be modified. The timeband setting has been designed to allow the higher layers to collect data in meaningful chunks.

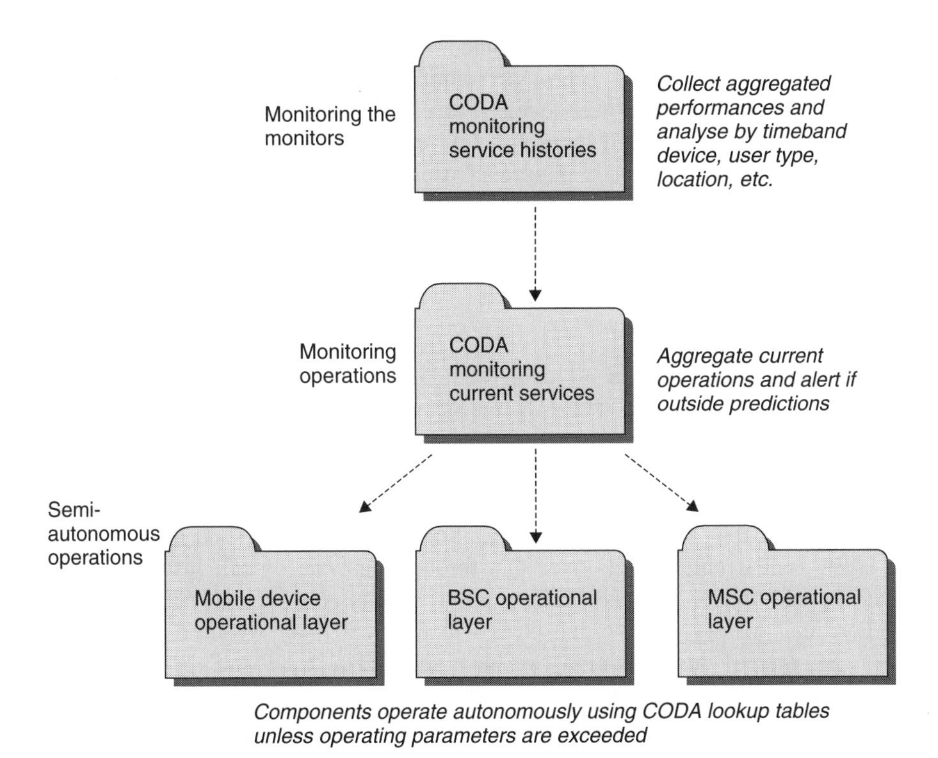

Figure 6.6 An overview of the CODA layers

Three-hourly timebands allow the CODA analysis software to deduce behaviour patterns typical of the timeband and to broadcast predictions for the next similar timeband.

The Network Manager and the Reconfiguration Controller both download copies of the updated information on the best performing configurations for the time of the day and the type of calls required. This is stored in lookup tables, which are closely monitored and regularly updated.

Under normal conditions, the agents and tasks at each level operate semi-autonomously making use of the CODA filters and successfully meeting contractual services. Tasks executing within the predefined CSFs need only report their status to the level above. They will not need to contact the layer unless they require assistance or to send or receive information from the levels above. Outside the normal predicted activities, interaction between layers usually occurs because performance falls outside pre-set tolerances. However, since the feedback loop is executed in both directions, it is perfectly possible for the upper levels to reschedule services and reconfigure resources if intelligent monitoring of the overall system predicts a bottleneck or some resource contention, which the lower levels do not anticipate.

Typical problems requiring higher-level intervention may involve comparison of current operations with past data. This may show an unexpected increase in resource requests in a timeband or a location. Several solutions may be available. It may be possible to allocate a narrower bandwidth to the resource (for example restrict the colour channels for graphic data, or to restrict users to those who pay more for services, to divert calls, or even to 'buy' the bandwidth and time from another provider). However, it may be an across-the-board increase which will lead to an imminent crash of the system. CODA should have sufficient historical information to provide solutions for this. However, the historical data may not be available to the lower levels. For example, the BST and MSC monitoring operations records are archived approximately every three hours. Thus the monitoring operations layer does not have a picture of user behaviour over time. The monitoring the monitors layers will have this information and should use it to make predictions and suggest reconfiguration.

6.4.1.2 The Operational Layer

The operational layer includes all data required for effective operation, as shown in Figure 6.7. The data stored in each MS is necessarily limited. Therefore the BSC stores details of each task activation and MS requesting a service unless the MS fails to reach a BSC, in which case the failed calls are archived and sent to the BSC at the next opportunity for later analysis.

For example, it may be that the user is failing to successfully send MP3 files. CODA upper layers will eventually discover this through analysis of call histories. This may result in the addition of a further 'help' option for this type of call. All calls are stored in the BSC call history archive and passed to the next layer within a specified period or event. By way of demonstration, we have selected a 'timeband' period and a 'type 2' reconfiguration failure as triggers to download call data, whichever is sooner. Up to 1 MB of memory may be needed to store calls in a BSC as service requests have priority, and are executed before archive requests so a BSC may have up to double the amount of data at very busy times.

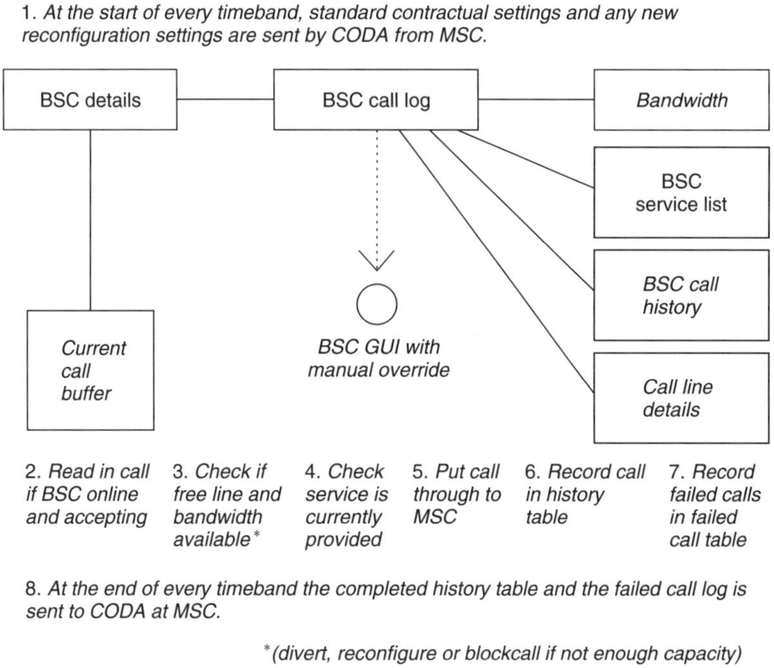

1. At the start of every timeband, standard contractual settings and any new reconfiguration settings are sent by CODA from MSC.

BSC details

BSC call log

Bandwidth

BSC service list

BSC call history

Current call buffer

BSC GUI with manual override

Call line details

2. Read in call if BSC online and accepting

*3. Check if free line and bandwidth available**

4. Check service is currently provided

5. Put call through to MSC

6. Record call in history table

7. Record failed calls in failed call table

8. At the end of every timeband the completed history table and the failed call log is sent to CODA at MSC.

**(divert, reconfigure or blockcall if not enough capacity)*

Figure 6.7 CODA operational layer components

The CODA system deals with the whole call as one operation. This is the level of granularity that has been selected for the demonstration. For the CODA demonstration, operating information, consisting of the contents of the call buffer for each call, is collected using a CODA 'wrapper' structure. This CODA wrapper collects a set of database entities and wraps them in a class structure which is denormalised and may contain repeating groups. More particularly, the wrapper contains status flags and dimensional data attributes which may trigger alerter conditions.

The call generator produces sample call data of this type by creating individual calls which are stored in a database. Data produced in this way must conform to the filters, which are in place for that timeband allowing us to use it to perform simple tests of the functionality of the layers above. Note that the MS can request a service provided the service requested is permitted by the user profile, the BS will let this call through, and provided there is sufficient actual and predicted bandwidth.

6.4.1.3 Monitoring the Operations Layer

The monitoring layers analyse operational performance and offer advice to fine-tune the operational layers. Since this is in the form of filters consisting of updated service tables and operational parameters, the CODA monitoring layers are transparent to the operational systems and there is a low network traffic overhead. The monitoring layers also collect and analyse aggregated historical data. The results of the analysis allow CODA to make decisions on the best operating tolerances and on the best performing services and

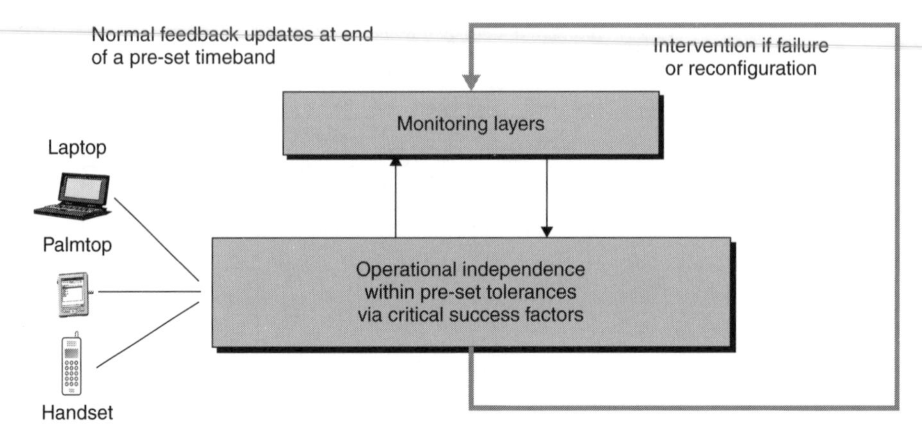

Figure 6.8 The monitoring operations

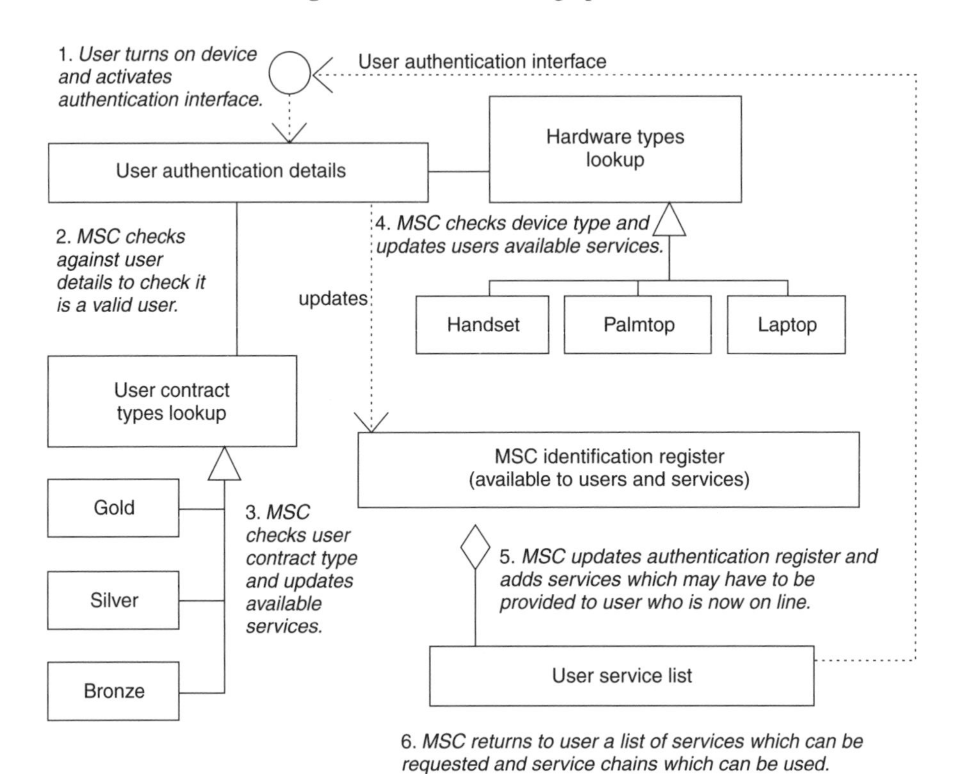

Figure 6.9 The authentication process using CODA filters

service chains. These are updated ordinarily at the end of a timeband or, extraordinarily, in the event of certain types of failure or reconfiguration. Figure 6.8 shows the monitoring operations.

Figure 6.9 shows the authentication process using CODA filters. Note that the user is able to own more than one device per SIM (provided there is some compatibility), but only one device may be activated at any given time. Once activated, the user is tracked automatically and does not need to re-authenticate when a service is requested. By using filters, the CODA monitoring operations layer is able to tell which services the user is able to leverage, since the services are further delimited by the user type and device capabilities. From a higher-layer perspective, CODA monitoring layers may be able to predict the system usage on this basis and send-users 'special offers' of services.

The call is passed to the MSC which checks the security, device type and user type and returns the timeband filter, special offers, and any help based on an analysis of the user profile. The process from the MSC perspective is shown in Figure 6.10.

Once the device is authenticated, no further authentication is needed when a service is requested. CODA is now ready to monitor timeband operations. At the BSC monitor operation layer, software is able to assess the call wrapper transactions passing through the network. If a transaction is successful, this is noted. However, failures and reconfigurations always alert the higher layers and may result in some immediate changes. For further improvement in performance, CODA also alerts the Network Manager and Reconfiguration Controller if there is any predicted shortfall in the provision of services and suggests possible reconfigurations based on analysis of past performance, current failures and current events, thus demonstrating the downward flow of the feedback loop. The MSC monitors the BSCs at all times, receiving regular status reports that all is well in those BSCs it is monitoring. Where a BSC is in condition 'amber' the MSC is able to either divert calls or restrict services.

Figure 6.11 shows the monitoring operations layer at the MSC. The MSC takes an overall view of all the operating functionality of all the BSCs at its disposal. If a BS hits condition 'amber' and another BS is experiencing a 'low' period as defined by CODA

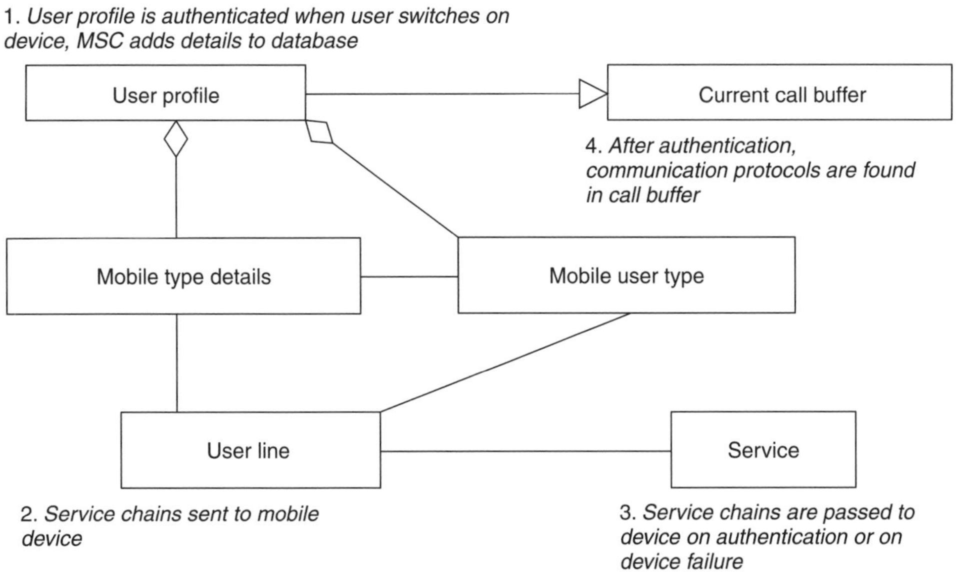

Figure 6.10 User authentication from the MSC perspective

MSC Details

MSC ID:	1	Time Band:	0900-1200	Tot Line Capacity:	100	Reconfig Status:	0
MSC Name:	CODA-TEL	Op Mode A:	Online	Tot Bwidth Capacity:	300	Alert Condition:	0
Location:	UK	Op Mode B:	Ready	Tot Lines Free:	100		
Status:	Active	Error Status:	0	Tot Bwidth Free:	300		

User Identification Register

	UserType	TelNo	Password	UserName	DeviceTy	MakeMod	Active	online	BS	sms	voice	wap	mp3
	silver	07896554	enitsirhc	christine	palmtop	Samsung	n	n	n	n	n	n	n
	silver	07896554	ave	eva	handset	NEC 300C	y	y	cambridge	y	y	y	y
	bronze	07896554	aderf	freda	handset	NEC 100C	y	y	canterbury	y	y	n	n
▶	bronze	07896554	lamek	kemal	handset	NEC 100C	y	y	cambridge	y	y	n	n
	gold	07896554	einnek	kennie	laptop	Dell Inspir	y	y	cambridge	y	y	y	y
	bronze	07896554	nitram	martin	handset	NEC 300C	y	y	cambridge	y	y	y	y
	[illegible]	[illegible]	[illegible]	[illegible]	[illegible]	[illegible]	[illegible]	[illegible]	[illegible]	[illegible]	[illegible]	[illegible]	[illegible]

BS Status Log

	Date	TimeBand	Location	Type	Nlines	sFree	locks	sFree	AlertStatus	Diversio
	16/09/2002	0900-1200	city ec1	down town	10	10	30	30	green	y
	16/09/2002	0900-1200	covent garden	down town	10	10	30	30	green	y
	16/09/2002	0900-1200	oxford st	down town	10	10	30	30	green	y
	16/09/2002	0900-1200	victoria	down town	10	10	30	30	green	y
▶	16/09/2002	0900-1200	cambridge	provincial	10	7	30	21	green	y
	16/09/2002	0900-1200	canterbury	provincial	10	10	30	30	green	y
	16/09/2002	0900-1200	edinburgh	provincial	10	10	30	30	green	y
	16/09/2002	0900-1200	highbury	suburb	10	10	30	30	green	y

Current Call Connection Log

Call ID	Sender BSC	Receiver BSC	Rec Status	Err Status	Reconfig Status	Err Type	Reconfig Type	CSF Value
33	Tottenham	Covent Garden	aok	0	0	0	none	0

MSC Connection History

	CallID	SenderBSC	ReceiverBSC	ErrStatus	ReconfigStatus	DeviceType	UserType

Figure 6.11 Demonstration of the monitoring operations layer at the MSC

analysis of histories, then the MSC will divert the next 10 calls to that BS and check the operating condition of the BS after this number of calls or after a period defined by CODA (again based on historical analysis of call flow). If the MSC is unable to find another BS and several BSs start to move into condition 'red', then the system may decide to set the status of the BSs to emergency service calls only. This will ensure that key services will still be available. A key feature of the monitoring operations layer is that it can be added to existing systems. The wrapper structure collects call data into a viable CODA component. This is stored by the operational layer and used by the monitoring operation layer to help CODA filter and analyse calls. At the end of each timeband or in the case of serious error, the operating data, which have already been aggregated, are sent to the next layer for analysis. Advice is returned to the monitoring operations layer in the form of new filters and new services and operating chains for a user or device. The upper layers are transparent to the existing systems and CODA can therefore be applied to any generic mobile network with minimal rebuild overheads.

6.4.1.4 Monitoring the Monitors Layer

The CODA monitoring the monitors layer collects the monitoring operations data from each BSC and from the MSC at the end of each timeband or after a serious failure. If a serious failure occurs, then this layer will be sent the operational data at the next available opportunity. This will be analysed and may result in operational layer being sent either new call chains or filters. If no serious failure occurs in a timeband, then monitoring the monitors proceeds as normal by analysing the call histories. The call data are used to make predictions about new services and service loads and fine-tune the operations as needed via global reconfiguration. Otherwise the priority at this layer is the analysis of failures, by cluster. As previously stated, failure is defined by severity range {0, 1, 2}; 0 = no error, 1 = user error, 2 = system or device error. Systems failures of any kind are a high priority as they mean that CODA has failed to predict failure to meet a service. CODA monitoring the monitors operates in both autonomous and operator-driven modes. If a severe (level 2) failure occurs, then normal operating mode is interrupted while CODA decides which type of failure has occurred and suggests a solution. In the case of a reconfiguration failure, CODA will suggest alternative chains.

CODA monitoring the monitors software uses the slice, dice and rotate operations available to a multidimensional warehouse storing dimensionalised and non-normalised data. In normal mode these operations are primarily used to discover failure clusters by dimensional type. The failure cluster may identify a device service failure, in which case it may be that a device has not been configured properly at the global level. Level 1 failures may result in the sending of a message to the user, as shown in the example of Figure 6.12. This is the result of cluster analysis showing consistent user failure in transmitting one type of service.

Monitoring the monitors, shown in Figure 6.13, requires a larger warehouse capacity than those of the monitoring operations layer. It should be large enough to manage a multidimensional database composed of data from many BSCs and one MSC.

Space and complexity considerations mean that this level should not perform forecasting and trend analysis. If we apply the CODA architecture, then forecasting and trend analysis should be performed by the control layer. At this layer controlling software allows the

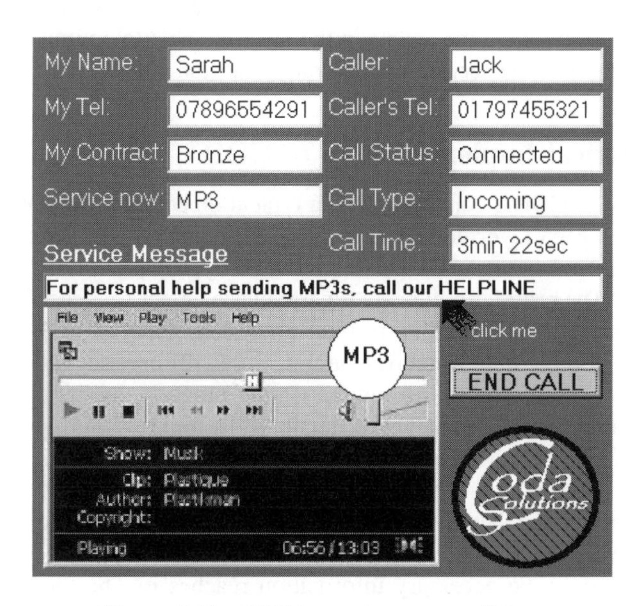

Figure 6.12 CODA response to user failure

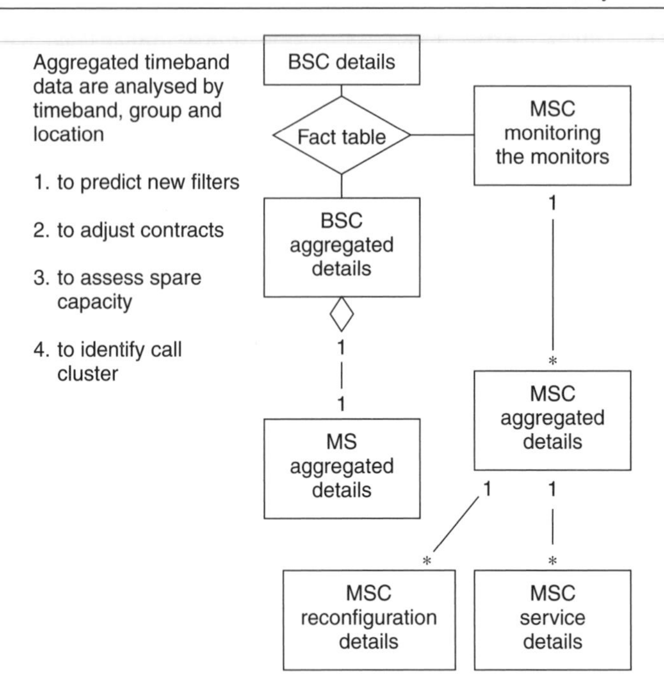

Figure 6.13 Monitoring the monitors layer UML model

network to make choices on the basis of past network performance and future predicted performance. CODA intelligence is based on the constant seeking of best performing chains. The software closely analyses failures to operate within the contracted tolerances as expressed in the critical success factors. On the basis of analysis it is able to offer advice on how to proceed. At the same time agents in the layer are able to optimise the performance of the system using average service-request rates, and totals from past performance.

6.5 The Practical Implementation of CODA

The physical view of CODA corresponds to a typical object-oriented (OO) layered architecture [20], where each layer denotes a level of the Viable System Model (VSM). Data sources, which can be data marts, data warehouses and legacy processing, must be attached to the right layer. The criteria to allocate a particular element to a certain layer (level) in the architecture are defined according to their processing type, as in the VSM described above. Observe that other data warehouse architectures [2] may suggest the separation of data into two levels according to the summarisation process; that is, lightly summarised data are data distilled from a low level of detail whilst highly summarised data are compact data. Unlike CODA, however, this levelling does not take into account the user of the data.

Interaction between the layers is achieved by two key elements of CODA, namely the filters and the feedback loop. As previously stated, each layer is separated by filters, which ensure that only the necessary information reaches it. The feedback loop allows components in a layer to only report unusual failure or success in meeting their CSF to the above layers. This means that tasks in the above layer(s) must take action. Basically the feedback loop corresponds to an event-based, implicit invocation style [7], where one layer generates alerting events, related to failures or successes, and the above layer registers the tasks to be invoked. Each layer is structured in a similar way, providing filtering, feedback and security capabilities.

6.5.1 The Internal Structure of a Typical CODA Layer

Two types of components usually provide data access in CODA: wrappers and filters. Object wrappers are widely used to integrate legacy or non-CORBA compliant applications with CORBA [21–23]. This approach provides universal and transparent access via the CORBA bus, using OO Application Programming Interfaces (APIs). This is also common in providing CORBA APIs for databases. More recently, this approach has been proposed to provide universal access to On Line Analytical Processing (OLAP) systems [24].

The filters are responsible for providing clean and secure data. The provision of clean and consistent data is a key property of any data warehouse architecture. In CODA, we use a hybrid data configuration, and filters perform reconciling and aggregation functions on data received from wrappers or other filters. This approach provides a highly distributed and scalable solution for warehouse creation. The structure of a typical CODA layer is presented in Figure 6.14.

6.5.2 Data Warehouse Architectural Considerations

The heart of CODA is the collection of data warehouses. Building a data warehouse requires complex systems integration to establish the architecture and tie together the

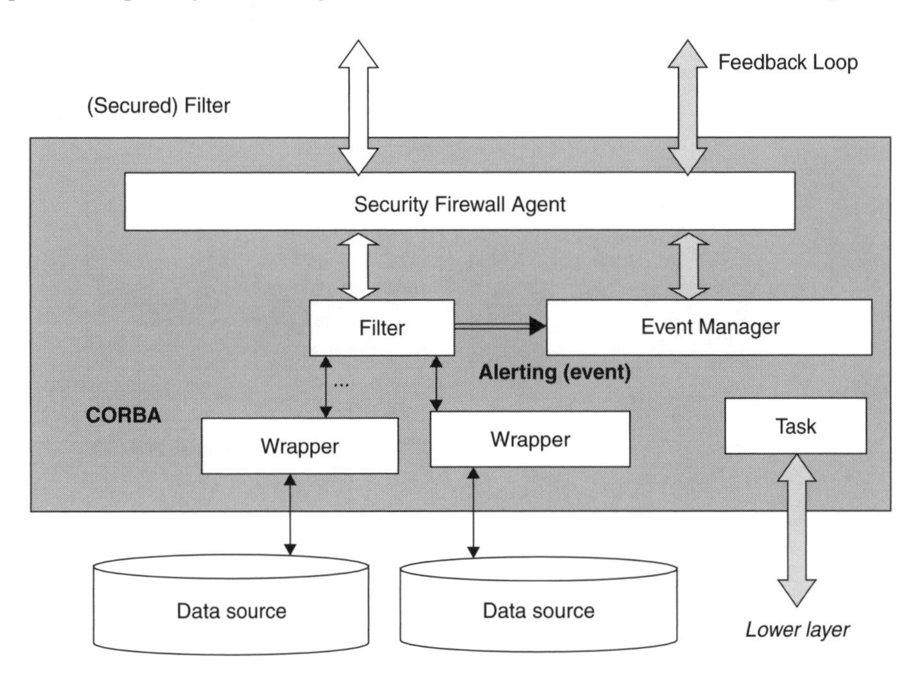

Figure 6.14 Coda typical layer

various components. CODA provides the bases for such integration, and defines strategies for modelling the overall data into its layered-system style. Most architectural decisions of the data warehouses are part of CODA. There are, however, a few aspects that are not covered by the system.

A data warehouse supports business analysis and decision-making by creating an integrated database with consistent, subject-oriented and historical information. A data warehouse is different from a (operational) database. Operational data are usually short-lived, change very often, are accessed at record-level, follow repetitive standard transactions and access patterns, and are updated in real-time. Data warehouse data, on the other hand, are long-lived, static, aggregated into sets, use *ad hoc* queries with some specific reporting, and are updated periodically with mass loads. This means, therefore, that the architectural and design issues are very different, since the main goal is not to get data into the database in the most efficient way but to access and summarise time and subject-based data using non-repetitive analytical processing.

6.5.2.1 Interface with External System

The best way of defining the interfaces in an architecture is through the concept of service. A service defines a contract between components at an abstract level. The interface of a component can be defined by the services it requires from and provides to other components. The interfaces between CODA components have been specified in earlier sections. We now have to establish what services the CODA components will provide to and require from components outside the architecture. The service architecture is described in Figure 6.15 where an arrow pointing out denotes a service provided and similarly an

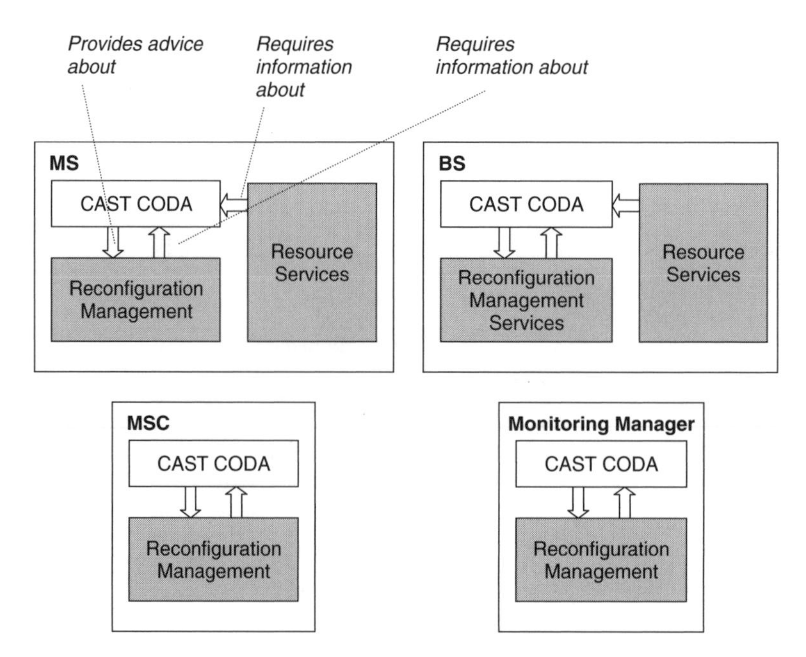

Figure 6.15 Coda services

arrow pointing in denotes a service required by the component. Since the reconfiguration takes place at both the MS and BS levels, this is where CODA obtains information about resources status in order to support the application service. This service is referred to as a *Resource Service*. The resource services must provide information to CODA and do not expect any information from CODA. In general, CODA will only request information about the latest application service and the resources used to support that service. CODA may also request information about the availability and status of resources on the MS or BS. The resource services could be implemented by the resource controller or the network management system. This decision, however, does not affect CODA.

Finally, it is also necessary to define the *connectors* between CODA and the external services. In Figure 6.15 the connectors are denoted by the arrows. A connector is an architectural element that mediates the interaction among components. It is essential in an architecture to specify how the components interact, especially external components. Since CODA is based on distributed object technology, more specifically using Java, it is expected that the components will use Java-based distributed object connectors. Java RMI (Remote Method Invocation) is used in the hardware controller and the reconfiguration controller.

Acknowledgements

The work described here is based on the research carried out in the EU-sponsored collaborative IST Framework V project CAST. The authors also gratefully acknowledge the contributions by George R. Ribeiro-Justo.

References

[1] H. Zuidweg, M. Campolargo, J. Delagado and A. Mullery Eds., 'Intelligence in services and networks', *Sixth International Conference on Intelligence and Services in Networks*, LNCS 1597 Springer, 1999.
[2] H.S. Singh, 'Data warehousing: Concepts, technologies, implementation and management', PTR, 1998.
[3] T. Karran, G.R. Ribeiro Justo and J. Zemely, 'An organic knowledge information management architecture', International Conference on Intelligent Systems and Control, Santa Barbara, California, 1999.
[4] G.R. Ribeiro Justo, T. Karran and J. Zemely, 'An organic architecture for distributed knowledge management', in R. Roy, ed., *Industrial Knowledge Management – A Micro Level Approach*, Springer, 2000.
[5] S. Beer, 'The viable system model: Its provenance, development, methodology and pathology', *Journal of the Operational Research Society*, **35**(1), 7–25, 1984.
[6] C.J. White, 'The IBM business intelligence software solution', Data Base Associates, Version 4, May 2000.
[7] M. Shaw and D. Garlan, *Software Architecture: Perspectives on an Emerging Discipline*, Prentice-Hall, 1996.
[8] P. Kruchten, *The Rational Unified Process: An Introduction*, Addison-Wesley, 1998.
[9] C. Hofmeister, R. Nord and D. Soni, *Applied Software Architecture*, Addison-Wesley, 2000.
[10] D. Garlan and J.P. Sousa, 'Documenting software architectures: Recommendations for industrial practice', Technical Report CMU-CS-00-169, School of Computer Science, Carnegie Mellon University, 2000.
[11] G. Booch, I. Jacobson and J. Rumbaugh, *The Unified Modeling Language Users Guide*, Addison-Wesley, 1998.
[12] H. Eriksson and M. Penker, *Business Modeling with UML: Business Patterns at Work*, Wiley, 2000.
[13] C. Bettstetter, H. Vögel and J. Eberspächer., 'GSM phase 2+ general packet radio service GPRS: Architecture, protocol and air interface', *IEEE Communications Surveys*, **2**(3), 1999.
[14] R. Morawek and H. Öczelik, 'UMTS: Basic network architecture', 2000, *http://www.unet.univie.ac.at*

[15] F. Waelchi, 'The VSM and Ashby's Law as illuminants of historical management thought, in R. Espejo and R. Harnden, eds, The Viable Systems Model: Interpretations and Applications of Stafford Beer's VSM', Wiley, 1996.
[16] W.R. Ashby, *Introduction to Cybernetics*, Chapman & Hall, London, 1965.
[17] R. Espejo, W. Schuhmsnn and M. Schwaninger, *Organisational Transformation and Learning: Cybernetic Approach to Management*, Wiley, 1996.
[18] M. Nyanchama and S. Osborne, 'The role graph model and conflict of interest', *ACM Transactions on Information Security Systems*', **12**(1), 3–33, 1999.
[19] CAST Project Web Site: *www.cast5.freeserve.co.uk*
[20] G.R. Ribeiro Justo and P. Cunha, 'An architectural application framework for evolving distributed systems', *Journal of Systems Architecture: Special Issue on New Trends in Programming and Execution Model for Parallel Architectures, Heterogeneous Distributed Systems and Mobile Computing*, **45**, 1375–1384, 1999.
[21] R. Zahavi, *Enterprise Application Integration with CORBA: Component and Web-Based Solutions*, OMG Press, 2000.
[22] A. Umar, *Object Oriented Client/Server Internet Environments*, Prentice-Hall, 1997.
[23] J. Hunt, *Java For Practitioners*, Springer, 1999.
[24] J.W. Buzydlowski, I.Y. Song and L. Hassel, 'A framework for object-oriented on-line analytic processing', DOLAP'98: ACM First International Workshop on Data Warehousing and OLAP, Washington, 1998.

7

Open APIs for Flexible Service Provision and Reconfiguration Management

Nancy Alonistioti, Spyros Panagiotakis, Maria Koutsopoulou,
Vangelis Gazis and Nikos Houssos

Communication Networks Laboratory, University of Athens

7.1 Introduction

7.1.1 Evolving Communications Environment

In second generation (2G) networks, services provided to mobile users were either rigidly integrated in network equipment or developed with proprietary tools by mobile operators or equipment manufacturers. In third generation (3G) mobile communications and beyond, an open and complex marketplace is expected to emerge: mobile devices with significantly different characteristics [e.g. computer laptop, Personal Digital Assistant (PDA) and cellular phone) should access a multitude of diverse applications and Value-Added Services (VAS), developed by Third Party Value-Added Service Providers (VASPs) that typically do not target solely mobile networks, using a variety of wireless link technologies (e.g. UMTS, GPRS, Wireless LAN, Bluetooth, etc.). Applications and services will be provided to end-users over various networking technologies and heterogeneous communication infrastructures, which should converge, interface and co-operate seamlessly and congruously to ensure consistent, transparent and ubiquitous service provision to end-users, hiding the involved network complexity.

Consequently, the prevailing design assumptions for mobile network, device and protocols must be re-evaluated in the light of the volatility and dynamism that dominate the raised wireless world. Design priorities should shift from vertical goals, such as optimum performance, toward more horizontal concepts, such as sustainable operation and service ubiquity. This implies the need for more flexible and intelligent networking architectures that are able to dynamically tailor their functionality and behaviour to the

Software Defined Radio: Architectures, Systems and Functions. Edited by M. Dillinger, K. Madani and N. Alonistioti
 ISBN: 0-470-85164-3

profiles of the services that are provided over them, the user and terminal profiles and the networking environments.

7.1.2 Evolving Management Paradigm

The third generation of mobile communication systems is the first that has been designed to flexibly support manifestations of the ongoing telecommunications, broadcasting and Information Technology (IT) sectors' convergence, by allowing for multiple delivery models (e.g. fixed, terrestrial and satellite) of any type of content (e.g. audio, video, data) to multiple classes of mobile devices (e.g. laptop computer, cellular phone). The most significant effect this proliferation of service provision options has is upon the prevailing management model. In the emerging era of mobile communications, managing network elements or even entire networks in isolation from the overall service provision context is simply not sufficient, for a number of reasons:

- The deregulation of the mobile communication sector marks a value chain shift from a monopolistic service provider model towards a more flexible and dynamic market structure, where various roles (e.g. connectivity provider, service provider, application provider, content provider, etc.) undertaken by multiple business entities must be managed efficiently to provide services to the customers [1–3].
- As the continuing increase in network bandwidth availability gradually shifts the value of the telecommunication service provision from plain connectivity toward more advanced and sophisticated service features, so must the respective management model [1–3].
- The proliferation of wireless access technologies (e.g. IEEE 802.11, Bluetooth, HiperLAN/2) combined with the enhancement of satellite data services clearly suggests that the future mobile user will enjoy globally available connectivity over a multimodal networking environment – a task that is evidently too complex for traditional network management approaches to tackle.
- User experiences from systems and services available from the fixed telecommunication sector (e.g. the Internet experience) increase the end-user's expectations, and makes them harder to be satisfied by employing network management approaches in isolation from other aspects of service provision (e.g. user preferences for content presentation).

For these reasons the traditional management models have shifted from network management to service management. Service management is not confined to the mobile network domain but spans the entire value chain, from the end-user terminal through the mobile network elements to the application-hosting servers and the content-controlling processes.

In the European Community (EC), the service management challenge is mostly addressed by the 'always best connected' vision for future mobile communications that includes dynamically reconfigurable systems and networks. Reconfigurable systems will change both in terms of structure (e.g. by downloading and instantiating a new radio interface and its associated protocol stack) and behaviour (e.g. by reconfiguring the operational parameters of a particular protocol or other software module). Reconfigurability management marries the network-level constituents that participate in the service provision with the higher-layer management intelligence required to enforce interoperability and consistency between network services and third-party applications over time and between different implementations.

7.1.3 *Implications for Software Defined Radio Aspects*

The vision of reconfigurability is considered to be a major technology enabler for the emerging dynamic and heterogeneous telecommunications environment described in the previous paragraphs. However, it is common sense that the application of Software Defined Radio (SDR) concepts in a context that a variety of different players contribute to the creation of VAS and the development of the supporting infrastructure and equipment becomes even more complicated.

For sure, managing and reconfiguring monolithic systems, where all elements (including the functional components of end-user applications) are developed by a single vendor and communicate with each other using proprietary interfaces, imposes fewer management requirements compared with a multi-technology, multi-vendor system where software components from different providers are interoperable and pluggable to a variety of possible configurations. The latter situation requires at a minimum the definition of generic, complete, easy to implement and use, open APIs that offer the functionality necessary for communication between different components and the control of the corresponding reconfigurability mechanisms. The following section concentrates on the provision of open Application Programming Interfaces (APIs), the requirements/challenges they should meet and the on going standardization efforts.

7.2 Open APIs

7.2.1 *Definition – Significance for Mobile Service Provision*

Traditionally, telecommunications equipment (e.g. network switches, terminals) is built according to the 'black box' paradigm: it incorporates specific, precisely defined functionality that is mostly not subject to modifications in the course of time. Interfaces for accessing and modifying the internals of network and terminal devices are, if available, very limited in scope, proprietary and based on low-level protocols that are very difficult to use.

Fuelled by the forthcoming heterogeneity and context volatility of future generations of mobile networks, there is nowadays a trend towards the denouncement of this paradigm and the introduction of programmability in networks through open APIs. With the latter term, we mean interfaces that provide access to network functionality at various levels, e.g. from an entire network down to a component of a network device, with the following characteristics:

- Independence from the underlying infrastructure, including vendor-specific hardware and software, networking protocols and technologies. This feature facilitates interoperability of different implementations and therefore the easier and faster introduction of new, optimised configurations of systems and services.
- Access to a significant subset of internal functionalities of the underlying systems, including the ability to dynamically modify their state.
- Interface accessibility through clients developed with the use of general-purpose software engineering tools and technologies (e.g. Java, CORBA, SOAP/XML). This implies

the definition and implementation of the APIs with IT methodologies and languages (e.g. object-oriented design, UML) that will enable a large number of apt programmers without necessarily strong telecommunications literacy to contribute to the development of high-quality telecommunications software.

The introduction of standardised open APIs is a major step towards a highly flexible mobile communications environment. Open systems and services are able to participate in the dynamic creation of network and application topologies and configurations that are optimal for the service provision chain at any point in time.

7.2.2 Applicability of Open APIs

By being technology agnostic, open APIs provide a common technological denominator that facilitates the interoperable implementation of functional specifications using different programming languages or implementation instruments and hosted on various operating system environments. Consequently, the application of an open API is in no way restricted by its nature, only by performance limits of available technologies.

Open API implementations would prove useful in various service provision contexts:

- In the mobile terminal firmware to allow trusted third-party control and management (e.g. by the mobile terminal manufacturer, or the SIM card issuer) over important mobile terminal features such as the runtime operating system and the network protocol stack.
- In the mobile network elements where local resources should be accessible and dynamically reconfigurable, so that higher-level management and control software is able to appropriately set parameters for achieving seamless operation and optimal performance in the entire network.
- In entire networks, where operation and maintenance policies are universally applicable to ensure the preservation of ultimately important features such as reliability and performance. That greatly simplifies the realisation of efficient end-to-end network management as required in future mobile communication systems.
- In VAS and applications to allow for service differentiation and manageability, accommodation of customer preferences and for paving the way towards the more ambitious goal of dynamic application composition, including runtime and on-the-fly component reallocation.
- In mobile environments to enable reconfigurability management. Reconfigurability management encompasses the triggering and control of reconfiguration actions, through reconfiguration policies, in various layers on mobile systems and networks, as depicted in Figure 7.1. That requires the support of open APIs by hardware/firmware and software components operating at all layers.

7.2.3 Advantages/Challenges

The advantages of introducing open access to network information and infrastructure can be summarized as follows:

- Open APIs facilitate the creation of a dynamic environment, where hardware and software entities will be able to adapt their configuration/behaviour on-the-fly to provide new levels of quality of network services and end-user applications.

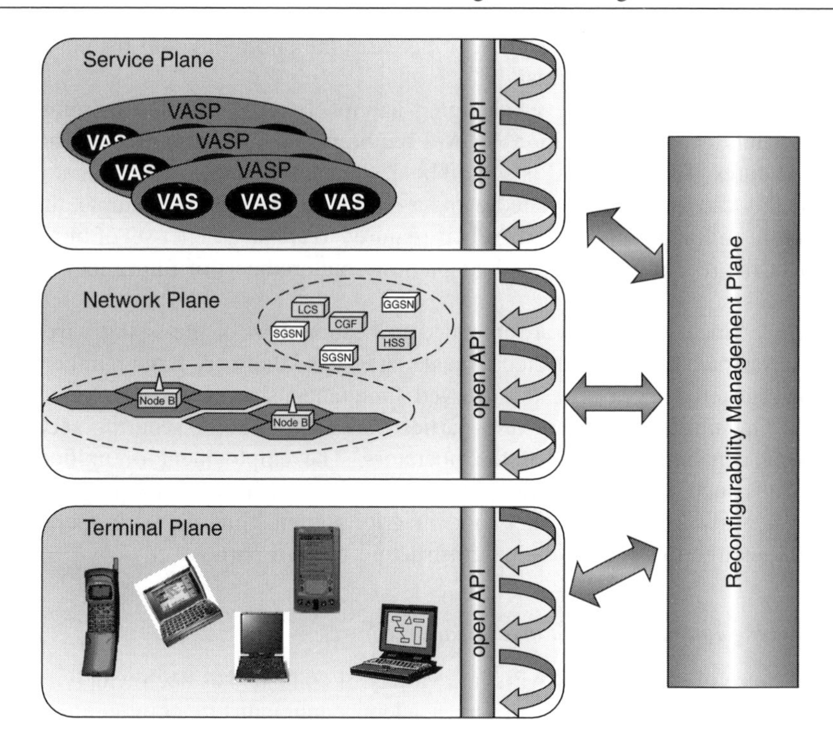

Figure 7.1 Reconfigurability management across various layers

- Open APIs hide the heterogeneity that is likely to dominate the forthcoming mobile communications era and thus enables the development of portable end-user and management applications as well as protocol and service components.
- The existence of open interfaces at various layers removes the monolithic approaches in system and service development and thus creates unlimited potential and business opportunities for enterprises of any size, which are now able to exploit their expertise for freely competing and contributing to the mobile communications and services market.
- Reusability of legacy infrastructures is facilitated through their wrapping by standardised open interfaces.

However, since interface definition is an inherently demanding task, several challenges have to be confronted by open API designers in order for the above benefits to be reaped. The most important of them include:

- *Generality/completeness:* Generality refers to universal applicability and independence from factors such as the underlying infrastructure and networking technology. This way reusability is achieved since the specified building blocks can be used in various contexts. Open APIs should include methods and provide access to the whole range of functionality and services that is frequently needed by the majority of client services/applications. That completeness is particularly important for cross-platform applications since the lack of many important functions leads to the need to write platform- and vendor-specific code that might be inconsistent in style, level or generality with the functions in the API. Hence, it increases the portability and maintainability of the client code among platforms.

- *Extensibility:* The dynamic, ever-changing nature of future mobile environments will most probably cause modifications in API requirements and, in particular, the addition of new features. Thus, open APIs should be designed to be extensible, so that additional functionality can be accommodated without excessive delays. For example, the work of various fora defining standardised APIs is required to create a richer set of interfaces that will cover the reconfigurability and adaptability requirements of future mobile systems and services.
- *Scalability:* Since open APIs are intended to become part of the rising wireless computing infrastructure, it is expected to scale largely, so that very large numbers of users (e.g. the registered services) can be served simultaneously.
- *Security:* Opening networks to third parties raises the issue of securing access to the underlying capability servers of the operators. The employment of unified security mechanisms such as those specified by specifications such as OSA/Parlay and JAIN is a major step towards ensuring network integrity and enabling network operators to be confident enough to open up their infrastructure to third parties.

7.2.4 Ongoing Research and Standardisation Efforts

The significance of standardised APIs in a variety of contexts has been widely recognised by the communications and networking academic and industry communities. Thus, a number of efforts and initiatives for defining open interfaces are under way by various organisations and fora.

Such efforts include the following:

- *Parlay* [27]. Parlay provides application developers with secure access to underlying networks and allows them to control a selected range of network capabilities. The Parlay APIs are network-independent; they can be provided over different types of networks (e.g. cellular, wired).
- *OSA* (Open Service Access) [28]. OSA is a Third Generation Partnership Project (3GPP) specifisation that essentially consists of a specialisation of the Parlay API, for the particular case of 3G mobile networks. Further API definition work is performed jointly by the corresponding subworking groups.
- *JAIN* (Java APIs for Integrated Networks) [29]. JAIN is a set of Java APIs that aim to enable the rapid development of portable, next generation telecommunications services on top of the Java platform. The JAIN family of specifications includes various levels of APIs: Java interfaces to network protocols (TCAP, MAP, SIP, H.323), Java Call Control (JCC) interfaces for the development of underlying protocol agnostic call-based applications as well as high level interfaces (JAIN Parlay API) to be used by third party applications for access to network functionality.
- *OSGi* (Open Services Gateway initiative) [30]. OSGi's goal is to enable the deployment of services, possibly over wide area networks, to local, mainly residential, networks. OSGi complements relevant local networking standards and initiatives (e.g. Bluetooth, HAVi, HomePN, HomeRF, Jini, UPnP, 802.11B) by defining open APIs for life cycle management, inter-service dependencies, data management, device management, client access, resource management and security.

- *SDR* (Software Defined Radio) Forum [31]. The SDR forum has defined a high-level generic framework, comprising specific functional blocks, and corresponding standardised APIs for facilitating the implementation of dynamic switching between different functional modules in SDR terminals and network nodes (e.g. base stations), as well as the communication between them.
- *IEEE P1520* [32]. This is an initiative that aims to facilitate programmability in networks of different types (e.g. IP, ATM, SS7). By introducing a layered reference model and specifying object-oriented interfaces between the different layers, it enables the development and dynamic introduction of new network services at various levels (e.g. from real-time stream management and synchronisation of different types of media streams down to routing algorithms and admission control schemes).

Despite their different scopes, all the above specifications share a common philosophy; namely the implementation details of the functionality offered by telecommunication devices and systems should not be standardised and rigidly incorporated in them. Instead, only generic frameworks and reusable capabilities should be predefined and integrated into the corresponding equipment. Those capabilities should be accessible through open interfaces to third parties that will thus be able to differentiate themselves in the market by introducing new, innovative services and functions that are portable to a variety of underlying infrastructures. This approach follows the model that has been successfully applied in the Personal Computer (PC) industry, where openness of operating system platforms has fuelled the creation of a wealth of applications by independent software vendors.

7.3 Service Provision and Reconfiguration Management

7.3.1 The Need for Adaptability and Reconfiguration in Various Layers

The upcoming world of fourth generation mobile communications, where reconfigurable mobile devices with significantly different characteristics (e.g. computer laptop, PDA, cellular phone) will access a multitude of independently developed applications and VAS, using a variety of wireless link technologies (e.g. UMTS, GPRS, Wireless LAN, Bluetooth), holds great challenges for the mobile communications community.

In ref. [6] the enormous range of operational parameters and Quality of Service (QoS) metrics that must be met in such heterogeneous and dynamic environments is identified as one of the most difficult challenges for future multiple wireless access communications. Consequently, the prevailing design assumptions for the mobile network, the terminal device and their protocols must be re-evaluated in the light of the volatility and dynamism that dominate the wireless world. Design priorities may shift from vertical goals such as optimum performance towards more horizontal concepts such as adaptable behaviour, sustainable operation and service ubiquity.

Adaptable systems capable of detecting significant changes in their operational environment (e.g. the short-term error bursts of the wireless link) and engaging the appropriate actions and counter-measures to maintain the desired QoS level are at the focus of current wireless research. However, adaptability cannot only be reactive; as the power supply or some other precious resource of the mobile device drains completely, all reactive strategies become unavailable. Therefore, adaptability must also be proactive, continuously monitoring the system's operational variables and predicting their future levels. Defining, monitoring, analysing and reacting to changes in the operational context of a system in an autonomous fashion is a property known and referred to as *context awareness*. Context awareness can be divided into extraction of context information, interpretation of context information, reasoning about the current contextual situation, and adaptation of application behaviour. Adaptability and reconfigurability encompass the entire service provision domain, extending from the mobile terminal through the network infrastructure to application and services. Reconfigurability procedures may cut across multiple system layers, and therefore may depend on and be triggered by various disparate factors:

- Service provision scheme
- Mobile device capabilities (e.g. cellular phone, PDA, laptop) and network profile
- User and application profiles [e.g. Virtual Home Environment (VHE) preferences]
- Applicable policy (e.g. security policy)
- Context and location

Ideally, the aforementioned factors should be readily accessible by any functional entity whose operation or performance depends on them. More importantly, access to that information should be realised in a technologically neutral manner, without imposing unnecessary technological constraints on its beneficiaries. Currently, open API approaches (e.g. IDL [7], CORBA [7], COM) present some of the most widely interoperable technologies available. Functional entities that communicate via a specific open API may be written in different programming languages, reside in different administrative domains, run on different host architectures and operating systems and, in short, be entirely independent of each other's implementation aspects.

Opening up reconfigurability features to service provision platforms and applications can create a dynamic, context-aware environment for the provision of personalised services. Reconfiguration functionality needs to be accessible via open, secure, standardised interfaces, such as the Parlay/OSA APIs [27, 28].

7.3.2 Open APIs for Flexible Service Provision and Service Adaptability

Technology-agnostic open API initiatives such as Parlay, OSA and JAIN abstract the underlying network infrastructure and all its complexities (e.g. communication protocol stacks, topological configurations), thus allowing application developers to focus on the development of core application features rather than the accommodation of all possible network-related events that may interact with the application logic – however remote these may be. Yet, despite allowing the exposure of (mobile) network functionality to the application domain, Parlay/OSA/JAIN leave it to the application developer to deal with all the context-specific cases that may arise in the course of the service provision process. Service provision platform approaches offload this burden – which in mobile environments may incur significant efforts due to the relaxed assumptions one can make about them – to the service provision platform operator. Consequently, the definition of an integrated framework for the support and management of service provision, service adaptability and reconfigurability aspects is a prerequisite for the emergence of advanced service provision in future mobile communications [12, 21]. A critical step in introducing such a framework is the provision of secure open APIs that enable third-party services and service providers to access network functionality and perform reconfiguration actions without compromising system integrity. In order to fulfil flexible service provision and adaptability requirements, open APIs to be introduced can be identified as follows.

7.3.3 Open APIs for Policy-Based Reconfiguration Management

Some basic requirements for a flexible reconfigurability management framework are:

- Support for flexible business models comprising multiple parties in different administrative domains
- Dynamic reconfiguration based on terminal/network, user and service profiles, application requirements, available network/device resources and according to user preferences
- Automatic negotiation of applicable service level parameters between mobile devices and access networks (i.e. bi-directional 'plug-and-play')
- Dynamic reconfiguration based on policy provision (e.g. charging, QoS)

Ideally, reconfiguration management concerning user, terminal, network and service-related entities should be exposed through a secure open API, thereby enabling third parties (e.g. application and service providers) to apply authorised reconfiguration actions to the mobile network.

In the definition and specification of an open API it is proposed to provide for policy-based management of mobile networks [13]. The proposed policy-based management architecture assumes that the communication of the involved Policy Decision Point (PDP) [1] with the associated network elements is built upon common protocols for policy enforcement such as the COPS [14] or COPS-PR [15]. The proposed API for the policy-based reconfiguration of the underlying network infrastructure includes methods that enable:

- the creation, modification and deletion of policy classes and instances;
- the activation and deactivation of policies;
- the creation, modification and deletion of policy-related events;
- the registration and deregistration for notification triggered by specific policy-based events;
- the handling of event notification from the network;
- the traffic flow monitoring of specific user sessions that concern VAS usage for charging purposes.

By employing the proposed API, authorised entities may create and apply context-aware policies to the mobile network as well as register for receiving notifications whenever specific events occur.

7.3.4 Open APIs for the Support of Service Adaptability

Service adaptability is an important technology enabler for mobile service provision over mobile systems and networks from 3G and beyond [1]. A task present in any type of service adaptation procedure is the decision on the most suitable (given the current context) transition to another state/behaviour of the adaptable entity [4, 5].

It is worth noting that typically the logic employed to make intelligent decisions through the processing of context information for a particular adaptation action may be useful in other cases where adaptation is required (e.g. for other adaptable entities). Thus, one could clearly identify the utility of a generic adaptation server, which can be used by end-user services as well as management applications (e.g. service provision platforms) for identifying the optimal choice for the state/behaviour of an adaptable entity. This way, the developed decision-making functionality can be re-used for adapting various types of entities in diverse environments.

An important observation is that intelligent decision-making may be essentially reduced to generic, entity-agnostic matching procedures. The aforementioned server implementation should be totally agnostic of the adaptable entities, the specific types of profiles processed during adaptation and the algorithms that are used for profile matching (the latter could be loaded dynamically at runtime and thus should not be a static part of the adaptation mechanism). The adaptation server could provide its functionality to clients through two simple operations [5]:

- Matching of two profiles of the same type, resulting in a Boolean decision on whether or not they are compatible with each other.
- Adapting a profile according to a profile of another type. After this operation the server returns the adapted profile to the client.

7.3.5 Open APIs for the Application of Reconfiguration Management Signalling

7.3.5.1 The Reconfigurability Process

At a high level of abstraction, the reconfigurability process manifests itself in the following distinct phases:

1. Identification of context. Owing to is generality and wide applicability the reconfiguration process must have identification of context as a starting point. Identification of context concerns an investigation procedure that spatially scopes the technological surrounding of the requesting entity by identifying affected elements in the communication and computing infrastructure. Owing to its inherently combinatorial complexity, without proper bounding measures the problem space of the aforementioned investigation procedure could potentially increase beyond tractable levels. Consequently, appropriate and efficient scoping mechanisms are necessary in any reconfiguration process so as to safeguard the existing operating infrastructure from 'reconfiguration storms'. For example, for a mobile terminal with a reconfigurable radio protocol stack and currently under Wireless Local Area Network (WLAN) and Universal Mobile Telecommunication System (UMTS) radio coverage, identification of context would isolate the particular network elements (e.g. Nobe B, RNC and/or WLAN base station) that will engage in reconfiguration-related signalling. In this case, technical capabilities would constitute the first applicable scoping variable, since the mobile terminal is reconfigurable with regard to radio protocols only. Mobile terminal location and proximity to particular network elements in the radio domain could serve as additional scoping variables for the identification of affected systems in the existing network infrastructure. Generalising, geographical location constitutes just one dimension in the multidimensional space within which context identification operates.
2. Identification of feasible alternative solutions. Before reconfigurability can be applied to a collection of systems, it is necessary to identify precisely the feasible alternative solutions that can be employed in a given context. For example, in a wireless

communication context that includes a mobile terminal with a reconfigurable protocol stack for the UMTS and HiperLAN/2 standards, which is currently under UMTS, GSM and HiperLAN/2 radio coverage, all the alternative solutions would be feasible combinations of UMTS and HiperLAN/2 modes. In a component-based application context that includes a media streaming application capable of employing different codec modules, the alternative solutions would comprise the set of all available codec modules that provide an interoperable interface to that particular application. The aforementioned examples implicitly reveal a notion of capability exchange and negotiation procedure under a specific policy that necessarily takes place within the identification of alternative solution(s) process. Furthermore, they expose the need to maintain and retrieve information about the set of all available solutions.

3. Decision on the solution and respective implementation. Once the feasible alternative solutions have been identified, the reconfigurability process proceeds to decide on the particular solution that will be deployed over the set of affected systems. As already mentioned, we consider an 'improved customer experience' as the primary goal of the reconfigurability process. However, the marginal improvement that a particular solution may render in the customer's experience will depend on a set of inherently subjective criteria that characterise each individual customer. Thus, the aforementioned decision process must necessarily incorporate user-related aspects about the set of feasible alternative solutions; that is, being able to take into account generic (e.g. user) preferences alongside strictly technical considerations. Building on our previously described example, a particular user may exhibit a preference for UMTS mode solutions with specific properties and, therefore, an efficient decision process should always consider such preferences, if available. Consequently, accommodation of generic user preferences regarding properties of the alternative solutions is a mandatory requirement for any efficient reconfigurability decision process.
4. Physical deployment of the solution. Following the decision phase, the reconfigurability process proceeds to the physical deployment of the actual solution implementations, since – especially in a wireless environment – one cannot assume that these will be readily available at all affected systems. Thus, the reconfigurability process must exploit existing infrastructure resources (e.g. the control plane and user plane communication protocols, physical memory and other storage resources, and system processing cycles) so as to download and install the actual solution implementations (e.g. network protocol stacks, application support components). Depending on the specific properties of the selected solution implementations and the respective technological context of target systems, it may be necessary to engage in certain post-download configuration actions before proceeding to the next phase. For example, downloading different implementations for a particular solution (e.g. a precompiled C language executable and a compiled Java class) will probably require undertaking different post-download actions and installation procedures at each affected system.
5. Activation of the solution. Before activating the installed implementations, the reconfigurability process may have to perform certain post-installation configuration actions (e.g. in the case of an IP protocol implementation, configuration of the DHCP feature). However, the details of these configuration actions will inevitably depend on the network and/or application context at the moment of activation, and therefore it should be possible to conduct such actions within the activation phase as well. In addition, since

more than one solution implementation may be activated at a particular system, the activation procedure must be able to account for interdependencies between solution implementations and adapt the activation sequence accordingly.

Clearly, the reconfiguration process deals primarily with the identification of the most suitable choice among a set of available solution implementations, thus revealing the notions of (solution) specification and (solution) implementation. By direct inference, any reconfiguration model must necessarily exhibit independence between specification and implementation as its cornerstone while addressing the aforementioned – or even additional – relationships between the two. Identification of the most suitable (solution) implementation for a particular (solution) specification suggests that persistent information about instances of these pivotal concepts should be readily available to the reconfiguration management process. Thereupon, a globally accessible reconfigurability registry service must necessarily be part of any reconfiguration management process.

For reconfiguration management to be flexible and efficient, it must be possible to materialise and manage the aforementioned relationship between the entities of a reconfigurable system at various granularities. Extending our previous example, it should be possible to reconfigure a specific protocol [e.g. Generic Tunnelling Protocol (GTP)] in a UMTS mobile terminal without requiring reconfiguration of the entire UMTS protocol stack.

In addition to the operational parameters of the functional entities or the adaptation of service delivery, an open reconfiguration management API should also provide for registration management of functional specifications (e.g. protocol specifications) and their matching implementations. In principle, that process involves two distinct roles: (a) organisational bodies (e.g. standardisation fora) that manage the registration of functional specifications, and (b) trusted third parties that manage the registration of functional implementations that have been certified against the respective specification. The class diagram in Figure 7.2 illustrates that functional implementation (TpModuleImplementation) instances bear a unique identifier (TpModuleImplementationID) that associates them with a single functional specification (TpModuleSpecification) instance that, in turn, bears its own unique identifier (TpModuleSpecificationID).

Typically, functional specifications are re-used across system specifications; for example, the MAP [16] specification is employed in GSM, GPRS and UMTS system specifications [17–19]. Considering that multi-mode wireless access will dominate future mobile communications, we understand that reconfigurability management must also support efficient grouping of functional specifications into distinct 'mode' objects that can be managed as a whole independently of each other. Therefore, the notion of mode specification (TpModeSpecification) and mode (TpMode) is introduced in Figure 7.3 as management containers for functional specifications and functional implementations, respectively.

As mentioned previously, part of the reconfigurability management process concerns the dynamic registration of functional specifications (TpModeSpecification and TpModuleSpecification) and the respective type-approved functional implementations (TpModeImplementation and TpModuleImplementation) to the reconfigurability registry service, as illustrated in Figures 7.4, 7.5 and 7.6.

The application of reconfigurability management signalling (i.e. the reconfigurability process) proceeds through a number of distinct stages, namely mode discovery,

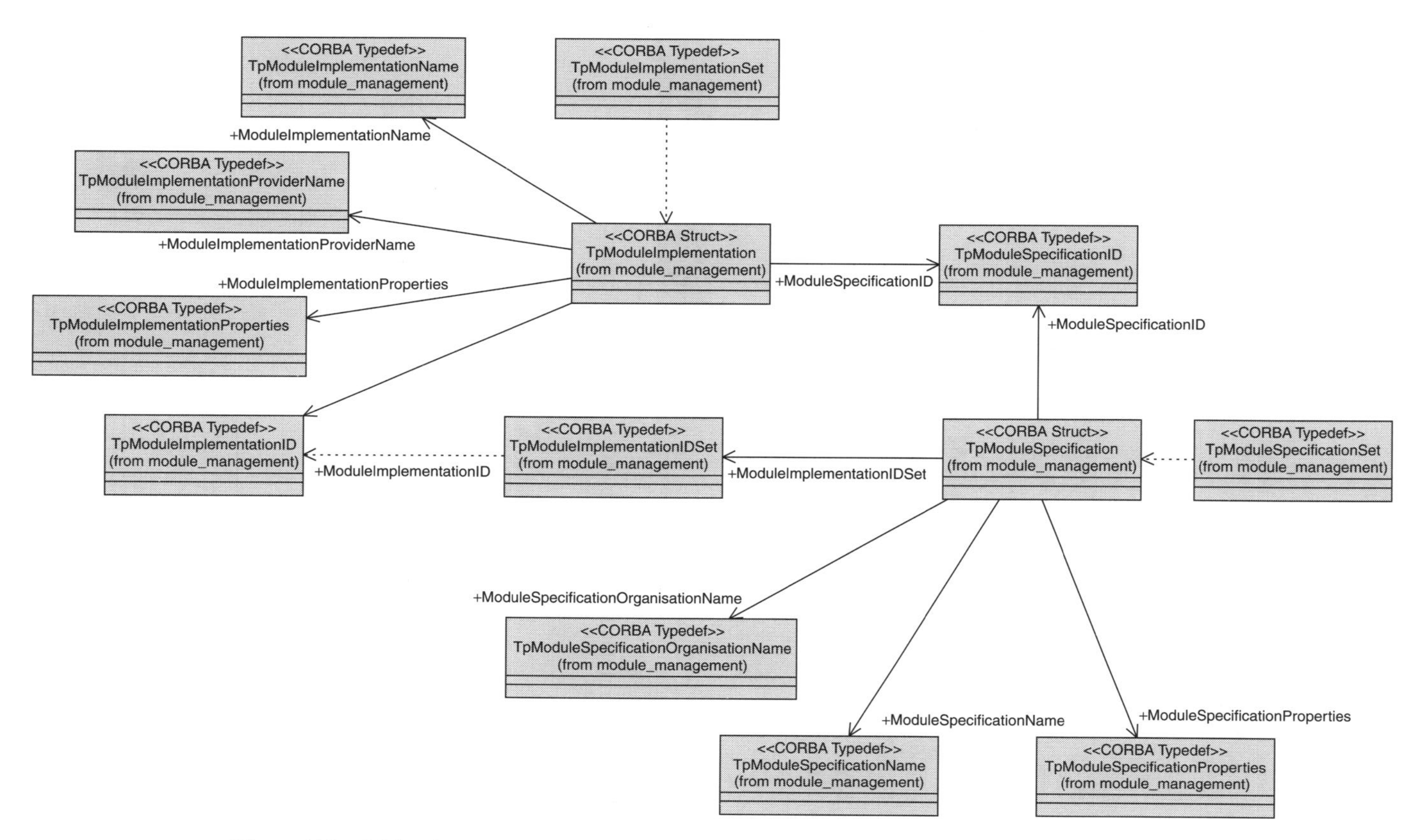

Figure 7.2 Main abstractions for reconfigurable functional specifications and implementations

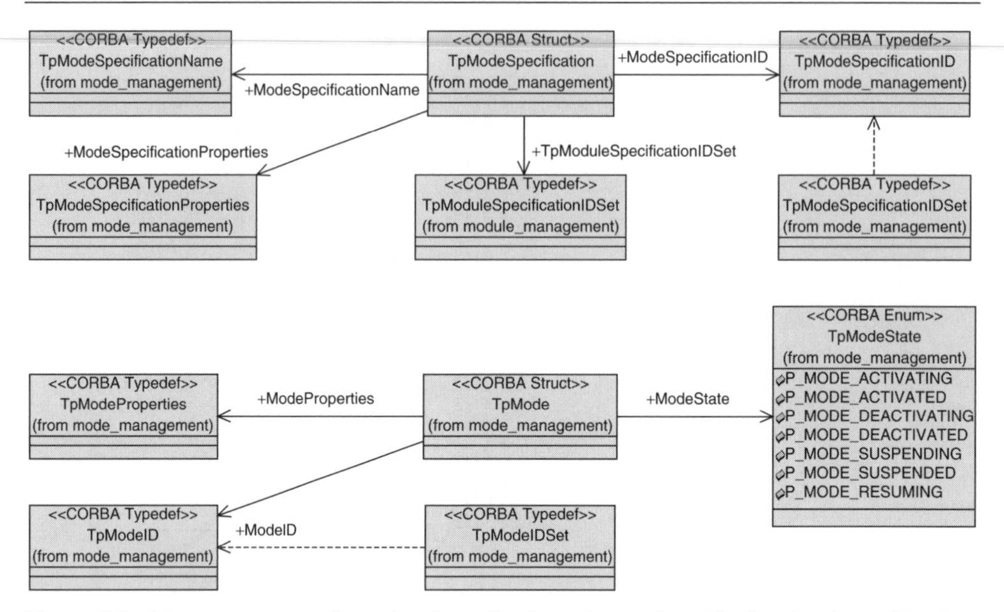

Figure 7.3 Management containers for the collection of reconfigurable functional specifications and implementations

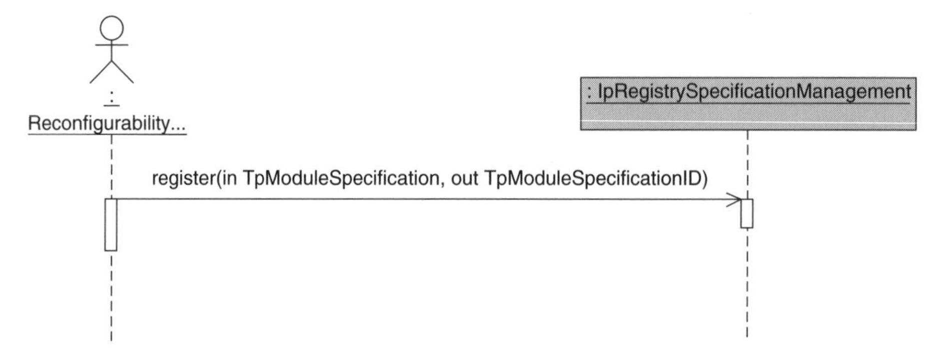

Figure 7.4 Registration of a TpModuleSpecification instance

Figure 7.5 Update of a TpModuleSpecification instance

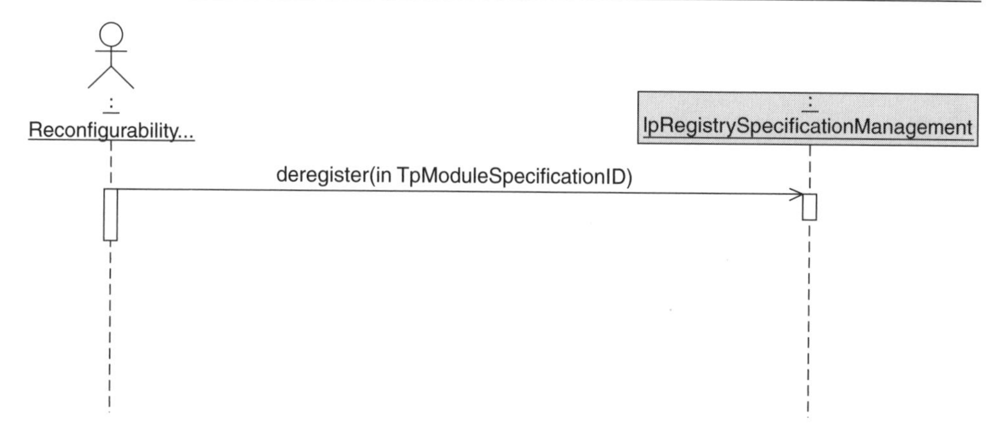

Figure 7.6 Deregistration of a TpModuleSpecification instance

resource discovery, capability negotiation, resource reservation, mode instantiation and mode activation.

As an example, and assuming that the desirable mode has been identified and all resources necessary have been successfully reserved, instantiation of the appropriate objects in a target reconfigurable system proceeds according to Figure 7.7. The depicted MSC assumes the availability of IpModeManager and IpModuleManager interfaces in the target reconfigurable system in order to preserve its configuration integrity. In addition, a factory design pattern [20] is applied to allow the creation of necessary instances (i.e. IpMode and IpModule) in a distributed manner, using remote factories (e.g. factories residing in the functional implementation provider administrative domain).

The IpModeManager and IpModuleManager interfaces provide for activation and life-cycle management of created instances within the target reconfigurable system. Figures 7.8–7.11 provide an illustration of reconfigurability management signalling as applied to the life-cycle management of functional implementations.

Although IpModule and IpMode instances may reside in independent threads of control, they maintain their life-cycle state synchronised with their assigned IpModuleManager and IpModeManager instances, respectively. By committing to an interface for state synchronisation purposes, any collocation requirements on the reconfigurability management model are removed, allowing for virtually any deployment combination, including terminal-centred, network-centred and fully distributed deployment models for the reconfigurability management signalling participants.

However, for reconfigurability management signalling to be efficiently applicable, it is essential that an unambiguous and universal specification for the behavioural aspects of the interfaces participating in reconfigurability management signalling exists. Such a specification should not only address the interface signature (i.e. the set of parameters and their respective data types) or the marshalling and demarshalling protocols, but more importantly, the underlying abstract state machine and its transition rules. Figure 7.12 provides an illustration of the abstract common state machine underlying the functional implementation management interface (IpModuleManagement) depicted in the previous diagrams. In order to efficiently manage a reconfigurable system regardless of the implementation details related to the set of functional implementations (e.g. device drivers,

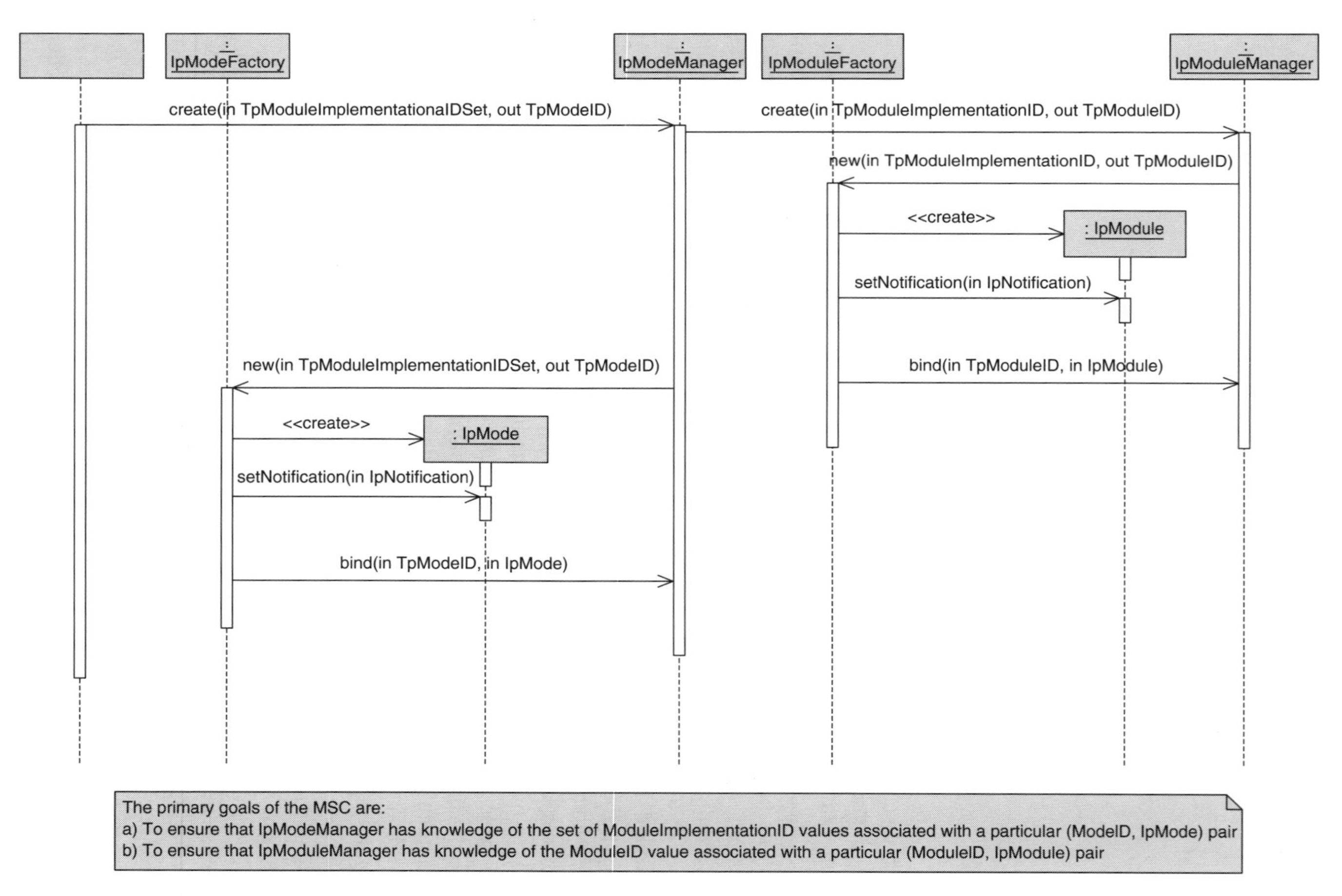

Figure 7.7 Creation of reconfigurable instances

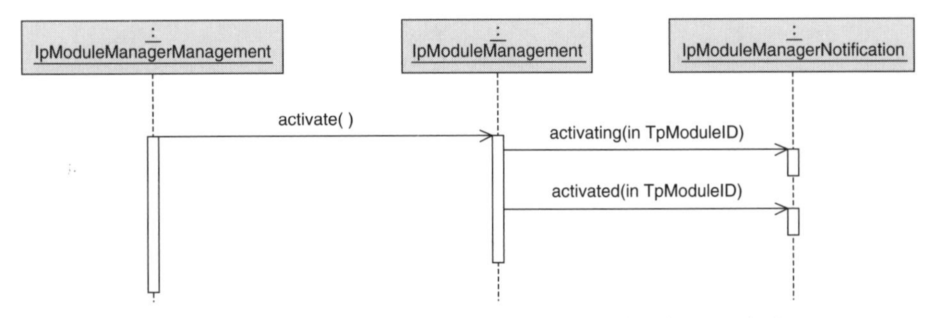

Figure 7.8 Activation of a module (functional implementation)

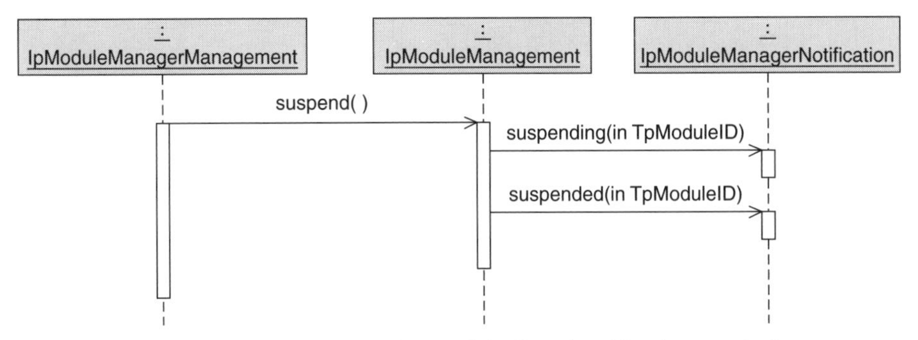

Figure 7.9 Suspension of a module (functional implementation)

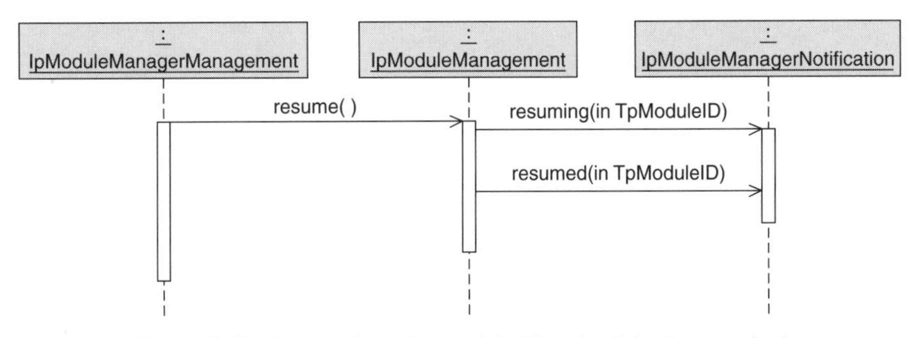

Figure 7.10 Resumption of a module (functional implementation)

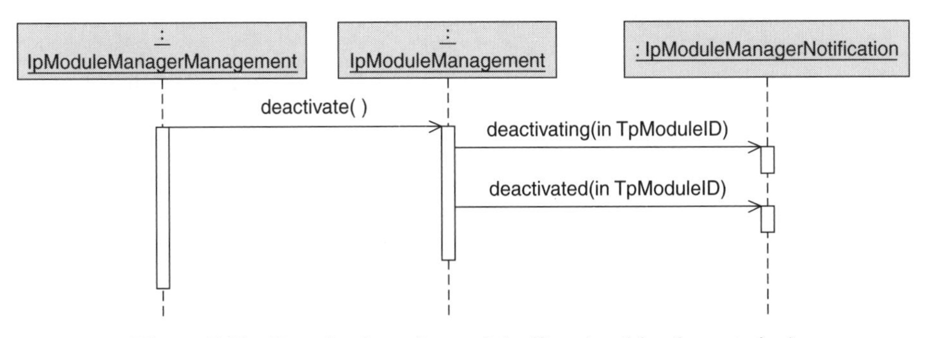

Figure 7.11 Deactivation of a module (functional implementation)

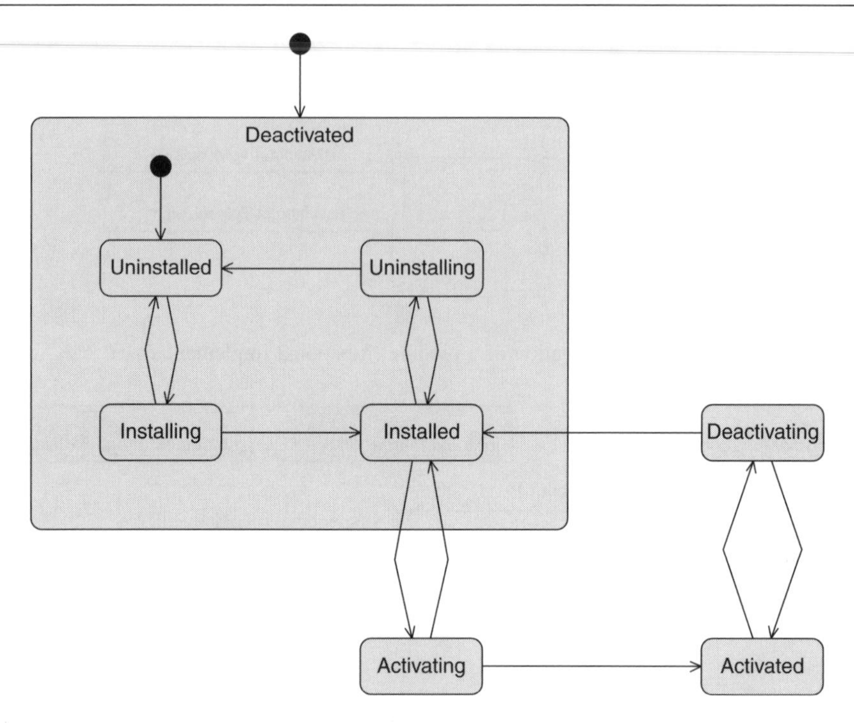

Figure 7.12 The abstract common state machine of a functional specification management interface (IpModuleManagement)

network protocols and system-wide services) active within it, it is necessary that all the management interfaces of all such type-approved functional implementations abide by a common abstract state machine throughout the life-cycle of the functional implementation.

7.3.5.2 Open APIs Providing for Location Information Management

The positioning technologies that have been developed recently for retrieving and tracking the location of mobile user offer truly customised personal communication services through typical cellular phones or other mobile devices [11, 23, 26]. Not surprisingly, several issues related to the management and exploitation of the location information in mobile communication systems are being studied by relevant standardisation bodies and fora [8, 9, 24, 25].

The exploitation of location information, as a prime factor for reconfigurability, presents a powerful new dimension to the range of information services that can be offered [22]. These include: location-sensitive user profiling, location aware service deployment, discovery and QoS management, location-based charging/billing, advanced mobility management providing for optimised network planning, location registration, paging and handover management, terminal device reconfiguration with software and protocol updates/upgrades, or performance enhancing mediation, network nodes' selection and reallocation, as well as conversion of multimedia services for content adaptation (e.g. voice to text, video to voice).

An open API providing for location information management should enable:

- Retrieval of the user's location with a specified accuracy. Location retrieval should be either immediate (in the case when the current location of the user is requested) or deferred (in case when the location of the user is requested when a specific event takes place).
- The creation/deletion/modification of specific location-based events, relevant to the current location or mobility of users.
- The registration of end-users and other third parties for receiving such location-sensitive event notifications.
- The notification of end-users and third parties of the available policies, restrictions, updates, tariffs, reconfigurations and other events that are associated with the current location of the user or induced due to the location updates that occur.
- The activation/deactivation of various special location-based services offered by the hosting platform (e.g. location-based charging/billing [26]).

Furthermore, the location API should ensure that location monitoring takes place with respect to the user privacy settings included in the user profile. The privacy rules that govern the tracking of a user's location by third parties are defined by end-users upon their registration with the location-monitoring platform. Hence, whenever an application/service requests the location of a user, the Location API Manager examines whether it is authorised by the user to do so. In the case when it is not, the end-user's preferences determine whether the request is simply denied or the end-user is asked for explicit Authorisation.

7.3.6 Open APIs for Resource Access and Management

Efficient resource management is of major significance for the operation of a network, since it can lead to the full exploitation of the network's capabilities, towards the goal of offering the best possible QoS to end-users. Network resources can be physical (directly corresponding to a physical component/device in the network) or logical. Physical resources are tangible entities such as switches, routers, Central Processing Units (CPUs), while examples of logical resources are point-to-point or end-to-end connections and routing tables. Network programmability involves the capability for efficient third-party access to network resources through standardised APIs. Open interfaces for resource management can be classified into two categories:

1. Interfaces for accessing local resources, namely entities whose state is expressed using only information local to a specific network node. The aim of these interfaces is essentially to provide programmable abstractions of low-level network element resources. Although these interfaces can be partly resource-specific, there should also be a number of methods common and applicable to all resource interfaces. Such methods could include functionality for:
   - Acquiring a resource (possibly following a prior resource reservation)
   - Reserving a resource for later allocation
   - Releasing a resource

   These interfaces can be used by network element management applications and/or functional modules undertaking distributed tasks such as routing and connection management.

2. Interfaces to resources (e.g. end-to-end virtual connections) whose state is distributed among multiple network nodes as well as entities performing distributed resource management tasks (e.g. connection managers). Interfaces in this category can be used by management and control applications that are used to map higher-level policies to the underlying network infrastructure. A case of particular interest is the possibility, through these open interfaces, of the dynamic loading and activation of distributed algorithms (e.g. for routing, admission control) in a particular network.

7.3.7 Open APIs for Terminal Reconfiguration Management

The mobile terminal traditionally has attracted the most attention in SDR reconfigurability activities. Although nowadays it is widely recognised that the application of SDR concepts in the network and application domain is a factor of major significance for flexible service provision, terminal reconfigurability is undoubtedly considered as an important technological enabler for next generation mobile environments. The introduction of reconfiguration management logic in the terminal that co-ordinates reconfiguration procedures typically in co-operation with managers in the network domain as well as the specified of open APIs in various layers of the terminal functionality can be clearly identified as a requirement for the development of advanced SDR terminals.

Figure 7.13 depicts a terminal architecture that has been introduced by the Information Society Technologies (IST) project SCOUT. The proposed framework divides terminal functionality into three layers, each of which exposes open interfaces to appropriate clients (other layers or end-users/carrier/manufacturer applications). The architecture aims to support the capability for dynamic determination and application of the optimal terminal configuration in various contexts, through the seamless co-operation of multi-vendor software and hardware components via standardised APIs.

The framework comprises the following parts:

1. *SDR Core Framework (SDR-CF).* This layer is placed on top of the operating system and is responsible for controlling the overall reconfiguration procedure, including tasks

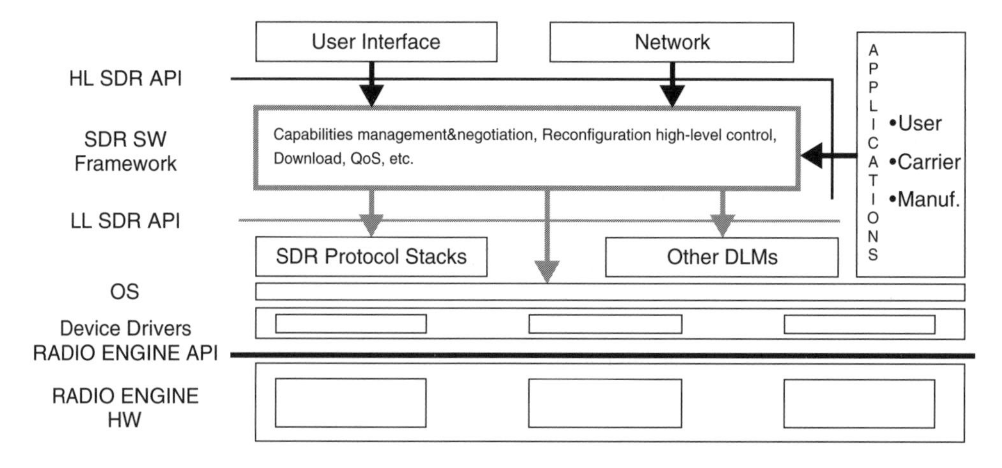

Figure 7.13 Terminal architecture for reconfiguration management based on open APIs

such as applications activation, QoS and download management. Applications may access its functions through the HL API.

2. *SDR Operating system (SDR-OS).* The SDR-OS implements important system functions that are used by the SDR-CF. More specifically, the SDR-OS:
   - controls current/future terminal configurations
   - supports domain managers and terminal agents
   - provides access between SDR-CF and HAL
   - provides data and resource management (LookUp Tables, Capabilities, Profiles, Databases, etc.)
3. *SDR Hardware Abstraction Layer/Radio Engine API (SDR-HAL).* The SDR-HAL is an abstraction layer used to enable open, vendor-independent access to reconfigurable hardware components.

7.3.8 Open APIs Providing for Advanced Charging Service Management

The introduction of a set of open APIs allowing dynamic reconfiguration actions related to charging is essential in order to provide applications with advanced charging services. Specifically, through open APIs the dynamic introduction of new pricing policies, modification of existing ones, one-stop billing, on-line provision of information concerning the service profits, etc. should be enabled.

Figure 7.14 illustrates the class diagram of the proposed interface for charging purposes. It is assumed that the client discovers the Charging, Accounting and Billing (CAB)

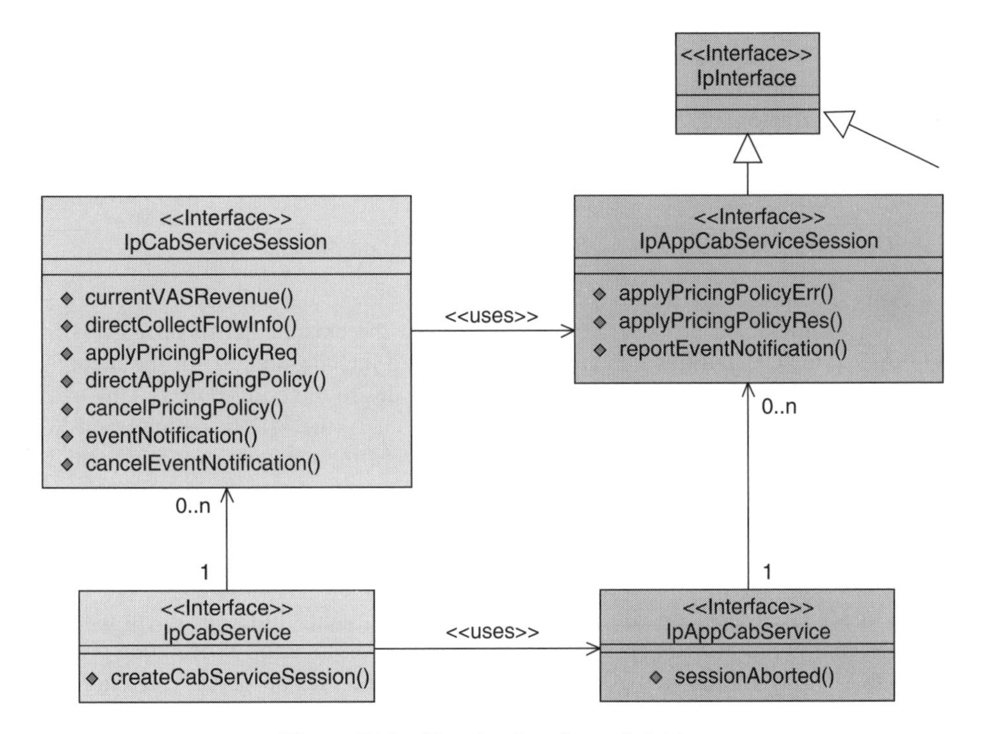

Figure 7.14 Charging interfaces definition

Service Capability Feature (SCF) applying the standard discovering procedures provided by the OSA/Parlay framework. As a result, the client receives a object reference to an object that implements the IpCabService interface and creates a local object implementing the IpAppCabService interface. The IpCabService interface provides the method `createCabServiceSession()` that is used in order to start a CAB service session. The introduced methods are supplementary to the existing methods provided by the standard OSA Charging API, since it enables applications to add charges to the user's bill only for the content.

Specifically, the introduced methods of the IpCabService interface enable:

- On-line retrieval of the status of the VAS revenue, implemented by the method `currentVASRevenue()`.
- Triggering of underlying network elements in order to gather real-time performance, metering and policy-related data, implemented by the method `directCollectFlowInfo()`.
- Dynamic application or cancellation of a specific pricing policy, implemented by the methods `directApplyPricingPolicy()` and `cancelPricingPolicy()`, respectively. Additionally, this functionality could be provided via asynchronous messages (such as `applyPricingPolicyReq()`, `applyPricingPolicyErr()` and `applyPricingPolicyRes()`). In the first case the application of the specific pricing model concerns only the current session, while the second case is more general and the specific pricing model should be applied whenever the provided conditions are met.
- Registration for receiving notifications by the CAB SCF whenever a specific event appears is implemented by the method `EventNotification()`.
- Cancellation of the aforementioned registration, implemented by the method `cancelEventNotification()`. Consequently, the client is notified through the method `reportEventNotification()`.

7.3.9 Open APIs Enabling Policy-Based Reconfiguration Management

Such APIs should include methods for applying policies to the underlying network elements. For example, we propose the enhancement of the existing set of open APIs [e.g. the OSA Service Capabilities Features (SCF)] [10], to include support for the reconfiguration of network metering entities according to the desire metering policy. This can be achieved through the introduction of a new SCF, namely the MDReconfiguration SCF. The proposed API enables the dynamic configuration of the network metering entities in order to monitor only the necessary flows and to gather only the required information, as defined by the applied metering policy for the specific flow, avoiding further processing of needless data.

The class diagram illustrated in Figure 7.15 depicts the proposed interface for the purpose of Metering Device (MD) reconfiguration. It is assumed that the client discovers the MDReconfiguration SCF by applying the standard discovering procedures provided by the OSA/Parlay framework. As a result, the client receives an object reference to an object that implements the IpMDReconfigurationManager interface and creates a local object implementing the IpAppMDReconfigurationManager interface.

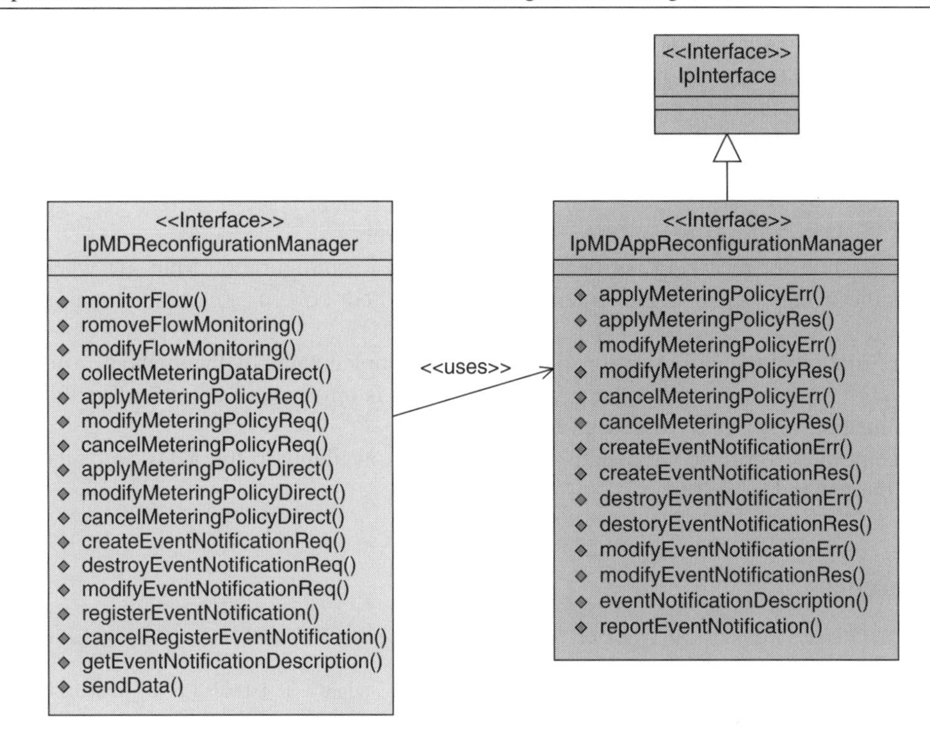

Figure 7.15 MD's Reconfiguration interface definition

The introduced methods enable:

- Dynamic configuration of the MDs in order to start or stop monitoring traffic flows (methods `monitorFlow()` and `removeFlowMonitoring()`). Flows typically refer to IP traffic and are identified by information such as IP address and port number. Modification of identification information of monitored traffic flows is also available to the API clients (`modifyFlowMonitoring()`).
- Triggering of the MDs to send immediately the metering data records concerning a specific service, a specific user, or a specific IP flow (`collectMeteringDataDirect()`.
- Application, modification or cancellation of a specific metering policy (e.g. metering granularity, frequency of metering records generation, type and format of generated records) for an application, a user or a combination of both, whenever certain conditions are met, implemented by the methods `applyMeteringPolicyDirect()`, `modifyMeteringPolicyDirect()` and `cancelMeteringPolicyDirect()`, respectively. Analogous functionality is included in the proposed API via asynchronous messages (`applyMeteringPolicyReq()`, `modifyMeteringPolicyReq()` and `cancelMeteringPolicyReq()`), targeted for cases where a client requests the application of a policy at some specific time in the future and thus does not block its execution waiting for a response. An appropriate response (e.g. `applyMeteringPolicyRes()` or `applyMeteringPolicyErr()`) are returned to the client upon the eventual completion of a request.

- Creation of new events (e.g. by a combination of existing events) that originate application notification by the MD Reconfiguration Manager whenever the specified conditions are met (`createEventNotification()`), for example in case when the applied metering policy has been modified. Deletion of an existing event as well as retrieval and modification of notification parameters are implemented by the methods `destroyEventNotification()`, `getEventNotificationDescription()` and `modifyEventNotification()`, respectively.
- Registration for receiving notifications by the MD Reconfiguration Manager whenever specific events occur is implemented by the method `registerEventNotification()`.
- Cancellation of the aforementioned registration is implemented by the method `cancelRegisterEventNotification()`. The client is informed of a notification through the method `reportEventNotification()`.
- Receipt of performance and statistical information available to the MDs, implemented by the method `sendData()`.

References

[1] UMTS Forum Report No. 9, 'The UMTS third generation market – structuring the service revenues opportunities', available from *http://www.umts-forum.org/*

[2] UMTS Forum Report No. 10, 'Shaping the mobile multimedia future – an extended vision from the UMTS forum', available from *http://www.umts-forum.org/*

[3] UMTS Forum Report No. 14, 'Support of third generation services in a converged network environment', available from *http://www.umts-forum.org/*

[4] N. Houssos, S. Pantazis and N. Alonistioti, 'Towards adaptability in 3G service provision', IST Mobile Communications Summit Thessaloniki, 16–19 June, 2002.

[5] N. Houssos, S. Pantazis and A. Alonistioti, 'Generic adaptation mechanism for the support of context-aware service provision in 3G networks', Fourth IEEE International Conference on Mobile Wireless Communication Networks (MWCN 2002), Stockholm, Sweden, 9–11 September, 2002.

[6] L. Kleinrock, 'On some principles of nomadic computing and multi-access communications', *IEEE Communications Magazine*, **38**(7): 46–50, 2000.

[7] 'Common object request broker: Architecture and specification, Version 2.6', available from OMG site *http://www.omg.org/*, document 01-12-01.

[8] '3GPP TS 29.198-1 V4.3.0 (2001-12); 3GPP; TSG_SA; Open Service Access (OSA); Application Programming Interface (API); Part 1; Overview', available from *http://www.3gpp.org/*

[9] Parlay Group, 'Parlay API Spec. 2.1', July 2000, available from *http://www.parlay.org/specs/index.asp*

[10] N. Alonistioti, N. Houssos, S. Panagiotakis, M. Koutsopoulou and V. Gazis, 'Intelligent architectures enabling flexible service provision and adaptability', Wireless Design Conference (WDC 2002), London, 15–17 May, 2002.

[11] S. Panagiotakis, N. Houssos and A. Alonistioti, 'Integrated generic architecture for flexible service provision to mobile users', IEEE 12th International Symposium on Personal, Indoor and Mobile Radio Communications (PIMRC 2001), San Diego, California, 30 September–3 October 2001, pp. 40–44.

[12] N. Alonistioti, N. Houssos and S. Panagiotakis, 'A framework for reconfigurable provisioning of services in mobile networks', Sixth International Symposium on Communication Theory and Applications (ISCTA 2001), Ambleside, Lake District, 2001, pp. 21–26.

[13] RFC 3198, 'Terminology for Policy-Based Management', available from *http://www.ietf.org/*

[14] RFC 2748, 'Common Open Policy Service Protocol (COPS)', D. Durham, J. Boyle, R. Cohen, S. Herzog, R. Rajan and A. Sastry, available from *http://www.ietf.org/rfc/rfc2748.txt*

[15] RFC 3084, 'COPS Usage for Policy Provisioning (COPS-PR)', K. Chan, J. Seligson, D. Durham, S. Gai, K. McCloghrie, S. Herzog, F. Reichmeyer, R. Yavatkar and A. Smith, available from *http://www.ietf.org/rfc/rfc3084.txt*

[16] 3GPP TS 29.002 V3.12.0 (2002-03); 3GPP; TSG_CN; 'Mobile Application Part (MAP) Specification', available from *http://www.3gpp.org/*
[17] 3GPP TS 21.102 V4.4.0 (2002-03); 3GPP; TSG_SA; '3rd Generation Mobile System Release 4 Specifications', available from *http://www.3gpp.org/*
[18] 3GPP TS 21.103 V1.2.0 (2002-03); 3GPP; TSG_SA; '3rd Generation Mobile System Release 5 Specifications', available from *http://www.3gpp.org/*
[19] 3GPP TS 21.104 V0.0.0 (2002-03); 3GPP; TSG_SA; '3rd Generation Mobile System Release 6 Specifications', available from *http://www.3gpp.org/*
[20] E. Gamma, R. Helm, R. Johnson and J. Vlissides, '*Design Patterns – Elements of Reusable Object-Oriented Software*', Addison-Wesley, ISBN-0-201-63361-2, 1995.
[21] M. Grech, R. McKinney, S. Sharma, J. Stanaway, D. Varney and K. Vemuri, 'Delivering seamless services in open networks using intelligent service mediation', *Bell Labs Technical Journal*, **July–September**, 2000.
[22] N. Alonistioti, S. Panagiotakis, N. Houssos and A. Kaloxylos, 'Issues for the provision of Location-dependent services over 3G networks', 3rd Generation Infrastructure and Services Conference (3GIS), Athens, Greece, July, 2001.
[23] S. Panagiotakis, M. Koutsopoulou and N. Alonistioti, 'Advanced Location Information Management Scheme for Supporting Flexible Service Provisioning in Reconfigurable Mobile Networks', IST Mobile Communication Summit, Thessaloniki, Greece, June, 2002.
[24] LIF TS 101 'Mobile Location Protocol' v2.0.0.
[25] 3rd Generation Partnership Project (3GPP), *http://www.3gpp.org/*
[26] S. Panagiotakis, M. Koutsopoulou, A. Alonistioti and A. Kaloxylos, '*Generic framework for the provision of efficient location-based charging over future mobile communication networks*', PIMRC, Lisbon, Portugal, September, 2002.
[27] Parlay Group, 'Parlay API Specification 3.0', December 2001, available from *http://www.parlay.org/specs/index.asp*
[28] 3GPP TS 23.127: 'Virtual Home Environment'.
[29] J. Keijzer, D. Tait and R. Goedman, 'JAIN: A new approach to services in communication networks', *IEEE Communications Magazine*, **January**, 2000.
[30] Open Service Gateway Initiative, *http://www.osgi.org*
[31] Software Defined Radio Forum, *http://www.sdrforum.org*
[32] IEEE P1520 Working Group, *http://www.ieee-pin.org*

8

Framework for Charging and Billing for Reconfigurable Services

Nancy Alonistioti and Maria Koutsopoulou
Communication Networks Laboratory, University of Athens

8.1 Introduction

In second generation (2G) networks, services provided to mobile users were either rigidly integrated in network equipment or developed with proprietary tools by mobile operators or equipment manufacturers. This situation led to the availability of a limited number of services that were tightly coupled with the type of network and the vendor of the equipment they were running on.

The convergence of the Internet Protocol (IP) and telecom worlds has a strong impact in the evolutions. In third generation (3G) mobile communications and beyond, an open and complex marketplace is expected to emerge: mobile devices with significantly different characteristics [e.g. computer laptop, Personal Digital Assistant (PDA), and cellular phone] should access a multitude of diverse applications and Value-Added Services (VAS), developed by Third Party Value-Added Service Providers (VASPs) that typically do not target solely mobile networks, using a variety of wireless link technologies (e.g. UMTS, GPRS, Wireless LAN, and Bluetooth) [1]. Applications and services will be provided to end-users over various networking technologies and heterogeneous communication infrastructures, which should converge, interface and co-operate seamlessly and congruously to ensure consistent, transparent and ubiquitous service provision to end-users, hiding the involved network complexity [2].

The Charging, Billing and Accounting schemes used in telecommunications and data networks have been quite simple until now. Users have been mainly billed at a flat rate, based on their subscription and/or the duration of their connection, for either making phone calls or accessing the Internet. Although the technology of these networks is quite different, users are mainly getting the same quality for the services they use. More specifically,

Software Defined Radio: Architectures, Systems and Functions. Edited by M. Dillinger, K. Madani and N. Alonistioti
 ISBN: 0-470-85164-3

telecommunication networks provide the same quality for any user call, while in the best effort environment of the Internet all users are treated equally when accessing an IP service. These schemes are expected to be altered soon as a consequence of the convergence of these two worlds. This alteration is accelerated due to the mobile telecommunications industry evolution.

During the recent past it has been well established that the IP will play a dominant role in the Universal Mobile Telecommunications System (UMTS) network. This has been decided in order to provide, in an efficient manner, a wide range of connectionless services to mobile users. The emergence of the UMTS and the convergence with the Internet world are key factors for the support of advanced business models. The requirements related to the provision of a number of services by independent application/service providers to users through a variety of network providers are now becoming feasible.

These new models introduce some difficulties in what relates to the execution of the charging process. It is expected that, contrary to existing approaches, the user should be able to access a plethora of services provided either by its home operator or by independent providers, without any additional contracts [3]. Innovative business models, encompassing the accommodation of the involved players, should enable the automation of procedures concerning sharing of the revenues and the introduction of flexible charging models. Furthermore, the evolution of mobile terminal and network reconfigurability as well as service adaptability aspects introduces additional requirements and complexity for the charging mechanisms to be applied in future communication systems [4].

Moreover, the introduction of the IP into mobile networks causes the design and adoption of new schemes for Quality of Service (QoS) provision that aim to support real-time services in a quality acceptable by the users. The deployment of such schemes signals the differentiation among users as well as the service flows and packets exchanged through the network. This differentiation creates the need for new mechanisms that will collect all information concerning chargeable events and, after appropriate processing, will impose flexible billing schemes on the users [5]. We should note that the records, containing all information related to the chargeable events, should possess adequate granularity to deploy advanced charging schemes such as content-based and location-based charging [6, 7].

Groups such as the Working Group 5 of the Third Generation Partnership Project (3GPP) [8], the UMTS Forum [9] in the telecommunication world, as well as the Authentication Authorisation Accounting ARCHitecture Research Group (AAAARCH) [10] of the International Research Task Force (IRTF) [11], are trying to establish the appropriate functionality.

Metering is the function of capturing all data related to network resources' consumption (e.g. volume of exchanged data) and is performed by network devices. After being collected by the network devices the resource usage data are sent to an Accounting Server for further processing.

The Charging function collects information related to a chargeable event from specific network nodes, which are able to generate such information. Since a chargeable event can be considered any activity utilising network resources and related services that the operator charges for, the charging information generated by network nodes is structured in the form of a Charging Data Record (CDR) and transferred via standard charging protocols. This is a well-established procedure since the establishment of a Global System for Mobile Communications (GSM) and, until recently, the abbreviation CDR stood for

'Call Detail Records'. The charging function is responsible for further processing and intermediate storage of CDRs, the correlation of partial records and the secure transfer of these records to the Billing System. The charging functionality is accomplished by two entities: the Charging Gateway Functionality (CGF) and the Charging Collection Functionality (CCF) [12].

The Billing function processes the records coming from the charging functional entity according to the respective tariffs [stored in the Home Location Register (HLR) or inside the Billing System] in order to calculate the charge for which the user should be billed. This function is performed by a Billing System that transforms charging records into bills requiring payment. The CDRs are transferred to the Billing System via a transfer protocol such as the File Transfer, Access and Management (FTAM) or the File Transfer Protocol (FTP).

In the case of roaming, the Accounting function is responsible for calculating the portion that is due to each operator. The billing record concerning a roaming user is forwarded to its home network operator. The Transferred Account Procedure (TAP) is the process that allows a visited network operator to send billing records of roaming subscribers to their respective home network operator. So, the billing records are converted and grouped in files under the TAP format. The transfer of TAP records between the visited and the home mobile networks may be performed directly, or via a Clearinghouse [13].

Charging, Accounting and Billing issues have been investigated in the framework of the MOBIVAS European Project [14, 15]. MOBIVAS produced an integrated system (named CAB) that collects charging information, generates a single itemised user bill and apportions the revenue among the involved players [16]. CAB has been designed and implemented taking into consideration the relevant approaches and recommendations of standardisation groups.

8.2 Involved Players in Service Provision and Charging Processes

Until recently the dominant business model in mobile networks was quite simple, since the only business players supported were the 'User' and the 'Mobile Operator'. Similarly, in data networks the only involved players were the 'Subscriber' and the 'Network Provider'. Nowadays the evolution of the UMTS has converged these different worlds to create an open market, where a large number of independent VASPs are able to offer their services to mobile users through a limited number of network providers. This evolution leads to the necessity for all the aforementioned business entities to be incorporated into a four-tier business model.

However, it is common sense that such a multi-party model will introduce some difficulties in terms of the charging process. The involvement of additional players in the service provision process raises the need that these players also be involved in the control and sharing of the cost of a provided service.

Figure 8.1 depicts the various players involved in service provision that might charge user/subscriber. Namely, these entities are:

- Value Added Service Provider (VASP): deploys services with added value (weather forecast, maps, on line stock exchange, etc.) to its subscribers; this role could be undertaken either by a network operator or by an independent provider.

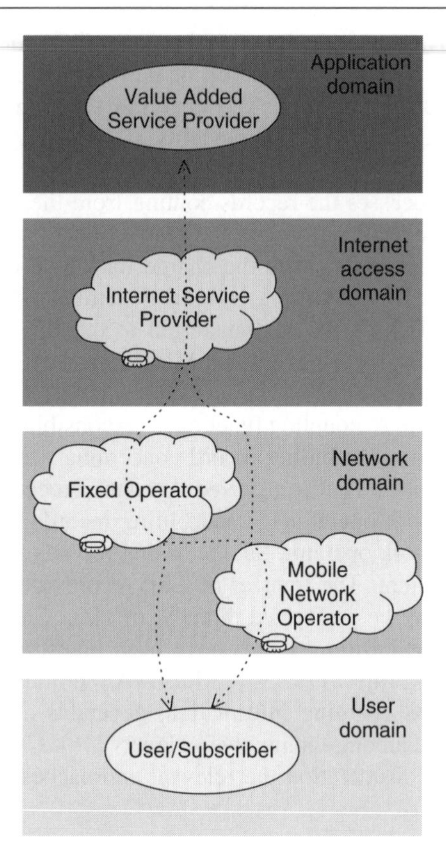

Figure 8.1 Involved business players in service provision

- Internet Service Provider (ISP): provides its subscribers with Internet access. This role could be undertaken either by a network operator or by an independent provider.
- Fixed Operator: provides telecommunications services to stationary (fixed) users but also its infrastructure could be used as a transport service by the other providers (ISPs, service providers).
- Mobile Network Operator: offers bearer and supplementary telecommunications services to mobile users.
- Subscriber: has a contract with an operator and/or provider and can be authenticated and charged by the operator and/or provider for services used. A subscriber could itself have the role of the user, or give to a number of different entities the ability to access services making use of the transport service provided by network operators.

To access a service a user/subscriber first has to come to an agreement with a network provider. This agreement enables the user to access services provided by the specific operator. Until recently, for Internet access, an additional contract between a subscriber and an ISP was required. Furthermore, the use of services provided by independent entities is either free of charge or is tied to an operator or an ISP. The latest technological developments in the telecommunication area have enabled the introduction of innovative business models. These models will enable users to access various networks and services regardless of the owner of the underlying network or the provider of a service. Specifically, a user will have the ability to make use of Internet access and a number of services with added value offered by third business entities, namely the ISP and the VASP, respectively, without the requirement of an additional subscription to them. To support such a scheme the existence of commercial agreements between network operators, ISPs and VASPs is necessary.

The involvement of new players in the service provision process and the introduction of a flexible model, which enables the service deployment without the need for additional subscriptions by the user, mean new charging requirements [17–19].

8.3 Requirements for Charging in Reconfigurable Environments

From the user/subscriber point of view, the main demand is the provision of 'One-Stop Billing' for using UMTS services. More specifically, users would like to receive a single itemised bill for using voice and data services offered by network operators and independent application/service providers. This requirement implies that the network operator would be responsible for collecting charging data from all players and billing the users. Another requirement is that the charging models should be in a form easily understood by the average user/subscriber. Also, the user/subscriber should be constantly aware of the charges to be levied for each chargeable event. This could be achieved through the deployment of a cost control mechanism that allows users to set and modify charging limits, as well as the deployment of near-real-time charging indication functionality.

The mobile operators, on the other hand, require a generic charging architecture that accommodates various charging models (e.g. time-, volume-, QoS-based, flat rate, one-off charge per service) in order to fulfil not only the traditional business models but also innovative ones. In addition, the selection of a specific charging model possibly could be based on user and/or the service profile parameters. Another important requirement, imposed by mobile operators, is the support of both pre-paid (a user has paid in advance an amount of money that is reduced depending on the usage) and post-paid (a user is charged regularly, e.g. every month, for the usage within this period) charging mechanisms. The provision of both mechanisms implies real-time or near-real-time processing of chargeable events, and the failure to meet such requirements obviously means a serious loss of incoming revenue for an operator.

In these networks a number of players have an active role in the service provision process. To bypass a complicated charging architecture, the operators would prefer a layered charging architecture approach [20]. More specifically, the chargeable events should be structured in three layers (i.e. transport, service and content) and processing of the relevant information should be made separately for each layer. Furthermore, it should be possible to apply different charging models at each charging layer.

From the independent application/service provider (ASP) point of view there is increasing demand that each authorised player should be able to apply dynamically the desired pricing policy for use of its services. The independent ASPs should be able to add or modify tariffs for the service and/or content part. This functionality requires the provision of an appropriate open Application Programming Interface (API) for the communication between the network operator (mainly considered responsible for the overall management of the charging and billing process) and the ASPs. In a flexible and open charging model, all the involved players should be able to apply the desired policy.

In terms of sharing the incoming revenue between the players (network operators, ASPs), it is necessary to introduce an automated process that apportions income based on the commercial agreements between them. Until now, only simplified mechanisms have been used for sharing revenues owing to practical considerations. However, in the upcoming UMTS environment, complex mechanisms that make use of information regarding resource allocation and usage could and should be possible to be applied.

Such mechanisms have already been provided even during the era of General Packet Radio Service (GPRS) [13]; however, their functionality was quite limited. The introduction of more complex functional systems such as the IP Multimedia Subsystem (IMS) in UMTS has made apparent that additional information (e.g. SIP session information) can be used in more advanced charging and billing systems [12]. Although, this advanced functionality, which is in the process of being adopted by standardisation bodies, seems ambitious to fulfil the aforementioned requirements, we believe that all requirements are still not met.

8.4 Open Framework for the Support of Advanced Charging and Billing Functions

An open framework will enable the sophisticated and reconfigurable support of charging, accounting and billing procedures as a discrete service, the CAB service. This framework sets a common charging, accounting and billing platform for all the involved players and enables the automatic apportioning of revenues among them ([i.e. Mobile Operators (MOs) and (VASPs)]. Figure 8.2 depicts the business entities involved in the charging process and their relationships in Unified Markup Language (UML) class diagram.

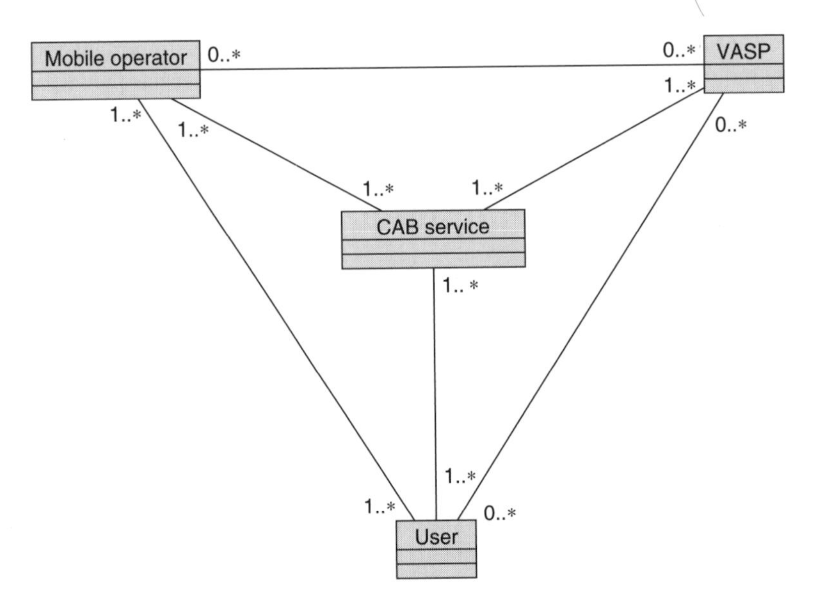

Figure 8.2 Business model for an open framework for charging and billing

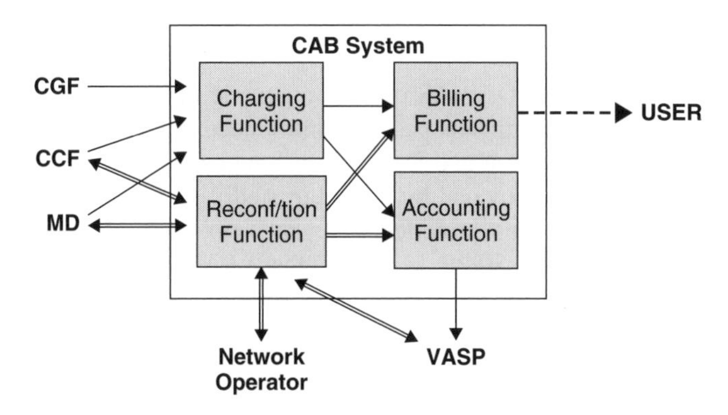

Figure 8.3 Basic functions of the CAB module and its interfaces

The CAB comprises four basic functions, as illustrated in Figure 8.3. Specifically:

- Charging is the function that receives and processes usage data from the network elements. In addition, Charging provides VASPs with an open API in order to be able to submit the usage data generated by their metering entities.
- Reconfigurability is the function that enables independent providers (i.e. VASPs) to define the metering and pricing policy dynamically, since it is able to reconfigure network elements through an Open Network Service.
- Billing is the function that performs usage-based charging models. It is able to apply different pricing policies on a user, service or session basis. Additionally, it provides authorised entities with the ability to add charges for value-added services or for content delivery. Moreover, a layered model is used for the calculation of the charges (transport, service and content layers).
- Accounting is the function that enables an automatic procedure for sharing of revenues. Specifically, the portion that is due to each player is calculated based on the agreement between involved players.

The combination of the functionalities provided by the aforementioned functions enables advanced charging services such as the support of one-stop billing for end-users as well as the separation of charging events based on transport, service and content usage information.

Furthermore, it provides a set of open APIs for the support of charging-related reconfiguration actions (e.g. pricing policy updates) and the deployment of advanced charging services (on-line charging indication, current balance of user billing, on-line provision of information concerning the service profits). Finally, this platform takes advantage of existing network components and their functionality and is in-line with the latest recommendations of the standardisation groups.

Additionally, the CAB service provides advanced charging services through open APIs in order to enable independent VASPs to configure the applied pricing policy dynamically, to retrieve statistical information concerning use of their services, to add content-based charges and to be informed about the current status of their VAS revenues.

8.5 Charging and Billing as an Intelligent Reconfigurable Service

In the framework of the IST MOBIVAS project, a generic management scheme for CAB in reconfigurable mobile environments has been introduced. Specifically, the CAB system has been designed in Specification and Description Language (SDL) and implemented as a discrete service that can be provided by a third trusted party, although it can also be an integral part of a service/transport provider. Relevant work takes place in standardisation groups, such as the WG5 of the 3GPP, as well as the AAAARCH of the IRTF. The CAB service has been designed to re-use the existing (or proposed) functions and network elements involved in the charging process [21]. Furthermore, it places a common basis for the homogeneous development of charging functionality for new services that can be used through a MO or an ISP.

In Figure 8.4, the MO offers the underlying UMTS infrastructure, while the ISP provides users with Internet access. The VASP could be an independent entity that makes use of the underlying network infrastructure as the transport means to provide their Value Added Services (VASs) to mobile users.

The existing and fundamental logical function of a UMTS network related to charging information collection is handled by the CGF [18] and the CCF [12]. These two provide

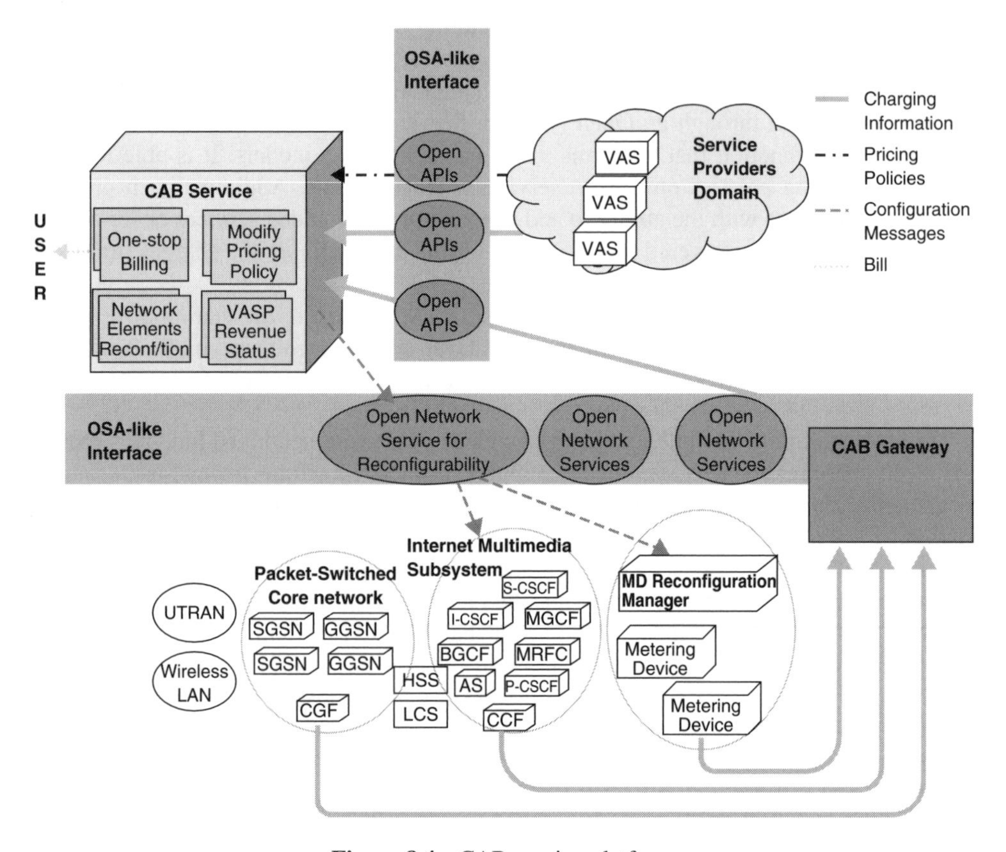

Figure 8.4 CAB service platform

mechanisms to transfer charging information, in the form of CDRs, from the network nodes to the Billing System. The prime network elements of the PS Domain and the IMS have the ability to generate charging records [18]. Additionally, the CGF and the CCF could act as a storage buffer for real-time CDR collection and may perform the consolidation of CDRs and pre-processing of CDR fields. The subscription-related information for home subscribers contained in the Home Subscriber Server (HSS) will be used in order to support an intelligent billing process.

In terms of an ISP, the usage-sensitive billing (i.e. the subscriber is charged according to the network resources utilisation) seems likely to dominate. This model requires the introduction of Metering Devices (MDs) for collecting and processing resources' consumption data [19]. The information collected by MDs is then transferred to an Authentication, Authorisation and Accounting (AAA) server, possibly together with Session Initiation Protocol (SIP) related information, which could play the role of a Billing System. In the proposed architecture shown in Figure 8.4 the MDs are depicted by dashed lines, since their introduction is still under consideration. Such an MD could actually be a Layer 4 SmartSwitch-Router [22] that processes traffic over the IP layer and could collect data about resource consumption in the network (e.g. transmitted volume). An alternative solution could be the deployment of the IPmeter approach [23] where all traffic is monitored by a device that is not involved in routing tasks.

Until now, to access VASs users made contracts with the VASPs and they were charged directly each time they used one of them. The VASP should include an AAA server necessary for the charging process and even a Billing System in order to calculate the bill for each subscriber. In this case the charging and billing process is tailored to the specific service.

To enhance the functionality of the aforementioned existing schemes, we propose to use open APIs among the players that will enable the configuration of network entities for the collection of all required information. For example, the Open Services Access (OSA) functionality [24] can be used between the CAB service and the MOs or the ISPs. A well-defined set of open APIs is also specified between the VASs and the CAB service.

We also propose to introduce MDs located at the edge of the MO in order to be able to process all traffic between VASPs and users. The information collected by the introduced MDs concerns VAS usage and is used to calculate service charges in the case of non-SIP services. To provide flexibility and efficiency, and contribute to avoiding bottlenecks at the edge of the network, the introduced MDs are dynamically configurable, which means that they collect metering data only for flows specified in the configuration policies. The functionality of MDs is under the supervision of the Metering Devices Controller (MDC), which is responsible for the policy-based configuration and reconfiguration of the MDs. The MDC mediates between authorised entities and MDs to securely grant the MD services to the respective parties. Since the reconfiguration of MDs is policy based, communication with the MDs is built upon the common protocols for policy provisioning such as the COPS [25] or COPS-PR [26]. The MDC maps all incoming calls on the aforementioned open interface to the appropriate COPS messages for interacting with MDs.

The CAB Gateway is responsible for collecting all the charging information concerning the transport usage of a MO or an ISP as well as service usage information from the IMS (through SIP session information) and the MDs. The network elements (CGF, CCF) send the charging records to the CAB Gateway using the standard protocols (GTP, Diameter) and interfaces, but also the CAB Gateway provides APIs to other trusted entities (MDs, AAA server) in order to submit their charging records. The proposed APIs implicitly indicate the basic information elements that should be included and standardised in charging-related records exchange. Note that content based charging is possible either through direct communications between VASs and the CAB service, or via appropriate method calls using the extended OSA architecture [24].

In summary, the platform that has been designed and implemented in the framework of the IST MOBIVAS project:

- Enables the sophisticated and reconfigurable support of CAB procedures.
- Sets a CAB platform for all involved players (i.e. MOs, VASPs and ISPs).
- Enables the automatic apportioning of revenues among the players.
- Implements the separation of charging events based on transport, service and content usage information.
- Generates a single itemised bill for each subscriber.
- Provides a set of open APIs for the support of charging-related reconfiguration actions (e.g. pricing policy updates).
- Sets up the deployment of advanced charging services (on-line charging indication, current balance of user billing, on-line provision of information concerning the service profits) offered through open APIs.
- Takes advantage of existing network components and their functionality (e.g. CGF, CCF).
- Is inline with the latest recommendations of the standardisation groups

8.6 The CAB Service Adapted to Each VHE – VASP Category

The Virtual Home Environment (VHE) [27] is a concept for personal service environment portability across network boundaries and between terminals. In the VHE concept, classification of different types of relationships between a VASP and a home or visited network are investigated.

The following types of relationships are identified:

1. The VASP is a Home Environment VASP (HE-VASP) to the home network. This implies that there is a privileged relationship between the VASP and the home network for specific services and users. This may be supported by OSA when the VASP is a third-party service provider. The level of privilege is likely to be higher than for a P-VASP.
2. The VASP is a privileged VASP (P-VASP) to the home or visited network. This implies that there is a privileged relationship between the VASP and the home or visited network for specific services and users. The level of privilege is likely to be lower than for a HE-VASP and may vary between different P-VASPs for the same or similar services.
3. The VASP is a non-privileged VASP (NP-VASP) to the home or visited network. This implies that there is no privileged relationship between the VASP and the home or visited network for the specific service and user. The user may or may not have

subscribed to the service, and has a direct business relationship with the NP-VASP. Moreover, the VASP does not use OSA for the specific user and service.

When the VASP is a HE-VASP, the home network is responsible for the charging and billing process. Therefore, the home network's components collect information not only for the use of its resources but also for the use of its service. The home network is also responsible for accounting for the revenue that is due to the HE-VASP.

In the case that the VASP is a P-VASP, the home network permits the service to use particular capabilities, but the full control of services remains with the P-VASPs. Thus, the service provider undertakes the charging and billing of the users for use of the service.

When the VASP is a NP-VASP, there is no privileged relationship between the home network and service provider. Therefore, the NP-VASP should calculate the charge for its service usage and obtain revenue directly from the user.

The assumptions that have been made during the design of the CAB system streamline with the requirements for flexible Charging models. In the first case, the home network provides charging and billing as a service to third parties (HE-VASP). The CAB Gateway collects charging records from the network's nodes (i.e. CCF, CGF, MD) and sends the required charging information to the CAB system in order to calculate separately the transport and the service part [20]. The home network bills the users and credits the HE-VASPs according to the agreements between them.

The service provision by a NP-VASP is possible over the proposed platform and the CAB calculates only the transport charge for the allocation and use of its network resources. The service bill is issued by the service provider, since the home network is not authorised to do that. In addition, the MD will not be configured to monitor the traffic to and from a NP-VASP, since there is no agreement between network operator and NP-VASP. Thus it cannot provide charging information to the CAB for such services usage.

In the case of a P-VASP, the network operator has a business agreement with the P-VASP, so the MD could monitor the incoming and outgoing traffic related to these services and generate charging records. Although the charging and billing is not responsible for the service charge, these charging records might be used, or alternatively discarded, by the MD. The CAB calculates the transport charge, while the service provider should estimate the service charge. In order for the P-VASP to be able to use flexible billing schemes for service use, metering devices such as a Layer 4 SmartSwitch-Router should be used to monitor the outgoing traffic from the P-VASP network. Another possibility is utilisation of the charging information provided by the network's components. In such a case, a new interface between the CAB and P-VASPs should be defined.

8.7 Location-Based Charging

Location-based charging [28] is a flexible, location-aware, charging model that takes into consideration location information from the network in order to provide subscribers with a personalised and customised charging/billing scheme. In order to enable the location-based charging service, one more module apart from the CAB module [6] is necessary. This is the Location Manager module [29], which is responsible for interacting with the location information's sources of the underlying network infrastructure (e.g. the LCS Server or the Presence Server) for tracking the location of the subscribers.

More specifically, with location-based charging the mobile provider enables subscribers to define their own Home Zones comprising some of their favourite places (e.g. home or work environment). Home Zones are included in the user profile of each subscriber. Location information is then compared against the Home Zones established for users, so that subscribers are charged, for example, the same low 'Home Zone' flat rate when they use their mobiles within Home areas. Outside the Home Zones they will be billed at a higher rate, while different tariffs may be applied in different zones based on the time of day or week.

Therefore, the descriptive document describing each VAS (probably in XML format) provided by VASPs upon VAS registration should include information relevant to the location-based charging offering. This information comprises a statement about whether the VAS is offered through the location-based charging service, along with the applicable pricing policies that are associated with VAS usage from within Home Zones or out of them. In addition, the location-based charging service will also be applicable on a group basis, which may be desirable, for example, for business groups. As subscribers move or cross predefined charging zones, or the applicable rate changes, it is required that individuals and group subscribers be notified of their location-based charging service offering indicating either 'in' or 'out' of zone, while a session is in progress or prior to its initiation.

The user preference related to the location-based charging activation/deactivation is included in the general user preferences part of the user profile, indicating consequently the user's general preferred policy relevant to this special service offering. So, in the case when the location-based charging feature has been enabled by a subscriber into his user profile, the CAB system will apply this advanced charging scheme to all activated sessions by the user that concern the use of the VAS. In such a case only the VAS explicitly associated with a different pricing policy (as included in the XML document describing the VAS) are excluded from the location-based charging offering. However, despite the user's acceptance of this feature, as implicitly expressed in their user profiles, prior to the location-based charging activation, an appropriate notification is sent to the involved subscriber asking his confirmation to enable location-based charging and tracking his location. Furthermore, location-based charging may be invoked upon initial registration with a VAS, while subscribers will be able to activate/deactivate this feature by using specific codes in their services subscription profiles kept in their user profile.

The location-based charging service is offered to the VAS and other authorised entities through the open API provided by the Location Manager. Hence, by invoking the ActivateLocationBasedChargingReq method, authorised entities are given the ability to request the Location Manager to activate this advanced charging feature onto a specified user, or session (that is per specific user and application). In the case when the location-based charging activation is requested on a per user basis, it results in the activation of this functionality for all the sessions the user is involved in, relevant to the use of the VAS. In such a case only the VAS explicitly associated with a different pricing policy (as included in the XML document describing the VAS) are excluded from the location-based charging offering.

The location-based charging offering can be disabled at any time by invoking the CancelLocationBasedCharging method that is offered to authorised entities by the Location Manager. Each cancellation requests should be sent by the same authorised entity that requested the activation of the location-based charging service and should explicitly specify the corresponding ActivateLocationBasedCharging request that it cancels.

To apply effective location-based charging, taking into account that a user may cross back and forth between zones multiple times during a session, and that a session may terminate in the zone it was originated from, or in a different zone, multiple interactions of the Location Manager with the Location Server [29] of the underlying network infrastructure may be required during a single session.

More specifically, interactions required to take place, in the case that location-based charging activation has been requested to the Location Manager, include:

1. Upon invocation by an authorised entity (e.g. an internal module, or a VAS) of the ActivateLocationBasedChargingReq method for a user or a specific session, the Location Manager first examines its caches to find out whether the location-based charging service has already been activated for the specified user. In the case that this is not so, an appropriate notification is sent to the specified user (by accessing the Push or messaging server of the underlying network through the open interface provided by the Basic Network Management Layer) asking his confirmation to enable location-based charging and tracking his location.
2. On positive response the Location Manager retrieves the current location of the user, by accessing the LCS Server of the mobile network (through the open interface provided by the Basic Network Management Layer), in order to identify the originating Home Zone of the user.
3. Then, the Location Manager invokes to the LCS server of the network a mobile terminated and deferred location request [29] specifying as response event the crossing of the applicable Home Zones by the targeted User Equipment (UE). This request includes all the parameters defined in refs [24] and [31] with respect to the user privacy settings. This information will enable the Location Manager to track the charging zones that the user visits while location-based charging is enabled.
4. Following the notification by the LCS Server (through the Basic Network Management Layer) that the requested UE has entered a new Home Zone or moved out of it, the Location Manager triggers the CAB service (calling the DirectCollectInfoReq method) to collect information regarding the specified session or user. Subsequently, the CAB service instructs the MD (calling the DirectCollectMeteringDataReq method through the open interface provided by the MD Reconfiguration Manager) to close and send to the CAB Gateway the charging records that are associated with the specified session or user and concern the usage of VAS within the previous charging zone. The Location Manager incorporates into the aforementioned request the geographical area (Home Zone or not) that the requesting user currently exited, along with the cause that triggered this method call (e.g. 'In Home Zone Billing' or 'Out Home Zone Billing') so that the MD includes this location-sensitive information into the records that it generates. The MD Reconfiguration Manager maps this incoming method call to the appropriate COPS-PR messages to interact with the MDs and stimulate them to close and send to the CAB Gateway the requested charging information. The CAB service, in turn, uses these records to charge the subscriber for the service usage in the associated Home Zone, based on the applicable rates for this zone.
5. By receiving the CAB service response, the Location Manager at first interacts with the Service DB to retrieve the applicable pricing policies of the active VAS on the

current Home Zone of the user and then notifies the subscriber about the new applicable pricing rates (by accessing the Push or messaging server of the underlying network).

6. Steps 4 and 5 are repeated until the authorised entity that activated the location-based charging feature in step 1 invokes the CancelLocationBasedChargingReq method provided by the Location Manager (specifying the ActivateLocationBasedCharging request that it cancels). Upon receipt of such a cancellation request the Location Manager interacts with the LCS server of the network (through the associated open API) to cancel the Mobile Terminated Deferred and Periodic Location Request invoked in step 3.

The Message Sequence Chart (MSC) in Figure 8.5 (derived from simulations that have been performed following the SDL specification) depicts the required interactions among components of the proposed architecture to accomplish location-based charging activation on a per user basis. In order for this procedure to work properly, it is required that the mechanism specified in [29], concerning the Mobile Terminated Deferred Location Requests, be enhanced to accomplish the new type of event (event = UE crosses Home Zone) described in step 3 above. Additionally, for privacy reasons, a notification message indicating the type of location request (e.g. current location) and the identity of the requesting LCS client should be sent to the targeting UE by the LCS server on a per location retrieval request basis.

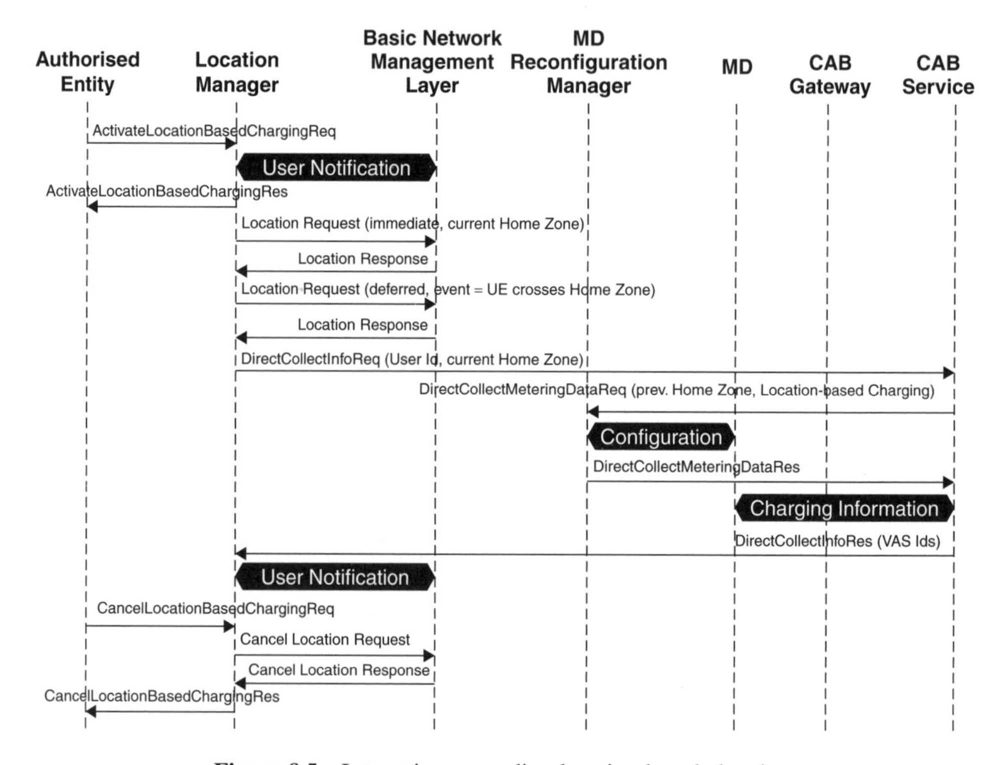

Figure 8.5 Interactions regarding location-based charging

References

[1] N. Houssos, V. Gazis, S. Panagiotakis, S. Gessler, A. Schuelke and S. Quesnel, 'Value added service management in 3G networks', 8th IEEE/IFIP Network Operations and Management Symposium (NOMS 2002), 15–19 April, Florence, Italy, pp. 529–545, 2002.
[2] A. Alonistioti, N. Houssos, S. Panagiotakis, M. Koutsopoulou and V. Gazis, 'Intelligent architectures enabling flexible service provision and adaptability', Wireless Design Conference (WDC 2002), 15–17 May, London, 2002.
[3] M. Koutsopoulou, C. Farmakis and E. Gazis, 'Subscription management and charging for value added services in UMTS networks', *Proceedings of the IEEE Semiannual Vehicular Technology Conference, Spring VTC2001*, **May**, Rhodes, Greece, 2001.
[4] SDR Forum, *http://www.sdrforum.org/*
[5] M. Koutsopoulou, E. Gazis and A. Kaloxylos, 'A novel billing scheme for UMTS networks', *Proceedings of the International Symposium on 3rd Generation Infrastructure and Services (3GIS 2001)*, **July**, Athens, Greece, 2001.
[6] S. Panagiotakis, M. Koutsopoulou, A. Alonistioti and A. Kaloxylos, 'Generic framework for the provision of efficient location-based charging over future mobile communication networks', PIMRC, September, Lisbon, Portugal, 2002.
[7] S. Panagiotakis, M. Koutsopoulou and A. Alonistioti, 'Advanced location information management scheme for supporting flexible service provisioning in reconfigurable mobile networks', IST Mobile Communication Summit, June, Thessaloniki, Greece, 2002.
[8] 3rd Generation Partnership Project (3GPP), *http://www.3gpp.org/*
[9] UMTS Forum, *http://www.umts-forum.org/*
[10] Authorization, Authentication and Accounting ARCHitecture Research Group, *http://iridal.phys.uu.nl/~aaaarch/*
[11] The Internet Engineering Task Force, *http://www.ietf.org*
[12] 3GPP TR 23.815 version 0.1.0 (2001-10), 'Charging implications of IMS architecture'. Available from URL *http://www.3gpp.org/ftp/Specs/Latest-drafts/*
[13] 3GPP TR 22.924 version 3.1.1 (2000-0), 'Charging and accounting mechanisms'.
[14] Downloadable mobile value-added services through software radio and switching integrated platforms (MOBIVAS). Available from URL: *mobivas.cnl.di.uoa.gr*
[15] N. Alonistioti, E. Gazis, M. Koutsopoulou and S. Panagiotakis, 'An application platform for downloadable VASs provision to mobile users', *Proceedings of the IST Mobile Communications Summit 2000*, **September–October**, Galway, Ireland, 2000.
[16] M. Koutsopoulou, N. Alonistioti, E. Gazis and A. Kaloxylos, 'Adaptive charging accounting and billing system for the support of advanced business models for VAS provision in 3G systems', invited paper at the PIMRC 2001, September–October, San Diego, 2001.
[17] Report 14 from the UMTS Forum, 'Support of third generation services using UMTS in a converging network environment', *http://www.umts-forum.org/reports/report14.pdf*, 2002.
[18] 3GPP TS 32.200 version 4.0.0 (2001-09), 'Charging management; charging principles'. Available from URL *http://www.3gpp.org/ftp/Specs/2001-12/Rel-4/32_series/*
[19] G. Carle, S. Zander and T. Zseby, 'Policy-based accounting'. Available from URL *http://search.ietf.org/internet-drafts/draft-irtf-aaaarch-pol-acct-04.txt*, IETF work in progress, March, 2001.
[20] V. Gazis, M. Koutsopoulou, C. Farmakis and A. Kaloxylos, 'A flexible charging and billing approach for the emerging UMTS network operator role', *Proceedings of the SCS ATS 2001 Conference*, **22–26 April**, Seattle, 2001.
[21] M. Koutsopoulou, A. Kaloxylos and A. Alonistioti, 'Charging, accounting and billing as a sophisticated and reconfigurable discrete service for next generation mobile networks', *Proceedings of the IEEE Semiannual Vehicular Technology Conference Fall VTC2002*, **24–28 September**, Vancouver, BC, Canada, 2002.
[22] 'Layer 4 switching: An overview'. Available from URL *http://www.enterasys.com/technologies/smart-switch-router*
[23] 'Ipmeter White Paper'. Available from URL *http://www.ipmeter.com/doc/whitepaper/t1.htm*
[24] 3GPP TS 29.198: 'Open service access (OSA); application programming interface (API); Part 1–12'. Available from URL *http://www.3gpp.org/ftp/Specs/2001-12/Rel-4/29_series/*

[25] D. Durham, J. Boyle, R. Cohen, S. Herzog, R. Rajan and A. Sastry, RFC 2748, 'Common open policy service protocol (COPS)'. Available from URL *http://www.ietf.org/rfc/rfc2748.txt*

[26] K. Chan, J. Seligson, D. Durham, S. Gai, K. McCloghrie, S. Herzog, F. Reichmeyer, R. Yavatkar and A. Smith, RFC 3084, 'COPS usage for policy provisioning (COPS-PR)'. Available from URL *http://www.ietf.org/rfc/rfc3084.txt*, IETF, March, 2001.

[27] 3G TS 23.127: 'Virtual home environment/open service access'. Available from URL *http://www.3gpp.org/ftp/Specs/2002-09/Rel-5/23_series/*

[28] 3GPP TS 22.071: 'Location services (LCS); service description, Stage 1'. Available

[29] S. Panagiotakis and A. Alonistioti, 'Intelligent service mediation for supporting advanced location and mobility aware service provisioning in reconfigurable mobile networks', *IEEE Wireless Communications Magazine*, **October**, 2002.

[30] 3GPP TS 23.271: 'Functional stage 2 description of LCS'. Available from URL *http://www.3gpp.org/ftp/Specs//2002-09/Rel-5/23_series/*

[31] 3GPP TR 23.871: 'Enhanced support for user privacy in location services'. Available from URL *http://www.3gpp.org/ftp/Specs//2002-09/Rel-5/23_series/*

Part IV

Profile and Radio Resource Management

9

Communication Profiles

Eiman Mohyeldin and Egon Schulz
Siemens AG

Michael Fahrmair and Christian Salzmann
Technical University of Munich

9.1 Introduction

Reconfigurable radio systems are expected to play a critical role in the area of mobile/wireless communications by increasing flexibility, reducing deployment as well as operation and maintenance costs, creating new business opportunities and employment, facilitating enhancement and personalisation, etc. [1].

In order to support fast reconfiguration or intersystem handover and the related mode negotiation procedure among heterogeneous networks, the co-operation between terminal and network should be highlighted. One important key of this co-operation is the information (user/terminal/service/network profile) needed by the mode negotiation and radio resource management for intersystem handover or any other adaptations. This information or profile data describes the capability of terminals, service, network and user preferences and presents an important decisive element in adaptation, such as intersystem handover decisions.

Two key features should be considered for mode switching or intersystem handover:

1. The database for system deployment supporting intersystem handover between coupled Radio Access Technology (RAT) sub-networks. The database can be organised as a table for storing network topology,
2. The management and storage of terminal (user/service) profile data in Radio Access Networks (RANs) and Cove Networks (CNs) depending on network interworking (e.g. tight vs. very tight coupling).

In this chapter the capabilities and status information of the relevant actors in a mobile scenario, i.e. the terminal, user, service and network, are used to introduce the concept of 'communication profiles'. After tackling all the communication profiles, similar features can be clustered together to form a classmark. Classmarks are shortcuts of the profile data.

Software Defined Radio: Architectures, Systems and Functions. Edited by M. Dillinger, K. Madani and N. Alonistioti
 ISBN: 0-470-85164-3

In systems with groups of a large number of similar devices, classmarks can describe the terminal efficiently and thus reduce the signalling overhead. Since classmarks should be used not only for reducing capability signalling network bandwidth but also for service trading decisions, the criteria for clustering profile data into classmarks are meaningful. We have introduced a strategy of abstracting the capabilities and status information of reconfigurable terminals into dynamic classmarks created during the course of a session. The concept of stage-based classmarks or context-based classmarks efficiently reduces the signalling overhead between the network and terminal.

In further optimisations, efficient methods to compress and code the profile structure are described. Also, to model the requirements for profile data elements in an exact way further attributes can be used, such as access frequency to features, how frequently the feature changes, and predication values of the profile data elements. This concept is generalised by adding meta attributes.

9.2 Communication Profiles

Terminal and network reconfiguration introduces additional functional perspectives to the classical management, control and user planes, therefore a new signalling system to control and manage the reconfigurations process is required [2]. Furthermore, the decision unit that controls the reconfiguration process needs to know information about the current context of the whole system (i.e. information about the main system entities such as user, terminal and network). Moreover, before the reconfiguration process takes place, the capabilities of the terminal must be known and checked regarding the target mode of operation (e.g. RAT). In addition, for mode negotiations or any other terminal adaptation the service, network and user profiles need to be retrieved before the decision process starts. Figure 9.1 illustrates the structure of the communication profiles.

In this section of the chapter we introduce the different entities of the communication profile, i.e. terminal, network, service and user profiles. In Section 9.3, the architecture

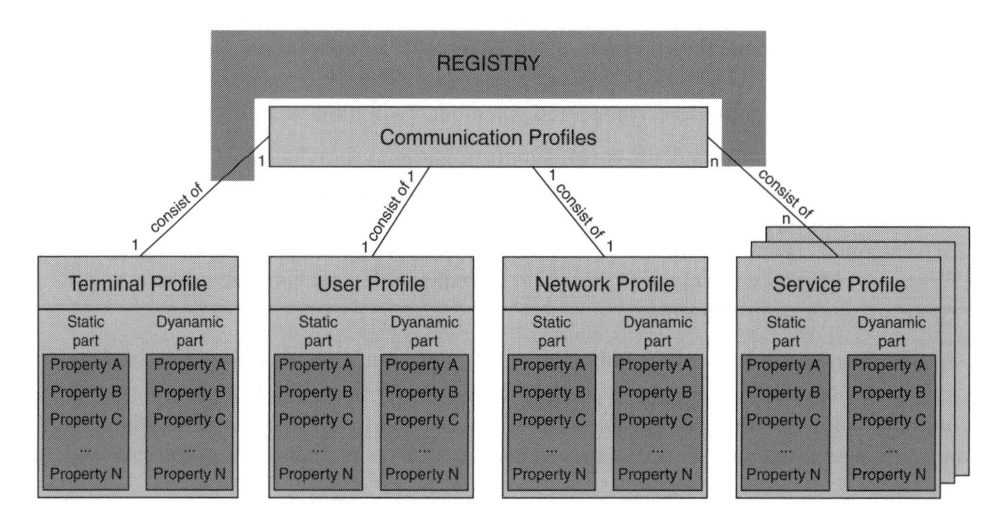

Figure 9.1 Communication profiles

for efficiently storing profile data information is described. A well-known tree structure, which uses XML to model and transfer the profile data between terminal and network, is presented. In Section 9.4, different management schemes for the communication profiles and signalling methods, such as efficient methods for compression and coding, are described. Also, dynamic classmarks in which the capabilities and status information of the reconfigurable terminal are abstracted into classmarks created during the course of the session so that a fast decision would be taken in the process of reconfiguration are introduced.

9.2.1 Terminal Profile

In the terminal profile attributes/features describing the user interface capabilities that are targeted for selecting a terminal suitable for certain applications are considered; see Section 9.2.1.2. Profiles are composed of static features that change occasionally or stay fixed for a certain period of time, such as the display type or screen size, and dynamic attributes that may change frequently or rapidly, such as current hardware status or the terminal radio capabilities and status. Changing the terminal radio capabilities, i.e. reconfiguring the terminal between different radio interfaces [Global System for Mobile Communications–Universal Mobile Telecommunications System (GSM–UMTS)] implies a reconfiguration at different levels: baseband, radio frequency (RF), protocols [4]. For example, in the baseband reconfiguration the functionality, behaviour and performance of the baseband transceiver chain will be redefined. It is envisaged that baseband reconfigurations may involve a change in modulation–demodulation, Forward Error Correction (FEC) encode–decode, and so on. This would be necessary if the terminal is to be re-configured to operate at a different standard to that currently in use. For example, if the terminal is to be reconfigured from GSM standard to UMTS-TDD, then the baseband sub-system, RF and protocols will need to be reconfigured so that the digital signal processing block of the physical layer would run the UMTS-TDD standard as supposed by the incumbent GSM. Reconfiguring the terminal from one standard to another implies that an important set of actions at the hardware level as well as at the software level will take place. In general these actions can be associated with the following principles:

- Adjusting the transceiver bandwidth to the bandwidth of the modulated signal to be handled.
- Setting up the synthesisers to provide the local oscillator frequencies that can result in a proper connection between the baseband and RF spectral positions.
- Adjusting the gain of the transmitter and receiver chain to adapt the terminal to the dynamic range of the particular standard.
- Controlling the performance of the transceiver according to the specification imposed by the particular standard.

9.2.1.1 Terminal Capabilities and Features

Terminal capabilities and resources are crucial for comprehensive decisions in terminal reconfiguration [5]. Naturally, only modes can be configured that are compliant with the capabilities of the radio transceiver(s) and the resources available at the terminal. But services depend on the capabilities and resources of the terminal as well. In what follows examples of the static and dynamic parameters of reconfigurable terminals are presented:

9.2.1.1.1 Static Parameters. Examples of the terminal static profile are:

- *General Terminal Capabilities:* Examples include the type of Central Processing Unit (CPU) (string), the performance of the CPU (benchmarks results), total size of memory space (Mbytes), total capacity of battery (mAh), display size (width in pixels × height in pixels) and colour depth of the display (bits per pixel).
- *Capabilities of the Transceiver(s):* Examples of transceiver-related terminal capabilities are the number of transceivers, supported RATs, terminal capabilities for specific RATs, terminal capabilities for specific RATs and maximum electric power consumption per mode. Depending on the degree of reconfigurability of the terminal, some capabilities may be hardware dependent, whereas others may be determined by software, and thus cannot be considered as terminal-specific in a strict sense.

9.2.1.1.2 Dynamic Parameters. Parameters that depend on the current usage pattern as available terminal resources or parameters characterising the usage of transceiver(s) are of a dynamic nature. They are being monitored by resource management entities such as a memory management unit or a power management unit. Examples of the dynamic profile are:

- *Available Terminal Resources:* Available (unused) terminal resources determine a particular service mode selection or terminal reconfiguration to a particular RAT; examples of resources are available processing power, free memory space (Mbytes) or available battery charge (mAh).
- *Transceiver Usage:* Information on the current mode of operation and usage of the transceivers is monitored by the terminal entity responsible for reconfiguration management issues. Combined with information on current service usage and power consumption, statistical data may be obtained that yield typical usage patterns and associated power requirements, such as the current mode of operation that describes the number of active transceivers and their configuration (RAT, frequency range, etc.) and current electric power consumption that is monitored to obtain statistical information on power consumption patterns for particular RATs.

9.2.2 Service Profile

One of the key points in the reconfiguration process is the service negotiation, in order to ensure service provision in the targeted system, since otherwise it would be a waste of time and resources to reconfigure to a system that is not capable of providing the expected user's services. Services are described by attributes that define service characteristics as they apply at a given reference point where the user accesses the service. In the service profile different aspects, which determine the final service provision to the user, are defined, such as service, service classification, service mapping, service negotiation and service allocation [3]. Service architecture provides standardisation of the service capabilities [8]. Figure 9.2 illustrates the service architecture. A number of bearer services

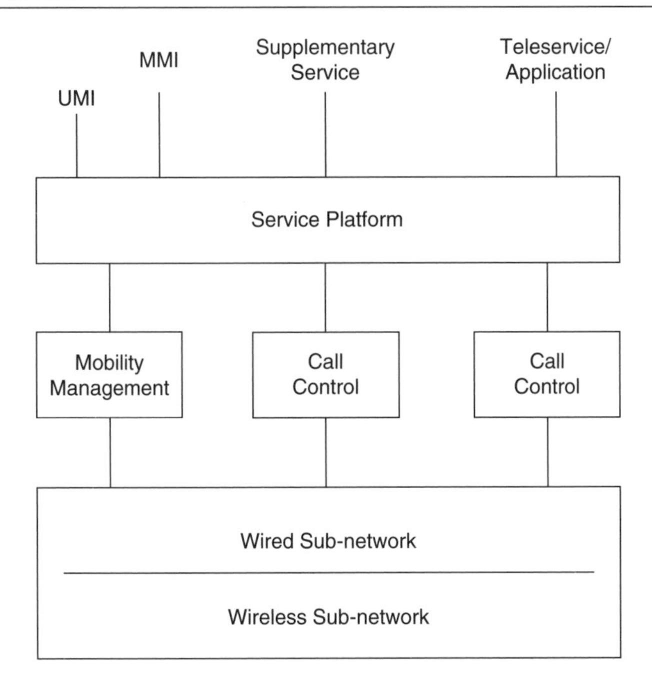

Figure 9.2 Service architecture

are provided that can differ in flexibility and offer different capabilities. Bearer services may be characterised by parameters such as 'throughput', 'delay tolerance', 'maximum bit error rate', 'symmetry', etc. These bearers enable information to be transferred by the appropriate provision of teleservices, and end-user applications generally, via sub-networks, which typically provide different specified qualities of service.

The assignment and release of bearer services is provided by the bearer control functions (see Figure 9.2). Provision should be made for several bearers to be associated with a call and for bearers to be added to a call and/or to be released from a call following call establishment. The bearer services is seen as independent of radio environments, radio interface technology and fixed wire transmission systems. Adaptation/interworking functions are required in order to take into account the differences between the bearers used for the provision of a teleservice/application in the fixed network and the mobile radio bearer services. Such adaptation functions are required that take into account the discontinuous and/or asymmetrical nature of most teleservices/applications.

The service platform provides interfaces (to serving networks and home environments) appropriate to the support, creation and control of supplementary services, teleservices and user applications. The service platform also provides interfaces enabling subscribers to control supplementary services, teleservices and user applications. Supplementary service provision and control are independent of radio operating environment, radio interface technology and fixed wire transmission systems.

As far as possible, the service platform is required to enable new supplementary services, teleservices and/or end-user applications to be supported at minimum cost, with minimum disruption of service and within the shortest possible time.

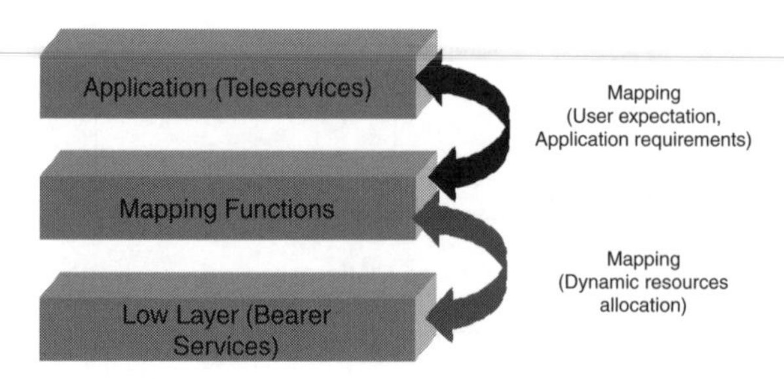

Figure 9.3 Service mapping

Services are classified into different groups according to the Quality of Service (QoS) classes they need; so far four classes have been identified [10, 13]:

- Conversational
- Streaming
- Interactive
- Background classes (services)

The mapping of services is the translation of higher-layer services (teleservices) into different levels of abstraction (bearer services), which implies that a mapping exists between parameters at different levels [11] (see Figure 9.3). Mappings are not usually one-to-one between parameters, but may be one-to-many, many-to-one or many-to-many. Although the mapping may be complex, the process of translation consists of simple arithmetic operations over a limited set of variables. The mapping is referred to as QoS parameters mapping [9, 14].

Service negotiation is mainly concerned with higher-layer features that applications may ask for in future services once reconfigured to the target system. The dynamic features of the service profile are the real-time status of the resources availability in the current access network that is subject to the cell load, whereas the static features of the service profile define the application offered that is subject to QoS.

9.2.2.1 Service Profile Parameters

Services such as Internet Protocol (IP) telephony, audio or video streaming, web browsing, etc. have their own specific QoS requirements. Usually, a service supports various different modes (e.g. different resolutions for video streaming) among which the user can choose according to his preferences. Each service mode requires specific capabilities and resources on the terminal side as well as in the network. The requirements of all active services must be considered in mode-switching decisions such that an optimal mode can be selected. In what follows, examples of the static and dynamic parameters of the service profile are presented:

9.2.2.1.1 Static Parameters. Examples of the static parameters of the service profile are:

- *Subscribed Services:* These are services that the user has subscribed to; they are listed in the service profile. These services are characterised by QoS parameters specifying the requirements of a service in terms of network and terminal capabilities and resources.

- *Service Preferences:* User preferences with relation to services are also stored in the Service Profile. User Service Preferences usually are specified per service or per service type. Different types of service preferences have been identified, such as Qualified services, which are treated in a special way if resources are scarce or some services have to be downgraded or dropped, Preferred service modes, Preferred service providers, Preferred RATs and Service degradation paths that contains a list of valid service modes ordered by decreasing QoS/resource requirements.

9.2.2.1.2 Dynamic Parameters. Dynamic service parameters are obtained as a result of monitoring procedures that may involve measurements, or via negotiation processes between terminal and network. Some examples are:

- *Supported/Unsupported Services:* The list of supported/unsupported services contains services that are supported (or not)/available in a particular network. This list can be obtained from a Service Management function residing in the network.
- *Active/Suspended Services:* The list of active/suspended service, currently active or not active. This list is kept locally in the terminal and updated whenever a change in the service status has taken place.
- *Delivered Service Quality:* The delivered service quality is obtained from a monitoring function, being either part of the application or of a separate entity responsible for QoS management.

9.2.3 Network Profile

In the network profile, different underlying networks need to be handled. The air interface defines the profile for every used protocol that describes the characteristics needed for accessing the network, such as QoS. The static features of the network profile are stored in a network database that defines the bearer services in each network. This database is an essential source for network capability negotiations and bearer services QoS assessment. The network bearer services (the static part of the network profile) provide the capability for information transfer between network entities and involve only low-layer functions (capabilities/attributes). The user may choose any set of high-layer protocols for his/her communication and the network does not ascertain compatibility at these layers between users.

In the general case a communication link between network entities provides a general service for information transport. The communication link may span over different networks such as the Internet, intranets, Local Area Networks (LANs) and Asynchronous Transfer Mode (ATM)-based transportation networks, having network-specific means for bearer control. Each network contributes to the end-to-end QoS perceived by the end-user. The requirements on bearer services are classified into two groups:

1. Requirements on information transfer, which characterise the network's transfer capabilities for transferring user data between two or more network entities.
2. Information quality characteristics, which describe the quality of the user information transferred between two or more network entities (e.g. the terminal and network).

It is possible to negotiate/renegotiate the characteristics of a bearer service at session/connection establishment and during an ongoing session/connection.

9.2.3.1 Network Profile Parameters

The network infrastructure in a network operator's domain provides static as well as dynamic information on its capabilities and available resources. In what follows some examples of the network profile are presented.

9.2.3.1.1 Static Parameters. The static network profile contains descriptions of available air interface modes and QoS bearer services. Examples of the static network profile are:

- *Available Modes:* The list of available modes provides information on the RATs supported by the network operator (GSM, UMTS/FDD, UMTS/TDD, Wireless LAN, HIPERLAN/2, etc.) and associated frequency ranges plus additional parameters. This information may be used by the terminal entity performing the mode monitoring in order to speed up the mode detection and identification process as scanning can be focused to the modes and frequency ranges specified in the network profiles of the local operators.
- *QoS Bearer Services:* The network profile contains a list of QoS bearer services available in a particular network; each QoS bearer service is characterised by a set of attributes. UMTS defines four traffic classes, each one with its own QoS bearer attributes.

9.2.3.1.2 Dynamic Parameters. The dynamic features of the network profile provide the current status of the cell load and other related informations, such as the current QoS of offered bearer services. Examples are:

- *Available Modes:* Information on available modes may also be acquired dynamically through a mode-monitoring procedure if not provided statically by the network operator. In this case, a wide frequency range has to be scanned to detect all available modes. Information on the quality of available modes is obtained via radio access stratum parameters (signal level, bit error rate, block error rate, etc.). These parameters are monitored by the terminal (for the down-link), and by the network (for the up-link). The combined information is used to determine if a given mode satisfies the requirements of the user and the active services in terms quality and reliability.
- *Load Information:* Monitoring of cell loads for local and neighbouring cells for different RATs yields information required for determining if a handover to a target cell can be performed. In addition, negotiations have to be carried out to check the actual availability of QoS bearers with specific attributes that are required to support the active services in the target cell. Cell load information and information on availability of QoS bearers are important criteria for the decision-making process.

9.2.4 User Profile

A user profile is separated into a static part, defining the invariant attributes like ID or address, and a dynamic part, which contains attributes that are likely to change during usage.

User preferences can be clustered by user classmarks, which are shortcuts of user profile data. This could include, for example, billing and application type preferences, as well as the amount of information and interaction possibilities. In systems with large groups of similar users, classmarks can describe user preferences without transmitting or even knowing the exact values of each profile element, thus saving bandwidth and increasing the accuracy of adaptation to specific users before gathering or configuring their preferences data. Users are classified into classmarks (i.e. basic user, advanced user, etc.) based on their preferences and certain usage characteristics. User classification can be used for decisions that involve a wide range of profile data but do not need exact values. Usually this concerns decisions about the user interface (e.g. amount of information and control possibilities displayed), application or mode usage or notification (which available services to use to announce the chosen QoS). This user classification will be helpful in designing the Graphical User Interface (GUI) for different purposes.

With technology advancing and usage patterns evolving, it is unlikely that the clustering criteria can be fixed on the basis of exact values of profile data. The criteria values that can be used to segment users into classes are more likely to change dynamically. However, since classmarks should be used not only for reducing the signalling load, but also for service trading decisions and initial user personalisation configuration, the criteria for segmentation should be both dynamic and meaningful. This can be achieved by defining abstract clustering criteria that can be dynamically populated with concrete profile data values during system run-time to reflect changes in user behaviour or allow for provider-specific configuration. The classification can influence the complexity of the configuration process. When a user classifies himself as an IT expert, there will be the possibility to configure the QoS settings for this user in a more detailed way.

9.2.4.1 User Profile Parameters

In what follows we present examples of a set of attributes for the static and the dynamic part of the profile.

9.2.4.1.1 Static Parameters. Examples of the user static profiles are:

- *ID:* This represents the unique ID of the user.
- *Cost-Related Preferences:* User preferences related to cost are of particular importance, because normally a user does not want to pay more than a certain maximum amount of money for a particular service or a combination of services. Cost-related user preferences may depend on the context, in particular on the role of the user. For instance, a user may be willing to ignore cost if on a business trip where high service quality is important, whereas in a leisure time situation, minimisation of cost may be desired.
- *Reliability-Related Preferences:* Reliability is an aspect of service quality that is not related to a particular service. For mobile communication, the availability of a connection or, put negatively, the risk of losing a connection, is an important issue. Another issue is the downtime of a terminal due to a handover process that leads to the disruption of services. Whereas the maximum duration of service disruption tolerated by a service may be service specific, the minimisation of service disruption by selecting appropriate target modes and reconfiguration strategies may be a general aspect of handover decision-making. Reliability-related user preferences might depend on the context – in particular on the role.

- *Provider-Related Preferences:* Provider Related preferences represent a convenient means to take into account good or bad service experiences, or low-priced vs. high-priced service offers when performing a mode-switching decision. Provider-related user preferences depend on the user's current context.
- *RAT Preferences:* A user prefers or exclude modes for quality or cost reasons. RAT-related user preferences depend on the user's current context. Two types of RAT preference are identified, namely the preferred modes and excluded modes.
- *Terminal-Related Preferences:* Terminal-related preferences request a preferred behaviour or configuration of the terminal that should be taken into account when mode switching is performed. Terminal-related user preferences could be, for example, to take into account the power consumption of different modes and try to minimise it (e.g. the user could specify or set a rule such that if the battery drops below 10%, then calls are not accepted).
- *User Notification Preferences:* In most situations, users prefer automatic decision-making and execution. In some situations they would like to be informed about what is going on or even be asked for confirmation if a particular decision is to be executed. This may be of particular interest if this involves an increase in costs or dropping active services. User notification preferences may depend on the current context. Most of the user notification options can be set to one of three different values *notify*, *confirm* and *silent*.

9.2.4.1.2 Dynamic Parameters. Dynamic parameters related to the user's context are obtained by automatic context determination or by manually setting the context. In what follows, examples of the user dynamic profiles are described:

- *Automatic Context Determination:* The automatic determination of context information is very convenient for the user, but requires additional devices and information bases that must be available either locally on the terminal or in the network. Different automatic contexts can be deduced, such as temporal, location, situation and role.
- *Manual Context Determination:* Manual context determination requires active participation by the user, but does not require any additional devices and information bases. This may be an alternative to automatic context determination in cases where the latter is error-prone or in cases where the user is deviating from the rules upon which context determination is based. Whereas it is very unlikely that the user will ever set the temporal context manually, a manual procedure is reasonable for the other three categories mentioned in the automatic context.
- *Provider:* The provider of the user for billing purposes.
- *Language:* The user's language(s) preferences.
- *Privacy Mode:* Indicates whether the user his/her their identity kept secret.
- *Accept Download:* Indicates whether the user will accept downloadable software.
- *Known/Already Visited Locations:* Is the user new to a given location?
- *Movement Speed:* How fast is the user moving (through the cells)
- *Appointments/Time Schedule:* Used to extrapolate dynamic requirements
- *Health Condition:* Heartbeat, breathing frequency, etc. might be used to monitor critical health situations, detect high priority emergency calls, etc.

9.2.4.2 Adaptive User Profile

In the previous section the user preferences were assumed to be more or less static over a longer time period. This is because user preferences that adapt automatically to the context of a user are highly complex topics. However, using communication profiles and a software defined system architecture for reconfiguration, a quite simple yet powerful mechanism can be defined that allows for both automatically changing user preferences based on the actual context of a user and adapting the terminal, network or applications features in use according to these changed preferences.

User preferences describe a set of requirements that is valid for a given user. As such they are part of the user profile that contains any information that describes a user and that is relevant to the interaction between the user and other entities in the communication system, such as the network, the terminal and application services.

Examples of user preferences could be very simple, such as the volume and sound of the ring tone. Other preferences might be more sophisticated descriptions of preferred cost and quality of different application services (see Section 9.2.4.1) or even basic service needs, such as the information that a specific user is on a journey in a foreign city and therefore is quite likely to look for a hotel to stay overnight.

These descriptions are embedded in the general user profile that contains also descriptions about the user that are not part of the preferences, for example the user's name, address and current location. However, this information that is not part of the preferences can be used to automatically derive user preferences. The address and current location are necessary to advise that the user is away from home and therefore might be in need of a hotel booking service. His location in a building together with some of his colleagues during working hours might indicate he is participating in a conference and therefore should not be disturbed by incoming calls, etc.

All the information in the user profile models the context of the user's current situation; that is, user preferences can be context-dependent. For example:

- Time-dependent: time-of-day, day-of-week.
- Role-dependent: on business or in leisure time.
- Situation-dependent: driving a car, on the train, waiting for a plane, attending a meeting.
- Location-dependent: in a particular city, in a megalopolis, in the countryside, in one's own country, in a foreign country.

The context of a user is monitored on the terminal or in the network and may be used to restrict user preferences to the specified context. The same type of user preferences may be specified for multiple contexts.

This is also true for the other profile types, such as network, terminal and service profiles. If the terminal battery is low and the user had scheduled a download with a high priority he is likely to prefer a faster but more expensive access network.

Context is, by definition, information about objects that are relevant for interactions between these objects [18]. The data stored in communication profiles therefore forms the context of a mobile wireless application, i.e. terminal, user and network profile data. This context can be used to derive user preferences that have impact on the reconfiguration of the system. In the above example the low terminal battery and a high scheduled download generate a user preference of finishing the download within the remaining terminal operation time and therefore a certain necessary bandwidth that can result in a reconfiguration request sent to the network for changing from GSM to UMTS bearer services for example. Similar to user profiles, adaptive terminal profiles can be envisaged.

There are many possible concepts of how the preferences can be derived. Examples are neuronal networks, rule-based expert systems or statistical evaluation. All these concepts, however, rely on access to context information on the one hand and a mechanism for adapting the technical system they are designed to control for on the other hand.

Therefore the existence of context information storage and management is a requirement for adaptive user preferences and adaptive communication profiles in general. The communication profiles described in this chapter can be used to provide this access to context information.

A working example of the adaptive user preferences mechanism described in this section is the CAWAR-Mobi@ project developed at the Munich Technical University [21] (see Figure 9.4). Mobi@ is a prototype for a Context AWare ARchitecture and uses both statistical and rule-based mechanisms to derive user preferences for a location-based wireless application service browsing portal that can operate in most RAT environments such as GSM, UMTS and WLAN that allow for network-based positioning of the reconfigurable terminals. With Mobi@ the user can search for different places and locations within a city, for example restaurants and shops, but also information services based on the current location can be found. We provide a simple example for the use of Mobi@. A user connects to the Mobi@ browser via GSM or UMTS. Mobi@ derives from the current time, position and previous behaviour of the user that he is most likely searching for a restaurant and therefore provides him with a list of all restaurants at his current site without requiring the user to enter a search term. Mobi@ determines the position of the user and searches for restaurants within the area, also taking into account the opening hours of each restaurant and the user preferences for fast food and vegetarian dishes. For each entry found Mob@ can offer additional services such as a map or

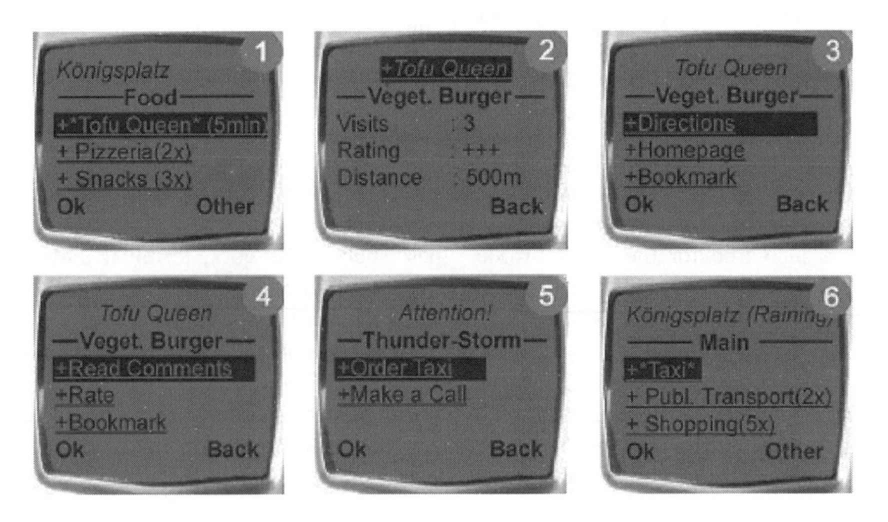

Figure 9.4 Adaptive user profile GUI

navigation service. During his meal the user is notified of a nearby thunderstorm and therefore offered to book a taxi instead of walking five minutes in heavy rain to the next subway station.

Although the Mobi@ example only covers adaptive user preferences for an intelligent selection of available application services it can show the clear benefit of using adaptive communication profiles, because the total number of cumbersome user interactions via a small screen and numeric keypad can be reduced to a minimum.

Since the application services in the Mobi@ example can be easily replaced by a more general concept of services, including network access technologies and QoS parameters, this is also a good example of how reconfigurable systems can profit from adaptive communication profiles.

9.3 Communication Profile Architecture

The profile data structure is basically a common tree structure that contains the attributes in the nodes. This gives us the needed flexibility to adapt the profile structure by exchanging nodes or branches. One of the most popular tree structures nowadays is defined by XML, the eXtensible Markup Language. We propose therefore in the following section an XML model for the profile data structure.

9.3.1 Profile Data Structure

For efficiently storing profile data information a communication profile architecture is described. A well-known tree structure, which uses XML to model and transfer the profile data between the terminal and the network, is used for the communication profile structure. The profile data distribution in the network and how to access it in an efficient mechanism will also be described.

XML provides a variety of advantages for accessing and processing the profiles:

- It is a standardised way to interchange data within heterogeneous systems.
- Data can be distributed over several documents and locations.
- It allows us to define an XML style for our specific needs.

In what follows brief examples of the general structure of an excerpt of the user and terminal profiles given in Sections 9.2.4.1 and 9.2.1.2, respectively, are modelled.

9.3.1.1 A Model of a User Profile in XML Format

```
<?xml version="1.0" encoding="UTF-8"?>
<profile type="user">
        <static>
                <firstname>Alfred</firstname>
                <surname>Hitchkock</surname>
                <userID>AH234711</userID>
                <address>Bates Motel - Bodega Bay CA</address>
                <preferences>
                       <cost>ignore</cost>
                       <reliability>
<linkLossAcceptance>0.01</linkLossAcceptance>
```

```
<handoverFrequency>low</handoverFrequency>
                         <handoverTime>0.5</handoverTime>
                  </reliability>
                  <provider>
                        <preferred
priority="8">Vodafone</preferred>
                  </provider>
                  <rat>
                        <preferredMode
priority="5">UMTS</preferredMode>
                         <excludeMode>GSM</excludeMode>
                  </rat>
                  <terminal>

<acceptCallBatteryLimit>10</acceptCallBatteryLimit>
                  </terminal>
                  <notification>
                         <mediate>confirm</mediate>
                         <modeChange>confirm</modeChange>
                         <serviceChange>notify</serviceChange>
                         <qosChange>silent</qosChange>
                  </notification>
               </preferences>
       </static>

       <dynamic>
             <provider> voicestream </provider>
             <language> english </language>
             <privacyMode> yes </privacyMode>
             <acceptDownload> yes </acceptDownload>
             <knownLocation> yes </knownLocation>
             <movementSpeed> 0.5 </movementSpeed>
             <visitedLocations>
                    <location date="30/11/01">D800001011</location>
             </visitedLocations>
             </schedule>
             </health>
       </dynamic>
</profile>
```

9.3.1.2 A Model of a Terminal Profile in XML format

```
<?xml version="1.0" encoding="UTF-8"?>
<profile type="terminal">
      <static>
              <type>ultrafone</type>
              <cpu>arm-10</cpu>
              <performance>120</performance>
              <memory>10</memory>
              <battery>12</battery>
              <screen x="200" y="320"c="4096">medium</screen>
              <standbyPower>1</standbyPower>
              <transceivers>2</transceivers>
              <transceivers id="1">
```

```
                <rat type="GSM">
                        <sub>GSM1800</sub>
                        <service>data</service>
                        <service>voice</service>
                        <service>fax</service>
                        <gprsClass>10</gprsClass>
                        <powerConsumption>5</powerConsumption>
                </rat>
        </transceivers>
        <transceivers id="2">
                <rat type="Bluetooth">
                        <sub>1.1</sub>
                        <profile>object</service>
                        <profile>pan</service>
                        <profile>lan</service>
                        <powerConsumption>1</powerConsumption>
                </rat>
        </transceivers>
    </static>

    <dynamic>
        <load>20</load>
        <freemem>8.123</freemem>
        <battery>80</battery>
        <mode status="active">
            <transceiver>1</transceiver>
            <service>voice</service>
            <powerConsumption>3.12</powerConsumption>
        </mode>
        <position>
            <city>D80000</city>
            <area>12345</area>
        </position>
    </dynamic>
</profile>
```

9.3.2 XML Structure

The XML profiles reflect the different properties of a terminal, a user, a network or a service. A profile is stored in a software module, called a registry (Figure 9.5), which takes care of the management of the profiles, such as accessing and efficient update.

To maintain a minimum level of redundancy and avoid data inconsistencies, most of the data will not be copied, but linked and accessed just in time. Take, for example, the user profile; the static part will physically be kept on the provider server. However, the dynamic part will be kept on the terminal since it is subject to frequent changes and therefore would cause high access traffic to the server. The averaged or filtered dynamic data can be sent, or copies of it mirrored, to the network either on demand or periodically, see Figure 9.6. The possibility of keeping dynamic data redundant might or might not be acceptable, depending on the synchronisation overhead, therefore the profile tree is logically split into two parts that are kept physically on each responsible Registry but are linked with each other.

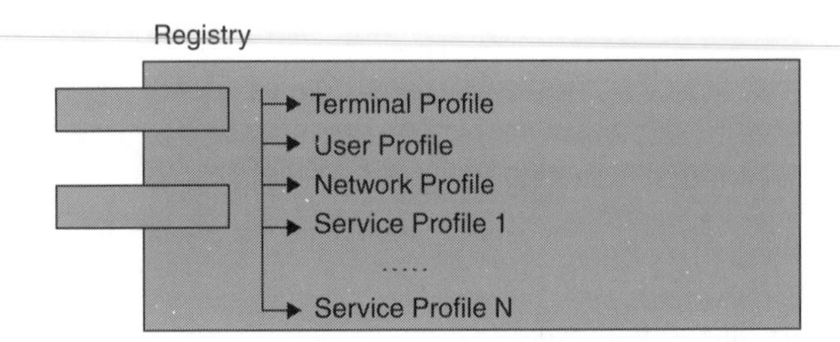

Figure 9.5 Registry structure

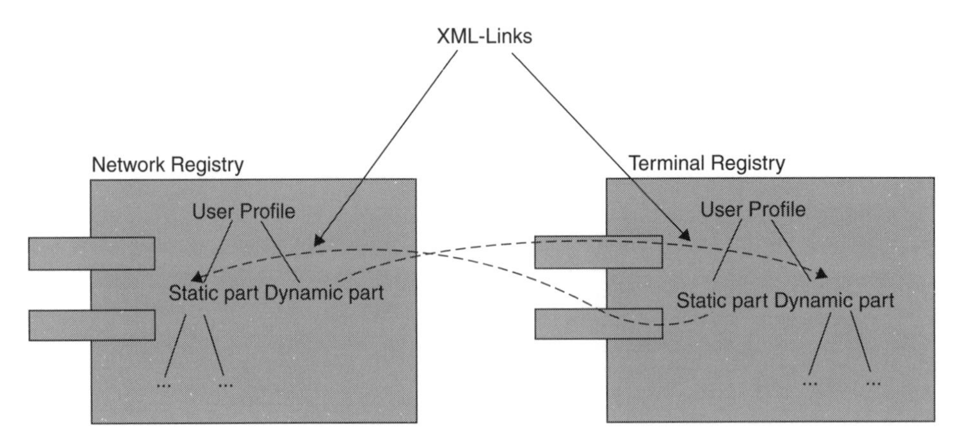

Figure 9.6 Tree structure of the profiles

The registry access the data using appropriately defined access methods that transparently resolve the links (XML-links) either by directly accessing the distributed data or by relying on sophisticated replication or proxy mechanisms. In the later case, the managing container "the registry" simulates to the system a locally stored profile. However, the profile may in fact be physically distributed in different locations.

9.3.3 Distribution of Profile Data

In the previous section the general architecture for a distributed profile management was discussed. The described profiles could of course be stored where the containing data originated, but in this section take a closer look at efficiency when it comes to deciding about the storage management location of each profile element.

In distributed systems, the origin of profile data can be spread all over the system (in the terminal and network), the same, as the places are, where the data might be used in the end. Sometimes there is no direct relation between the logic profile data classes and the location of the origin of the different elements. The origin of the network profile's data, for example, is distributed all over the system and user subscription data are stored somewhere in the network, usually in the HLR in the case of the GSM and in the HSS in the case of the UMTS.

The efficient distribution of profile data can mean several things. Regarding a reduction of the used bandwidth, data that are frequently accessed should be stored at or close to the location where it is consumed. Data that are often changed should be kept at the location where they are produced in order to reduce the signalling load. If the data are to be accessed by a large number of consumers at the same time they need to be replicated and therefore consistently synchronised should they change.

In the general approach, usually profile data are stored where they originate. Links are used to form a completely distributed data model. We can also define the possibility to plug in different methods to access distributed elements while having only one logic distribution model using XML links. The default access method is simply to resolve each link every time the profile element it links to is accessed via the registry container. Since this approach contains no data replication, no synchronisation is necessary. This general approach, however, does not address some important requirement conditions within wireless mobile networks and, moreover, has some disadvantages regarding used network bandwidth. Wireless networks are unreliable by nature and terminals might lose their connection from time to time (in a connected mode). Services that rely on profile data stored in the terminal could of course use a lease-based mechanism to prevent unnecessary resource allocation as well as precipitate aborting a running transaction. This, however, implies that the service needs to wait for the terminal to reconnect before it can continue even if the requested profile data are not very likely to change during the time of not being connected.

A more sophisticated approach in such cases would be to store static profile elements at a location within network entities and dynamic elements on the terminal. Furthermore, by filtering the dynamic data, the result might be stored in the network. Since static profile data (e.g. screen resolution) never, or rarely, change at predetermined points, synchronisation is no real issue. This way services can access some of the terminal's profile data even if the terminal is temporarily not connected (see Figure 9.7). The same is true for static and dynamic user profile data such as, for example, user subscriptions and 'do not disturb' preferences that are stored on the terminal/smart card and the provider's network, respectively.

Service profiles only use this separation if the service is located on the terminal. Web services and network profiles are completely located within the provider's network.

9.3.4 Access to Profile Data

Another efficiency issue regarding synchronisation and distribution of profile data is the efficient synchronisation of profile data between locations where the information is produced, stored, consumed or replicated.

With the default access method of direct resolving links on each access, profile elements could be distributed on different locations and are transparently synchronised on each access by resolving pointer connections where necessary. This can be very inefficient if the profile data consumer needs to query (poll) each time he wants to check whether or not an element has changed. It is even more inefficient if the polled element is in fact a sub-tree of the profile data structure with several children, again distributed on other remote locations, and if the consumer is only interested in a small portion of the sub-tree.

Two techniques for different access methods that can be plugged into the registry container can improve the general approach in this area. First, it needs to be possible

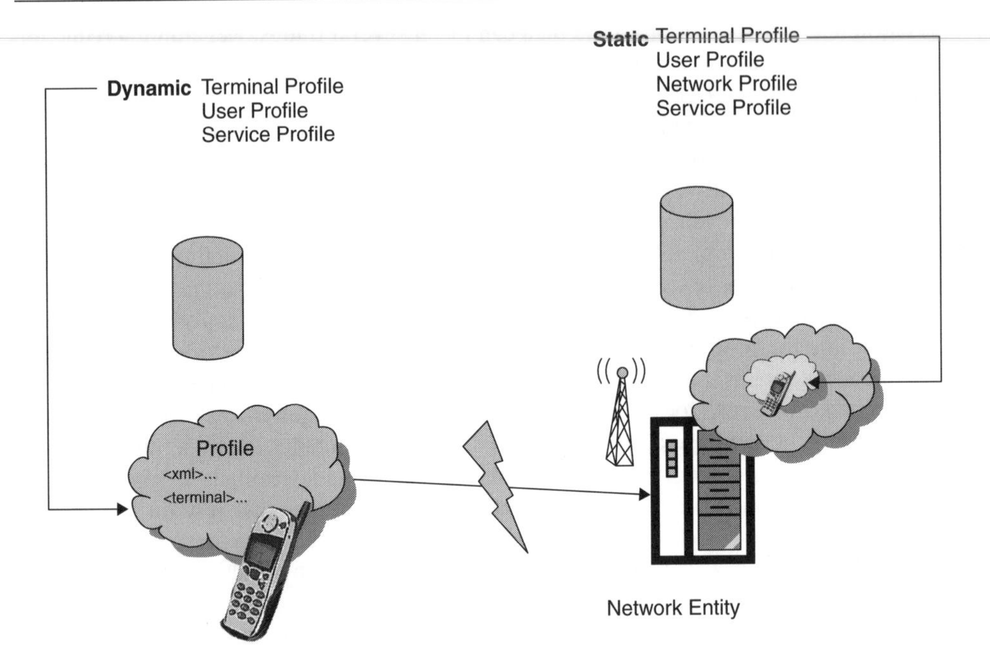

Figure 9.7 Profile distribution

to specify in more detail the part of the profile the consumer is interested in. Second, the fact that part of the profile data has changed should be propagated to consumers who registered their interest in the affected area, along with the exact point the change occurred. In some cases it might be useful to send the changed data element along with the event notification, especially if the affected data is of reasonable size. Not all consumers would decide to poll the changed data after receiving the notification and not all consumers would do it at exactly the same point in time. This way load peaks can be prevented in one producer/many consumer situations. Consumers who need to query every change, e.g. central storages, should be sent the changed data along with the event notification. Addressing parts of the profile data tree should contain addressing single elements without child elements, sub-trees, depths and sub-trees with exclusions, see Figure 9.8.

9.4 Management of Communication Profiles

During the establishment, configuration, observation or control of the current connections and radio controlling procedures, the network should know the up-to-date information about the terminals, e.g. power, potential load added to the network and priority. All the required information depends on the radio connection status, states and network types. On the other hand, the inhomogeneity nature of the reconfigurable terminal and the considerable amount of expected information stream needed in order to fulfil the required information from the network side greatly decreases the spectral efficiency of the actual payload. In this section detailed requirements regarding the efficient storage and management of profile data are described.

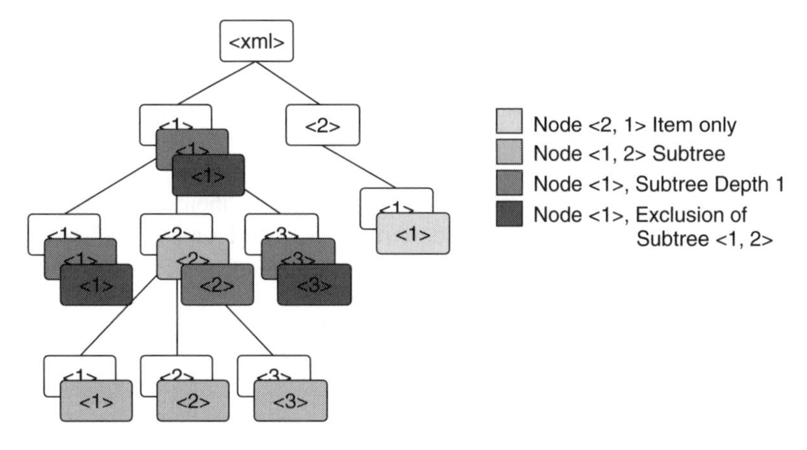

Figure 9.8 Addressing profile data elements

In the first stage a novel classmark concept for the communication management called the 'Dynamic Classmark for Reconfigurable Terminals' will be discussed, which greatly increases the signalling efficiency between terminal and network. Then strategies for the efficient coding and compression of the proposed profile structure will be discussed that will have an impact on both the amount of resources needed for storage, the processing power to access and the bandwidth necessary to transmit profile information within a distributed environment.

9.4.1 Communication Classmarks

There exist a variety of terminals from complex terminals that are expensive to produce and to buy to cheap ones that offer fewer features and restricted functionality. It should be possible to categorise the capabilities of reconfigurable terminals according to how advanced they are and which functionality they are able to offer. Terminal classmarks are used to categorise mobile devices based on their performance and capability. These criteria could include, for example, processing power, memory, display properties and screen size, as well as interactive capabilities.

Such a classification would be beneficial in that it would provide end-users with a standard representation of the capabilities and features of a terminal in terms of reconfiguration, and also the network and service provider need to be aware of the capabilities of each terminal. This places a slightly different emphasis on the framework. Whereas a descriptive, behavioural classification system would satisfy the architectural framework requirements, it would be of little use to the network.

Here classmarks are also part of the profile data. In systems with groups of a large number of similar devices classmarks can describe terminal capabilities without transmitting the exact values of each profile element, thus saving bandwidth. These classmarks have evolved alongside the evolution of mobile networks and tend to become more complicated as the number of possible features increases.

So far, there are several classmarks used in mobile communications systems. In the following sections we define some of them.

9.4.1.1 The GSM Classmark

The GSM [6] defines three different classmarks; the higher classmark contains all information contained in the lower classmark and provides additional information. The important information already contained in Classmark 1 is the radio frequency (RF) power capability. However, also information about available encryption algorithms is provided. The higher classmarks extend this information by giving details about additional capabilities, e.g. support for different frequency bands within the GSM or positioning capabilities. There is also the GPRS multi-slot classmark which denotes the combination of receive and transmit slots that the terminal can support [7].

9.4.1.2 UMTS

The concept of simple classmarks has been abandoned in UMTS in favour of a more flexible concept, as the variety of parameters to be communicated with the network increased considerably. The interface between user equipment and UTRAN uses instead 'UE capability information' which consists of a mandatory part containing radio access capabilities for UMTS and an optional part specifying the capabilities of other RATs, e.g. the GSM. A detailed description of these data can be found in ref. [15], so here only a short summary is given.

The radio access capabilities specify the capabilities of the terminal including support for different channel capacities for physical channels, e.g. 600 up to 76 000 bits per time interval for downlink channels, as well as transport channels, TDD hot-spot capabilities, and other technical details concerning the terminal radio capabilities. This also includes information about the measurement capabilities of the terminal, indicating for instance whether the compressed modes are to be used for carrying out measurements, e.g. on UMTS FDD or GSM. In addition to the radio-technology-related information, it also contains security-related information and specifies the support for positioning methods, e.g. GPS, by the terminal.

Information about other available RATs at the terminal is also given. For instance, for GSM, this includes the traditional GSM classmark.

9.4.1.3 MExE Classmark

In the MExE the capabilities of the terminal in terms of screen size, input capabilities, memory and processing power are provided [12].

9.4.2 Dynamic Classmarks for Reconfigurable Terminals

The purpose or aim of the dynamic classmark scheme is to provide the network, in an abstract manner, with the terminal's capabilities without specifying or constraining future implementation. A refined dynamic classmark scheme filters the terminal's dynamic features depending on the network's estimation of feature usage. The responsible network entity can estimate/predict the status of the dynamic feature, in this way the transmission of the classmark will be restricted only to dynamic features that are hardly to be predict and their frequency of change is much higher, thus saving bandwidth and reducing the signalling overhead. One example of dynamic feature behaviour prediction is the status of the battery; knowing which application the terminal is currently running and what kind of battery is used, the network can determine the battery stage of charge information. In the network, similar dynamic features behaviour can be stored and the network can be trained to forecast or predict future statuses of terminal resources or of other dynamic characteristics. Hence, the information to be reported to the network can be reduced during the course of a session. The dynamic classmark scheme can be described based on a finite number of stages that are specified for the classification of classmarks. At each stage, the value of the classmark has a different meaning.

The stages of classmarks (classmark-stage) can be upgraded according to the priority, upgrade frequency, access frequency and the accuracy of attributes of the features required by the network. The features can be clustered independently of the Open Systems Interconnection (OSI) layers.

Once the network requests the information from the reconfigurable terminal, the reconfigurable terminal sends the classmark in the current stage. With the information decoded from the classmark, the network will determine its next action, or even trigger the reconfigurable terminal to change to the required stage. However, the stage is independent of the reconfigurable terminal's actions; it is only related to the classmark and the value of the classmark when it is to be generated.

The features of the reconfigurable terminal consist of hardware and software capabilities, the capability of accessing potential RATs, the user preferences, the applied service type, etc. All these features should be input parameters to the radio resource control functions, in order to assign a reasonable radio link, reconfigure an on-going radio link, multiplex the number of connections, or even release or discard an unsatisfied connection. In order to effectively increase the signalling efficiency between the reconfigurable terminal and the network, the features should be classified into different groups, where only one group of features appears at one stage when the network requires the information corresponding to that stage.

The principle of classifying features is based on their properties. The static ones should be retrieved first, e.g. user identification and basic terminal capability, and then the more dynamic features should be requested. However, there are number of features that are not necessary to be retrieved, e.g. the maximum display size, although it is static. If the user only applies for the voice service, then it is not necessary to transmit this feature to the network. So, in the design procedure, the 'maximum display size' should be a feature requested after the 'user preference' and 'service application'.

In order to model the stage-based classmark clearly, a finite-state-chart illustrating the stages and the related features and classmarks is needed, as shown in Figure 9.9. Since it is designed that the network may trigger the reconfigurable terminal to change to arbitrary stages by leaving any stage, a default idle stage is introduced in Figure 9.9. However, for each stage (context) of the communication procedure a template is agreed between the terminal and the network for signalling specific features that the network needs for its decisions.

The network triggers the reconfigurable terminal by transmitting the stage-trigger command. Once the reconfigurable terminal receives the command, it sends the classmark corresponding to the current stage, e.g. once the reconfigurable terminal receives '001' shown in Figure 9.9, it should submit its up-to-date features in stage 1 represented by the classmark to the network.

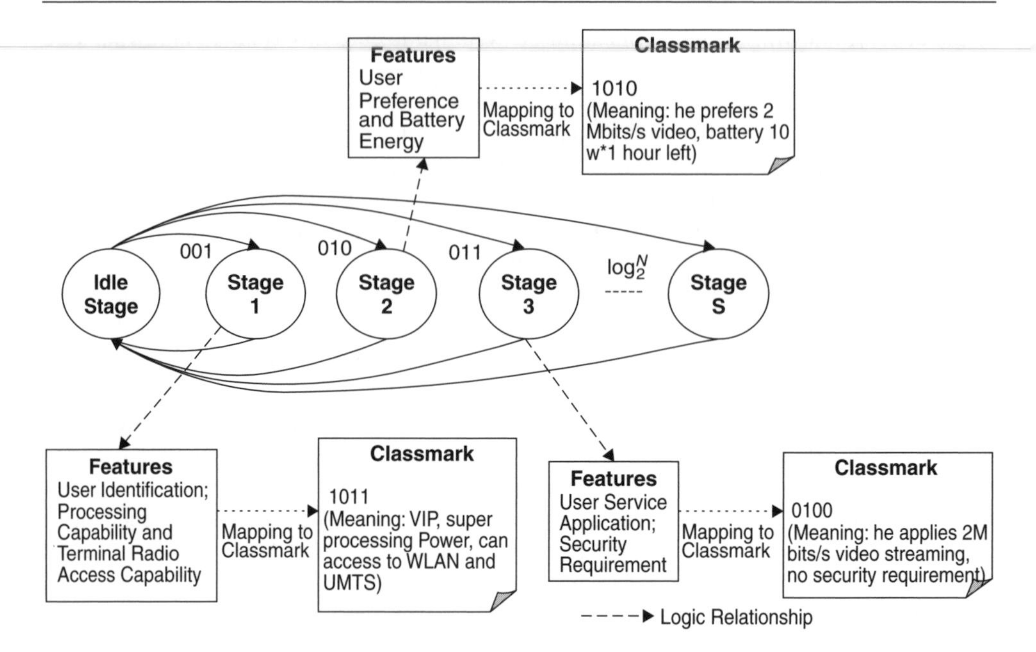

Figure 9.9 Classmark described by stages (context) and their representative features

The network must take into account the user preferences for assigning a possible radio link to a connection. The reconfigurable terminal can be connected to a single RAT only, or can even be simultaneously connected to more than one RAT. In this case, the traffic-splitting command, the joint load control and joint scheduling algorithm described in ref. [16] should manage simultaneous connections, as described in Section 10.3. In the second case, the reconfigurable terminal must have the ability to access the relevant RATs determined by the 'terminal radio access capability' features. During the communication procedure, a number of features will be frequently requested. Either the reconfigurable terminal or the network can abstract the frequently requested features to cluster them into a new stage-defined classmark, as depicted in Figure 9.10. However, before the classmark value is sent to the network, the definition or template of the classmark and the meaning of the classmark must be agreed by both the network and the terminal.

The background theory of the dynamic classification of the features is outlined in what follows. Let **A** represent all the features; it has certain possible realisations.

Lemma 1. *A complete generic set* **A** *is composed of S subsets, namely* $\mathbf{A} = \{A_1, A_2, \ldots, A_S\}$. *All subsets are orthogonal to each other, i.e.* $\langle A_i, A_j \rangle = 0$, *with* $i \neq j$. *Let* **A** *have N different realisations, and* A_i *have a total of* n_i *realisations. Owing to the orthogonality of all the subsets of* **A**, $N = \prod_1^S n_i$.

Lemma 2. *If and only if all realisations of the subsets are equal to each other, the efficiency of channel usage reaches an optimal point.*

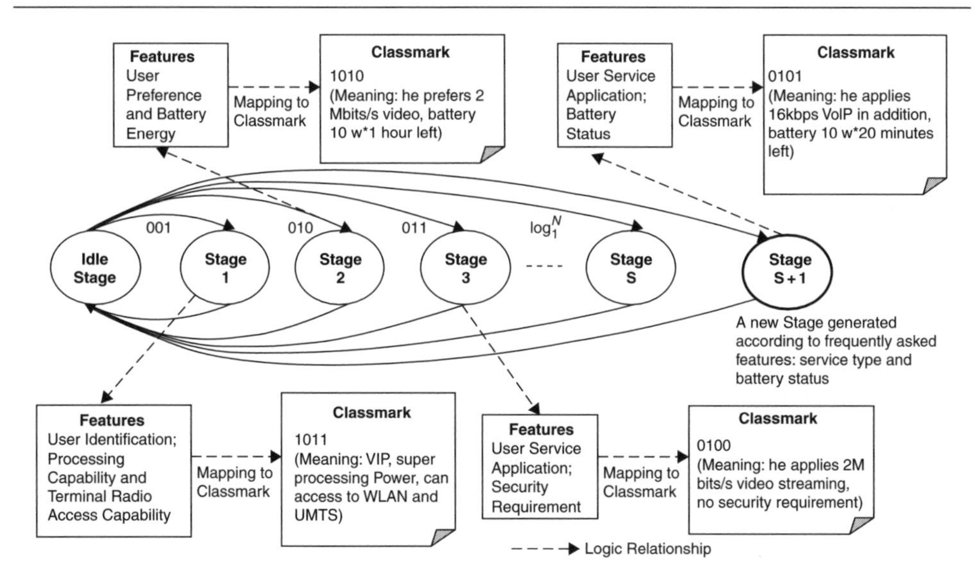

Figure 9.10 Dynamic generation of new classmark

Proof. Let C represent the channel capacity [bits/s], which carries classmarks in each call from the network. Let the number of bits carrying the possible realisation be $B = \log_2^N$, which in each subset should be $b_i = \log_2^{n_i}$. In order to have correct receptions, the number of bits transmitted over the required time T should have the following relationship with C, as $C \cdot T \geq b_i$. Let n_m be the highest number in the set $\{n_1, n_2, \ldots, n_S\}$. The channel capacity should be selected to be at least $C = b_m/T$. So, in each call from the network, the efficiency of channel usage is $\eta = \log_{b_m}^{b_i} \leq 1$. So, only when $n_i = n_j, \forall i \leq S, \forall j \leq S$, does the channel usage reach the highest efficiency, namely 1.

Suppose the network asks the mobile terminal to submit classmarks at a rate R. If this stage is partitioned, the mobile should submit the value of all the features to the network. In this case, the bits need to be transmitted in a unit of time $B \cdot R$. It is self-evident that $B \geq b_i$. So after a call from the network, the terminal needs only to transmit b_i bits. The maximum spectrum saving is immediately $G = \log_2^N/\log_2^{n_i} = \log_{n_i}^N = S$. However, the bigger S is, the more difficult it is to establish orthogonal subsets.

9.4.3 Compression and Coding

Efficient coding and compression of the proposed profile structure will have an impact on both the amount of resources needed for storage, the processing power to access, and the bandwidth necessary to transmit profile information within a distributed environment.

At first glance the text-based XML approach that was introduced because of the requirement for extensibility might look inefficient. It seems feasible to improve at least the resources necessary to store or transmit profile data by introducing a less redundant binary coding scheme to structure the profile data. Considerable research effort has been put into this subject. There already exist standards for 'binarising' an XML-data structure [17]. However, first reports indicate that the difference in overall response time between a serving compact HTML (iMode) to an iMode client browser versus a binary coded WML to a WML browser is negligible and thus not worth the effort of optimisation [19]. The consensus at the moment is that binary encoding strategies may be suitable for problems where the format and data are known in advance [19]. This, however, contradicts the requirement of flexibility, especially the extensibility of the profile structure presented in this chapter. Moreover, a text-based XML data structure would support any integration aspects with third-party software services. Furthermore, applying compression to reduce redundancy in the stored data in addition to a binary coded data structure format seems advisable (see Figure 9.11).

Another approach to save bandwidth and storage resources would be first to use a general binary or text-based compression mechanisms instead of specialised binary optimisation, either for the complete profile data or alternatively only for certain elements that are especially suitable for compression. Then to limit these steps to only parts of the distributed system where the bandwidth is especially expensive compared with a slight increase of processing power. This is the case for transmitting profile information between the end-user terminal and the base station. Table 9.1 shows the results of a very simple experiment using zip compression on XML files. The numbers show that even not XML optimised compression technology can achieve very good results on reducing necessary bandwidth to transmit profile data for medium size profile data while retaining full flexibility, extensibility and integration aspects of the initial text XML based approach. However, this approach does not yield good results for small portions of XML structured data, but the concepts for efficient addressing and signalling of only relevant and statistically rare

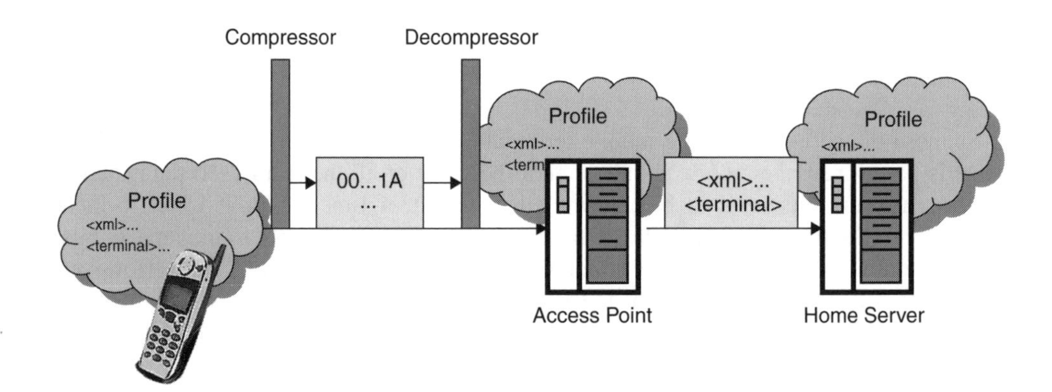

Figure 9.11 Profile compression

Table 9.1 XML compression (zip)

	Size (bytes)	Compressed
smallXML	317	**301**
medXML	**9500**	**2500**

parts of the profile data, as proposed with dynamic classmarks, usually results in small size chunks of XML data.

An approach that is better suited for small size XML data is to separately compress the structure, i.e. the XML tags, and content information in XML documents as proposed in ref. [20].

More sophisticated experiments comparing standard compression technology with an XML optimised tag/content compressor can be found in ref. [20] and show that the XML optimised separate compression of structure and content information is significantly more efficient for small size profile data up to 5 kb, while retaining XML flexibility and extensibility features (e.g. ignoring unknown tags) even without decompression.

The conclusion is that it is not necessary, and may even be disadvantageous, to manually optimise the coding of text-based XML to describe profile data by converting the XML structure to a fixed binary data format thereby loosing flexibility, especially extensibility. Instead, the already existing fly compression technology applied separately to the structure and content information should be used to reduce necessary storage or transmission bandwidth and only where necessary (e.g. transmitting profile data between terminal and base station); see Figure 9.11. This way the original advantages of an XML text-based approach (flexibility, extensibility and interoperability) can be preserved.

9.4.4 Meta Profile Data

Efficient signalling and coding introduced the requirements to describe certain properties of the profile elements and profile data, such as the frequency of occurrence of certain profile values within the network (for dynamic classmarks) or within one profile (for compression). This concept can be generalised by adding meta-attributes, attributes that describe the information stored within a profile in the form of profile elements. Further meta-attributes such as Change Frequency, Access Frequency, and Accuracy Attributes can be used to model requirements for profile data elements more exactly. This will allow for a more refined distribution of profile data between mobile devices, network entities and the proxy reconfiguration manger as well as more sophisticated propagation of changes. Some optimisations based on this information are described briefly in the following sections.

9.4.4.1 Transmission Delaying, Event Delaying and Grouping

The transmission of profile data updates to storage points not on the same location as the data origin, or the transmission of update events, can be delayed under certain circumstances:

- If transmission bandwidth is limited and could be used for other purposes. For example, from terminal to base station during video transmission with variable bit rate, the transmission of a profile data update could be delayed until video frames with a lower bit rate are sent.
- If the communication mechanism only allows for sending information at given slots, e.g. using the IMEI to transmit information during network logon.
- If information can only be sent in packets of a certain size. The data update could be delayed until more updates can be grouped.

- Rapidly or continuously changing profile data that need not be accessed in real time, for example battery status, can be delayed and grouped to save both bandwidth and processing power on the consumer side.

Which profile data elements are suitable for delaying can be derived from their describing attributes, e.g. static, priority, change frequency, access frequency and accuracy attributes.

9.4.4.2 Prediction Values of Profile Elements

Some profile data elements that change often in a predictable way (e.g. battery status) can be replaced by functions describing their predicted development from a given starting point, stored in a prediction function meta-attribute. The function can be hosted on the fixed network and therefore would be available even if the data origin (e.g. the terminal) is temporarily not connected.

9.4.4.3 Restricting Change Possibilities

Introducing a meta-attribute marking static, session static and dynamic profile data can achieve a further refinement of the static/dynamic classification. It separates the profile data into three classes. Static elements rarely or never change. If they change, they change at predefined points, such as, for example, reconnecting the terminal after being switched off. Typical examples of static elements in this classification would be the user's address, date of birth, the terminal's screen resolution, etc.

Session static elements never change during a session. If they change, they either change at predefined points, or it is not necessary for already bound services to take these changes into account during an active session. Examples of session static elements could be firmware version, additional connected hardware, etc. Session static elements usually are checked when the session starts.

Dynamic elements can change at anytime and services need to know about this also during active sessions, e.g. battery status.

Whereas static and session static data can be polled at predefined points, changes in dynamic profile data need to be propagated by the storage location to all consuming entities using an event mechanism. This optimisation reduces network traffic, because polling consumers can reduce polling frequency for certain data to a second predefined change point. This optimisation can be further refined by modelling more then two change points, however at the cost of increasing profile data size.

References

[1] W. Tuttlebee, *Software Defined Radio: Origin, Drivers and International Perspectives*, Wiley, 2002.

[2] K. Mossener, S. Gultchev and R. Tafazolli, 'Software defined radio reconfiguration management', PIMRC, 2001.

[3] T. Farnham, R. Haines, N. Olaziregi, R. Falk and D. Bourse, 'Novel solutions on system aspects of reconfigurable terminals: Mode identification and switching, configuration management and software download security', IST Summit, 2002.

[4] L. von Allmen, J.A. Garica, J. MacLeod and M. Mehta, 'Capability characterisation of reconfigurable terminals', IST Summit, 2001.

[5] C. Niedermeier, R. Schmid, E. Mohyeldin and M. Dillinger, 'Handover management and strategies for reconfigurable terminals', SDR Forum General Meeting, September, Edinburgh, UK, 2002.
[6] 3GPP TS 04.08 v 7.16.0, 'Mobile radio interface layer 3 specifications', December, 2001.
[7] 3GPP TS 05.02 v 8.10.0, 'Multiplexing and multiple access on the radio path', August, 2001.
[8] 3GPP TS 22.101 v 5.4.0, 'Service aspects; service principle'.
[9] 3GPP TS 22.105 v 5.0.0, 'Services and service capabilities'.
[10] 3GPP TR 22.925 v 3.1.0, 'Quality of service and network performance'.
[11] 3GPP TS 23.002 v 3.3.0, 'Network architecture'.
[12] 3GPP TS 23.057 v 4.4.0, 'Mobile execution environment (MExE); functional description', December, 2001.
[13] 3GPP TS 23.107 v 5.3.0, 'QoS concept and architecture'.
[14] 3GPP TS 23.110 v 3.4.0, 'UMTS access stratum services and functions'.
[15] 3GPP TS 25.331 v 4.2.1, 'RRC protocol specifications'.
[16] J. Luo, M. Dillinger and E. Mohyeldin, 'Radio resource management schemes supporting reconfigurable terminals', 2nd Karlsruhe Workshop on Software Radios, March, 20/21, 2002.
[17] WAP Binary XML Content Format, W3C NOTE, 24 June, 1999. *http://www.w3.org/TR/wbxml/*
[18] A. Dey, 'Providing architectural support for building context-aware applications', PhD Thesis, College of Computing, Georgia Institute of Technology, December, 2000. *http://www.cc.gatech.edu/fce/ctk/pubs/dey-thesis.pdf*
[19] L. Dodds, 'Intuition and binary XML', 18 April, 2001. *http://www.xml.com/pub/a/2001/04/18/binaryXML.html*
[20] Marc Girardot and Neel Sundaresan, 'Millau: an encoding format for efficient representation and exchange of XML over the Web', 9th International World Wide Web Conference, Amsterdam, Netherlands, May 2000, *http://www9.org/w9cdrom/154/154.html*
[22] *www.cawar.de*

10

Radio Resource Management in Heterogeneous Networks

Jijun Luo and Markus Dillinger
Siemens AG

Lucas Elicegui and David Grandblaise
Motorola Labs, European Communications Research Lab, Paris, France

10.1 Introduction

10.1.1 Definition of Radio Resource Management

Software defined radio (SDR) technologies enable the future network to meet challenges of high quality of service requirement by supporting high mobility and throughput for multimedia services with heterogeneous user requirements. Mobile users expect services that will not only depend on a traditional single traffic type, but multiple traffic types even supported by adaptive simultaneous connections from different network types, e.g. cellular, *ad hoc*. However, the freedom of co-existing heterogeneous networks raises further questions about how to manage the traffic in networks in an efficient way. From a system management point of view, a simple physical layer processing restricted system cannot meet the co-existing multiple radio access technologies (RATs). A reconfigurable environment, flexible and scalable network and traffic management is of high interests and expectations to emerge. It should support the complete network with convergence towards an Internet Protocol (IP)-based network and ubiquitous, seamless access among second generation (2G), third generation (3G), broadband and broadcast wireless access schemes, augmented by *ad hoc* networks schemes and short-range connectivity between intelligent communication applications. The ability offered by the emerging 3G WCDMA system, namely 384 kbps for urban areas and 2 Mbps for local-area coverage with a single band, can be obtained. Because of the physical characteristics of cellular radio networks and the requirements of users, the data rate of an ongoing radio link will also vary, demanding more advanced joint radio resource management over heterogeneous networks covering cellular and decentralised networks.

Software Defined Radio: Architectures, Systems and Functions. Edited by M. Dillinger, K. Madani and N. Alonistioti
 ISBN: 0-470-85164-3

In order to handle radio access properly, efficient spectrum allocation, access control, session flow control, traffic management and power control mechanisms are required. The management of all or part of these elements falls under the term Radio Resource Management (RRM).

In this chapter, after the general concepts and issues related to RRM have been being addressed, new RRM challenges in the context of heterogeneous radio environment are introduced and the concept of Joint Radio Resource Management (JRRM) is developed. Finally, an overview of JRRM performance bounds is presented and the functions needed for a common RAN control thereby allowing simultaneous connections on radio level for better QoS and connectivity for users are introduced.

Definition of RRM: *Network-controlled mechanisms that support intelligent admission of calls and sessions; distribution of traffic, power and the variances of them, thereby aiming at an optimised usage of radio resource and maximised system capacity. These mechanisms can also work over multiple radio air interfaces with the necessary support of reconfigurable terminals.*

10.1.2 Radio Resource Units over RRM Phases

RRM stands at the interface between the Radio Access Network (RAN) and the terminal and aims to optimise the use of radio resources of the air interface. Radio resources being limited, the challenge of RRM is to apply schemes in order to reach high system performances in terms of spectral efficiency (i.e. the appropriate use of the limited spectrum resource). These schemes would rely heavily on the RRM functions (i.e. principally defined by the network operator management policies) and reflect the priorities and rules the operator wants to establish in his network. The following describes the main RRM concepts and principles for a single radio communication system.

Radio resources are clearly not limited to the spectrum itself. Depending on the envisaged system they could be decomposed into several basic Radio Resources Units (RRUs). In order to establish communications between most radio communications systems, elementary RRUs are required for the uplink (reverse) and downlink (forward). For example, an RRU can typically consist in

- One frequency carrier
- One time slot
- One orthogonal code
- An amount of power

As a result, the radio channel is a combination of one or several of these required RRUs and allows the establishment of a proper communication, as depicted on Figure 10.1. Table 10.1 gives some examples of radio channels for several Radio Access Technologies (RATs). Some more examples and details of radio channels can be found in [4]. Once the RRUs are defined, the main RRM issues and the derived concepts terminology will be discussed.

In the context of a radio wireless communication system, from the initial time when a user applies to set up a communication to the final time the user decides to end his call, different communication phases between the user terminal and the network can be

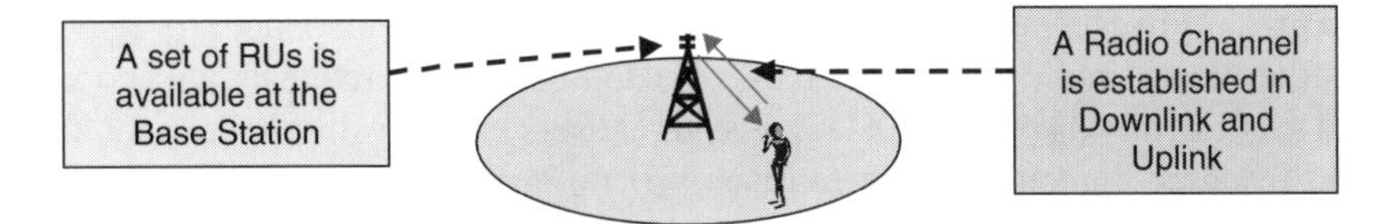

Figure 10.1 Mapping between RRUs and the radio channel

Table 10.1 Examples of RRUs

GSM	Frequency carrier + Time slot + Power
UMTS FDD	Frequency carrier + Code(s) + Power
UMTS TDD	Frequency carrier + Time slot + Code(s) + Power

identified. Each of these steps addresses specific issues and requires appropriate mechanisms to ensure the system operation. An overview of the main RRM phases according to the nature of the service handling procedure of a radio network are depicted in Figure 10.2 and described in what follows for the specific case of a mobile communication system.

10.1.2.1 Initial Access Phase

In this phase the user attempts to initiate a communication with the network. The user applies for a particular radio channel so that a connection with the network is established

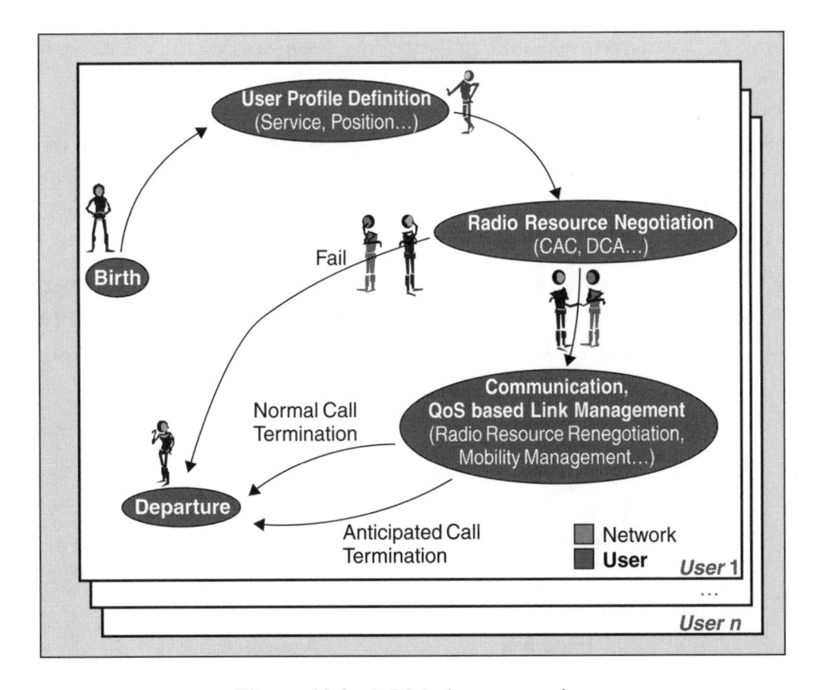

Figure 10.2 RRM phases overview

and supported. However, at the same time, many other users independently may also try to access the limited radio resources. It is expected that users' requests will be served as fairly and successfully as possible. Having several users to serve simultaneously, this leads to a radio resource contention between users. If no appropriate mechanisms are carried out, some unfair and non-optimised radio resource usages can be expected, penalising both users and operators.

In this access phase the main issue to be faced is *blocking*. Blocking appears when users are facing the impossibility of accessing the radio resource during the initial access phase. It is the most critical one in terms of quality of service (QoS) perceived by the user, as it is impossible for a user in this case to establish the communication. This phenomenon logically becomes evident when there is a lack of RRUs (e.g. frequency carrier or time slot). At this step, *hard* and *soft* blocking may occur. Hard blocking occurs when a mandatory RRU is unavailable (e.g. one frequency carrier or time slot) to establish a link. Connection establishment in this case is definitively impossible. Soft blocking occurs when the 'mandatory' RRUs are available (e.g. frequency carrier or time slot) but some others RRUs are not available due to the limited RRUs set (e.g. a large amount of power might be needed to maintain the quality of the users' radio links experiencing bad radio conditions). To overcome these initial phase issues, RRM has developed appropriate radio resource negotiation schemes [under the terminology Call Admission Control (CAC)]. Those management rules provided by RRM schemes help in globally organising the access to RRUs at the system level (i.e. for a given part of the network in a given geographical area).

10.1.2.2 Ongoing Communication Phase

Once the link between terminal and network has been successfully established during the initial access phase, the user faces new issues when moving across the radio cellular networks (Figure 10.3). Owing to the limited size of the various RRU pools (e.g. codes or power), the challenges rely on the most appropriate accommodation of the RRUs among users competing for resources. Various phenomena mainly linked to queuing theory appear, representing the major issues that RRM algorithms would have to face and overcome. The main phenomena can be captured under the scope of the following indicators.

- *Blocking*. Mainly due to the movement across the mobile cellular network, the user is led to leave the serving cell coverage he is being connected to. Therefore, a new connection with a neighbouring cell must be set up to ensure (seamless) handover.

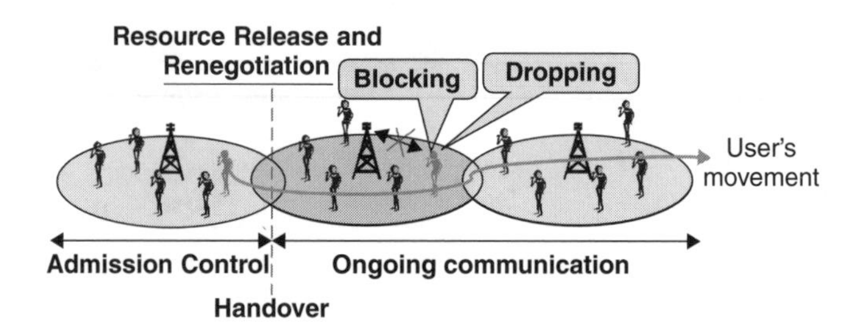

Figure 10.3 Ongoing communication phase

To perform the handover process as appropriately as possible, careful radio resource re-negotiation (mechanisms captured under the terminology 'handover') must be carried out to ensure communication continuity with the same QoS. Failing in this re-negotiation is the Blocking phenomenon. In the meanwhile, some users may also initiate a call in that cell in the same way as described in Section 10.12.1. This introduces some new radio resource contention issues regarding priority management rules between different user profiles (e.g. requested services). Of course, several alternatives exist, such as pre-emptive mechanisms allowing anticipation of the handover and effectively triggering it, if it has been previously made possible to reserve required resources.

- *Dropping.* Appears when users are dropped during the already established communication. It is also critical since consequences on perceived QoS are immediate as communication is cut. The main reason for the emergence of dropping would not be based on a lack of resources, as a link is already established, but on the insufficient capacity for the system to provide the user with a reliable link, i.e. the link quality becomes degraded during a consecutive time period that is not clearly acceptable. Different RRM strategies in terms of resource re-negotiation are required to overcome these issues.
- *Bad Quality Call.* Appears when bursty time periods of degraded link quality [e.g. in terms of a bad frame error rate (FER)] occur due to the difficulty in maintaining link reliability. For applications being not time critical, users are not dropped. Hence, appropriate RRM schemes are also needed to overcome these critical situations.

Close monitoring of these quality indicators contributes heavily in evaluating the global QoS perceived by each user and transposing it at the system level to get a better insight into global satisfaction. Of course, it could be envisaged to derive or monitor other indicators, which could help in getting a sharper picture of the system status.

10.1.3 RRM Challenges and Approaches

10.1.3.1 RRM Problems in Conventional Networks

10.1.3.1.1 Radio Resource Contention. This section generally describes the problems that should be solved by RRM algorithms based on the previously introduced basic concepts. This is discussed with respect to RRM functions, performance and analytical models.

As mentioned above, RRM policies rely heavily on the chosen perspective. Basically, they can be classified into two main categories: user and system perspectives. These perspectives could be outlined as in Table 10.2.

The implemented RRM policies will be impacted directly by the choice to favour the user's QoS (i.e. user perspective) or the overall system capacity (i.e. system perspective). Depending on the balance between these two perspectives, differently implemented RRM policies can therefore be envisioned. What can be highlighted from these two perspectives is the dilemma that exists between them. Indeed, these points of view are in opposition. For instance, privileging the user's QoS would impact the system capacity (because supporting the QoS may require many RRUs thereby preventing new users from entering into the system) and vice versa. Thus, the unavoidable trade-off must be analysed when tackling RRM, as depicted in Figure 10.4.

Table 10.2 User and system perspectives for RRM

User perspective	1. Service requirements: best QoS, continuity of the service 2. Radio user profile: experienced radio conditions
System perspective	1. Operator point of view: capacity as important as possible 2. Provide the best service to the user independently of the number of users 3. Consider different user profiles

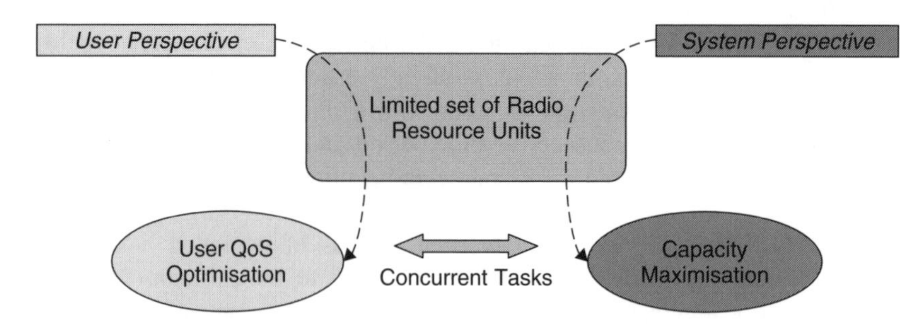

Figure 10.4 Single user and overall system perspectives – concurrent tasks

In order to study RRM performance it is necessary to derive an abstraction of the considered radio system. If, for instance, a cellular radio system is considered, this could be considered from a radio system point of view to be a whole system where its evolution depends entirely on the occurrence of asynchronous events over time [5]. Therefore, such a system can be modelled as a stochastic timed discrete event system [more simply denoted as a Discrete Event System (DES)] or a Finite State Machine (FSM). It provides analysis and representation tools for the events along with time. Moreover, it provides potential bridges with queuing theory formalism. More details on RRM problem modelling and resolution are outlined in Section 10.1.4.

Figure 10.5 illustrates a temporal representation of events as they appear in a realistic operating mobile radio communication system. This figure also captures the dynamic timely resource contention during some of the illustrated events [e.g. Call Admission Control (CAC) or radio resource renegotiation]. In order to tackle these events, it is

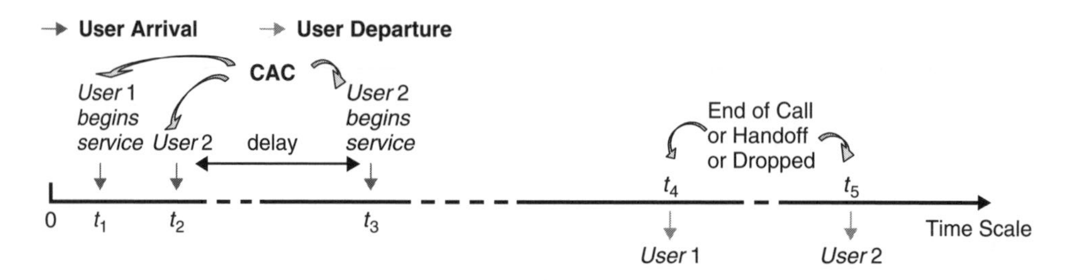

Figure 10.5 Event and temporal representation

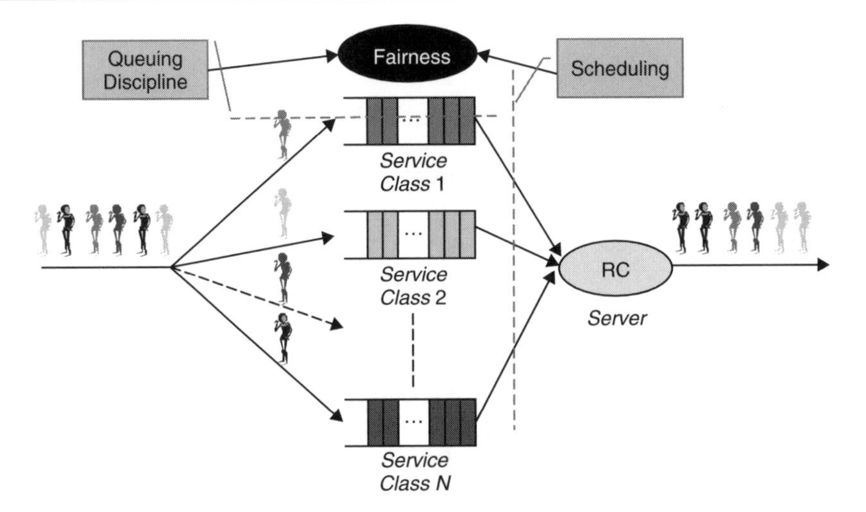

Figure 10.6 Radio resource contention illustration based on the queuing theory formalism

required to introduce specific mechanisms, e.g. scheduling, queuing discipline and traffic load management.

On the basis of the queuing theory formalism, Figure 10.6 illustrates radio resource contention between users when they request different services in the same operating network, where the typical DES system can be modelled by a queuing system [5, 7]. Queuing modelling is very useful when a limited set of resources is available since a call-waiting state can be considered.

In wireless communication systems, typical problems of resource allocations policies and related trade-offs are: (a) each individual user's request must be satisfied as far as possible, (b) radio resource units must be distributed fairly among all users belonging to the same service class as far as possible, and (c) complexity must be as low as possible. Depending on the way these three parts are managed, different operating policies can be designed. In addition to the CAC mechanism discussed previously for the initial access phase, scheduling policies and queuing disciplines are other specific mechanisms tackling RRM issues. The *scheduling* process enables which service class must be completed first and which service class to be processed next. Pre-emptive, no-pre-emptive and other derived policies are possible. For a given service class, the *queuing discipline* process aims at specifying the order in which the server selects the users to be processed first and which user to be processed next. Depending on the system features, different selecting schemes are possible: first come first served (FCFS), last come last served (LCLS), random, round robin (RR) and last finished first served (LFFS) are some examples.

In mobile communication systems, different channel allocation strategies have been proposed, which take spatial-temporal varying traffic loads into account.

- The fixed channel allocation (FCA) scheme provides a fixed RRU pool for each cell. This scheme performs better when the offered traffic load is high.
- The dynamic channel allocation scheme (DCA) provides both a fixed RRU pool for each cell and an RRU pool to be shared between different cells.

- The hybrid channel allocation scheme (HCA) is a mix of FCA and DCA to adapt dynamically to low and high offered traffic loads.

This section has described a high-level view of RRM introducing typical RRM issues and challenges for a single wireless communication system. Further information on the main challenges and outputs of the research activities performed in the field of RRM can be found in the literature [1–3]. Before the RRM functions are modelled, a further analysis of RRM problems is needed.

10.1.3.1.2 Environment Constraints Influencing RRM Behaviour. Loss of energy over a wireless connection leading to the signal degradation due to path loss is very considerable over large distances. The cell coverage can be based on signal coverage or on traffic coverage. Signal coverage can be predicted by coverage prediction models and initially is applied for network planning. Prediction models, especially the widely used point-to-point model, is widely used to illustrate the problem of path loss [8]. In order to solve the path loss problem, proper cellular network design and antenna techniques are needed. In addition to this, diverse technologies, channel coding schemes, joint detection techniques, and power control algorithms are additional powerful RRM techniques to maintain the required QoS of single links.

Owing to the 'Brownian dance' of electrons in all materials and electronic components, thermal noise cannot be neglected. Assuming a certain level of noise power, error-free transmission or within the tolerable bit error rate (BER) requires a minimum signal-to-noise ratio. In real communication system, multiple access interference will considerably add to thermal noise power, which adds more problems to a reliable communication system and, as a consequence, more advanced RRM functions are required, as depicted in Section 10.3.

From the central limit theorem we know that the envelope of summed signals from multipaths without line of sight is represented by a Rayleigh distribution and the received signal is the product of multipath fading and transmitted signal with noise added on. Lognormal fading is another fading parameter, also called shadowing. Besides diverse technology, a channel coding solution and even a joint detection technique, power control algorithms work on single links for maintaining the required QoS.

In future multi-standard environments, co-operation among different operators will bring operational gain for spectrum efficiency. The benefit of JRRM will be discussed in Section 10.2.

10.1.3.1.3 Multiple Access Interference. Multiple access techniques allow the efficient use of scarce radio resources. Furthermore, we differentiate between dedicated usage and shared resource usage. This usage can be supported respectively by circuit switched and packet switched approaches. It is known that the circuit-switched approach allows a guaranteed/reserved capacity for a connection and the packet-switched approach allows a higher utilisation due to a non-guaranteed link capacity allocation.

For an interference-limited system, additional RRM functions must be considered, e.g. outer-loop and inter-loop power control methods must be applied. High data rate services mostly affect system capacity, i.e. the soft blocking bound of system capacity.

10.1.3.1.4 Link Adaptation. When applying link adaptation in flexible systems, e.g. channel coding rate and modulation scheme, two tendencies can be observed. For instance, in HIPERLAN/2 system, it encounters either long time interference to existing users when a lower mode is selected (lower rate), or high interference when the higher mode is selected (higher data rate). As depicted in refs [9–11], there are seven different modes, which can be selected by the CAC, and a link adaptation function during the session/call.

On the basis of the theoretical assumption, the interfered user wants to select a link type to achieve the maximum throughput at a given signal-to-noise ratio. The link adaptation algorithm is carried out in each active link in the interfered cells [12]. It has been investigated that the smaller is the amount of candidate (newly admitted) user traffic, the higher mode should be selected by the CAC [13]. In other words, small bursty traffic should be transmitted at a high mode (high order modulation scheme) to keep the experienced interference low over time to other users.

10.1.3.2 Challenges from Heterogeneous Networks

Current research on wireless communication systems addresses more and more critical RRM issues. The problems related to this and to be solved by RRM in the converged heterogeneous networks context are discussed in the following subsections.

10.1.3.2.1 Challenges for JRRM. The intelligent collaboration between multiple RANs potentially enabled by the emergence of reconfigurable equipment is a promising approach to improve the overall wireless systems capacity and QoS combining the capabilities of each system. The convergence of heterogeneous networks is promising since it extends existing wireless systems capabilities while introducing new services, which could not be supported by individual systems if they were considered independently.

However, this inter-operability approach addresses new RRM issues compared with the case where only one system is managed. Until now, the main RRM issues have dealt with intra-RRMs since existing systems operate independently. With the emergence of collaboration capability among existing independent networks, new challenges have been introduced at both the radio (inter-system RRM) and network architecture levels. Different inter-operability levels can be envisaged depending on the degree of inter-operability at the radio or at the network level. Subsequently, RRM issues in a heterogeneous radio environment (also called multi-radio environment) not only focus on radio level issues but also raise some issues at the network level to suitably manage radio resources jointly and optimally. In the heterogeneous scenario, the joint scheduling algorithm, joint load control and intersystem handover must be further applied in order to increase the overall system capacity. This topic is discussed in Section 10.3.

In this context, the main RRM research work aims to find solutions at least related to the following issues:

1. How do we choose the most suitable RAT (following which criteria?) among multiple co-existing technologies delivering similar or competitive services in a given geographical area?
2. Is intra RRM impacted by inter-system RRM?

3. How do we ensure that the single system performance is not disturbed by the operation of inter-system handovers?
4. How are inter-system RRM rules designed to support scalability, reliability, flexibility (introduction of a new RAT for example)?
5. What are the required equipment modifications needed to support inter-system handover?
6. What is the role of the regulator in the inter-system RRM rules design?
7. How far is inter-system RRM impacted by the operators' policies? What is the flexibility provided by the operators? What are the operators' wishes?

These are some very preliminary questions that require much attention to ensure appropriate inter-system RRM rules design. In addition to purely technical issues (points 1–5), some specific topics dealing directly with operators/regulatory bodies and impacting the way inter-system RRM schemes are designed is more widely developed in Chapter 11. Points 3 and 4 are discussed in Section 11.3.

10.1.3.2.2 JRRM Topics. The transition from a single wireless system management to the joint management of heterogeneous wireless systems implies the re-definition or the re-thinking of RRM basis rules. This transition not only consists in merging the different rules developed for each of the systems independently, but consists in combining harmoniously most of the features of existing RRM schemes. In addition to this, new RRM rules must be designed to overcome the incoming new services specificities (high bit rate, high system capacity) through new business models.

This is the challenge of the intra-system RRM to inter-system handover transition. All the mechanisms relative to RRM on a joint basis between heterogeneous networks are included under the terminology JRRM. These related topics are also discussed in [6]. As opposed to the single operation, when a user requests a service (application) in a heterogeneous environment, several networks can be available to support the request. Basically, at least five levels of contention can be distinguished, as depicted in Figure 10.7: application, bearer, transmission mode, RANs and spectrum.

In this context, a user is provided with one or several bearers capable of supporting the required service. Each of these bearers can be transmitted by one or several transmission modes (such as unicast, multicast or broadcast). Then, each transmission mode can be supported by one of several RANs. For each RAN, different spectrum management strategies are possible. The introduction of flexible spectrum management schemes can also enhance the RRM capabilities. This last topic is more widely discussed in Chapter 11.

For each contention level, a set of units is available (e.g. the 'Transmission Mode' set is composed of the 'unicast', 'multicast' and 'broadcast' units). This raises specific issues and leads to a careful choice between the units of each set to ensure user satisfaction while optimising the impact on the overall system (i.e. avoid the waste of radio resources and maximise the number of active users in the system).

In addition to the issues relating to each contention level, it is also necessary to address the issues relative to the interfaces between those levels (denoted by ❶, ❷, ❸ and ❹). This also is of concern when designing appropriate JRRM schemes.

In the following subsections, some more detailed discussions on the sets, contention levels and interfaces are provided.

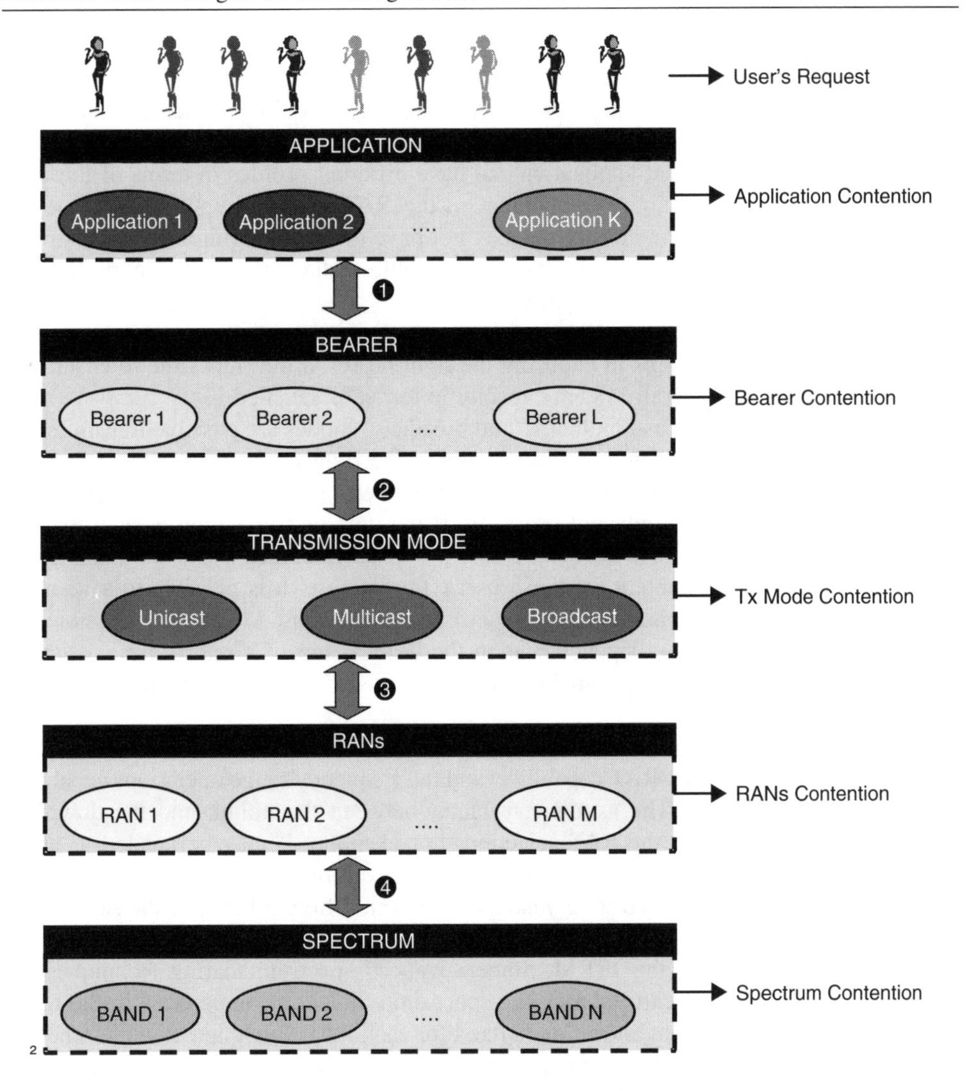

Figure 10.7 Contention levels of JRRM

10.1.3.2.3 Input Parameters for the JRRM Function. In this subsection, issues related to the previous contention levels, sets and interfaces are discussed.

- *User and terminal profiles.* One of the main features of the terminal profile is the radio user communicating equipment capability and features explained in Chapter 1. The user equipment may support one or several RATs and may operate on one or several frequency bands for each RAT. In addition to this, it can be imagined that several links can be operated simultaneously with the same RAT or between different RATs (radio multi-homing). This directly impacts on how the service can be provided to the user and how the radio resources can be managed. The joint scheduling approach shown in Figure 10.13 reveals further details on radio multi-homing.

An important feature of the user profile is the internal user equipment capabilities. For example, the service requested by the user (e.g. visualisation of images on the terminal) must be compliant with his equipment capabilities (e.g. a colour screen is required to display images) to avoid unnecessary use of radio resources.

- *Application.* The requested application can have different profiles in terms of throughput and real-time constraints. Depending on the RAT used to support the requested application, the application constraints may not have the same requirements. Moreover, the requested service may be asymmetric (for multimedia services it is expected that more traffic is offered in the downlink).
- *Bearer services.* Different combinations of bearer service can support the application. The main challenge remains in choosing the right bearer at the right time to ensure the service QoS and the overall system capacity in terms of satisfied users.
- *Transmission mode.* Unicast, multicast and broadcast modes are possible transmissions modes to support each of the previous bearers. The right choice can be based on the commonalities of the service required by a set of users. For example, if an application is requested by a large number of users and if the type of application is such that the content is common to most of these users, it can be imagined using only one radio channel to deliver the content for those users. In that case, it is possible to save radio resources by the use of multicast or broadcast transmissions. Moreover, the choice of transmission mode can be dependent upon the need or not of a return link (typically, continuous information streams application – e.g. 'pushy' traffic information – does not need a return link).
- *Radio Access Technology (RAT).* The appropriate choice of RAT depends on the user terminal capability (multi-RAT capability) and the frequency range operations capability (multi-band capability). The asymmetry factor between the uplink and the downlink can also be an input for the RAT selection. For example, it may be needed to use a bi-directional unicast RAT if no information is sent on the return link.
- *Spectrum.* The way the spectrum is managed between different RATs of the same operator or the way the spectrum can be managed between different operators can provide additional flexibility for the JRRM. Some advanced spectrum-sharing techniques (as depicted in Chapter 11) can improve the spectrum efficiency compared with the traditional case where spectrum allocation is fixed for each RAT. Such new flexible schemes are promising to cope with asymmetric traffic management expected for the multimedia services to be delivered in the near future.

10.1.4 RRM Modelling and Investigation Approaches

Operating wireless communications systems can be formalised (see Section 10.1.3) as a whole finite state machine (FSM) where the system state space (SS) is discrete and where the SS evolution depends entirely on the occurrence of asynchronous discrete events over time. Therefore, such a system can be modelled as a DES. Such a DES is defined as being a system in which the SS is a discrete set and the State Transition mechanism is the action of asynchronous discrete events over time; this mechanism is said to be event-driven (ED) [5].

Depending on which RRM issues are to be solved, different approaches for RRM problem modelling and resolution are possible (analytically, or numerically with simulations).

The main challenge remains in the right choice of the method for the particular issue so that RRM scheme design or analysis is appropriately performed.

10.1.4.1 Analytical Approach

In a wireless communications context, the analytical approach expectations are typically used to provide asymptotic system performance. The analytical approach for RRM issues modelling and analysis is usually based on control and system theory. Typically, birth/death Markov chains are appropriate tools to assess preliminary system performance boundaries.

On the basis of Markov chain properties, queuing theory provides powerful tools when assessing more accurately radio resource contention issues, e.g. between different access points (i.e. servers) and users (i.e. customers). Typical simple RRM issues can be modelled and solved with this approach. In general terms, three following basic elements define a queuing system: (a) customers: these are input entities waiting in their quest for resources; (b) servers: these are resources entities offering the requested service and for which the waiting is done; and (c) queue: this is the space where the waiting is done. The queue is entirely specified by its capacity (i.e. the maximum number of customers that can be accommodated in the actual queuing space) and its discipline (i.e. rules according to which the next customer to be served is selected from the queue). In the specific case of wireless communications systems: (a) Customers are user arrivals requesting one or more services in the single or multi-radio environment (b) Servers are the limited radio resources units (codes, slots, frequency carriers, etc.) specific to the RAT. Here, servers are the combination of one or several RRU to define radio channels. The channel is viewed as a delay block, which holds a user for some amount of service time. (c) The queue is a finite length buffer.

As discussed in Section 10.3.2.1, the way the different user classes can be managed can lead to different scheduling policies. Similarly, the way the users can be managed for a given class can lead to different queuing disciplines.

However, this approach is only suitable for a given class of stochastic DESs and is it not generally applicable to all classes. This approach is more dedicated for RRM scheme description than for RRM scheme performance evaluation. The limitations of the analytical approach based on Markov chains are that it is difficult to include radio conditions constraints in the modelling. Therefore, performance is somehow independent of some realistic systems behaviour. Moreover, the previous analytical tools are mostly based on the Markovian property. Thus, when the wireless system under investigation does not conform to the Markovian property, fewer tools are available and modelling is less straightforward. In addition to the pure queuing theory approach, some more advanced techniques have been proposed to deal with specific resources contention issues [5]. Typically, a stochastic timed automata language has been introduced to cope with low complex DESs. At the same time, stochastic timed Petri nets is a powerful tool based on a language enabling to cope with more complex DES. Depending on the issues to be modelled/solved and the DES properties under consideration, different Petri nets families have been developed to deal with specific issues. Petri nets offer a large plethora of adapted tools to model, analyse, design and synthesise wireless communications systems.

10.1.4.2 Computer Simulation Approach

To model more realistically any relevant wireless communication systems, including radio conditions, the use of ED (or simultaneously timed and ED) methodology based computer simulations is needed. This approach enables to:

- consider systems more deeply in term of complexity;
- cope with systems that do not conform to Markov chains and queuing theory assumptions;
- perform relevant parameter sensibility analysis; and
- assess accurately a given scenario performance.

In addition to this, this approach provides powerful simulation capabilities to specify and design accurate and 'implementable' algorithm into products.

However, in the same time, this approach is often dedicated to specific scenario investigations (not so much a generic as an analytical approach, which can cover a wider spectrum of scenarios) and requires subsequently a huge amount of computer development effort (which may also require some non-negligible additional time).

Is there an ideal approach? There is no straightforward ideal approach when investigating radio contention issues for wireless communications systems. The ideal method is the one that best suits the addressed problem in terms of modelling and resolution.

However, a powerful approach could consist of combining both analytical and computer approaches so that modelling, analysis, design, synthesis, control and optimisation capabilities are available. Chapter 11 gives detailed examples of JRRM over UMTS FDD and TDD systems and simulation results of Joint Call Admission Control (JCAC) and intersystem handover procedures.

10.2 Investigations of JRRM in Heterogeneous Networks

With the emergence of co-existing different radio interface and protocols, meaningful measures dealing with such a complicated system must be considered. In this section the concept of JRRM dealing with managing radio resources in heterogeneous networks is introduced. A link to detailed explanations is also given. Here, a theoretical proof shows that JRRM leads to lower packet loss and higher spectrum efficiency. In the theoretical analysis part, we have not included the intersystem handover phenomenon. Nevertheless, the number of users leaving a system is equal to the number of arrivals, and thus the bounds for capacity gain are valid.

10.2.1 Measuring Gain in the Upper Bound Due to JRRM

In this section the basic system capacity gains due to JRRM are analysed. There are two basic layers of JRRM that are of great interest, namely the joint call admission control (JOCAC) layer and the joint radio resource scheduling (JOSCH) layer. A different evaluation approach is described according to the packet-switched and circuit-switched nature of JOCAC and JOSCH.

System capacity can be evaluated in the JOSAC/JOCAC layer, where the incoming traffic is not split, i.e. the session/messages cannot be split over different networks, but can be admitted alternatively to different sub-networks. In the case when the incoming traffic can be split over sub-networks with the support of joint scheduling algorithm in radio frame level, extreme gain can be obtained. References [14] and [15] depict the dimensioning of the scenarios and specify in more detail the theoretical scenarios, concepts and definitions.

10.2.1.1 Switching Algorithms: Packet-Switched Services and Circuit-Switched Services

If the dedicated channels are assigned to services, a circuit-switched-based model can be used to evaluate the performance. UMTS supports both the packet-switched service and circuit-switched connections [16, 17]. WLAN, i.e. HIPERLAN/2, only supports packet-switched connections (shared channel) in the radio frames. All LCH channels are shared, which is only indicated by the FCCH channel in the header of the radio frame, where the location of LCH for particular users is indicated [18, 19].

10.2.1.2 RRM Control Layers: JOCAC and JOSCH

JOSAC does not provide detailed traffic-splitting to sub-networks; it only can give a gain due to traffic routing, i.e. alternatively divert incoming traffic into different systems. The joint scheduling algorithm provides a refined traffic split, which further improves the throughput of the collaborating systems.

10.2.1.3 Interference Limited Systems

System capacity greatly depends on the load, fading and reception qualities of the network. Different loads on the network constrain the traffic management algorithms. It is described in Section 10.3.2.3 that a higher loaded network likely cannot accept further traffic. Basically this follows the water-filling theorem, which is well known in information theory [20]. In order to minimise the system load, incoming traffic should be diverted or split into less loaded systems with a higher volume of traffic.

10.2.2 Circuit-Switched System

For networks with different capacities, the gain in circuit-switched traffic can be obtained by the well-known Erlang B formula, which is applied in conventional telephone systems [7, 21]. In Figure 10.8 the upper figure is the queuing model for two independently operated systems. If we assume the incoming traffic is based on a Poisson distribution with arrival rate λ and one basic channel has the capability to process the incoming call with rate μ, then the call-blocking probability is calculated as:

$$p_{\mathrm{i}} = \frac{(\lambda/\mu)^{M_{\mathrm{i}}}/M_{\mathrm{i}}!}{\sum\limits_{k=0}^{M_{\mathrm{i}}} (\lambda/\mu)^k/k!}, \tag{10.1}$$

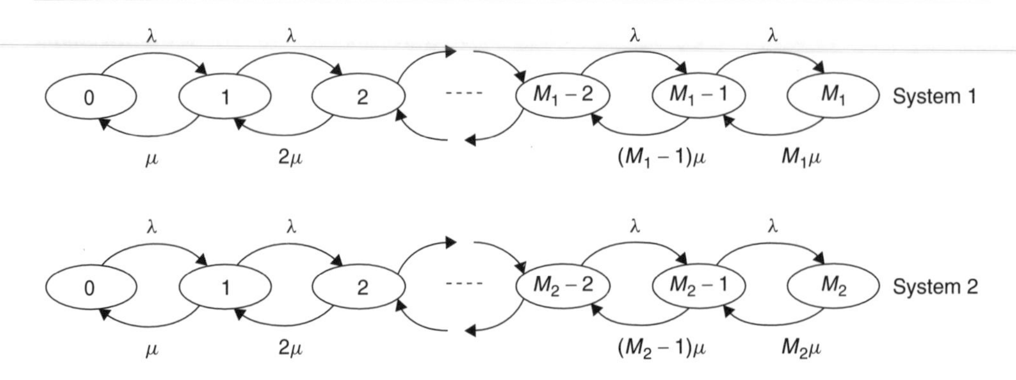

Figure 10.8 Two independently operated systems

where M_i is the maximum number of servers in sub-network i. In this example, $i \in \{1, 2\}$. The blocking probability for the overall system is then calculated as the average of the blocking probability in each sub-system.

10.2.2.1 Gain Due to Resource Sharing with JOSAC Involvement

The sub-networks co-operate with each other, i.e. the incoming calls can be admitted by any sub-system and only stay in the CAC/SAC level. Therefore, in this case, the call-blocking probability is:

$$p = \frac{(\lambda/\mu)^{\Sigma_i M_i} \sum_i M_i!}{\sum_{k=0}^{\Sigma_i M_i} (\lambda/\mu)^k / k!}. \tag{10.2}$$

The system capacity can be evaluated by the maximum system load once the call-blocking probability (system grade of service parameter) is given, i.e. since the conventional Erlang B formula is monotonically characteristic, the higher the number of server, the less the blocking probability. The gain of co-operating systems versus stand-alone systems can be evaluated by a comparison of the maximum admitted loads by each system. It is shown in the traffic-scheduling layer that more gain can be achieved, as depicted in Section 10.2.2 (see Figure 10.9).

10.2.2.2 Gain Due to Traffic Splitting with JOSCH Involvement

Suppose the incoming traffic can be split according to the splitting factor, γ, i.e. the incoming traffic is split over two sub-networks, $\gamma = 2$. Suppose the traffic arrival rate to the sub-networks is λ, we then calculate the processing rate in both sub-networks according to traffic split. If the same unit of traffic load is split over γ sub-networks, then the processing rate through the birth–death chain is calculated as $\mu' = \gamma \cdot \mu$. For a conventional traffic unit split by γ, the maximum number of basic servers will increase γ times. The model is depicted by a hierarchy structure. The main stages are identical to

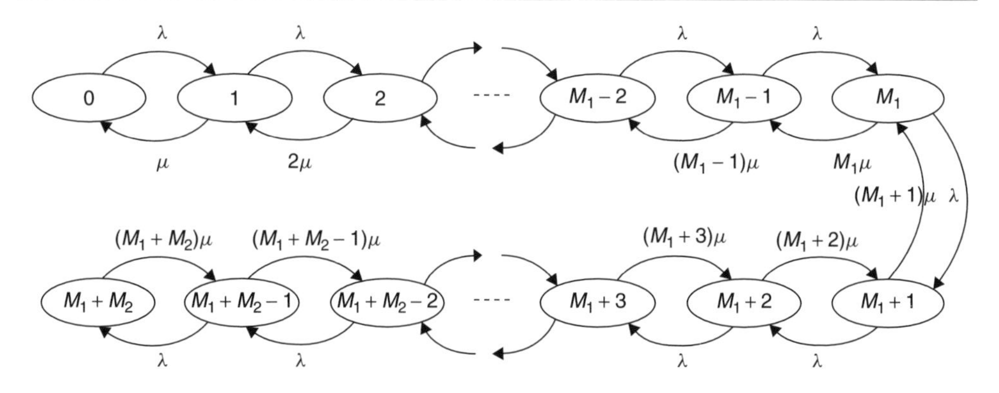

Figure 10.9 Two co-operative systems (CAC layer)

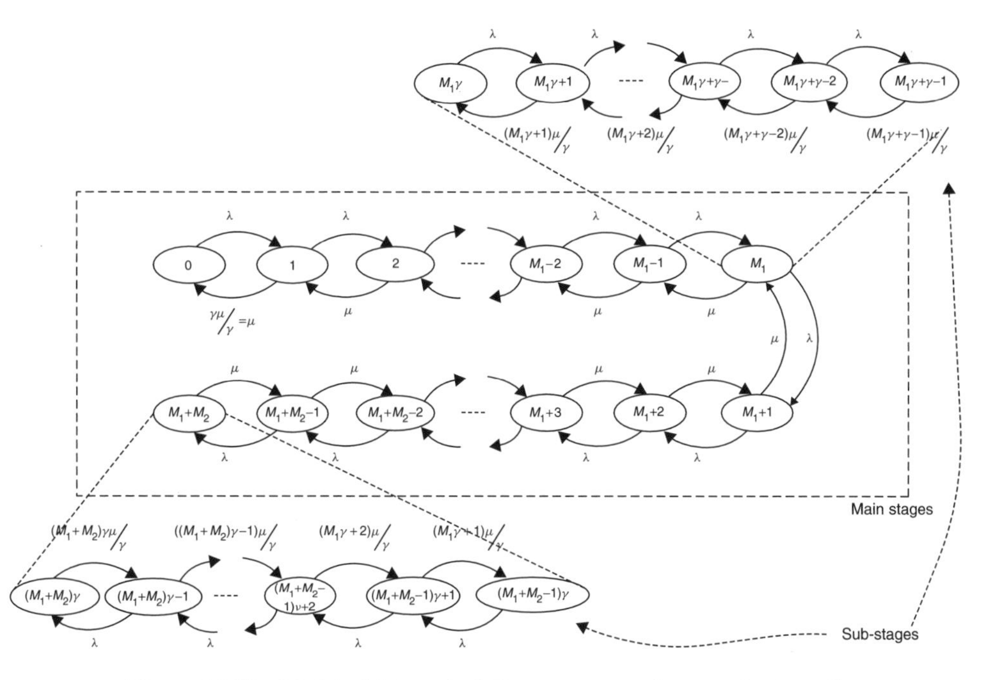

Figure 10.10 Model of joint scheduling over two sub-networks ($\gamma = 2$)

the previous models. The difference with the joint admission case is that each stage in the joint admission case has γ sub-stages in the joint scheduling case; therefore there are $\gamma \cdot (M_1 + M_2)$ stages (Figure 10.10).

The probability that the system is in the kth basic stage is:

$$p_k = \begin{cases} p_0 \left(\dfrac{\lambda}{\mu}\right)^k \dfrac{1}{k!} & k \leq (M_1 + M_2) \cdot \gamma, \\ 0 & \text{else.} \end{cases} \tag{10.3}$$

This equation is derived in the same way as equation 3.11 in ref. [21]. The probability of the system being in the initial state is calculated as:

$$p_0 = \left[\sum_{k=0}^{(M_1+M_2)\gamma} \left(\frac{\lambda}{\mu}\right)^k \frac{1}{k!} \right]^{-1}. \tag{10.4}$$

The call-blocking probability, i.e. that all channels are busy, is:

$$p_B = \frac{\left(\frac{\lambda}{\mu}\right)^{\chi} \Big/ \chi!}{\sum_{k=0}^{\chi} \left(\frac{\lambda}{\mu}\right)^{k} \Big/ k!}, \tag{10.5}$$

where the complete system availability (complete number of basic channels) is modelled by $\chi = (M_1 + M_2) \cdot \gamma$.

It should be pointed out that the traffic split must consider the user perspective, e.g. traffic must be split according to application requirement. For instance, if a user downloads HTTP traffic through a number of air interfaces – UMTS, DVB and wireless LAN – the application is divided into different sub-streams. The most important sub-stream should be routed through the most stable system with the highest coverage. The remaining streams are routed through the other networks. For HTTP traffic, the frame/structure is more important and the in-line objects are less important.

The principle of this model is to consider that all the split packets are of the same priority. Therefore, the call-blocking probability is identical to the one defined in equation (10.1). The blocking probability p_B defines the grade of service (GoS). From the GoS value the maximum traffic ρ, i.e. the system capacity, that can be handled by the system is calculated. The gains due to different JOSAC and JOSCH are calculated according to the corresponding capacities compared with the non-co-operating networks. The results of the maximum gain from the scheduling algorithm and the simulation parameters are listed in Table 10.3.

Table 10.3 Parameters of the analytical model

Parameter	Characteristics
Number of sub-networks	$\gamma = 2$
Scenario 1	Non-co-operating systems
Scenario 2	Joint admission controlled system (co-operating systems only at call/session setup, i.e. no traffic split during session). Furthermore, no vertical handover during the session
Scenario 3	Joint scheduling system (cooperating systems for data link level. The incoming traffic belonging to the same session is split over γ different sub-networks)
Basic channels	Varying
GoS (blocking probability)	0.01

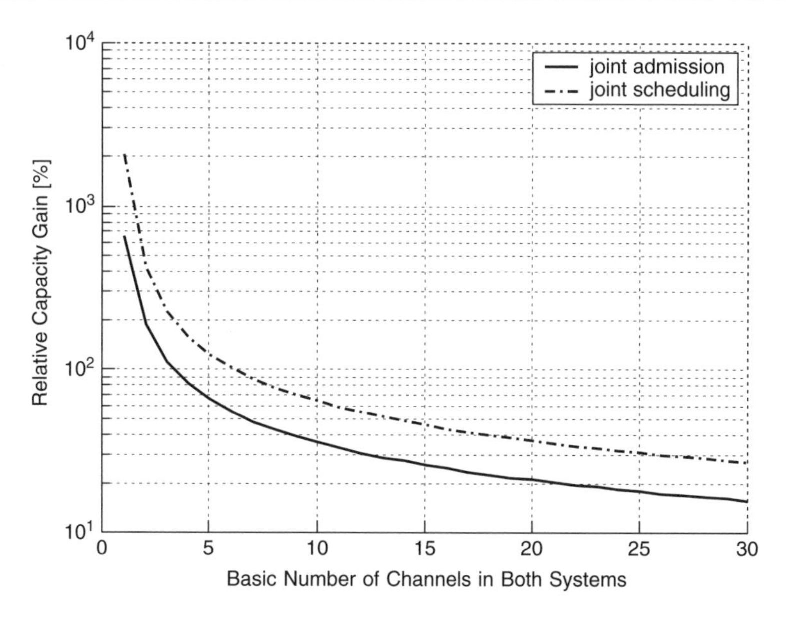

Figure 10.11 Relative gain due to JRRM in circuit-switched networks

The current analytical work is based on a finite number of servers. In this case, the gain of JOSAC/JOCAC and joint scheduling models can be obtained by numerical analysis and system-level simulation (Figure 10.11).

10.2.3 Packet-Switched System

In this section, the gain of JRRM is modelled by the M/G/R PS model, which was first described by Lindberger [22], where 'M' stands for Markov process, 'G' stands for generic processing rate and 'R' is the number of reserved servers according to the peak rate of incoming traffic. The term 'PS' stands for packet-switched service. The expected transfer time for the M/G/R PS model is given by the following equation [15, 23]:

$$T(x) = \frac{x}{r_P} \cdot \left(1 + \frac{E_2(R, R\rho)}{R(1-\rho)}\right) = \frac{x}{r_P} \cdot f_R, \tag{10.6}$$

where the term f_R determines the delay factor, x is the size of the file, ρ denotes the utilisation of the link ($\rho = \lambda_e \cdot \overline{x}/c$), with $\overline{x}$ the average value of the file size, c the channel capacity and λ_e the traffic arrival rate. E_2 is the Erlang C's formula defined as:

$$E_2(R, A) = \frac{\dfrac{A^R}{R!} \cdot \dfrac{R}{R-A}}{\displaystyle\sum_{i=0}^{R-1} \frac{A^i}{i!} + \frac{A^R}{R!} \cdot \frac{R}{R-A}}. \tag{10.7}$$

As the previous analysis shows, the gain of the joint admission control is determined by the shared capacity; the gain of the joint scheduling algorithm is achieved by the traffic split over the involved sub-networks.

It is assumed that no error on the links occurs, i.e. no Automatic Repeat Request (ARQ) mechanisms are required. The M/G/R PS model can be used to calculate the gain achieved by the co-operating networks. If the peak rate of incoming traffic is smaller, the gain will be bigger. It can be stated that the occupation factor of the traffic in the radio link is highly related to the joint admission control algorithm.

With traffic splitting and sharing of radio resources, higher capacity gains are obtained. Since the average file size is reduced due to traffic splitting, the sojourn time can be decreased according to the changed traffic model in sub-networks thanks to the traffic split (Table 10.4).

On the basis of Table 10.4, the following results can be obtained. It can be seen from Figure 10.12 that for a given delay upper bound for the in-coming traffic, the highest arrival rate of the in-coming traffic can be afforded when the joint scheduling algorithm is applied.

It can be seen from Figures 10.11 and 10.12 that the gain depends heavily on the number of available channels and the system bandwidth. For sub-systems with very high bandwidths or much higher capacity than the traffic demand, the gain is negligible.

Table 10.4 Analytical model of the joint scheduling algorithm

Parameter	Characteristics
Number of sub-networks	$\gamma = 2$
Peak rate (original traffic)	5 Mbits/s
Average file size	10 kbytes
Utilisation factor ρ	[0, 1]

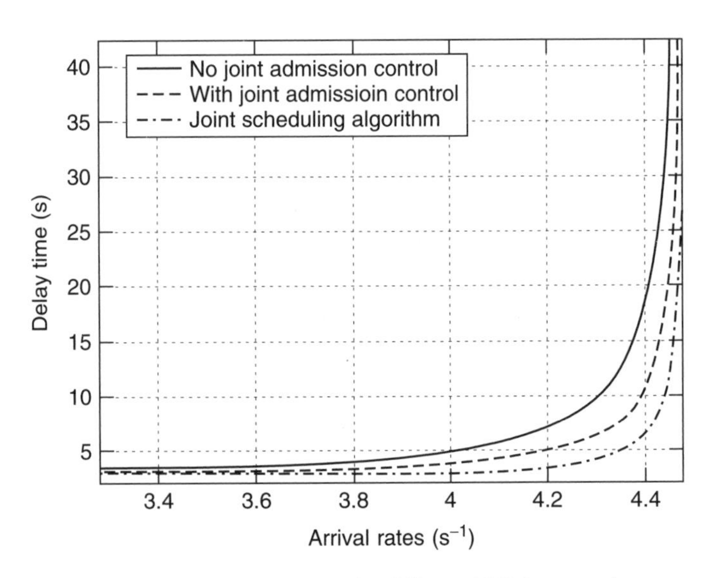

Figure 10.12 Performance for different RRM approaches

10.3 Functions and Principles of JRRM

10.3.1 General Architecture of JRRM

10.3.1.1 Interworking Among Different Sub-Networks

In what follows we discuss the architectural components needed for JRRM. The interworking between different RATs requires new protocols defined for convergence reasons. It should also offer IP packet-based convergence sub-layers to networks to guarantee QoS. Owing to the heterogeneity of co-existing different networks, many different policies are conceivable for joint management functions, in particular when considering legacy and new network types. Systems in different generations are equipped with different functionalities, protocols and management requirements. Thanks to future reconfigurable terminals providing simultaneous connections to different RATs, a very tight interworking between networks is important to benefit from soft handover or load balancing algorithms. In general, the loosening of very tight coupling schemes between different network types must be considered for the provision of multiple connections, as detailed in Chapters 1 and 5. For a possible very tight coupling between a UMTS subsystem and a wireless LAN or 4G system, one must consider the restrictions in each sub-systems, e.g. the transport block size and minimum transmission time interval for each are defined differently according to the specifications. Very tight coupling allows the joint scheduling of traffic streams between involved networks and terminals. Joint interworking must also take into account the user, service, network and terminal profiles, which consist of static and dynamic features, as outlined in Chapter 1.

A two-stage admission control approach is shown in the following subsection (see Figure 10.14). It is introduced to define the handling of network operations related to the static and dynamic networks, service, terminal and user profiles. The first stage handles the static features, whilst the second stage handles the dynamic features, e.g. the current QoS of each user. In a co-operating environment with different co-existing subsystems, the JRRM working with different scales will enhance the utilisation of resources. The spectrum can be shared by a dynamic allocation scheme; the load can be balanced by a joint admission control and load control scheme; an even lower layer resource management function e.g. power control, resource scheduling (RS)algorithms, can jointly contribute to the spectrum efficiency.

The conventional CAC is designed for each access system to work independently of co-existing RANs. In the heterogeneous co-operating environment, a joint session/call admission control must be defined. The neighbouring RAT system load is taken into account by the joint session admission control (JOSAC), as shown in Figure 10.13. The traffic stream can be routed alternatively through the co-operating sub-systems according to the restriction and load conditions of each. For instance, for wide coverage reasons we prefer a cellular system, e.g. GSM or UMTS, whereas very high transmission rates can be obtained by using wireless LAN. With the information of estimated load in all the sub-networks (dynamic network profile), the joint load control entity (JOLDC) located together with JOSAC will distribute the traffic based on the characteristic of the co-existing RATs (static and dynamic network profiles), the QoS requirements for the service and the number of applicants for the software download service to determine the software download

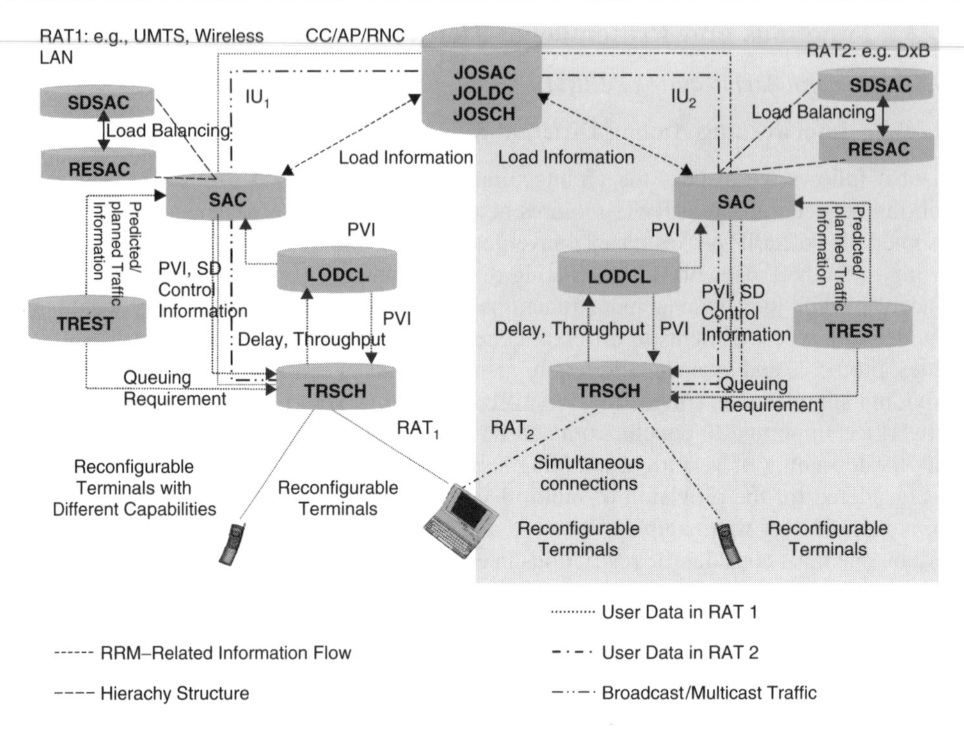

Figure 10.13 JRRM architecture

strategy, i.e. which RAT and channels with sufficient capacity should be selected. The joint resource scheduler (JOSCH) is important for terminals working with simultaneous connections to different networks. JOSCH is responsible for scheduled traffic streams being split over more than one RAT. It helps to optimise the utilisation of radio resources in the whole system. It also synchronises the stream being split, e.g. a video stream, with the basic layer and the enhancement layer being transmitted over different air interfaces individually or separated into main object and inline objects of HTTP service belonging to the same session, etc.

The SAC entity in each sub-network consists of a software download traffic control entity (SDSAC) and a regular traffic control entity (RESAC). On the basis of the time schedule and QoS requirement of the SD traffic the SDSAC assigns an optimal broadcast or multicast bearer service to distribute the download to the terminals, when many terminals are affected. The load generated by this push service traffic and regular traffic should be jointly balanced by SDSAC and RESAC by taking an appropriate SD strategy to determine the needed common channel capacity defined by the JOSAC.

In the case when the single traffic is split into two streams carried by two RATs in a very tight coupled manner, the joint admission control must admit both connections to the systems. In this tighter coupled case, queues for each sub-system are needed for the amount of data coming from the same traffic stream.

10.3.1.2 Interworking between Different Layers

Because of the requirement for heterogeneous services, the queuing of data stream and priority arrangement should be executed by TRSCH. Some important input information required by CAC must also be offered by TRSCH. The trade-off between maximising the utilisation of the system and reducing the rate of dropping or negatively affecting the QoS demand of users must be considered to design an efficient CAC algorithm. The interworkings between CAC/SAC and RS are shown in Figure 10.14.

The incoming traffic for the system is divided into different traffic types after the first stage. The traffic types with different QoS requirements follows, i.e. the real-time requirement, the throughput requirement, etc. The location of users is part of the dynamic user profile information to decide on the bearer services to be assigned. Therefore, SAC selects the transmission physical mode of the bearer service or drops the application in the case when the network cannot provide the requested service. On the basis of the chosen static service and network profile in the first stage (JOSAC), the first stage of the CAC assigns a certain range for the scheduler priority weights (PW) defined for the service types based on the network, terminal and user profiles, which are offered to the second stage in the SAC. In the case when the incoming calls ask for service beyond the static restrictions, the JOSAC should reject them. The very tight-coupled traffic stream over two RATs should be scheduled by the JOSCH, which works between the JOSAC and

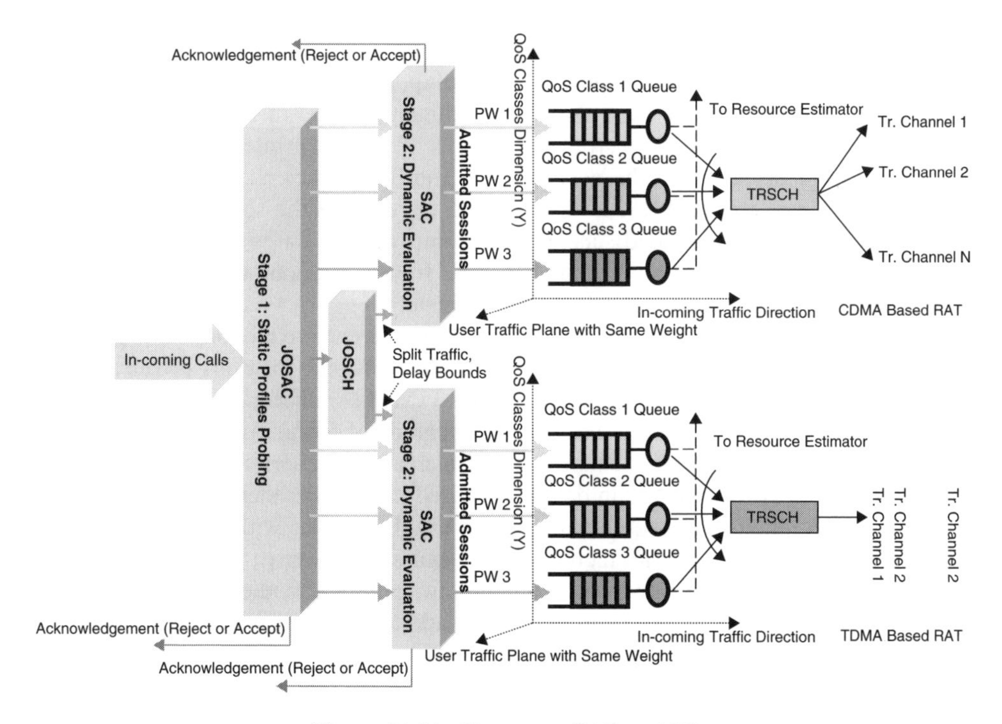

Figure 10.14 Two-stage CAC and RS

the SAC. The split traffic after the JOSCH is forwarded to the individual SAC in each RAN. With the offered control information from the JOSCH, the SAC should map the split traffic into conventional traffic types with a concrete PW.

As introduced in Figure 10.14, the QoS class dimension (Y) of the scheduling allocates resources for different QoS classes. The cost function upon which the CAC is based requires the PW for each traffic type. In this case, the cost function parameter will be harmonised with the Y-dimension of the scheduling algorithm. The principle of the generic process-sharing algorithm (GPS) [24, 25] can be applied to guarantee the resources for the sessions with different priority; i.e., a minimum service rate can be obtained as the ratio between the chosen weight and the sum of all weights. In addition, the PW can also be applied for the calculation of allocated resources for the chosen service type according to the weights assigned for this type. The PW can be handled by the network operator to offer dedicated radio resources to particular users.

10.3.1.3 Interworking Between Different Services

As explained in Chapter 1, the software download managed by the software download module is one important issue for reconfigurable systems. The impact of mass downloads on regular traffic in wireless networks is an important issue, which is tightly related with the peer-to-peer (P2P) QoS. For a soft blocking problem restricted system, the impact between download-related traffic and regular traffic can be modelled by an interference value to evaluate the impact of the chosen software download strategy. In general, we must consider point-to-point and point-to-multipoint software downloads. Point-to-multipoint can be carried out by a shared or common channel in a network. For many users a broadcast channel with a certain data rate is of particular interest here for the cell grouping strategy discussed in the next chapter. The SDM takes charge of either downloading the software from a PRM in the network, or retrieves the software from another terminal (decentralised download schemes) or from local databases in the terminal, e.g. libraries, which may contain the required module from a previous reconfiguration. It is also responsible for forwarding detailed SD parameters.

The download channel utilisation and download interference reduction are of particular interest when choosing a download strategy [26]. The probability of using broadcast channels for downloading software to a mobile terminal (MT) is rather high when many users want to receive the same information or a mass upgrade of terminals takes place. Broadcast channels are not fast power controlled and increase the interference to surrounding cells. If the broadcast with a certain data rate is used in all cells, then the cell capacity is further decreased due to the cell coupling in the CDMA system with a frequency re-use one. Cell coupling means that an increase in transmission power in one cell will lead to an increase in transmission power in neighbouring cells due to the need for maintaining the target SIR for services. To minimise the additional interference due to software download when using, for example, cell broadcasts, a specific download management for all involved cells in a geographic area must be applied. By spreading download traffic, a cell-grouping strategy will greatly decrease the interference to the regular traffic and the download traffic in neighbouring cells. Also, parallel downloads for different cell groups can be applied. In other words, in cells with a low cell-coupling factor, e.g. cells in a

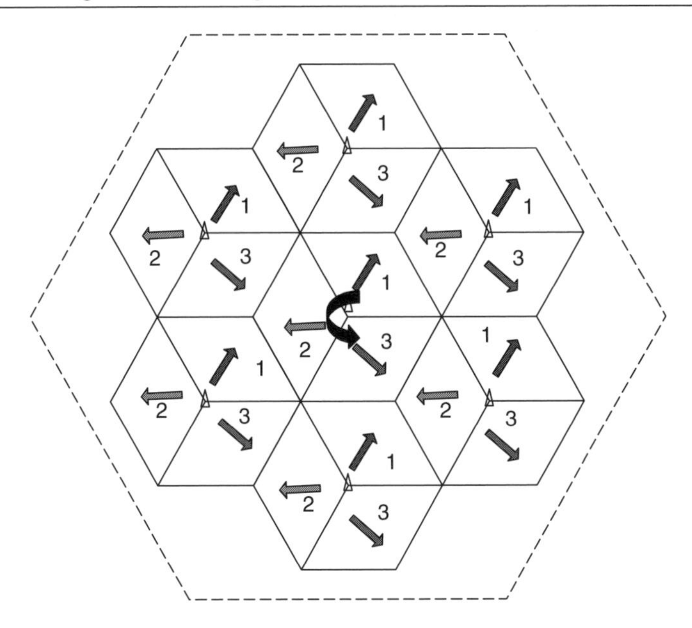

Figure 10.15 Cell-grouping algorithm in the WCDMA macro cellular system

macro environment, where their 120° sector antennas are oriented towards one geographic direction, can be clustered into one download group, as shown in Figure 10.15.

Figure 10.16 illustrates the interference reduction due to software download using this cell-grouping approach [27, 28], where the interference in the figure means interference to regular traffic. It can be seen that a certain gain has been obtained by the cell-grouping strategy. A higher download probability can be achieved by repeating download broadcast sessions.

10.3.2 Detailed RRM Functions in Sub-Networks and Overall Systems

Detailed functions of the JRRM will be tackled in this section. The load information and traffic information are required to be shared by co-operating networks. Each RAN needs efficient interworking between traffic volume, the measurement (prediction) function, the scheduler, the load control unit and admission control. We will show how CAC takes into account the scheduler with an efficient PW assignment.

To admit new users, the objective is different to dropping an already ongoing call (active user), which is generally perceived to be more annoying than blocking a new call request (new user). Hence, if an overload situation occurs because of the assignment of a new user to the channel, an appropriate resource management decision is to refuse the new user. By doing so, one will favour the already active users in the system.

A new user requires a base station and a channel, and the CAC has to make trade-offs between channel quality, low outage and blocking probabilities. In general, several users may concurrently seek admission. For the QoS requirement, the delay tolerance according to different services should also be considered besides the as well as the Bit

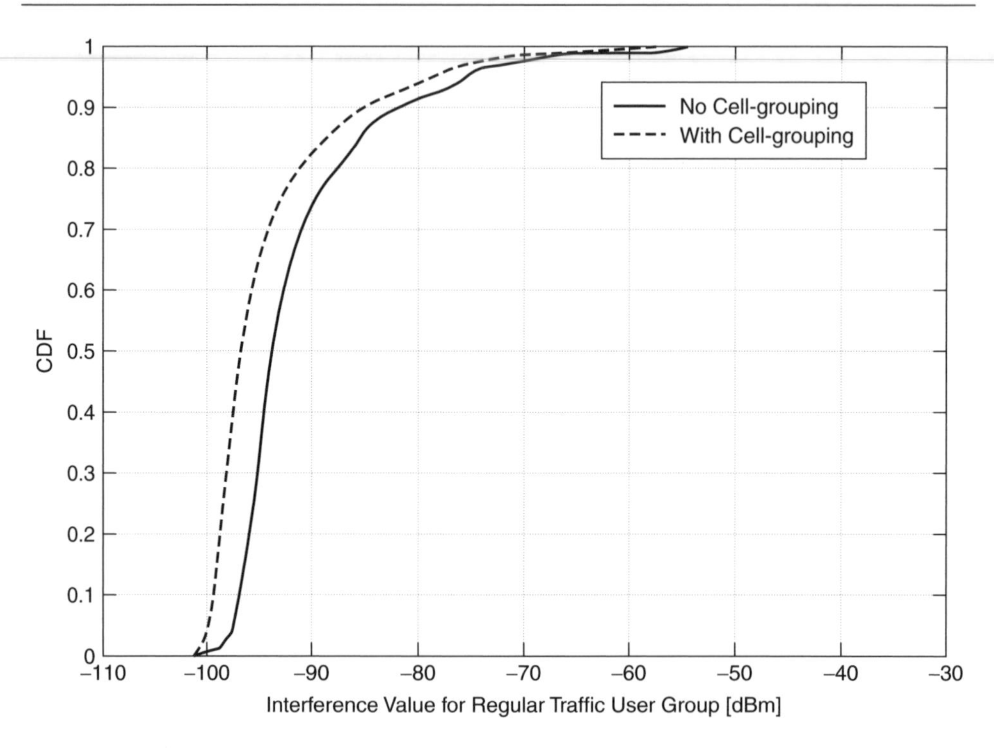

Figure 10.16 Interference reduction with the cell-grouping algorithm

Error Rate (BER). This can be addressed either by a non-reprioritised admission type algorithm, or by an algorithm that considers the user's priority.

10.3.2.1 Principles of Scheduling Algorithms

Radio resource scheduling works in the Radio Link Control (RLC) layer, controlling the scheduling of transmission units with different priorities. The radio resource allocation is executed and realised by scheduling algorithms. In this subsection, the radio resource scheduling algorithm is analysed and examples of scheduling algorithms in the TDMA and CDMA systems and the relevant performances are given.

10.3.2.1.1 Overview of Scheduling Algorithms. Along different protocol stacks in wired network and radio systems, queues corresponding to different traffic classes are defined in different layers, e.g. the IP layer, TCP layer or RLC layer are established. Therefore, each one of these queues utilises a concrete scheduling algorithm according to the configuration or traffic nature in that layer. The Transmission Time Interval (TTI) value related to radio delivery is also explained, which is a very important parameter for scheduling algorithms. The TTI value corresponds to a parameter setting for each transmission delivery, i.e. it gives the minimum transmission time. In brief, the performance of scheduling algorithms depends on the basic disciplines, e.g. First Come First Served (FCFS), Round Robin (RR), Last Come First Served (LCFS), whereas the scheduling delay requirements for each transmission are bounded by the TTI value specified for the bearer services in radio systems.

FCFS Algorithm. The FCFS scheduling algorithm describes the sequential arrival of data (packets) in a queue, in such a way that these traffic units are delivered in the same order as they arrived. This kind of scheduling algorithm is intended for both low-data, rate-real time traffic, e.g. voice over IP traffic. The earlier arrived IP packets, i.e., talk-spurts, are scheduled with higher priority.

RR Algorithm. The RR algorithm is a well-known and widely used scheduling algorithm. The services are being scheduled based on a time-shared delivery manner. Quantum is the main parameter for this algorithm, which is defined as the time unit when the customer is served (transmitted). Once the service time defined by Quantum expires, the radio resource resources are assigned to the next customer according to the Quantum size (it has close similarity to a circular queue). In addition, when the traffic unit belonging to the customer being served finishes before its Quantum time is over, the system waits to serve the next customer until the next serving cycle arrives. From the time-sharing nature of the RR algorithm, if the Quantum size defined is very large, RR shows similar behaviour to FCFS.

Modified RR Algorithm. The main difference between the modified RR algorithm and the conventional RR algorithm is that in the modified RR algorithm, when the Quantum is released the resources are given to the next customer who has data stored in the buffer. In such a way the modified RR algorithm presents a better utilisation of resources with a great delay time improvement. This algorithm is applied in the simulations for HTTP using dynamic scheduling in DCH, which are described in the following sections.

LCFS Algorithm. The LCFS algorithm is mainly intended for high-bit-rate, real-time services. The best appropriate traffic for this scheduling algorithm is video traffic. The method of polling users in the same QoS class with successive transmissions requires the recalculation of the virtual finishing time for the service traffic. The user traffic, which could be the last one to finish, must be served first. The high bit rate reduces the impact of the quantisation effect resulting from packetising.

10.3.2.1.2 General Traffic Scheduling for Different Air Interfaces. 1. Scheduling in a typical CDMA Based System – UMTS/FDD. This section gives an overview of relevant system functions to clarify the assumptions and the performance of the RS algorithm, where some results are discussed.

System-level simulations consider both the transmission and the receiver with multiple co-existing radio links in a cellular system. The scheduling algorithms defined in the Radio Network Controller (RNC) and the User Equipment (UE) are modelled and some performance curves are also shown. Owing to the multiplexing process, interleaving scheme, the data for an active link cannot be decoded within a Transmission Time Interval (TTI) period. For this reason the TTI defines the minimum delay time for the scheduling algorithms [29–36].

Point-to-point connections cannot allow delay with arbitrary value, where the overall delay is composed of the transmission delay, processing delay, scheduling delay and coding delay. 3GPP Rel. 99 has specified the delay component where the scheduling delay is mainly constructed by multiplexing and interleaving delays. On the other hand, ITU-T recommendations G. 173 [37] and G. 174 [38] say that, 'the total one way delay of the PLMN for voice services should be kept within 40 ms'. However, the demand mentioned in ref. [39] that the maximum overall transfer delay for RT service should be kept within 300 ms is more realistic.

Since the RS generates a considerable delay for real-time services, as the 3GPP TR documentation outlines, we need to evaluate the maximum delay due to scheduling algorithms in different air interfaces. Owing to the long-tail distribution of Internet traffic, the QoS can be evaluated by the effective throughput compared with the incoming traffic itself, as depicted in Figure 10.18 below.

The most important values for a scheduling algorithm, i.e. the schedule delay, the traffic throughput, the impact on spectrum efficiency and channel occupation must be accessed to evaluate the performance of the scheduling algorithm.

An example of scheduling HTTP traffic in a DCH channel with parallel sessions for each user is given in Figure 10.17. For users running two sessions, S1 and S2, the mechanism does not change as compared with the previous scenarios explained, except that a new dynamic scheduling algorithm is applied in order to maintain the fairest share of the resources between sessions. This means that the resources are shared dynamically. The two upper charts in Figure 10.17 show the independent HTTP traffic for S1 and S2. One way to manage the scheduling is to create a vector where every element (every session) contains the number of bits that will be sent every TTI. Equation (10.8) shows the relationship between the amount of data that must be transmitted within a TTI period:

$$\left(b_k \geq \frac{R_a \cdot T_\mathrm{T}}{N_\mathrm{S}} \right) \Bigg|_{k=1,\ldots,N_S} . \tag{10.8}$$

Here, N_S is the total number of sessions, b_k is the data accumulated in the buffer for session k, R_a is the data channel capacity assigned by the UMTS/FDD between the Medium Access Control (MAC) and Physical Layer (PHY), and T_T represents the TTI value.

A fast approach to obtaining the static scheduling vector of sessions during each TTI can be estimated by $\varepsilon = \lfloor R_a \cdot T / N_S \rfloor$. However, this does not yield the exact number of data bits transmitted during a TTI, before the scheduler finds the optimal Quantum to be chosen for each session.

In Figure 10.17, the middle chart shows the scheduling procedure according to the data accumulated in both buffers. If the buffer for a session is empty, then all resources are given to the other session. If the last session does not have enough data to transmit, padding bits are embedded. In the case when both sessions have no data to transmit, the multiplexing chain would work in the same way as in the first scenario. One result is selected as an example to show the performance of the scheduling algorithm with a spreading factor (SF) of 64 in the downlink, as shown in Figure 10.18.

2. Scheduling in a typical TDMA-based system – HIPERLAN/2. HIPERLAN/2 is a strong representative of the wireless LAN, which is designed to provide a flexible platform for a variety of business and home multimedia applications that can support a set of bit

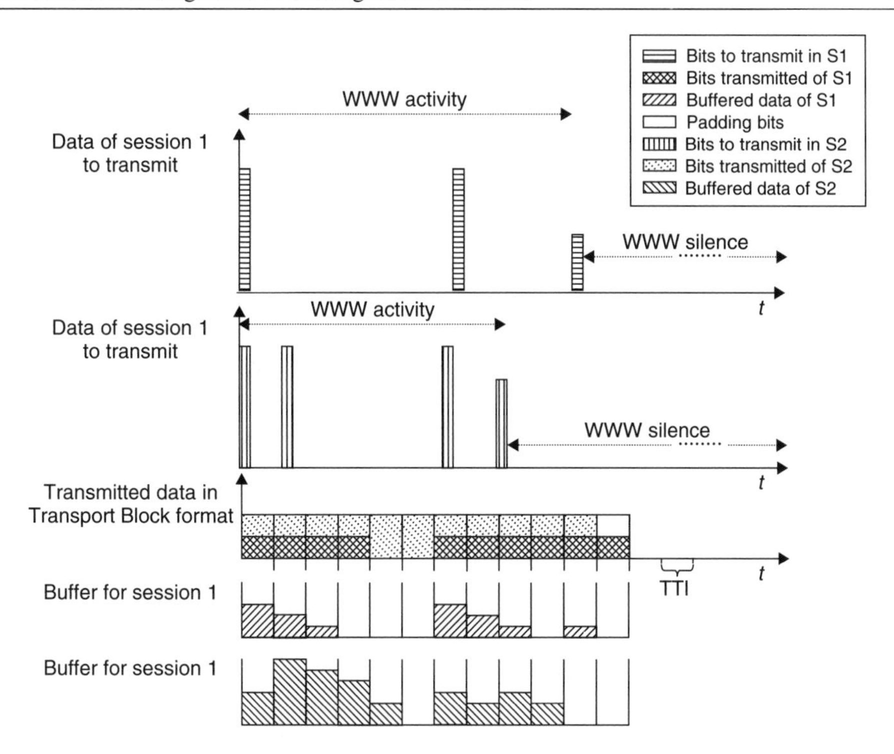

Figure 10.17 HTTP traffic scheduling for two sessions/users in the Data Channel (DCH)

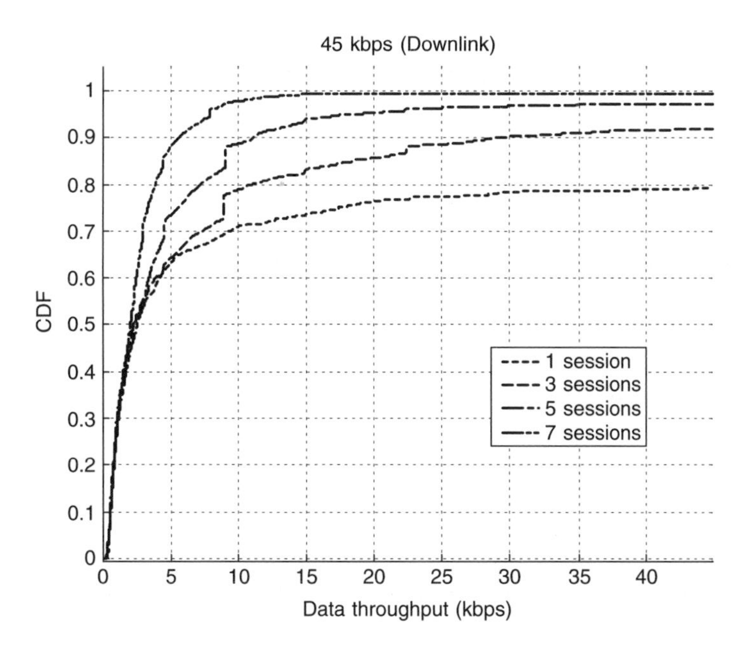

Figure 10.18 45 kbps DCH in the downlink with SF = 64

rates up to 54 Mbits/s [10, 9]. It has to support multiple classes of services with widely different traffic characteristics and QoS requirements.

In order to analyse the performance of the RS algorithm, a proper traffic model of the HIPERLAN/2 system needs to be set up. HIPERLAN/2 supports the QoS demands of IP applications and we particularly investigate the voice service and draw conclusions for low up to medium data rate services. Although designed for a very high data rate service, the HIPERLAN/2 system must be able to receive services from other mobile radio systems carrying lower data rate services in order to cater the end-to-end connectivity. In addition, many more applications of wireless LAN can be found in person-to-machine and machine-to-machine communications, e.g. on-line games, industry surveys, etc. However, the low data rate services with real-time constrains are of importance to should also be tested. The purpose of this subsection is to investigate scheduling algorithms in order to increase the number of low to medium data rate users with QoS demands, e.g. bit rate, delay, etc. In addition, the mixture of services case is also studied.

As voice over IP is real-time traffic, which is usually assigned the first priority, we model it and investigate the performance of FCFS scheduling for this type of traffic. The problem of low efficiency for the radio resource utilisation of HIPERLAN/2 spectrum is revealed and algorithms of how to improve the efficiency are proposed. To fulfil the dynamic association, a number of Short Transport Channels (SCHs) containing the resource request message and a number of RCH channels containing the resource grant message are required. In addition, one Long Transport Channel (LCH) Protocol Data Unit (PDU) can only be occupied by a unique user, which offers a much higher intermediate data rate than is required for voice traffic. This is the main reason for a low HIPERLAN/2 spectrum efficiency in voice services.

It is specified that LCHs only carry UDCH data, i.e. 54 bytes PDU packets are generated for transmitting voice traffic. We assume the basic mode in HIPERLAN/2, i.e. 6 Mbps due to BPSK modulation and a coding rate of 1/2 for convolutional codes. Owing to the nature of voice services, uplink and downlink transmissions are symmetric. In a single sector environment, the preamble of the control channel is 16 μs long, with variable FCH length according to the number of established links; the preamble for downlink phases are 2 OFDM symbols per user, whereas in the uplink there are 3 OFDM symbols; preambles for RCHs are 3 OFDM symbols; the number of SCHs is assumed to be equal to the number of users in both directions. One OFDM symbol duration time is 4 μs long. One RCH channel exists in the MAC frame. In addition, the guard period is 6 μs; the guard time between UL/DL bursts is set to be 2 μs; whereas guard times among BCH/FCH, FCH-ACH/DL, UL/RCH are set to 800 ns. The frame structure in one MAC frame is shown in Figure 10.19.

Scheduling Algorithm for Low Data, Rate-Real Time Services. Shorter PDUs are embedded into the MAC frame in order to provide better spectrum efficiency. The size of the PDU should be designed according to the traffic characteristics. The higher the data rate, the bigger the PDU. However, due to the cell relay feature of the HIPERLAN/2 system, we cannot reach the maximum spectrum efficiency when using variable PDU sizes. The PDU size should be known by the system before the user data is assigned in the MAC frame. Basically, the LCH is designed for high data rate services, e.g. video streaming, videoconferencing. Here, we choose the optimal PDU size for a low data rate

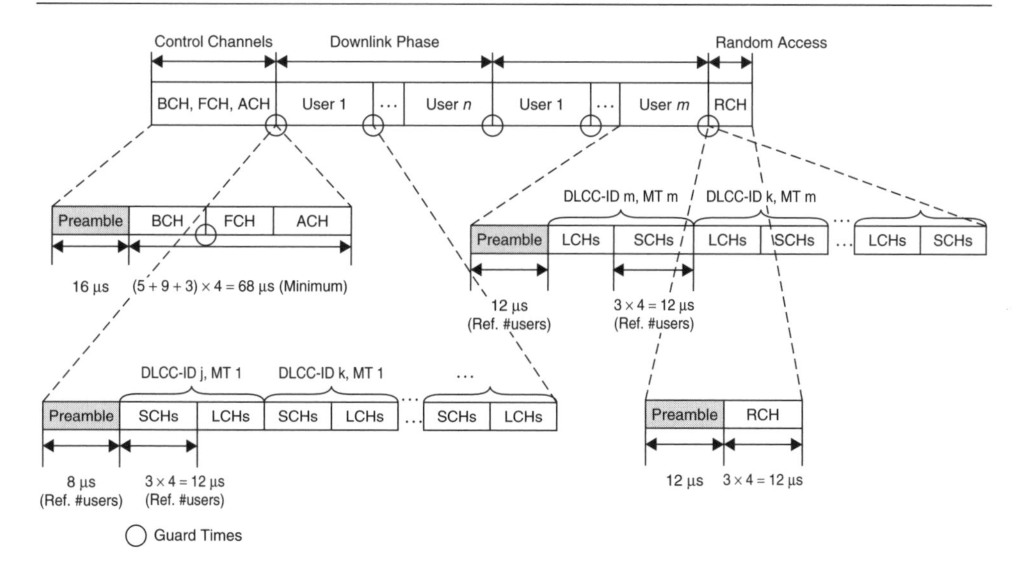

Figure 10.19 Frame structure for HIPERLAN/2

service such as voice traffic. The following analysis shows the policy of assigning different PDU sizes according to different real-time services. It must fulfil $T \leq T_{WM}$, where T_{WM} is the maximum waiting time for the real-time service, R is the data rate of service and T defines the time the user must be scheduled. According to the FCFS algorithm, the user that arrived first has higher priority with the scheduled time T. In general, variable PDU sizes optimised for data rate and maximal tolerable delay yield higher spectrum efficiency [40].

In order to admit more users without changing the standardised system protocol, i.e. without changing the PDU size, an alternating MAC frame utilisation algorithm (AA) can also be implemented for a low rate, real-time service. It is based on the alternating factor (AF) to schedule the user traffic in a round robin (RR) manner, where RR scheduling is one of the fairest scheduling algorithms designed especially for time-sharing systems. The difference to the alternating scheme is that the RR scheme requires a fixed Quantum size assigned to each service waiting in the circular queue [9, 40]. In our proposal, the scheduler checks the user data queue in all AF frames. By the requirement of real-time service and the varying bit rate nature, the AA algorithm does not need to assign the bandwidth to an individual user if its traffic has not arrived. It can also assign a bigger Quantum size to the user if its data rate is high enough. The AF value controls the schedule routine, and the size of the individual queue is affected by the Quantum size.

It can also be shown that the throughput for those two schemes is similar, which results in similar spectrum efficiency for the HIPERLAN/2 system. It is the user delay time that evaluates the performance of the AA and adaptive PDU usage approach.

Example of Mixed Services in HIPERLAN/2. It is investigated that if only HTTP users are admitted in the HIPERLAN/2 system, the basic Radio Resource Unit (RRU) with two LCH PDUs is the optimal Quantum size to implement the RR algorithm. In order to compare the scheduling algorithm in a mixed service scenario, as suggested in ref. [16], we assume at least 10% of the dedicated user throughput for satisfied user criteria, if each user can be admitted with one LCH channel in each MAC frame. In a single AP-controlled cell, 400 satisfied HTTP users meet the system upper capacity. In the case that both voice users and web browsing users exist in the network, we assume the voice users have higher priority, and 10% of voice users have also simultaneous HTTP sessions. The scheduler performance is shown in Figure 10.20.

The left-hand side of Figure 10.20 shows the system throughput vs. the number of voice users in a one-cell HIPERLAN/2 system. The right-hand side shows HTTP user satisfaction with respect to the number of voice users. It can be seen that voice users dominate the radio resource, which greatly degrades the QoS of the HTTP service.

10.3.2.2 Handover Principles

In this section the concept of inter-system handover supported by reconfigurable terminals and inter-cell handover are explained. Investigations on soft/softer handover and system capacity gain due to soft inter-system handover (soft vertical handover) are illustrated.

10.3.2.2.1 Inter-Cell Handover. Owing to the advanced transceiver technologies, soft/softer handover can be applied in UMTS FDD system. Soft/softer handover is applied for UE-RNC connections on dedicated channels, which serve at least one bearer service. The admission control at call set-up or when a new bearer service is added/removed to a connection decides whether soft/softer handover can be applied. For complexity reasons, a selective combining scheme is applied in the soft handover case when more than one single site is involved. On the other hand, signals are combined as maximum ratio combinations in the softer handover case because of single site involvement [41–43] (see Figure 10.21).

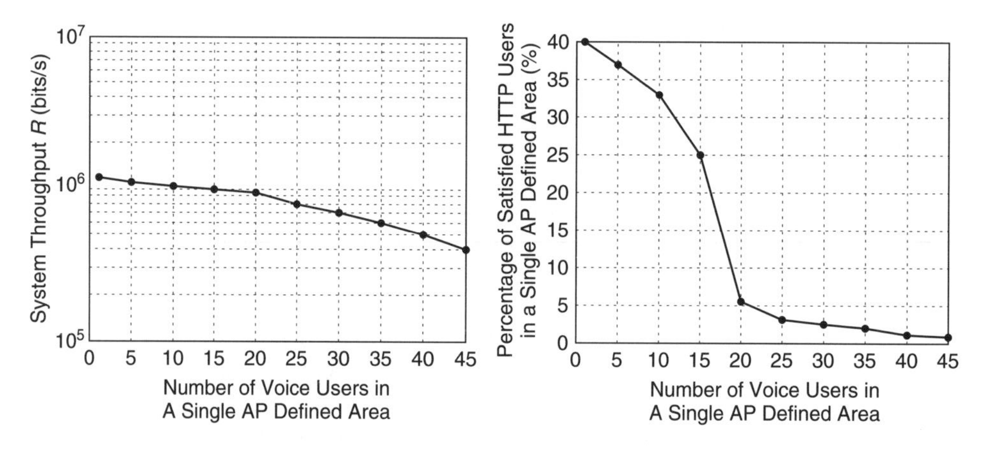

Figure 10.20 System throughput and percentage of satisfied users in the mixed service case

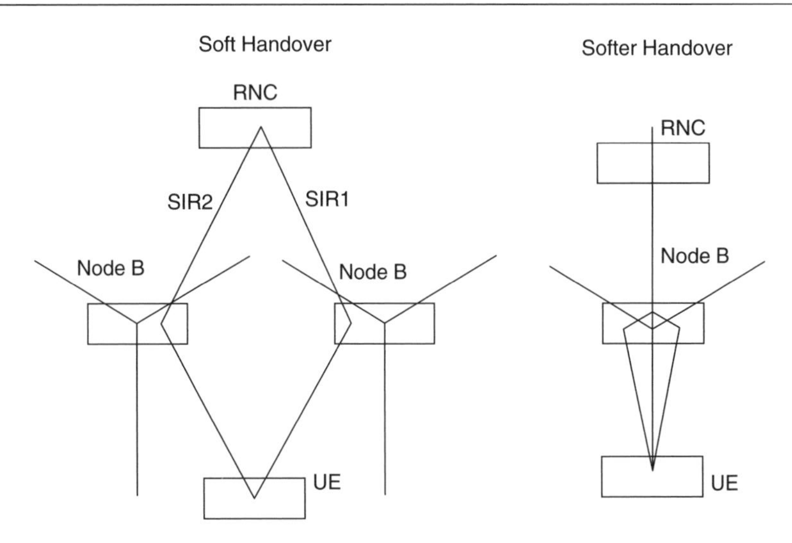

Figure 10.21 Soft handover and softer handover in UMTS FDD

In addition to the received signal strength at the UE, the UE decodes the Physical Common Control Channel (PCCPCH) and reads the required Signal to Interference Ratio (SIR) of the base station and its uplink interference level for the initial access (RACH) which is broadcast on the Broadcast Control Channel (BCCH). Based on the knowledge of these measurements, the UE chooses the sites requiring the lowest transmission power.

Having established the candidate cells by the UE, it is up to the network to decide, when the UE has to enter the Soft Handover (SHO) mode. The active set is the number of sites the UE has active links to. The allowed set is defined by three parameters: handover margin, hysteresis for inclusion and hysteresis for replacement. The handover margin requires the best sites (usually the serving sites before entering the SHO) and all other sites must be within this handover margin for establishing a link. Inclusion hysteresis delays the decision on the inclusion of the sites in question until this hysteresis has been defined. New sites must have a higher level to get into the active set and can afford some additional attenuation before being removed from this table. Replacement hysteresis demands that new sites must have a certain level higher than the worst site before the worst site is replaced. The replacement parameter is needed when the set is already full.

10.3.2.2.2 Inter-System Handover. As explained in Section 10.3.1, JRRM architecture and algorithms are established. It also appears in ref. [44], where the feasibility of JRRM over heterogeneous networks is studied. In ref. [44] we proposed an approach to traffic splitting in simultaneous connections over tight-coupled sub-networks. In this subsection we go one step further with a soft vertical handover approach: 'GPS [24, 25] based vertical (inter-system) soft handover', to show the benefit of soft inter-system handover and the approach of its realisation. Here the term '*soft*' means that the link in the old sub-network is not released during the handover process. That is to say, during the '*soft handover*' process, the terminal has more than two simultaneous connections over different sub-networks, as Figure 10.13 shows. On the basis of the capability information in both system, the current load of both systems and the load due to the handover process in the old system, the resource required by the connections in each sub-system can be handled by the PW.

1. Schematic comparison between soft and hard handovers. The inter-system handover (vertical handover) can be managed in two basic ways: hard handover and soft handover. The future reconfigurable multi-mode terminals guarantee the feasibility of inter-system handover, where hard handover means the connection affiliated to the old sub-network is released before a new connection to the new network is established. A schematic illustration of the difference between the two approaches is shown in Figure 10.22.

2. Introduction to the GPS approach. As shown in Figure 10.13, the Priority Weight (PW) carries the weight of each traffic type corresponding to the traffic class. The TRaffic SCHeduler (TRSCH) allocates resources for different QoS classes. The cost function the CAC is based on the PW requirements for each traffic type. In this case, the cost function parameter must be harmonised with the scheduling algorithm. The principle of GPS [45] can be applied to guarantee the resource for the sessions with different priority, i.e. a minimum service rate can be obtained according to the ratio between the assigned weights and the sum of all. In the case when the same services types exist, by

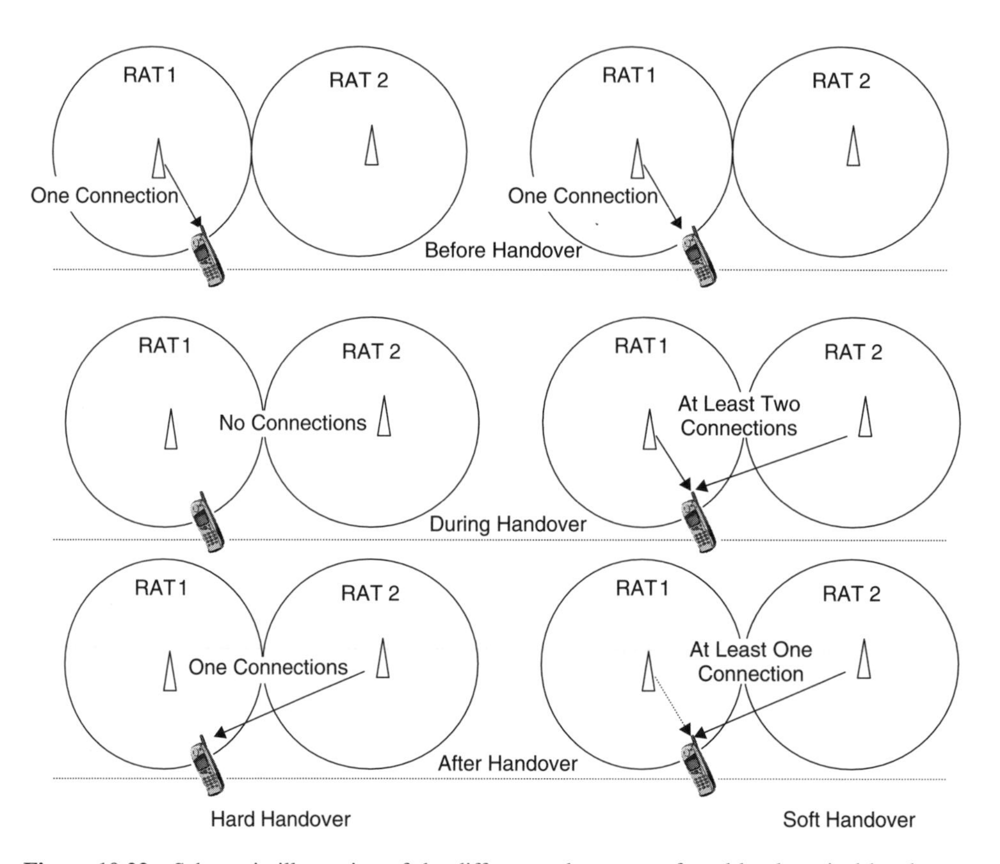

Figure 10.22 Schematic illustration of the differences between soft and hard vertical handovers

assigning different weights to individual users special demands for the radio resources can be implemented, e.g. the handover process has higher priority than users even they have the same service type.

In addition, the vector consisting of PWs belonging to different service types can also be applied in the calculation of allocated resources for the committed services. The network operator can handle the PV to offer dedicated radio resources to particular users and particular vertical handover processes.

The gain of the soft handover depends on the number of vertical handovers in the network. Since the overall arrival rate λ of the system has a linear relationship to the load ρ, we model the amount of inter-system handovers with respect to its contribution to the system load by means of its weight ϕ_{H} in the PV. Here, ϕ_{H} corresponds to the weight of the handover process assigned by the admission function in the sub-network; the sum of weights belonging to all other processes in the sub-network is denoted by $\phi_{\overline{\mathrm{H}}}$. The weight ϕ_{H} is handled by JOSAC and JOSCH, as shown in Figure 10.13. JRRM functions should assign the value of ϕ_{H} to the handover processes in each sub-network. As our assumptions require, the sum of load in each sub-network belonging to this process should be constant, so that the handover process maintains the same QoS during the handover procedure. Therefore, a comparison of the impact on the overall Grade of Service (GoS) of sub-networks is bias-free. The idea of changing weights for handover processes during the soft vertical handover phase is illustrated in ref. [45]. Changing the load due to the handover process can be designed by assigning different shapes to the PW changes over time. The optimal PW shape over time results in the lowest impairment during handover. The load change $\Delta_{\eta}(t)$ needs to be controlled by the time-variant PWs. It can be shown that different system performances will result from different PW changes over time.

The load of the system can only be changed in a discrete manner due to the limitation of the radio resource control, i.e. the fastest load changes are based on the minimum transmission time unit, namely TTI values for bearer services in UMTS. On the basis of the theoretical analysis and simulation in ref. [45], we can conclude that the more frequently a handover takes place, the higher is the benefit of using the soft vertical handover. A slow start soft handover is preferred to allow the RRM in the new system enough time for radio resource adaptation. If the two systems have quite different capacities, then a soft vertical handover is also beneficial. Moreover, due to the soft handover, the QoS for individual users is better than applying a hard handover.

On the other hand, a hard handover will bring a sudden interference increase in a CDMA based system or a high time slot resource occupation in a TDMA based system. The power control functions or radio schedulers need some time to bring the system smoothly back to a stable state [46]. In Table 10.5 comparisons of soft vertical handover and hard vertical handover are summarised.

3. *Realisation* of a soft vertical handover between HIPERLAN/2 and UMTS. The time variant PW can be mapped onto time slots in 2G GSM, onto frames in shared channels in UMTS [47], onto spreading codes in WCDMA and a number of LCHs in WLAN systems, e.g. HIPERLAN/2 [9, 48]. The mapping table can be accessed by the JRRM entities. According to the shape of the change of the PW that each sub-system is based on, the JRRM assigns the weight and maps the weight to the concrete resource in order to carry out the soft handover procedure.

Table 10.5 Comparison of soft and hard vertical handovers

	Soft vertical handover	Hard vertical handover
QoS of handover services	Seamless handover	Sudden QoS changes. There would be a sudden break in the service; for a real-time service this is especially important. (Video services have their own importance)
QoS for the existing network (control aspect)	Stable control system, no sudden changes	Large interference during the transition phase. Potentially unstable
GoS for the existing network (load aspect)	Less load with much less variance	Higher load with much higher variance

During the handover process from UMTS to HIPERLAN/2, the terminal is taken smoothly from the old to the target system. Figure 10.23 shows the time sequence in the vertical direction during the intersystem handover procedure. The left-hand side shows the connections in the UMTS system; the right-hand side shows the connections in HIPERLAN/2 system. On the basis of the information on system type, load and handover load, the JRRM should decide the strategy to change the PW to meet the optimal shape of diverting handover load over time; different handover strategies previously stored previously as a reference table in the JRRM entity.

10.3.2.3 Load Control Between Sub-Networks

In order to investigate the benefit of load control over sub-networks, two co-existing systems are modelled in this section. Once a new session arrives, there are two possibilities: the first one is that all the incoming traffic should be added to a single system, the other is to split the session into both sub-systems with certain load contribution. A model of the load (noise rise) for a single system can be produced by the exponential noise arise formula: $L = \mathrm{e}^{an} - 1$, where n is the number of traffic units added to the network and a is the weight of noise arise. Different sub-systems have different noise-rise curves due to their air interface specification.

The load contribution can be distributed and the traffic can be split, i.e. $m = m_1 + m_2$. For the individual sub-system, the load increase of the system where all the traffic is added to is: $L'_i = \mathrm{e}^{a(n_i+m)} - 1$, with i the index of the sub-network. In the following description, a prime is used to denote a non-traffic split, i.e. incoming traffic with unit m is only added to a single system; a double prime is used to show that the traffic is split over two sub-networks. For the traffic split case, the load value in sub-systems is calculated as:

$$L''_1 = \mathrm{e}^{a(n_1+m_1)} - 1 \quad \text{and} \quad L''_2 = \mathrm{e}^{a(n_2+m_2)} - 1. \tag{10.9}$$

Let $n_1 = n_2$, then the traffic added to the single system is calculated as:

$$\Delta' = L'_i - L_i = (\mathrm{e}^{a(n_1+m)} - 1) - (\mathrm{e}^{a(n_1)} - 1). \tag{10.10}$$

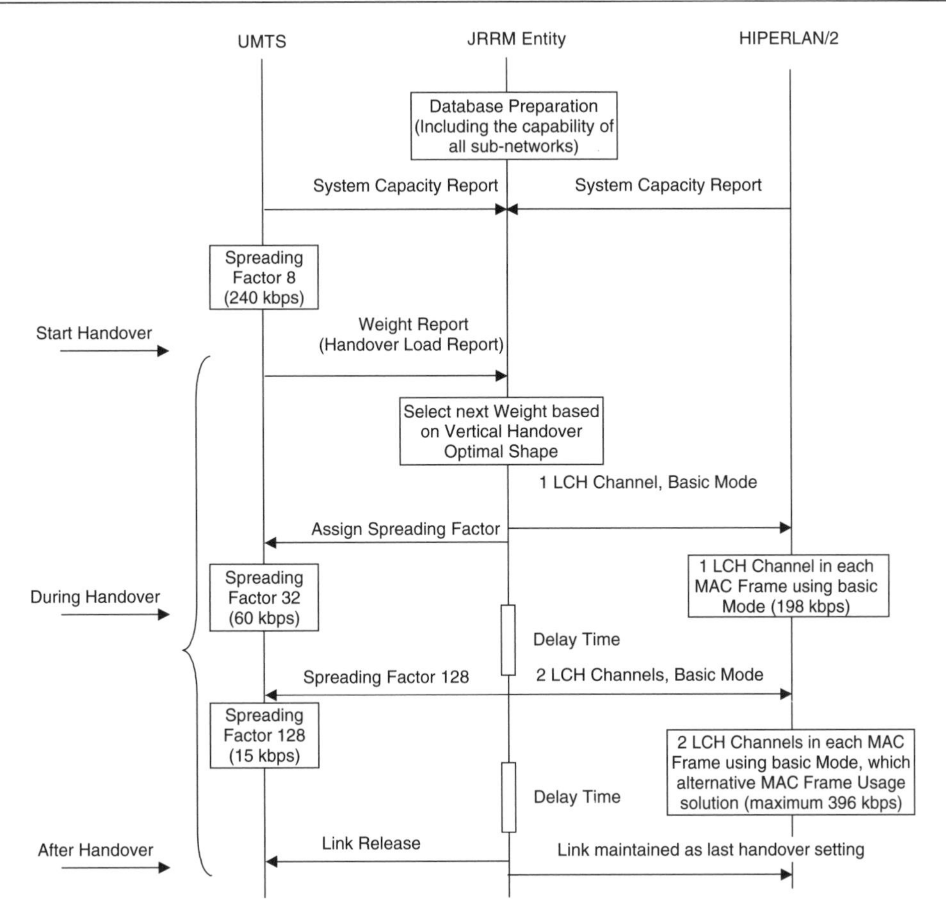

Figure 10.23 Example of soft vertical handover in UMTS and HIPERLAN/2

For the case when traffic is added to both systems by splitting the traffic, the load contribution is given by:

$$\Delta'' = \sum_i L_i'' - L_i = \sum_i [(\mathrm{e}^{a(n_i+m_i)} - 1) - (e^{a(n_i)} - 1)]. \tag{10.11}$$

With the assumed conditions $n_1 = n_2$, $m = m_1 + m_2$ and $m_1 = m_2$, the ratio between the added load with load balancing due to traffic split and the added load to a single system is:

$$R = 2 \cdot \frac{\mathrm{e}^{a(n+m/2)} - \mathrm{e}^{an}}{\mathrm{e}^{a(n+m)} - \mathrm{e}^{an}}. \tag{10.12}$$

The gain for the traffic split is:

$$G = 1/2 \cdot \frac{\mathrm{e}^{a(n+m)} - \mathrm{e}^{an}}{\mathrm{e}^{a(n+m/2)} - \mathrm{e}^{an}} - 1 = (\mathrm{e}^{a \cdot m/2} - 1)/2. \tag{10.13}$$

Since $a > 0$ and $m/2 > 0$, then $G > 0$.

In the case when two sub-networks are non-equally loaded and with different noise-rise curves, namely a *dynamic constellation*, which depends on the current load of the sub-networks and the noise-rise factor:

$$R_i = \frac{\sum_j e^{a_j(n_j+m_j)} - e^{a_j n_j}}{e^{a_i(n_i+m)} - e^{a_i n_i}}, \quad \text{with respect to } \sum_i m_i = m, \tag{10.14}$$

from the Kuhn–Tucker minimisation theory, by intelligently assigning m_i, the minimum load poured into the overall system can be obtained [49]. For the two sub-networks case, traffic can be split according to the optimal constellation, which obeys the principle of pouring higher traffic load into the less loaded system.

10.3.2.4 Principles of Admission Control

In this section the basic principles of admission control and admission control approaches for different air interfaces during the reconfiguration process embedded in the reconfiguration module are discussed and an example of a Call Admission Control (CAC) in a WLAN (HIPERLAN/2) is given.

The admission control function decides whether a new radio link can be admitted to a particular cell or not. Furthermore, it reserves capacity and code resources for the new radio link. The following functions can be distinguished that relate to admission control:

- Admission calls and sessions
- Code allocation (downlink scrambling code set, downlink channelisation code, defragmentation) (UMTS FDD/TDD)
- Time slot allocation (UMTS TDD and GSM)
- LCH PDU assignments (WLAN)

A well-designed CAC not only enhances the utility of the system resources, but also reduces the probability of dropping calls. An appropriate CAC covers consideration of session scheduling according to QoS requirement and current system load [50, 51]. On the other hand, the RS function in the HIPERLAN/2 system needs to deal with heterogeneous traffic in a proper way. It is important to emphasise that to deal with the CAC algorithm according to the MAC, the layer constrains in a TDMA-based WLAN system are more significant than the interference-based CAC strategy in a CDMA system. Therefore, interworking with resource scheduling and the load control module must be seriously considered, as illustrated in Section 10.3.1.2.

The objective of admitting new users is different from dropping an already ongoing call (active user), which is generally perceived to be more annoying than blocking a new call request (new user). Hence, if an overloaded situation occurs because of the assignment of a new user to the channel, an appropriate resource management decision is to refuse the new user. By doing so, admission control must favour the already active users.

A new user requires a base station and a channel, and the CAC has to trade off between channel quality, low outage and blocking probabilities. In general, several users may concurrently seek admission. For the QoS requirement, the delay tolerance according to different services should be highly considered besides the BER. This can be addressed either by a one-by-one admission type algorithm, or by a multiple admission type, i.e. one-by-one and multiple oriented algorithms.

In the HIPERLAN/2 system, the conventional CAC can be designed for each AP working independently with neighbouring APs. In a cellular WLAN network, a more efficient cell loading can be achieved with a centralised CAC scheme, i.e. by taking neighbouring cell loads into account. The centralised and decentralised CACs are shown in Figure 10.24, where the decentralised CAC is defined inside each AP, and the centralised CAC is defined in the APC, which needs preliminary knowledge of all cells' loads to inform the affiliated APs.

To efficiently utilise the limited radio spectrum and maximise system capacity, the cell size of the HIPERLAN/2 network tends to be small. As a result, a mobile user's connection could handover from one serving base station (BS) to a neighbouring BS several times during the entire session period. The trade-off between maximising the utilisation of the system and reducing the rate of dropping or negatively affecting the QoS demand of users should be considered to design a CAC algorithm.

Admission control approaches for future coupled networks is different from conventional CAC algorithms. Since admission control must jointly work with load control and the resource scheduler, it is one of the most important tasks of JRRM. In a co-operating radio network, co-existing with a number of radio sub-networks, a new radio link has to be admitted in the JOSAC, which is aware of the resources of the corresponding sub-networks. The admission control is triggered via the sub-network CAC/SAC entity, which sets up, adds, deletes or reconfigures a radio link or requests resources on a common transport channel. Depending on the type of procedure, e.g. new arrival sessions/calls or sessions/calls being handed over from other sub-networks, resources have to be either allocated or de-allocated.

The capacity requirements of the new bearer depend on the actual load and the resource requirements of the new bearer. Because of the requirement for heterogeneous services, the queuing data stream, based on the PW assigned by the JOSAC and SAC modules, resource allocation must be carried out by the RS module.

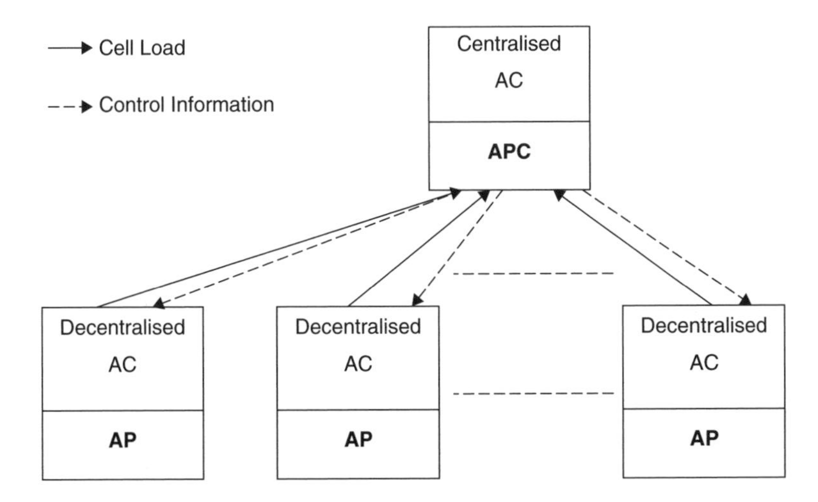

Figure 10.24 Centralized and decentralized CACs

On the other hand, owing to the requirement of multi-media services and the requirement of reconfiguration, software download management becomes very important. Therefore, more features must be added to admission control in order to partially control the software download procedure. The bearer services types, e.g. unicast, multicast or broadcast, should be selected by admission control with an appropriate transmission format. The optimal link mode, e.g. one proper mode from different basic transmission modes, must also be determined by the CAC/SAC.

References

[1] I. Katzela and M. Naghshineh, 'Channel assignment schemes for cellular mobile telecommunication systems: A comprehensive survey', *IEEE Personal Communications Magazine*, June, 10–31, 1996.
[2] J. Zander, 'Radio resource management–An overview', IEEE Vehicular Technology Conference, 1996. Mobile Technology for the Human Race., IEEE 46th, Vol. 1, pp. 16–20, 1996.
[3] J. Zander and S.-L. Kim, 'Radio resource management for wireless networks', Artech House, Mobile Communications Series, 2001.
[4] T.S. Rappaport, *Wireless Communications, Principles and Practice*, Prentice-Hall, 1996.
[5] C.G. Cassandras and S. Lafortune, *Introduction to Discrete Event Systems*, Kluwer, Dordvecht, 1999.
[6] J. Zander, 'Trends in resource management future wireless networks', IEEE Wireless Communications and Networking Conference, 2000, Vol. 1, pp. 159–163, 2000.
[7] D. Gross and C. M. Harris, *Fundamentals of Queuing Theory*, Wiley Series in Probability and Statistics, Wiley, 1998.
[8] W. Lee, *Mobile Cellular Telecommunications Systems*, McGraw-Hill, 1989.
[9] J. Luo, M. Dillinger and E. Schulz, 'Investigations of scheduling strategy for low data rate real time IP based service in wireless LAN', IEEE ICCCAS, June, Chengdu, China, 2002.
[10] ETSI TR 101 031 V2.2.1, 'High performance radio local area network, type 2, requirements and architectures for wireless broadband access', 01, 1999.
[11] ETSI, 'Brandband radio access network (BRAN); HIPERLAN TYPE 2 functional specifications; data link control (DLC) layer', Part 1 and Part 2, 11, 1999.
[12] J. Luo and M. Dillinger, 'Ideal power control for HIPERLAN/2 system', IST Summit 2000 October, Galway, Ireland, 2000.
[13] J. Luo, M. Dillinger and B. Walke, 'Link adaptation criteria for downloading software to reconfigurable terminals', IEEE 3G wireless 2002, San Francisco, USA, 2002.
[14] A. Riedl, M. Perske, T. Bauschert and A. Probst, 'Dimensioning of IP access networks with elastic traffic', 9th International Telecommunication Network Planning Symposium (Networks 2000), September, Toronto, 2000.
[15] A. Riedl, M. Perske, T. Bauschert and A. Probst, 'Investigation of the M/G/R processor sharing model for dimensioning of IP access networks with elastic traffic', First Polish–German Teletraffic Symposium PGTS 2000, September, Dresden, 2000.
[16] ETSI: Universal mobile telecommunications system (UMTS); selection procedures for the choice of radio transmission technologies of the UMTS. UMTS 30.03, v 3.2.0, November, 1997.
[17] ETSI SMG2: UTRAN Architecture Description, UMTS ZZ.01, v 0.1.0. January, 1999.
[18] B. Walke, et al., 'IP over wireless mobile ATM–Guaranteed wireless QoS by Hiperlan/2', *Proceedings of the IEEE*, January, 2001.
[19] A. Doufexi, et al., 'A Comparison of HIPERLAN/2 and IEEE 802.11a', VTC, 2000.
[20] J. G. Proakis, *Digital Communications*, 3rd edn, McGraw-Hill, 1995.
[21] L. Kleinrock, *Queueing System*, Vols. I and II, Wiley, 1975.
[22] J. Charzinski, 'Fun factor dimensioning for elastic traffic', *Proceedings of the ITC Specialist Seminar on IP Traffic*, September, Monterey, CA, USA, 2000.
[23] E. Kreyszig, *Advanced Engineering Mathematics*, Wiley, 1975.
[24] Abhay K. Parekh and R. Gallager, 'A generalized processor sharing approach to flow control in integrated services networks: The single-node case', *IEEE/ACM Transactions on Networking*, **1**(3), 344–357, 1993.
[25] Abhay K. Parekh and R. Gallager, 'A generalized processor sharing approach to flow control in integrated services networks: The multiple-node case', *IEEE/ACM Transactions on Networking*, **2**(2), 137–150, 1994.

[26] T. Farnham, G. Clemo, R. Haines, E. Seidel, A. Benamar, S. Billington, N. Greco, N. Drew, T.H. Le, B. Arram and P. Mangold, 'Reconfiguration of future mobile terminals using software download', IST Mobile Communications Summit 2000, Galway, Eire, 2000.
[27] M. Dillinger, J. Luo and E. Mohyeldin, 'Circular cell broadcast for software downloads in WCDMA', IEEE 3Gwireless 2002, June, San Francisco, USA, 2002.
[28] M. Dillinger, J. Luo and E. Mohyeldin, 'Software download management for broadcast channels in WCDMA system', 2nd Karlsruhe Workshop on Software Radios, March, Karlsruhe, Germany, 2002.
[29] ETSI SMG2, 'MS-UTRAN radio interface protocol architecture', UMTS YY.01, v 1.0.3, January, 1999.
[30] ETSI SMG2, 'UMTS XX.20', v 0.02, December, 1998.
[31] ETSI SMG2, 'The ETSI UMTS UTRA ITU-R RTT candidate submission', SMG2 260/98, 1998.
[32] ETSI SMG2, 'Layer 1 – General requirements, UMTS YY.02', v 1.1.0, January, 1999.
[33] ETSI SMG2, 'Description of UE states and procedures in connected mode', UMTS YY.03, v 0.4.0, January, 1999.
[34] ETSI SMG2, 'UTRA FDD: Transport channels and physical channels', UMTS XX.03, v 1.3, 1999–2001.
[35] ETSI SMG2, 'UTRA FDD: Physical layer procedures', UMTS XX.07, v 1.1, 1999–2001.
[36] ETSI SMG2, 'UTRA handover', UMTS XX.15, v 0.3.0, January, 1999.
[37] ITU-T G. 173, 'Transmission planning aspects of the speech service in digital public land mobile networks'.
[38] ITU-T G. 174, 'Transmission performance objectives for terrestrial digital wireless systems using portable terminals to access the PSTN'.
[39] 3GPP 22.105, 'Services and service capabilities'.
[40] J. Luo, M. Dillinger and E. Schulz, 'Radio resource scheduling algorithms for mixed VoIP and HTTP traffic in HIPERLAN/2 system', IST SUMMIT 2001, September, Barcelona, Spain, 2001.
[41] A. Viterbi, 'CDMA – Principles of spread spectrum communication', April, 1995.
[42] J. Luo, M. Dillinger, E. Schulz and Z. Dawy, 'Optimal timer setting for soft handover in WCDMA', IEEE 3Gwireless 2000, June San Francisco, USA, 2000.
[43] J. Luo, M. Dillinger, E. Schulz and Z. Dawy, 'Probability estimation for soft handover in WCDMA', IEEE 3Gwireless 2000, June San Francisco, USA, 2000.
[44] J. Luo, M. Dillinger and E. Mohyeldin, 'Radio resource management schemes supporting reconfigurable terminals', 2nd Karlsruhe Workshop on Software Radios, 20/21, March, Karlsruhe, Germany, 2002.
[45] J. Luo, et al., 'Generic process sharing based soft vertical handover over multiple radio networks', ICT June, Beijing, China, 2002.
[46] E. Mohyeldin, J. Luo and M. Dillinger, 'Performance analysis of adaptive fuzzy outer loop power control with comparison to basic qualcomm algorithm in W-CDMA system', Africom 2001, May, Cape Town, South Africa, 2001.
[47] H. Holma and A. Toskala, *WCDMA for UMTS*, Wiley, 2000.
[48] ETSI, Broadband Radio Access Network (BRAN), 'HIPERLAN TYPE 2 functional specification; data link control (DLC) layer, Part 2: radio link control (RLC) sublayer', September, 1999.
[49] E. Mohyeldin, J. Luo and M. Dillinger, 'Joint admission control and scheduling algorithm in tightly coupled heterogeneous networks supporting reconfigurable terminals', WWRF, June, London, 2002.
[50] Z. Dziong, M. Jia and P. Mermelstein, 'Adaptive traffic admission for integrated services in CDMA wireless-access networks', *IEEE Journal on Selected Areas in Communications*, **14**, 9, pp. 1737–1747, 1996.
[51] Z. Liu and M. El Zarki, 'SIR-based call admission control for DS-CDMA cellular systems', *IEEE Journal on Selected Areas in Communications*, **12**, 4, pp. 638–644, 1994.

11

An Efficient Scheme for JRRM and Spectrum-Sharing Methods

Didier Bourse, Lucas Elicegui, David Grandblaise and Nicolas Motte

Motorola Labs, European Communications Research Lab, Paris, France

Reconfigurable Software Defined Radio (SDR) equipment is seen as the key technology in the evolution of telecommunications, terminals being able to support several Radio Access Technologies (RATs) and provide seamlessly and transparently to users a large panel of applications that could only be available today with different devices, each being dedicated to a specific RAT. Currently, RATs are available through different systems such as Cellular (i.e. GSM, IS-95, UMTS), Broadcast (i.e. DVB-T), or WLAN (e.g. HIPERLAN/2, IEEE 802.11) systems. Each of these systems has been designed initially for specific services and applications in identified operating areas. The future of telecommunications will be characterised by the convergence of those technologies (beyond the 3G – B3G perspective) and the possible introduction of new specific RATs (e.g. 4G, UWB). The future challenge will consist in building harmoniously and conjointly on those heterogeneous technologies. Therefore, SDR technology has been identified as the means to achieve the desired inter-operability. However, to reach the full benefit of this telecommunication evolution, more flexibility will also be needed in current spectrum management rules, and practices will need to evolve and accompany the SDR technological evolution. Thus, an additional challenge will also reside in the design of new rules for a more flexible spectrum allocation and radio resource management.

This chapter addresses first the current spectrum allocation and regulation. The flexible and dynamic spectrum allocation schemes are then studied in detail, from the perspective of research followed in some European projects to the complete formalisation of the flexible spectrum management encompassing spectrum sharing and spectrum pooling. The chapter Section 11.3 focuses on the joint management of radio resources in heterogeneous networks and details the operation of a Joint Radio Resource Management (JRRM) scheme proposal. Finally, specific emphasis is placed on future research to be followed on the evolution of Radio Resource Management (RRM) in a hybrid *ad hoc* environment.

Software Defined Radio: Architectures, Systems and Functions. Edited by M. Dillinger, K. Madani and N. Alonistioti
 ISBN: 0-470-85164-3

11.1 Spectrum Management

The recent emergence of the control of radio functions by software algorithms embedded in future smart communications devices will directly affect the manner in which these devices will use the spectrum [1, 2]. Indeed, with the advent of reconfigurable SDR equipment, changing between different modes to adapt to specific operating conditions will become possible, bringing in flexibility and technology independence for both mobile users and network operators. This flexibility will improve network performances as well as the Quality of Service (QoS) delivered to the user. However, this cannot be fully exploited with the traditional spectrum practices. Advanced spectrum allocation and utilisation techniques have to be investigated to incorporate this flexibility and to enhance further the performance of reconfigurable SDR systems.

The decision to undertake spectrum sharing is highly dependent on policies and rules dictated by various regulatory bodies (ITU, CEPT, ARIB, FCC) having significant influences in shaping spectrum utilisation. This is detailed in Section 11.1.2. Traditionally, rigid rules apply in spectrum management, imposing heavy restrictions in frequency utilisation. Indeed, frequency channels are well defined and their allocation operation is limited only to the allocated spectrum. Nevertheless, fixed and standardised planning of the spectrum provides several advantages: ordered planning of frequency spectrum use encourages the careful choice of technology, both for maximum spectrum efficiency and for the profitability of operators. Strict standardisation, both in network equipment and end user equipment, coupled with normalised practices in network design and implementation, further assist in the most efficient use of the frequency spectrum and at the same time minimise the interference. The main drawback of the conventional approaches is the lack of flexibility in resource allocation, which does not take into consideration the heterogeneous users' requirements and the wide range of traffic profiles.

Thanks to the future flexibility accessible through the telecommunications convergence and the introduction of reconfigurable SDR equipment, operators will have the opportunity to diversify their services while optimally distributing traffic loading across different radio networks and maximizing capacity gains. Operators (in its general meaning) will be able to exploit in the most optimal way the different portions of the spectrum and spectrum collaborations would imply inter-operability between technologies as well as inter-operator collaborations.

11.1.1 Spectrum Allocation

The spectrum allocation process aims to define globally the frequency bands a radio system can use to operate and distribute the scarce radio spectrum among services. As such, depending on existing systems within a geographical area that could be as wide as a region, a country or a continent, frequency band plans among countries could be very different. The allocation process takes into consideration emerging systems that require initial allocation but also aims, for examples to manage existing and obsolete systems that are not used any more. In such a case, this is an opportunity to engage in re-allocating the freed spectrum for other services.

Moreover, one of the major challenges in the frequency allocation process is to ensure spectrum coexistence, whether with services operating in the same band or with systems operating in adjacent bands. Adapted spectrum engineering techniques make it possible to evaluate the required isolation between the radio systems. Of course, owing to the diversity of the RATs and the possible frequency arrangements, engineering techniques have to deal with numerous co-existence scenarios.

The combination of the fact that historically all regions have not used the same radio systems and that all emerging technologies are not necessarily deployed worldwide leads to a non-harmonised configuration in terms of frequency band plans. As can be observed in Figure 11.1 [2], the non-harmonised bands for 2G systems together with the diversity of 2G technologies (GSM, PHS, IS95, etc.) highlight the need for SDR equipment development to consider spectrum use in an optimised manner.

Nonetheless, with emerging radio technologies being considered more and more worldwide, there is a global tendency to reach harmonised frequency allocation, at least for these new systems. As a consequence, if focusing on public radio systems, specific frequency bands appear to be used intensively for public applications. The diversity of the presently available RATs for the end-user contributed to reaching a situation of patchy frequency band patterns where it would be possible to find various radio systems for different types of final applications covering a wide range of users' requirements.

Basically, these radio applications that drive public telecommunication growth can be categorized in to three main types:

1. Cellular systems (GSM, UMTS, CDMA2000, etc.)
2. WLAN/PAN systems (HIPERLAN/2, 802.11, Bluetooth, etc.)
3. Broadcast systems (DVB, etc.)

Figure 11.2 represents from a macroscopic point of view the occupied bands by these radio system categories, it has to be noted that for the considered applications, the figure is fairly valid in every ITU Regions, namely Japan, the United States and Europe. These allocations aim to provide the necessary bandwidth for these radio applications. For emerging

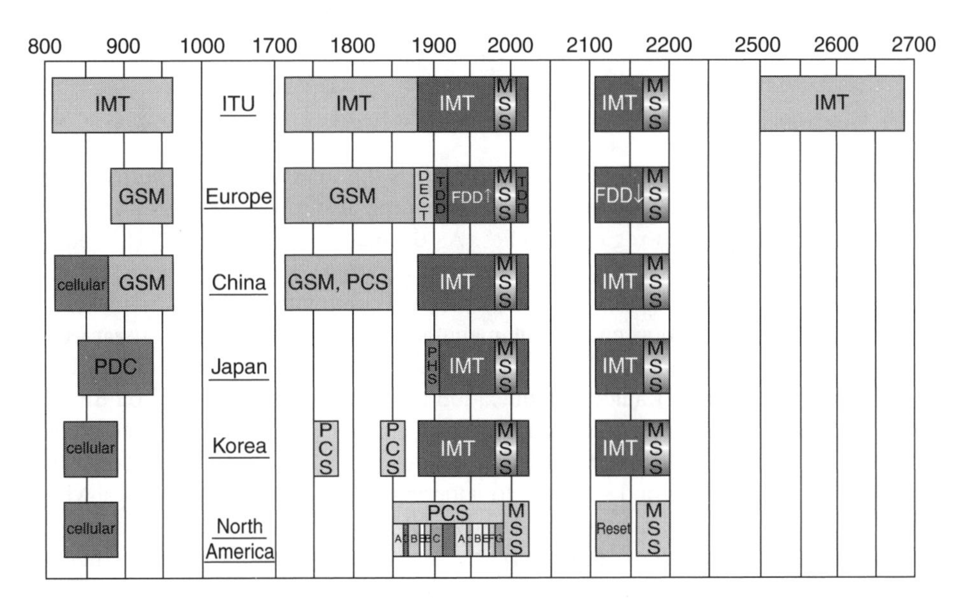

Figure 11.1 Spectrum allocation – 1800–2200 MHz [2]

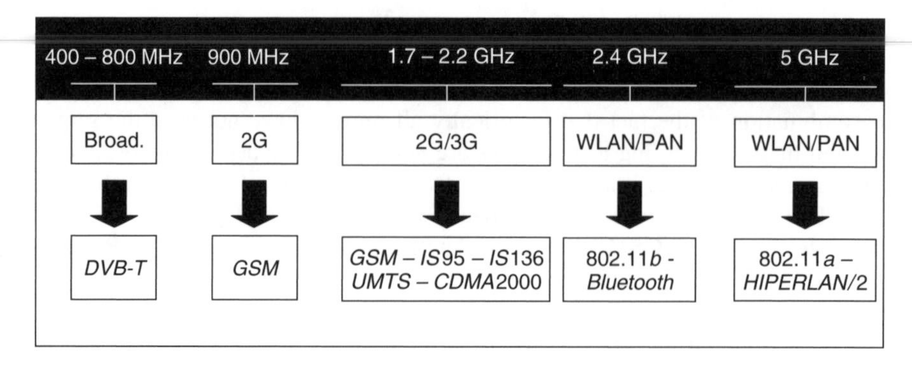

Figure 11.2 Macroscopic view of some wireless allocations

technologies such as WLAN or 3G systems, the required spectra have been assessed based on the initial demand and forecasted penetration rates of these new services. Nonetheless, technology maturity, in addition to the adoption by customers of these emerging services, could lead to envisaging extension bands to operate. Owing to the nature of these radio applications, the evolution of the allocation could for instance consider assigning an asymmetric spectrum (i.e. not the same amount for downlink and uplink communications) in order to better fulfil traffic patterns. Indeed, it is fairly predictable that due to targeted applications for 3G systems, for instance, the demand will become more important in the downlink. Thus, it could be seen as a first step in the optimisation and rationalisation of the spectrum allocation.

As an example, Figure 11.3 [4] illustrates the requirements in terms of asymmetry for UMTS. Following this high level view of the commercial and spectrum allocation aspects of wireless communications systems, next section focuses on the regulatory aspects of spectrum allocation.

11.1.2 Spectrum Regulation

The International Telecommunication Union (ITU) is the international authority on spectrum allocation matters. The ITU operates by providing answers to specific questions and making recommendations to regional bodies. As detailed in ref. [5], the ITU, created in 1865, is founded on the principle of co-operation between governments and the private sector, with a membership grouping comprising telecommunication policy-makers and regulators, network operators, equipment manufacturers, hardware and software developers, regional standards-making organisations and financing institutions. The ITU is composed of three sectors: Radiocommunication (ITU-R), Telecommunication Standardisation (ITU-T), and Telecommunication Development (ITU-D). The three sectors' activities cover all aspects of telecommunication, from setting standards that ease seamless inter-working of equipment and systems, to adopting operational procedures for wireless services and designing programmes to improve telecommunication infrastructure. Each of the three sectors works through conferences and meetings, where members negotiate the agreements, which serve as the basis for the operation of global telecommunication services:

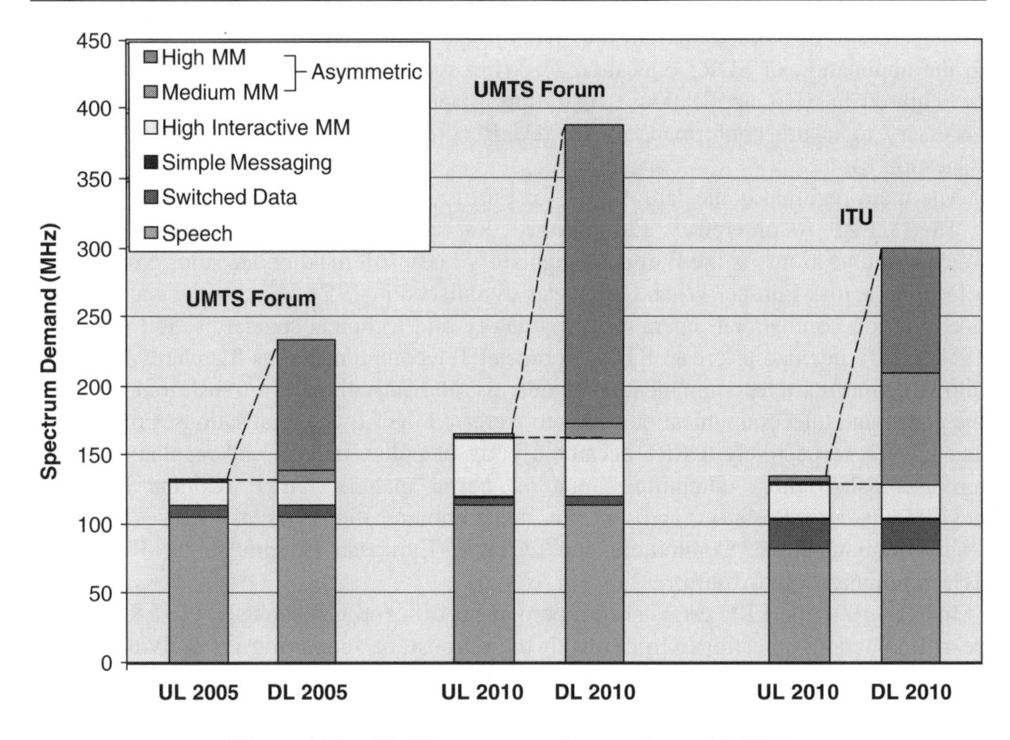

Figure 11.3 UMTS asymmetry forecast figure 11.33[4]

- *ITU-R:* This sector draws up the technical characteristics of terrestrial and space-based wireless services and systems, and develops operational procedures. It also undertakes the important technical studies that serve as a basis for the regulatory decisions made at radio communication conferences.
- *ITU-T:* In this sector, experts prepare the technical specifications for telecommunication systems, networks and services, including their operation, performance and maintenance. Their work also covers the tariff principles and accounting methods used to provide international service.
- *ITU-D:* In this sector, experts focus on the preparation of recommendations, opinions, guidelines, handbooks, manuals and reports, which provide decision-makers in developing countries with 'best business practices' relating to a host of issues ranging from development strategies and policies to network management.

The ITU is based on 24 Study Groups (24 SG: 8 in ITU-R, 14 in ITU-T, 2 in ITU-D), made up of experts drawn from leading telecommunication organisations worldwide, who are carrying out the technical work and preparing the detailed studies that lead to authoritative ITU Recommendations. The 8 ITU-R Study Groups comprise 10 Task Groups and 32 Working Parties.

The ITU address the specific question of the Software Defined Radio (Question ITU-R 230/8) [5], considering that SDRs may facilitate spectrum efficiencies in complex mobile radio configurations and that recommendations on SDR design would be complementary to other ITU-R Recommendations on mobile telecommunications. Among the questions

ITU decides to be studied on SDR are: What frequency band considerations are important to the application of SDR? (Question 2); What special interference considerations may be required in SDR applications? (Question 3); and What technical considerations are necessary to insure conformance with ITU Recommendations and Radio Regulations? (Question 6).

The main regional bodies are: CEPT for Europe, FCC for US and ARIB for Japan.

The CEPT (Conference Européenne des administration des Postes et des Télécommunications) is the European regulatory body [6], making decisions on spectrum allocation across Europe. When CEPT was established in 1959, its activities included co-operation on commercial, operational regulatory and technical standardisation issues. In 1988, CEPT decided to create ETSI (European Telecommunications Standards Institute), into which all its telecommunication standardisation activities were transferred. In 1992, the postal and telecommunications operators created their own organisations, Post Europe and ETNO, respectively. CEPT became a body of policy-makers and regulators. CEPT has established three committees, one on postal matters, CERP (Comité Européen des Régulateurs Postaux) and two on telecommunications issues, ERC (European Radiocommunications Committee) and ECTRA (European Committee for Regulatory Telecommunications Affairs).

In May 1991 the ERC established a permanent office in Copenhagen, the ERO (European Radiocommunications Office), with the purpose of supporting the activities of the committee and conducting studies for it and for the European Commission. In September 1994, ECTRA also established a permanent office in Copenhagen, the ETO (European Telecommunications Office) for the same purpose. CEPT has reorganised in order to meet the challenges of convergence in the radio and telecommunications market, and the two offices, ERO and ETO, have merged, a new ECC (Electronic Communications Committee) having been established.

It should be noted that individual member countries have their own regulatory authorities (e.g. ANFR in France, RegTP in Germany and the Radiocommunications Agency in the UK). The national regulatory authorities have the right to allocate spectra as they please. However, owing to the possibility of interference in border areas, CEPT recommendations are widely accepted.

Considering SDR in Europe, the Telecommunications Conformity Assessment and Market Surveillance Committee (TCAM) [7] is the Standing Committee assisting the Commission in the management of the Directive 99/5/EC [8, 9]. This Directive establishes a regulatory framework for the placing on the market, free movement and putting into service in the Community of radio equipment and telecommunications terminal equipment. Article 3 of this Directive (essential requirements) states that radio equipment shall be so constructed that it effectively uses the spectrum allocated to terrestrial/space radio communication and orbital resources so as to avoid harmful interference. The TCAM was set up to facilitate the exchange of documents between Member States and the Commission and the TCAM created the specific sub-group TGS (TCAM Group on SDR) on SDR regulation with respect to the R&TTE Directive.

The FCC (Federal Communications Commission) [10] is an independent US government agency, directly responsible to Congress, that runs spectrum allocation for private/commercial systems in the United States and operates in close co-operation with NTIA. The FCC was established by the Communications Act of 1934 and is charged with regulating interstate and international communications by radio, television, wire, satellite and cable. The FCC includes six operating Bureaus (Consumer and Governmental Affairs, Enforcement, International, Media, Wireless Telecommunications, Wireline Competition) and 10 Staff Offices (Administrative Law Judges, Communications Business Opportunities, Engineering and Technology, General Counsel, Inspector General, Legislative Affairs, Managing Director, Media Relations, Plans and Policy, Work Place Diversity). The Wireless Telecommunications Bureau (WTB) handles nearly all FCC domestic wireless telecommunications programmes and policies. Wireless communications services include Amateur, Cellular, Paging, Broadband PCC, Public Safety and more.

Regarding SDR, in March 2000 the FCC published a Notice of Inquiry (NoI) [10], requesting comments on (1) SDR technology status, (2) inter-operability, (3) improving spectrum efficiency and spectrum sharing and (4) equipment approvals (ET Docket No. 00-47). In the NoI, the FCC asked for comments about how SDR could result in improved spectrum efficiency and spectrum sharing (e.g. would SDR enable greater flexibility in access to open frequencies, could such equipment be designed to include some 'intelligence' that would let it monitor the spectrum to detect usage by other parties and then transmit on open frequencies?). The FCC asked about the implications of SDR with regard to the current spectrum allocation model. In December 2000, the FCC released a Notice of Proposed Rulemaking (NPRM), focusing essentially on equipment authorisation, rule changes based on inter-operability or spectrum efficiency having been judged premature. The FCC adopted its first Report and Order (R&O) on SDR in September 2001. In June 2002, the FCC Spectrum Policy Task Force issued a Public Notice (ET Docket No. 02-135) seeking public comment on various spectrum policy issues.

The NTIA (National Telecommunications and Information Administration) runs spectrum allocation for public service systems in the United States [11] and operates in close co-operation with the FCC. The Office of Spectrum Management (OSM) of NTIA comprises three divisions (Spectrum Engineering and Analysis, Computer Services, Spectrum Support) and Spectrum Plans and Policies (SPandP). The Interdepartment Radio Advisory Committee (IRAC) within the Spectrum Support Division consists of a main committee, four subcommittees, a group for notifying frequencies to the ITU, and 12 *ad hoc* working groups that consider various aspects of spectrum management policy. The basic functions of the IRAC are to assist the Assistant Secretary in assigning frequencies to US government radio stations and in developing and executing policies, programmes, procedures, and technical criteria pertaining to the allocation, management and use of the spectrum.

The CITEL (Inter-American Telecommunication Commission) [12] endeavours to make telecommunications a catalyst for the dynamic development of the Americas by working with governments and the private sector under the auspices of the OAS (Organisation of American States), which resides in Washington, DC, and has 35 Member States and over 200 Associate Members. CITEL is based on a committee organisation and its objectives include facilitating and promoting the continuous development of telecommunications in the hemisphere. CITEL has a Permanent Executive Committee (COM/CITEL) consisting of 11 members, and two Permanent Consultative Committees (I) Standardisation Telecommunication, (II) Radiocommunications including Broadcasting, whose members are all Member States of the Organisation, Associate Members that represent various private telecommunications associations or companies, permanent observers and regional and international organisations.

The ARIB (Association of Radio Industries and Business) is the Japanese standards authority, part of the MPHPT (Ministry of Public Management, Home Affairs, Posts and Telecommunications of Japan), chartered in May 1995 [13]. The objectives of ARIB are to conduct investigations, research and development (R&D) and consultation of utilisation of radio waves with the view to develop radio industries, and to promote realisation and popularisation of new radio systems in the field of telecommunications and broadcasting. Regarding investigations and R&D on radio spectrum utilisation, the ARIB initiatives encompass: the study of spectrum utilisation, R&D on spectrum utilisation and IMT-2000 technical studies.

The APT (Asia-Pacific Telecommunity) [14] was established in May 1979 as a Regional Telecommunication Organisation by an inter-governmental agreement. APT now has 32 Members, 4 Associate Members and 51 Affiliate Members. The objectives of the Telecommunity shall be to (i) correlate the planning, programming and development of telecommunication networks, (ii) promote the implementation of all agreed networks, (iii) assist in the development of national components of efficient networks, (iv) foster coordination within the region of technical standards and routing plans, and (v) seek adoption of efficient operating methods in regional telecommunication service. APT Study Groups (SGs) conduct studies on telecommunication issues that are of concern to members (four SG dealing with 19 questions for the period 2001/2002). The output of the studies are brought out in the form of inputs to ITU, recommendations, handbooks, guidelines and reports.

The TELEC (Telecom Engineering Center) is, since 1998, the successor to MKK, which was founded in 1978 to provide a technical regulations conformity certification service and a calibration service for measuring devices under designation by the MPHPT pursuant to the Radio Law of Japan [15]. The TELEC has a Tokyo Telecom Research Park which is an open-house type of R&D facility that supports regional research and the development of telecommunication technology.

Regarding SDR in Asia, the TELEC is commissioned by MPHPT since 2000. An SDR Study Group was established in January 1999 by IEICE (Institute of Electronics, Information and Communications Engineers) and a first workshop on SDR activities in Korea, Japan and Taiwan, sponsored by RCBC (Radio Communications Broadcasting Committee) was organised in Korea in April 2000. In October 2001, an IEICE SDR workshop was organised by IEICE, TELEC, MPHPT, SDR Forum and ITU Japan.

11.2 Flexible Spectrum Allocation (FSA) in a Reconfigurable Radio Context

11.2.1 Vision for New Flexible Spectrum Management – European Projects

This section introduces some new spectrum and RRM approaches, investigated in the European IST projects TRUST, DRiVE, SCOUT and OverDRiVE.

11.2.1.1 TRUST [16]

The TRUST (Transparently Reconfigurable UbiquitouS Terminal) project was the largest European IST project on SDR, partly founded by the European Commission. It started from the user requirements from the perspective of the terminal and investigated enabling technologies and the system aspects of SDR. Within the system aspects, a specific research focused on spectrum-sharing techniques and investigations were done on the advantages of flexibility in spectrum engineering practices in the context of SDR applications [37]. Spectrum research on TRUST focused on (i) the evolution of standards and regulatory bodies to enable the use of SDR and spectrum-sharing techniques, (ii) the definition and description of possible RRM techniques relevant to flexible spectrum management, (iii) UMTS FDD/TDD compatibility studies for static scenarios, (iv) Joint Radio Resource Management (JRRM) (Figure 11.4) and (v) spectrum-sharing and collaboration between operators.

11.2.1.2 DRIVE [17]

The European research project DRiVE (Dynamic Radio for IP Services in Vehicular Environments) aims to enable spectrum-efficient high-quality wireless IP in a heterogeneous multi-radio environment to deliver in-vehicle multimedia services. DRiVE addresses this objective on three system levels:

1. The project investigated methods for the coexistence of different radio systems in a common frequency range with dynamic spectrum allocation contiguous Dynamic Spectrum Allocation (DSA) in both temporal and spatial configurations (Figure 11.5). DRiVE DSA performance (by simulation) has shown that, for temporal and spatial DSA, a typical gain over fixed spectrum allocation could be around 30% for both cases, although the exact value achievable depends on the traffic patterns seen for the services [23].
2. DRiVE developed an IPv6-based mobile infrastructure that ensures the optimised interworking of different radio systems (GSM, GPRS, UMTS, DAB, DVB-T) utilising new dynamic spectrum allocation schemes and new traffic control mechanisms.
3. Furthermore, the project designed location-dependent services that adapt to the varying conditions of the underlying multi-radio environment.

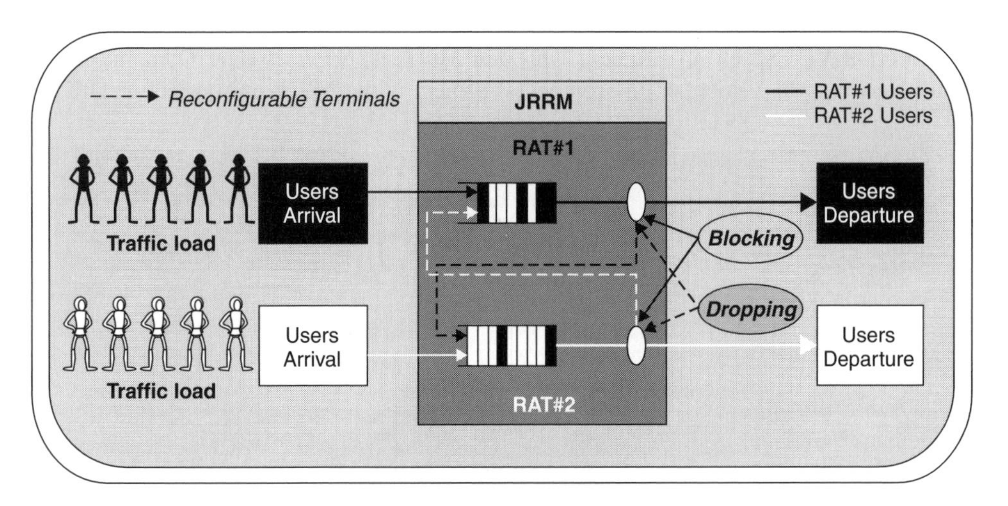

Figure 11.4 JRRM in TRUST

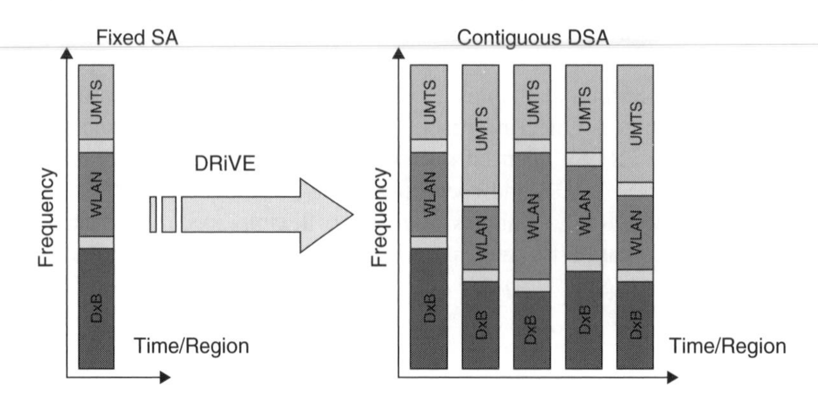

Figure 11.5 DRiVE temporal and spatial DSA

11.2.1.3 SCOUT [18]

The SCOUT (Smart user-Centric cOmmUnication environmenT) project is the follow-up project of TRUST, placing a specific focus on the evolution of spectrum management in an SDR context. Spectrum research analyses and proposes efficient solutions for RRM in the highly complex context of reconfigurable terminals, supporting multiple RATs, in different networks topologies (hierarchical and decentralised), being potentially managed by different operators. The main research thematics address (i) the evolution of radio spectrum regulation and management in a reconfigurable context, (ii) the evolution of regulator rules/impact of the R&TTE Directive, (iii) spectrum management of asymmetric regular traffic and software downloading impact on overall system performance, (iv) analysis of the inter-system handover measurement and criteria and inter-working between heterogeneous networks, and (v) flexible spectrum allocation. The highlights of the asymmetry/download and hybrid *ad hoc* RRM are presented in Figure 11.6.

11.2.1.4 OverDRiVE [19]

The OverDRiVE (Spectrum Efficient Uni- and Multicast Services over Dynamic Multi-Radio Networks in Vehicular Environments) project is the follow-up project to DRiVE.

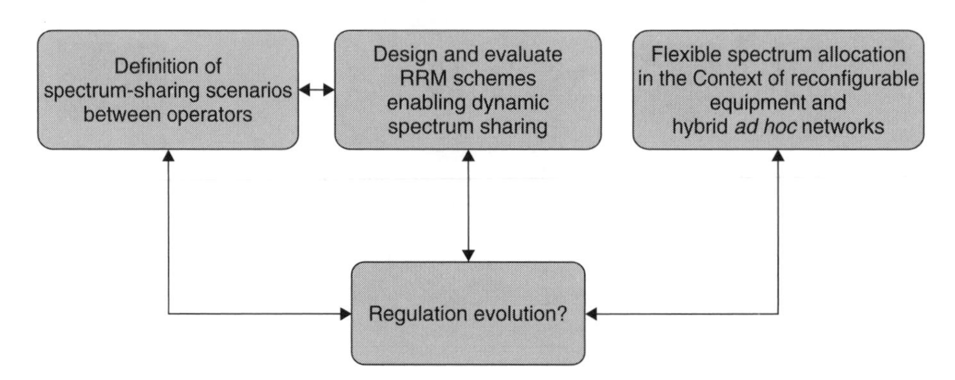

Figure 11.6 Highlights of SCOUT spectrum research

OverDRiVE aims at enhancing UMTS and combining it with existing radio networks into a hybrid network to ensure spectrum-efficient provision of mobile multi-media services. An IPv6 based architecture enables inter-working of cellular and broadcast networks in a common frequency range with DSA. The project objective is to enable and demonstrate the delivery of spectrum-efficient broad- multi- and unicast services to vehicles. The project introduces more flexibility into the achievable gain of spectrum efficiency in studying fragmented DSA schemes, based on the contiguous DSA algorithms developed in DRiVE. The spectrum research of OverDRiVE focuses on (i) system coexistence and DSA, (ii) DSA requirements on reconfigurability, (iii) multicast over UTRAN and (iv) asymmetric UMTS. The fragmented DSA is illustrated in Figure 11.7.

11.2.2 Flexible Spectrum Management and Spectrum Sharing

Until now, only static spectrum allocation has been considered (Section 11.1.1). However, there is a need for more advanced approaches such as dynamic or even flexible spectrum allocations that may usefully benefit next generations of wireless communications systems. This section highlights these new spectrum allocation perspectives.

11.2.2.1 Motivation and Challenges

In order to provide the optimum delivery of the user's wanted service via the most appropriate air interface available in a multi-radio environment (composed of systems supporting uni-, multi- and broadcast transmission modes), the handset is assumed to be provided with multi-mode and multi-band capabilities. This diversity of air interfaces can be achieved through reconfigurable air interface functionalities in future equipment. SDR is expected to be a key enabling technology providing these new flexible reconfigurable capabilities at the radio segment level.

SDR equipment is seen as a key technology in the evolution of telecommunications, terminals being able to support several RATs and provide seamlessly and transparently to users a large panel of applications that could only be available today with different devices, each being dedicated to a specific RAT. The reconfigurability capacities of multi-band multi-RAT SDR terminals will dramatically change the current vision of

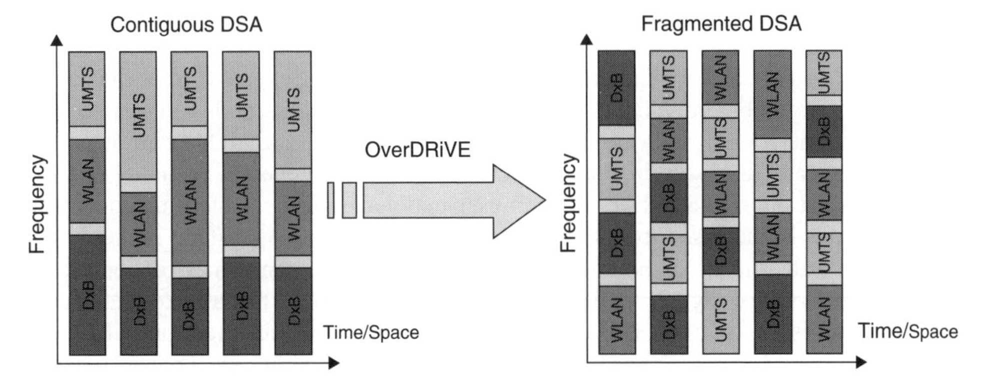

Figure 11.7 OverDRiVE DSA spectrum research

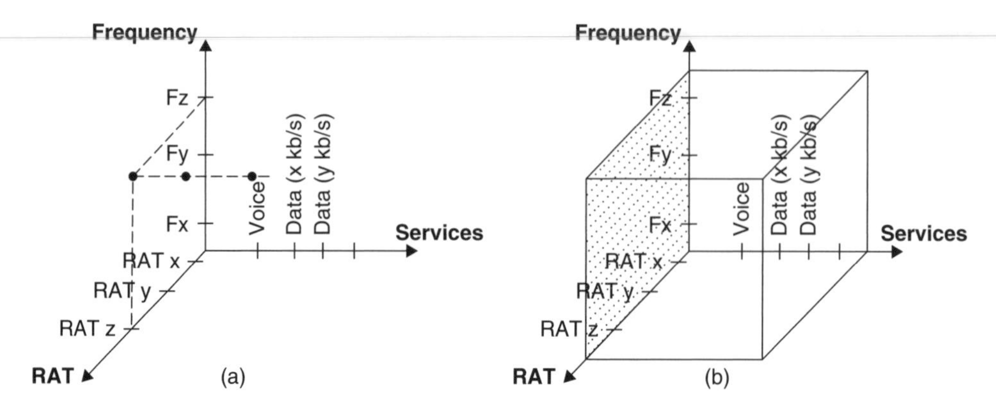

Figure 11.8 Evolution of spectrum management in an SDR context: flexible spectrum allocation (FSA)

spectrum allocation. Today communication equipment can be characterised by the triplet (Services, RAT, Frequency), where generally the projection in the RAT/frequency plan [Figure 11.8(a)] corresponds schematically to one point (or a couple of points, if the device is multi-band). The introduction of SDR terminals will extend the communication capabilities into a volume, the spectrum allocation then being potentially stretched to the overall RAT/frequency plan [Figure 11.8(b)]. One of the main challenges in a future spectrum and RRM will be to succeed in (re)thinking the spectrum allocation, considering the potentials brought by the SDR equipment.

With regard to this challenge, spectrum-related perspectives moving smoothly from today's fixed spectrum management (traditional approach) to tomorrow's full flexible spectrum management (FSM).

Indeed, the introduction of reconfigurable air interfaces allows allocating different users to the most appropriate RAT in a multi-radio environment. Therefore, FSM can introduce flexibility into/RRM in the sense that it enables us to provide the needed radio resources to peer entities (terminal, network) as required while maximising Quality of Service (QoS) and/or overall system capacity. Therefore, the use of radio resource units is more closely performed in connection with actual identified needs. It ensures the efficient and effective use of the scarce radio spectrum resource while avoiding any waste of expensive spectrum resources.

A gain in system capacity and/or spectrum efficiency can be expected from the deployment of reconfigurable equipment (terminal or any other access points in the network) only if appropriate FSM rules are designed. These FSM rules are highly correlated with RRM rules already discussed in Section 11.1 and further developed in Section 11.3. 'Providing the most appropriate RAT' to each user may have different meanings depending on whether this is addressed from the end-user's or the network's (system) point of view. Those views can sometimes be complementary but also can sometimes be opposite. In addition, FSM rules design is all the more challenging in that it has to take into account both technical issues (current 2G technologies deployment has matured, but what about the emerging technologies?) and economic issues (current actors are known, but what about their future roles and the roles of the future emerging actors?).

To outline this hot topic, the following basically illustrates how these rules could be driven at least from the end-user's terminal and the network's perspectives in the context of a multi-radio environment:

- *User-centred perspective.* The way the service is provided to the user is dependent on the end-user's terminal capability. Because future terminal generations might not be equipped with the same reconfigurable air interface capabilities, all services might not be delivered to each user in the same way. Therefore, users' required QoS might not be provided with the same reliability between all users. In addition, in the case when several operators (and/or service providers) offer different service proposals (with different rates) to deliver the user's requested service, a user-centred decision (manually on the terminal) can lead to the selection of one or several RATs. Therefore, FSM should take into account users' preferences for the design rules.
- *Network-centred perspective.* The way each user's requested service is managed at the radio system level is dependent on: (i) The overall system features and requirements (available RATs in the operating geographical area, offered traffic load to be supported by each RAT and all RATs, time and space varying nature of the traffic, interference conditions occurring on each technology, asymmetric traffic, etc.). Therefore, FSM rules must tackle the overall traffic and spectrum distribution/splitting between the different RATs. (ii) The user's behaviour (mobility, location, experienced propagation conditions, etc.). (iii) For each user, network criteria or any combination of them could be used to dynamically reconfigure the spectrum allocation. Thus, flexible spectrum management gives more opportunity to balance the offered traffic load (and consequently the necessary amount of spectrum allocated to each technology) between the available radio systems while maintaining users' requirements, maximising the spectrum efficiency (and/or overall system capacity) and minimising the interference. Different approaches are presented and discussed in the following subsections.

11.2.2.2 Concepts

Motivations for new spectrum allocation schemes have been discussed in Subsection 11.2.2.1. This subsection discusses and illustrates how air interface reconfigurability enabled by SDR technology introduces new perspectives for the design of new spectrum management schemes in the context of inter-operability/co-operation capability between heterogeneous radio access systems. As stated in Subsection 11.2.2.1, the re-definition/re-thinking of spectrum management impacts the way radio resources are managed. Therefore, these new flexible spectrum management schemes raise new RRM challenges already addressed in Section 11.1. An interesting algorithm tackling inter-system handover in an FSM context is discussed in Section 11.3.

In what follows, short, medium and long-term worldwide research community effort in the design of new concepts enabling flexible spectrum management is illustrated through different initiatives. Today, different flexible spectrum management philosophies are envisaged and some of them are under investigation. Different schemes can be introduced gradually (cf. the horizontal bars in Figure 11.9) to reach full flexibility:

- Fixed spectrum allocation scheme (but including inter-system, inter-operability capability)

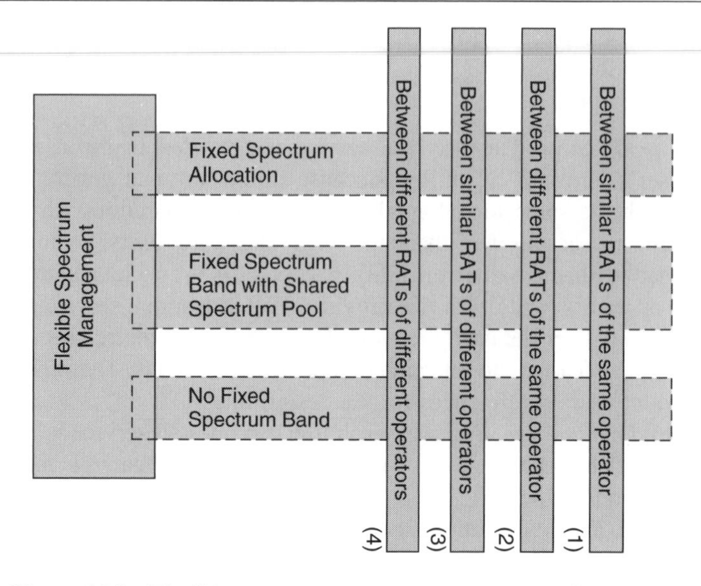

Figure 11.9 Flexible spectrum management concept and approaches

- Fixed spectrum band with a shared spectrum pool scheme
- No fixed spectrum band scheme (full flexibility)

These schemes could be considered as the most reasonable steps to reach the full reconfigurability status in terms of deployment and evolution path.

Different scenarios can be considered for each of these schemes. Those scenarios are defined by different combinations of the 'operator' and 'RAT' entities (Figure 11.10):

1. A single operator operates several RATs of the same family (e.g. mobile cellular enabled handsets)
2. A single operator operates several RATs of different families (e.g. mobile cellular enabled handsets, broadcast receiver enabled handsets, WLAN enabled handsets)
3. Several operators operate several RATs of the same family (e.g. mobile cellular enabled handsets)
4. Several operators operate several RATs of different families (e.g. mobile cellular enabled handsets, broadcast enabled receiver handsets, WLAN enabled handsets)

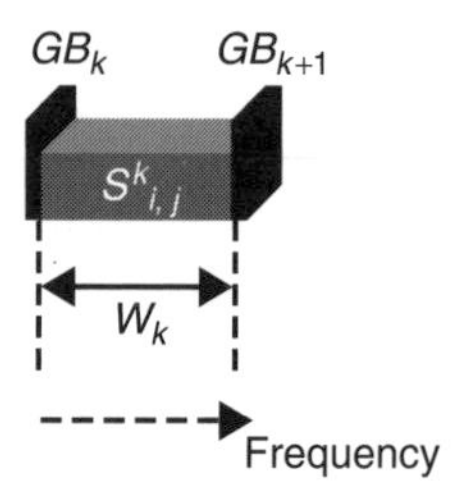

Figure 11.10 Spectrum band formalism

In what follows, to make the description more tractable let $S_{i,j}^k$ be the spectrum block k of bandwidth W_k used by operator i and operating RAT j on band k (Figure 11.10), where $k \in K = \{k_1, k_{2,\ldots,} k_l\}$; K is a finite set indexing (identifier number) the different defined frequency bands (blocks) in the spectrum. $i \in I = \{i_1, i_{2,\ldots,} i_m\}$; I is a finite set indexing (identifier number) the existing (and potentially future) identified operators in a given geographical area. $j \in J = \{j_1, j_{2,\ldots,} j_n\}$; J is a finite set indexing (identifier number) the existing (and potentially future) identified RATs in a given geographical area. GB_k and GB_{k+1} denote respectively the lower and the upper guard band of the spectrum block k with its neighbours.

11.2.2.2.1 Fixed Spectrum Allocation. This is the natural evolution from the current fixed spectrum allocation as regulatory bodies have proposed it. In this scheme (Figure 11.11), a fixed amount of spectrum $S_{i,j}^k$ is still allocated to each operator i. The bandwidth W_k of each $S_{i,j}^k$ is constant over time and space $\forall k \in K$, $\forall i \in I$ and $\forall j \in J$. However, each user is allowed to operate on any RAT j thanks to reconfigurable terminals. This basic scheme can be applied to all the above mentioned scenarios 1–4. Each of these scenarios introduces new perspectives for a more flexible RRM. However, this flexibility raises new issues to be resolved. In what follows, some of these topics are outlined.

(1) *Radio Resources Sharing.* In this fixed spectrum allocation scheme, the same operator (or different operators) owns one or several frequency blocks in which different RATs can be operated. Here, allocated spectrum bands and RATs are not operated independently since a user can switch from one RAT to another one while changing spectrum band. Hence, radio resources are managed on a joint (shared) basis for the call admission process for both the initial access phase and the inter-system handover phase for on-going communication. The introduction of inter-system handover capabilities allows to balance the offered traffic load between technologies and Spectrum Bands and thus to overcome congestion. This is the first step toward a more flexible spectrum management.

As the first approach of short-term research (considering dual UMTS FDD and TDD modes terminals of the same operator), the performance of this scheme has been assessed in terms of capacity gain and QoS improvement compared with the case when no inter-system handover is performed. A significant gain can be achieved (up to 20%). More details on this are available in Section 11.3.

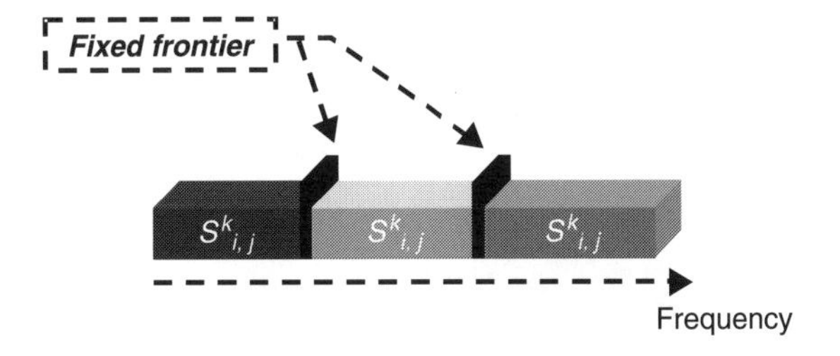

Figure 11.11 Fixed spectrum band scheme

The specific case of inter-system handover between two or more operators raises additional issues in term of fair radio resources sharing. Therefore, some rules and new mechanisms must be specified to ensure operator inter-operability. This is discussed in Subsection 11.2.2.3.1 with the introduction of the concept of a 'meta-operator'.

(2) *Spectrum Pooling.* Spectrum pooling illustrates a second strategy enabling flexible spectrum management between different operators in the context of fixed spectrum allocation. This scheme was initially proposed by Dr Joseph Mitola [20], who defines Spectrum Pooling as:

> 'an arrangement under which spectrum owners agree to rent it to each other for a time as brief as one second'. 'In a spectrum pooling, the licence owner of the spectrum has the highest priority to access the spectrum and allows priority specified candidate renters to use his spectrum until he needs his spectrum himself. It is therefore a way to make sporadically used frequencies available for commercial purpose without a drawback for the licence owners'.

Spectrum pooling is all the more efficient when it is based on an awareness of the user's context in terms of location, mobility, environment, etc.

In this original version, the motivation for pooling was the rental of public and government spectra by the present owners to cellular service providers for a time as brief as one second since federal, state and local governments could generate revenue streams by renting channels that were not currently in use. Spectra that could be made available for pooling include all the bands allocated for mobile terrestrial uses, and not satellite, radar and radio navigation bands.

The concept of spectrum pooling is illustrated in Figure 11.12. The licence owner of the 'grey' spectrum can rent his spectrum to the 'dark' operator [a dark user can use a part of the grey spectrum (phase 1)]. When a grey user (with the highest priority) needs some of his grey spectrum (phase 2) and if there is no grey spectrum left for his own operation, the dark user must leave the grey to release spectrum (phase 3). Table 11.1 gives an example of a possible priority arrangement between different user classes to access the under-used spectrum.

Spectrum pooling seems to be more suitable for non-real-time applications since at each time the renting resource must be released. Cognitive Radio is expected to be a key technology enabling spectrum pooling. The 'Cognitive Radio' concept is more widely described in Subsection 11.2.2.3.2.

In conclusion, this basic scheme does not offer the full expected flexibility because the amount of spectrum allocated to each of the technologies is still fixed. In addition, in this

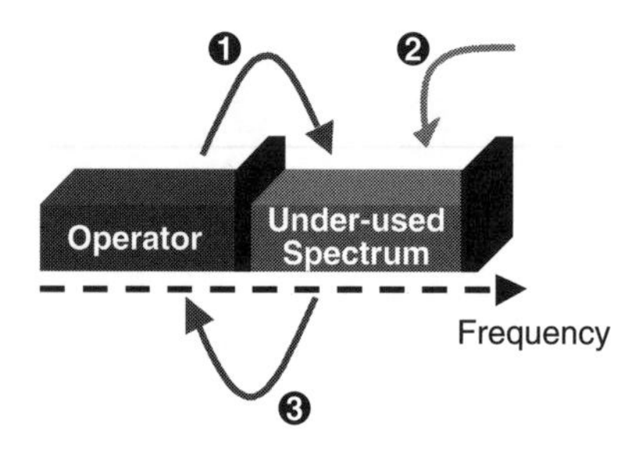

Figure 11.12 Illustration of the spectrum pooling concept

Table 11.1 Example of priority between different user classes

Priority	User classes
1	Emergencies
2	Government
3	Public interest
4	Commerce
5	Other

context of inter-system inter-operability, more signalling is expected with respect to the static and independent resources allocation.

11.2.2.2.2 Fixed Spectrum Band with Additional Spectrum Pool Sharing. In this scheme (Figure 11.13), a minimum fixed amount of spectrum $S^k_{i,j}$ is still allocated to each operator i. The minimal bandwidth $(W_k)_{\min}$ of each $S^k_{i,j}$ is constant over time and space $\forall k \in K$, $\forall i \in I$ and $\forall j \in J$. Each user is allowed to operate on any RAT j thanks to reconfigurable terminals capabilities. The major difference with respect to Scenario described in Figure 11.11 and noted (a) is that now a portion of the spectrum is in the pool and can be accessed by any RAT j, while the radio compatibility between neighbouring RATs is still provided. The scheme developed in (a) is considered as a sub-class of this scheme. Indeed, graph (3) of Figure 11.13 is the scheme discussed in (a).

This new flexible spectrum allocation scheme offers more degrees of freedom for a more efficient spectrum management. This scheme enhances the previous fixed spectrum allocation scheme in the sense that the amount of spectrum allocated to each technology can be increased or decreased depending on the traffic needs. Likely in scheme (a), the load can be balanced between RATs. However, in addition to this and conversely to (a), this new scheme provides the capability to balance the spectrum resource itself. Therefore,

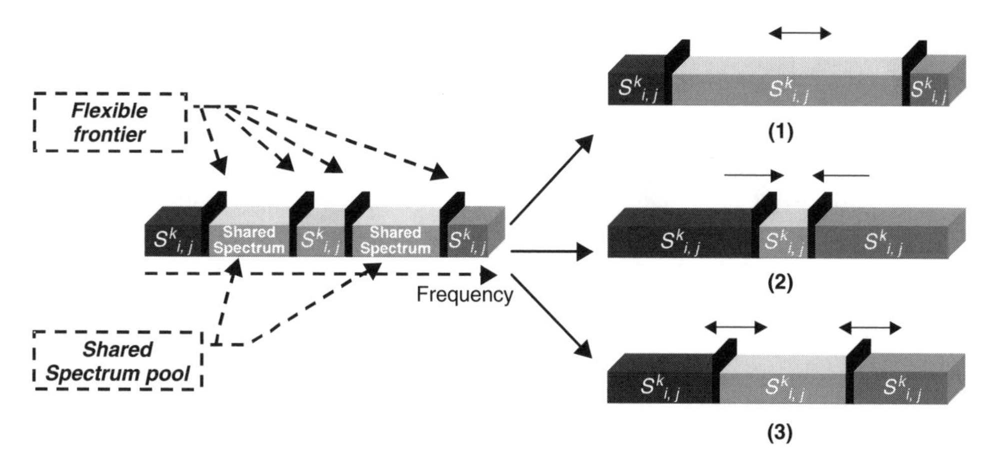

Figure 11.13 Fixed spectrum band with a shared spectrum pool

it becomes possible to provide a RAT with more spectrum band if necessary. Hence, because operators can access additional shared parts of the spectrum, this can introduce radio resources contention between operators for a fair access to the spectrum resources. Consequently, some appropriate rules and new mechanisms must be specified to ensure operators' inter-operability on spectrum sharing. Regarding this topic, the concept of a 'meta-operator', discussed in Subsection 11.2.2.3.1, has been developed.

It is noticeable that the fixed bandwidth for each technology guarantees a minimum amount of resources for each technology. This is a first step towards more spectrum efficiency. Conversely to (a), this scheme enables switching between one RAT and another one without changing frequency carrier.

Of course, the introduction of this new scheme raises some stringent requirements and increases complexity. For example, it is expected that more signalling information on the new spectrum allocation scheme will be transmitted to the RAN and end-user terminals. At the same time, radio resource management schemes might be more complex.

11.2.2.2.3 No Fixed Spectrum Band. In this scheme (Figure 11.14), there exists no predefined part of the spectrum $S_{i,j}^k$ allocated to each operator i to operate RAT j. Here, the spectrum allocation is totally flexible: there is no specific bandwidth W_k allocated to any of the RATs.

The bandwidth W_k of each $S_{i,j}^k$ varies dynamically over time and space $\forall k \in K, \forall i \in I$ and $\forall j \in J$. As in schemes a) and b), which is described in Figure 11.13, each user is allowed to operate on any RAT j thanks to reconfigurable terminals. The scheme developed in (b) is considered to be a sub-class of this scheme. Indeed, scheme (c) (see Figure 11.14) can support scheme (b).

This scheme introduces full flexibility in the sense that there is no pre-defined RAT and no frequency carriers (spectrum bands) allocated to each operator i to deliver each user's service request. Hence, the user's terminal can be fully reconfigured (RAT + frequency carrier) if needed. In this framework, different RATs can share the same frequency carrier at different points in time in the same geographical area. However, the implementation

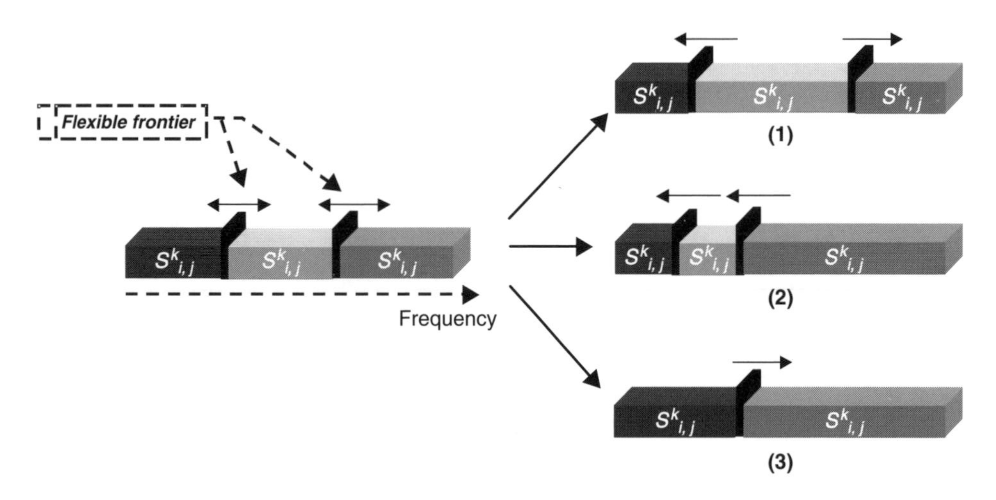

Figure 11.14 No fixed spectrum band

of this scheme may require more complex radio resource management. Higher signalling is expected when compared with the previous solutions.

As mentioned in Section 11.2.1, active research on advanced flexible spectrum management schemes is being carried out though different European initiatives. This additional work completes the FSM overview in the context of the No Fixed Spectrum Band scheme. Currently, the European Commission has sponsored one of these initiatives (namely the OverDRiVE project) to investigate spectrum-efficient temporal and spatial DSA schemes in multi-radio environment systems (UMTS, DVB-T and WLAN), where several transmission modes (uni-, multi- and broadcast) are possible. The overDRiVE Project builds on the findings of the successful DRiVE project [17]. The following paragraphs outline the motivations for DSA and discuss some interesting developed approaches to tackle the identified issues. In the OverDRiVE project, different DSA strategies are under investigation regarding the spectrum blocks $S_{i,j}^k$ (of width W_k) distribution between each operator i and each RAT j into a given spectrum range. An overview of this project can be found in refs [19] and [21]. More specific details dealing with DSA research are available in ref. [22].

(1) *Temporal Dynamic Spectrum Allocation.* This paragraph discusses the motivations for temporal DSA and describes the developed concepts in the DRiVE and OverDRiVE projects. More details can be found refs [22] and [23].

Temporal DSA motivations are based on the time-varying utilisation of the spectrum (Figure 11.15). This means that during some time observations windows, the spectrum may be under-used. For illustration, two different radio access networks are considered: UMTS delivering speech service and DVB-T delivering multicast video streaming. When observing over a 24-hour period, this graph depicts three curves:

- The traffic demand (and therefore the spectrum demand) for UMTS [curve (a)]
- The traffic demand (and therefore the spectrum demand) for DVB-T [curve (b)]
- The cumulative value of the spectrum demand for UMTS and DVB-T [curve (c)]

To make the illustration more tractable, both spectrum demand peaks are normalised to 1. In the case of a fixed spectrum allocation scheme, enough spectrum is allocated to each RAN to support their respective peak demands at any time. This leads to an overall amount of allocated spectrum set to 2. However, it is observed the peak demands for UMTS [peak

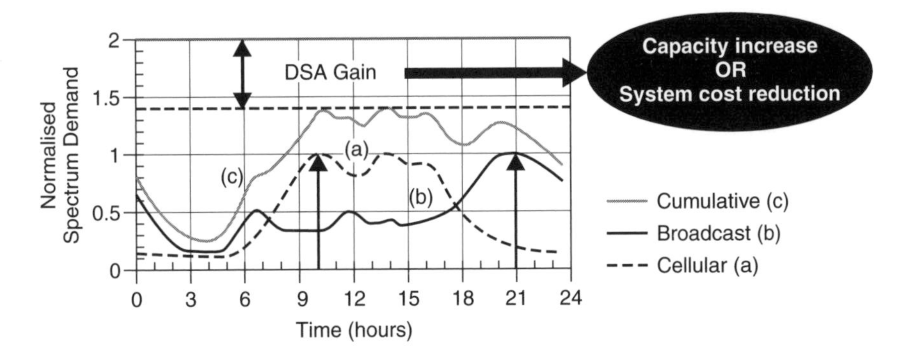

Figure 11.15 Temporal traffic distribution

of curve (a)] and DVB-T [peak of curve (b)] do not occur simultaneously. Hence, some gains can be made from the combined use of co-ordinated networks and DSA. In the ideal case of a perfect DSA scheme being used, only as much spectrum as required for the traffic demand would be allocated to the RAN. This would mean that sufficient spectrum would be required to support the cumulative traffic demand from UMTS and DVB-T at any time. This corresponds to the peak of curve (c). This spectrum demand peak is below the amount of spectrum required for fixed allocation. As a consequence, the gain can be expressed into two different ways: either the same amount of traffic could be supported in less spectrum bandwidth or, conversely, more offered traffic could be supported in the same amount of spectrum with DSA as a fixed allocation. DSA gains will be all the more important if UMTS and DVB-T traffic peaks occur at different times, i.e. traffic patterns have negatively correlated patterns [22].

Two schemes, namely contiguous and fragmented DSA, have been identified (Figures 11.6 and 11.7) for the temporal DSA [23]. The goal is to define the spectrum blocks $S_{i,j}^k$ (of width W_k) distribution (in the time domain) between each operator i and each RAT j for a given spectrum range. In other words, the purpose here is to specify $S_{i,j}^k(t)$ and $W_k(t)$ as a function of time t.

The *contiguous DSA* scheme was investigated during the DRiVE project [23]. This scheme can be seen as the evolution between the fixed and dynamic assignments. The contiguous DSA scheme allocates a single contiguous block of spectrum to each RAN in the system. However, the width of the spectrum block assigned to each RAN is allowed to vary in order to allow for changing demand. This technique will only allow the spectrum partitioning to change at the expense of the adjacent spectrum allocations. In the case of UMTS and DVB-T, with the traffic patterns shown in Figure 11.15, it has been shown that the contiguous DSA can increase system capacity by 30% up to 98% satisfaction compared with the fixed spectrum allocation [22, 23]. A more theoretical analysis is detailed in ref. [25].

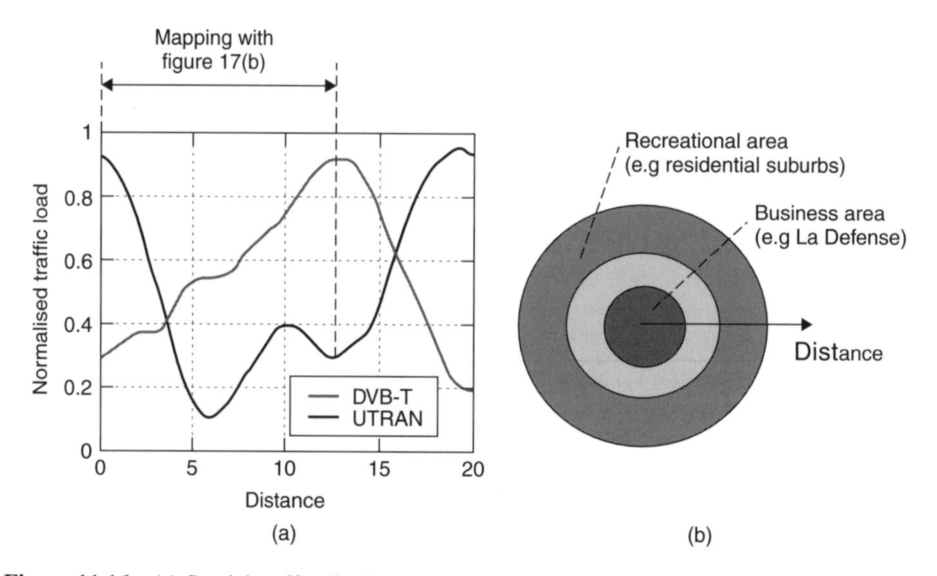

Figure 11.16 (a) Spatial traffic distribution vs. distance; (b) services and spatial area mapping

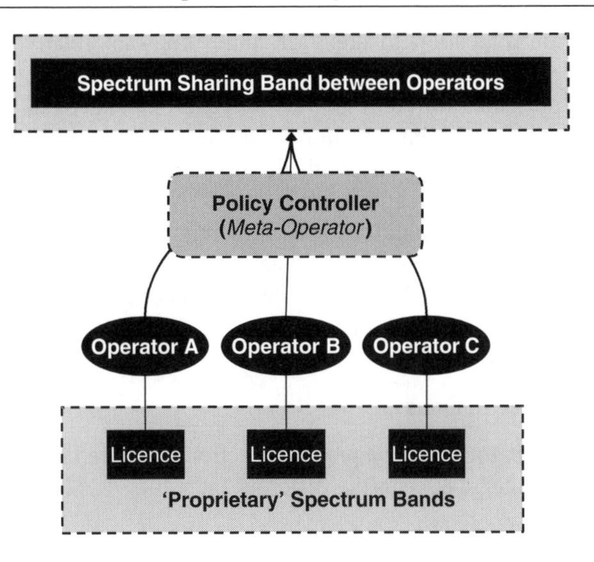

Figure 11.17 Architecture for dynamic inter-operator spectrum sharing

Current and ongoing OverDRiVE research activity is investigating some additional DSA schemes. One such scheme is the *fragmented DSA* concept. The fragmented method is seen as an enhancement of contiguous DSA. This scheme allows RANs to be allocated blocks of spectrum from anywhere in an overall spectrum pool. Here, the blocks do not need to be contiguous, i.e. there can be several blocks of spectrum allocated to one RAN that are spread out over the spectrum with other RANs spectrum between them. In this way, the size of the access network's spectrum is not dependent on its adjacent spectrum neighbours [22, 23].

(2) *Spatial Dynamic Spectrum Allocation.* The goal is to define the spectrum blocks $S^k_{i,j}$ (of width W_k) distribution in the overall spectrum band between each operator i and each RAT j in the space domain. In other words, the purpose here is to specify $S^k_{i,j}(s)$ and $W_k(s)$ as a function of space.

Apart from being temporally variable, Figure 11.16(a) and (b) shows how demand for services is also spatially variable. Two different RANs are also considered: UMTS delivering speech service and DVB-T delivering multicast video streaming. Over a given geographical area [Figure 11.16 (b)], Figure 11.16 (a) depicts two curves:

- The traffic demand (and therefore the spectrum demand) for UMTS
- The traffic demand (and therefore the spectrum demand) for DVB-T

The spectrum demand for UMTS (delivering speech service) is expected to be larger in the business area, whereas the spectrum demand for DVB-T (delivering multicast video service) is expected to be larger in recreational areas, such as residential suburbs. However, with this current approach, the geographical areas with the largest traffic demand typically determine the spectrum demand for spectrum allocation. Therefore, without spatial spectrum allocation adaptation, the spectrum is often wasted in most regions. Therefore, motivations for new spatial spectrum partitioning are clear. In this context, new spatial DSA strategies have been proposed [24, 26]. Spatial DSA aims to confine the uniform spectrum allocation to areas of rather constant relative traffic density by establishing DSA regional areas.

This section has presented different flexible spectrum management schemes. However, it is important to note that the introduction of these new schemes will be possible only if investigations are performed jointly with the operators' views to have a better understanding of users' needs and to capture more precisely the business models that impact directly on those schemes. In addition, spectrum regulatory bodies' views on a potential open spectrum and new spectrum co-existence rules design will also directly impact the feasibility of these schemes. Therefore, flexible spectrum management research and success is at least a three parties effort.

11.2.2.3 Enabling Mechanisms

11.2.2.3.1 Meta-Operator. In the previous sections, the need for a mediator (a kind of high level entity controller) between the different operators was identified for two cases:

1. In the case when a fixed amount of spectrum is allocated to each operator but where each user can be allowed to switch from one system to another one (the fixed spectrum allocation scheme).
2. In the case when the spectrum itself can be shared dynamically between operators (the fixed spectrum band with an additional spectrum-sharing pool and the no fixed spectrum band schemes).

This second scenario (spectrum-sharing scheme) illustrates the concept of spectrum loaning and renting in the sense that additional spectrum is provided to each operator and users might switch from one system to another through inter-system handover. Therefore, this enables us to face high and low peaks of traffic by dynamically adjusting spectrum allocation depending on traffic needs and conditions. Smart load balancing can therefore be performed. However, this requires mutual agreement between operators with appropriate mechanisms ensuring the agreed rules. To tackle that, the concept of a higher-level entity has been introduced to deal with radio resources contention between operators and to ensure fair resources access. This can lead to the introduction of different levels of co-ordination between competitive systems.

The conceptual architecture of dynamic inter-operator spectrum sharing is depicted in Figure 11.17. On the one hand, each operator, namely A, B and C, is allocated with 'proprietary' spectrum. On the other hand, these operators agreed to flexibly share some frequency carriers. In this case, each network can potentially use up those frequency carriers. The shared channels access management consists in setting some pre-defined priorities between operators. This can be performed with the introduction of an additional entity in the network playing the role of the decision, policy controller manager. The policy controller could be implemented as:

1. A single operator: an operator hosts the policy controller, but it is not necessarily the master.
2. A distributed decision process manager between the operators.
3. An external operator: 'operator of operators', namely a meta-operator.

This policy controller has to be accessible by all interested operators. The policy controller and the operators can communicate through wired or wireless links (in that case, a 'common channel' would be needed).

The policy strategy can be derived from different pre-defined rules agreed between operators. The design and implementation of such mechanisms require some information exchange. This raises some issues relating to security and confidential information exchange between operators. In addition, there is a need to investigate an appropriate architecture (centralised, de-centralised, hybrid centralised/de-centralised) supporting these mechanisms. For more details on this topic, the reader is referred ref. [27] regarding a proposed method and an architecture enabling such frequency carrier sharing between different operators. The pros of this method and architecture are to dynamically adjust a set of thresholds without the need for load information exchange between the operators. In this scheme, priority levels can be modified in time, which in turn makes it possible to automatically update the allocated threshold values according to current needs. That means that if operator B has lower priority than operator A, and is starting to be overloaded (for example, by the arrival of high-bit-rate services) and while operator A's resources are under-utilised, operator B's access priority can be increased temporarily. The policy controller will arbitrate between concurrent needs.

11.2.2.3.2 Cognitive Radio. As seen in Subsection 11.2.2.2, spectrum pooling is a novel approach for better managing the spectrum. The goal is that users who need spectrum at a given time and location, rent this spectrum from entities who do not need it at the same time. There are many technical difficulties to be solved before allowing spectrum pooling. The user must first detect that the spectrum is not in use, negotiate the renting with the legacy user and detect the entering of a legacy user in order to free the bands. J. Mitola introduced the Cognitive Radio concept [20] enabling all necessary actions to be performed. A Cognitive Radio is defined as an agent controlling the Software Radio. It is capable of adapting its radio resources use as function of the user context. So the radio has to be context aware. In order to detect its context, J. Mitola described [20] how it interacts with the external world thanks to the cognition cycle: the Cognitive Radio continually observes the environment, orients itself, creates plans, decides, and then acts. In addition, machine learning can take place in these different phases in order to enhance the robustness when the environment rapidly evolves. In order to enable spectrum pooling, a new class of protocol, named Radio Etiquette, has been envisaged [20]. This protocol provides for the advertising and rental of spectrum. Etiquette includes the spectrum-renting process, assured back-off to authorised legacy radio, assured conformance to precedence criteria and an order-wire network. A complex de-centralised solution enabling Spectrum Pooling has been proposed, this referring to the Cognitive Radio.

11.2.2.4 Flexible Spectrum Management and Reconfigurability

11.2.2.4.1 Joint FSA and Reconfigurability Research. Previous sections have discussed different flexible spectrum management proposals and the potential derived mechanisms. However, the flexible spectrum management operations in a heterogeneous wireless network environment require the definition, or even the re-thinking, of new functionalities in existing and future systems to ensure the feasibility and integration of these new schemes and mechanisms. One of the expected new features of these functions is the reconfigurability capability. Therefore, both research on FSM schemes and research on SDR equipment design (terminals and entities in the network to support them) are required to be investigated jointly to ensure an adequate synergy between FSA schemes and SDR equipment implementation. This research work is currently carried out within the European project OverDRiVE [19, 28], where dynamic spectrum allocation scheme investigations are performed jointly with reconfigurability [29]. The main research challenges and expectations on this topic are depicted in the following paragraphs. More details on the project framework dealing with this topic are described in ref. [29].

11.2.2.4.2 Reconfigurable Systems Architecture Model. Research on reconfigurable functionalities supporting DSA is closely connected with research on the design of potential architectures supporting those functionalities. In this reconfigurability-based, flexible spectrum-engineering context, the main challenges are to identify the functions and entity classes relevant for supporting a flexible spectrum scheme. To properly carry out this study, an architecture model is needed to support and capture the interactions between the objects composing the architecture. The architecture is expected to be composed of functions, entities, components and interface objects. Motivations for research on a reconfigurable systems architecture model are that current worldwide reconfigurable systems architecture model proposals [30–33] often encompass numerous issues (related to all aspects of terminal and network) but do not explicitly address specific flexible spectrum management issues. Hence, it is very difficult to have a clear visibility on how flexible spectrum management is impacted in those generic models. Alternatively, other models focus on specific issues (e.g. RF, Base Band, security) and miss some relevant points needed for an appropriate reconfigurability model of dynamic spectrum allocation.

Therefore, the main OverDRiVE challenges relevant to these investigations are to design an appropriate model enabling the capture of the overall relevant flexible spectrum management reconfigurability features into an adapted architecture, and to investigate how far this architecture model can inter-operate with/complement existing ones.

In order that reconfigurable functions are supported and operated, the identification of different entities is needed. It is expected that several different types of entities will be designed or enhanced to support the reconfigurable functions. Typically, the physical entities can consist of the end-user terminal, a RAN, a core network, backbone or any specific part of the system. In the meanwhile, the identification of some new entities supporting the reconfigurable functions will be investigated if needed. For instance, if necessary, the introduction of new physical entities will be considered and/or new concepts enabling new logical entities (e.g. virtual entities) will be developed.

Then, regarding the reconfigurable systems architecture model, one of the goals is to map each reconfigurable function between the different entities. One of the research challenges is to investigate how far each function (control, monitoring, decision, etc.) should be centralised within a single (or alternatively distributed between several) entity(ies). The motivation for these choices should be rationalised by investigating appropriately the potential strategies for splitting the different reconfigurable functions. Those investigations should be achieved considering different approaches and including terminal-centred,

Table 11.2 Preliminary DSA requirements

Spectrum requirements
1. Time spectrum management
2. Spatial spectrum management
3. Spectrum efficiency optimisation management
4. Scalability management
Additional requirements
1. Mobile and seamless connections
2. Coverage complementarity between RANs
3. Service complementarity between RANs
4. Downlink complementarity between RANs
5. Individual and System QoS optimisation

Table 11.3 Potential reconfigurable functions

Physical level functions
1. Multi-radio access operations
2. Multi-band operations
3. Radio Access Technology (RAT) and frequency carrier detection
4. RAT and frequency carrier identification
5. RAT and frequency carrier monitoring
6. Variable duplex operations (asymmetric spectrum)
System level functions
Spectrum
1. Spectrum availability detection
2. Fair frequency carrier distribution between RATs
3. Appropriate channel bandwidth management
4. Temporal spectrum allocation/release
5. Spectrum negotiation between operators
6. Information exchange between operators (including signalling)
7. Spectrum coexistence (guard band management in space and time)
8. Spatial area dimensioning
Vertical handover (VHO)
1. Transmission modes (unicast, multicast, broadcast) identification – spectrum needs for each mode
2. Simultaneous UL and DL spectrum management
3. VHO between operators – VHO between private and legacy operators
4. Group creation and management for multicast
5. Frequency planning/coordination and coverage management
Traffic awareness
1. Temporal traffic load statistics
2. Spatial traffic load statistics
3. Mobile nodes location
4. Mixed traffic load and spectrum needs prediction

network-centred or combined terminal- and network-centred perspectives. In this context, different architecture models can be envisaged depending on the degree of coupling between different RANs. Hence, in addition to this, the research activities will be led by both loose and tight coupling approaches relevant to the flexible spectrum management framework.

11.2.2.4.3 Flexible Spectrum Management and Derived Reconfigurable Functions.

With regard to the OverDRiVE scenarios under investigation [34], DSA concepts (Subsection 11.2.2.2) and, more generally speaking, the flexible spectrum allocation schemes, the first three steps of the project framework are:

1. Identification of the flexible spectrum management requirements.
2. Identification of the reconfigurable functions derived from previous requirements.
3. Mapping between requirements and functions.

Regarding specific OverDRiVE scenarios and DSA concepts, some very preliminary requirements can be identified, as illustrated in Table 11.2. Two high-level sets of requirements are listed in this table: specific spectrum requirements and additional requirements directly impacting flexible spectrum management. As listed in Table 11.2, scalability management is important to enable the current investigated flexible spectrum management schemes to be operated by future emerging technologies and equipment. Given these flexible spectrum management requirements, some initial reconfigurable functions can be derived (Table 11.3).

Different types of functions can be derived from multi-radio access and multi-band switching operations. For instance, depending on the current allocations set by the DSA, the user equipment may need to be able to retune to different frequencies and RATs. Moreover, additional functions can be considered in the case when several RATs are supported simultaneously by the same Tx/Rx device.

11.3 Characteristics of the JRRM Scheme

Section 11.2 outlined the challenge and potential future progress for new spectrum management schemes to be rolled out in the short, medium and long term. As a first approach of short-term research, the performance of a simple joint RRM scheme has been assessed in the case of the 'fixed spectrum allocation' scheme (Subsection 11.2.2.2.1) supporting inter-system handover capability between several RATs.

Here we present an efficient JRRM scheme between RATs optimising the use of radio resources. The advantage of having N RATs in the same geographical area must not only be considered as a possibility to basically handover, but as a means to increase the overall system's performance (RAT#1 $+ \cdots +$ RAT#N). The introduction of the JRRM scheme results in a QoS and capacity improvement for the overall system. This novel scheme considers radio conditions experienced by the user as well as overall system's behavior in order to maximise the inter-system handover success probability while minimising the impact on the performance of the hosting RAT. The JRRM algorithm is accurately described in the following paragraphs. As a first approach of short-term research, the performance of a simple joint RRM scheme has been assessed for two RATs. Gain results generated by simulations are then presented. Some of this work has been investigated in the context of the IST TRUST project and is resumed in ref. [37].

11.3.1 Enabling the JRRM Operation

In cellular communication systems the radio conditions experienced by users depend on his location, speed and the environment where he starts out. For instance, the traffic load of his current cellular network plays an important role. In order to maintain a certain QoS through the network and to permit a high mobility for users, a process called handover was developed. This latter triggered by the network jointly with the mobile consists in searching a better Base Station (BS) (in term of pathloss, or in term of loading) inside the network in order to provide the user with a better quality. However, no handover solutions could be available if the network were heavily loaded in the user's geographical zone. In this case, the user could lose his communication, i.e. he might be dropped. Thus, another user trying to access this network in the same geographical zone probably would be blocked. Other users having a communication in this zone would maintain it but with poor quality (e.g. high FER) caused by the interference of other users. In the near future, more RATs will be available for users. Nevertheless, in the context of multi-technologies and multi-bands, the problem described will still exist. Given multi-technologies and multi-bands capabilities of future terminals, a user experiencing poor radio conditions in his RAT could be ordered to perform an intersystem handover to another RAT, aimed at increasing both user and system QoS. This intersystem handover can be seen as a Radio Resource Management (RRM) rule, and has to be more precisely analysed in order to optimise its performance. The following describes and presents an optimising scheme to jointly manage the radio resources of two RATs: the JRRM scheme. In what follows, 'JRRM module' refers to a subpart of the 'JRRM scheme'.

11.3.2 Theoretical Models

11.3.2.1 High-Level Description of the Process

A high-level description of the process is represented in Figure 11.18. The launch of the JRRM module will be motivated by the fact that one user experiences poor radio conditions or blocking and that no solution are currently available in his native i ($i \in$ {RAT#1, RAT#2}) since only two RATs are considered here).

In order to trigger this algorithm, some user's quality indicators have to be monitored and transmitted to the module in the case of trouble (1). After launching the process and before ordering the inter-system handover, the probability that the user experiences better radio conditions or can access radio resources in the new mode without degrading the performance of this mode is evaluated. To do so, some system's indicators have to be monitored. The JRRM module requires those values about both RATs as inputs (2). Those system's indicators are then provided to the JRRM (3). After performing these phases, the module notifies (4) the user and both RATs of its decision (accept/refuse the inter-system handover).

Before providing a complete description of the JRRM algorithm, the 'Individual User' and 'Overall System and User' indicators have to be more precisely defined.

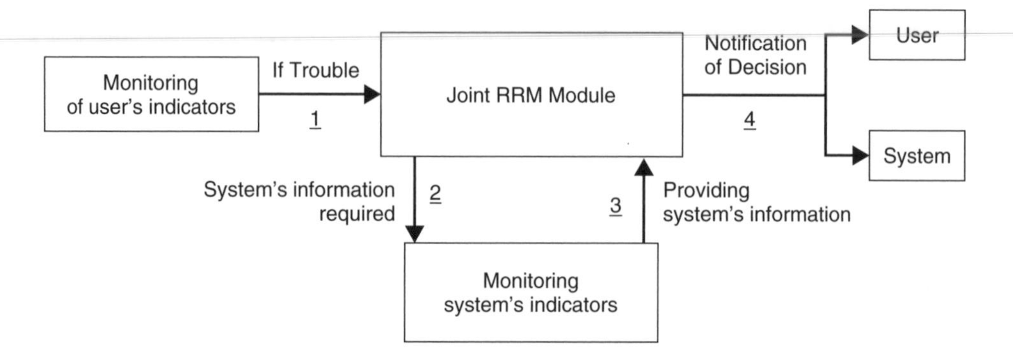

Figure 11.18 Macro-description of an inter-system handover

11.3.2.1.1 Monitoring and Assessing an Individual User's Indicators. Some parameters characterising the communication quality of a user have to be monitored in his native i, $i \in \{\text{RAT\#1}, \text{RAT\#2}\}$. Three different parameter sets are relevant $\mathbf{Pu}_1$, $\mathbf{Pu}_2$, $\mathbf{Pu}_3$; $\mathbf{Pu}_1 = \{Pu_1\}$ concerning the Dropping (D), $\mathbf{Pu}_2 = \{Pu_{2,1}, Pu_{2,2}\}$ the Bad Quality Call (BQC) and the last $\mathbf{Pu}_3 = \{Pu_3\}$ the Blocking (B).

An ongoing call user will be dropped if the time Pu_1 during which all consecutive frames are lost represents a time higher than a given time threshold T_{dropped}. If Pu_1 becomes higher than a threshold $T_1 (T_1 < Pu_1 < T_{\text{dropped}})$, problems are detected for the user, and the JRRM procedure is triggered.

An ongoing call user can also loose frames, which are distributed along the communication. If the FER (Frame Error Rate) is relative high, the user will be dissatisfied due to the overall BQC. Such users experiencing bad radio conditions also require higher power levels. $Pu_{2,1}$ is equal to the FER at a given time instant during the user's communication. $Pu_{2,1}$ indicates a problem if it becomes higher than a threshold $T_{2,1}$ (a). $Pu_{2,2}$ is equal to the power required for a BQC user. $Pu_{2,2}$ indicates a problem when $Pu_{2,2}$ becomes higher than a threshold $T_{2,2}$ (b). The user sends a message to the JRRM module indicating he is experiencing problems during his communication if both (a) and (b) are experienced.

The last major problem a user can experience is blocking. A user tries to access radio resources, but the network refuses his communication because no resource is available. Pu_3 is the consecutive time during which a user tries to access radio resources (if $Pu_3 > T_{\text{Block}}$, the user is blocked). If Pu_3 becomes higher than a threshold $T_3 (T_3 < Pu_3 < T_{\text{Block}})$, problems are detected and the JRRM module is contacted to try to perform an inter-system handover.

For the set-up of the different thresholds T_1, $T_{2.1}$, $T_{2.2}$, T_3, a compromise has to be found in order to maximise the probability that the user is experiencing real problems and so minimise the probability of triggering the JRRM module if not really needed.

11.3.2.1.2 Monitoring and Assessing the Overall System and Users' Indicators. These indicators refer to both the overall system and user indicators, i.e. monitored and assessed from the system perspective. From the system perspective, 'overall user indicators' refers to overall user QoS monitoring and assessment (statistics performed over all the users). From the system perspective, 'overall system indicators' refers to both. Traffic monitoring and assessment includes both native traffic and inter-system handover traffic, and overall system QoS monitoring and assessment.

11.3.2.1.3 Traffic Monitoring and Assessment. Before performing an inter-system handover, the JRRM scheme has to be aware of the radio conditions in $i(i \in \{\text{RAT\#1}, \text{RAT\#2}\})$ in order to avoid that a user experiences more degraded radio conditions after his inter-system handover from i to j ($i \neq j$ and $(i, j) \in \{\text{RAT\#1}, \text{RAT\#2}\}$). The first parameter to monitor and to assess is the current traffic load ($A_{\text{current}\,i,k}$) per BS in the two systems at the time $t = k$ and where $i \in \{\text{RAT\#1}, \text{RAT\#2}\}$. To estimate this, the traffic load can be divided into two parts. $A_{\text{native}\,i,k}(i \in \{\text{RAT\#1}, \text{RAT\#2}\})$ due to the birth of users in the native i at time k. The other traffic is $A_{\text{HO}\,i,k}$ taking into account the arrivals and departures of users due to inter-system handover during the time interval $[k-1, k]$ in i, where $i \in \{\text{RAT\#1}, \text{RAT\#2}\}$. In this definition, $A_{\text{HO}\,i,k}$ refers to the handover traffic considered from i (i.e. RAT#1 or RAT#2) point of view. Given this definition, the traffic $A_{\text{HO}\,i,k}$ can also be either positive or negative since $A_{\text{HO}\,i,k}$ is a traffic variation. The equivalent traffic is then calculated as follows:

$$A_{\text{current}\,i,k} = A_{\text{native}\,i,k} + A_{\text{HO}\,i,k}, \tag{11.1}$$

where $i \in \{\text{RAT\#1}, \text{RAT\#2}\}$.

Typically, since these traffic estimations will be processed periodically (between two consecutive assessments instants $t = k-1$ and $t = k$), a recursive approach can be used in order to process only the differential between $t = k-1$ and $t = k$. The formalism based on recursion enables us to obtain estimated values with a higher likelihood since estimated values at $t = k$ take into account the system memory state, and enables us to minimise the processing loading within the network. This approach is described in the following.

For native traffic:

$$A_{\text{native}\,i,k} = A_{\text{native}\,i,k-1} + \Delta_{\text{native}\,i,k}, \tag{11.2}$$

$$\Delta_{\text{native}\,i,k} = \delta_{\text{native}\,i,k} * T_{\text{MEAN native}\,i} / N_{\text{BS}\,i}, \tag{11.3}$$

where $i \in \{\text{RAT\#1}, \text{RAT\#2}\}$, $\Delta_{\text{native}\,i,k}$ stands for the $A_{\text{native}\,i}$ traffic variation on $[k-1, k]$ per BS, $\delta_{\text{native}\,i,k}$ stands for the call arrival rate variation of users on $[k-1, k]$, $T_{\text{MEAN native}\,i}$ is the mean users' communication time in i (based on well-known voice traffic model), and $N_{\text{BS}\,i}$ is the number of BSs for i (this number is well known within the considered coverage area). With this approach, non-stationary traffic in the native network can be taken into account by $\delta_{\text{native}\,i,k}$ (arrival rate).

For the inter-system handover traffic, we then define:

$$A_{\text{HO}\,i,k} = A_{\text{HO}\,i,k-1} + \Delta_{\text{HO}\,i,k}, \tag{11.4}$$

$$\Delta_{\text{HO}\,i,k} = \delta_{\text{HO}\,i,k} * T_{\text{MEAN HO}\,i,k} / (T_{\text{OBSERVATION}} * N_{\text{BS}\,i}), \tag{11.5}$$

$$\delta_{\text{HO}\,i,k} = N_{\text{ARRIVALS}\,i,k} - N_{\text{DEPARTURES}\,i,k}, \tag{11.6}$$

$$T_{\text{MEAN HO}\,i,k} = [(k-1) * T_{\text{MEAN HO}\,i,k-1} + \nu_{i,k}] / k, \tag{11.7}$$

where $i \in \{\text{RAT\#1}, \text{RAT\#2}\}$, $\Delta_{\text{HO}\,i,k}$ stands for $A_{\text{HO}\,i,k}$ traffic variation on $[k-1, k]$ per BS due to inter-system handover for i, and $\delta_{\text{HO}\,i,k}$ stands for the variation between the number of user arrivals and number of user departures on $[k-1, k]$ per BS for i due to inter-system handover. Note that $\delta_{\text{HO}\,i,k} = -\delta_{\text{HO}\,j,k}$ with $i \neq j$ and $(i, j) \in \{\text{RAT\#1}, \text{RAT\#2}\}$, $N_{\text{ARRIVALS}\,i,k}$ is the number of inter-system handover users entering i on $[k-1, k]$, $N_{\text{DEPARTURES}\,i,k}$ is the number of inter-system handover users leaving i on $[k-1, k]$, $T_{\text{MEAN HO}\,i,k}$ is the mean remaining communication call time assessed at $t = k$ for users who have joined the hosting i (new RAT after inter-system handover), $T_{\text{OBSERVATION}}$ is the time window observation, $v_{i,k}$ is the remaining communication call time assessed within time window $[k-1, k]$ for users who have joined the hosting i after inter-system handover, and $N_{\text{BS}\,i}$ is the number of BSs for i.

11.3.2.1.4 Overall System/User QoS Monitoring and Assessment. QoS indicators in each native mode also have to be monitored in order to assess the inter-system success likelihood from the native i to the target j, $i \neq j$ and $(i, j) \in \{\text{RAT\#1}, \text{RAT\#2}\}$. This likelihood depends on the performance from both the system and user perspectives. Therefore, some specific indicators have to be monitored in each of them to carry out the final inter-system handover decision with the highest likelihood. Both system and user cases are described below.

(1) Overall system QoS monitoring and assessment. For each $A_{\text{current}\,i,k}$ traffic as described above, the following indicators also have to be monitored and calculated. These parameters characterise each native RAT overall system performance. They provide a general trend on performance from the system point of view. They are considered as first insights into performance assessment, but are not considered as sufficient.

The mean total transmit power per BS on $[k-1, k]$ for i is:

$$P_{\text{BS tot}\,i,k} = [(k-1) * P_{\text{BS tot}\,i,k-1} + \delta_{\text{BS tot}\,i,k}]/k, \tag{11.8}$$

where $\delta_{\text{BS tot}\,i,k}$ is the mean total transmit power per BS on $[k-1, k]$ and $i \in \{\text{RAT\#1}, \text{RAT\#2}\}$. The mean number of available codes per BS on $[k-1, k]$ for i is:

$$N_{\text{Available Codes}\,i,k} = [(k-1) * N_{\text{Available Codes}\,i,k-1} + \delta_{\text{Available Codes}\,i,k}]/k, \tag{11.9}$$

where $\delta_{\text{Available Codes}\,i,k}$ is the mean number of available codes per BS on $[k-1, k]$ and $i \in \{\text{RAT\#1}, \text{RAT\#2}\}$.

(2) Overall user QoS monitoring and assessment. For each $A_{\text{current}\,i,k}$ traffic, the following indicators also have to be monitored since they provide a more accurate analysis on individual user performance and his own link requirements. This ensures that the inter-system handover scheme takes into account the availability of the radio resources needed by the user in target i. Each indicator is an average of all individual link qualities in each native i. The mean transmit power needed per MT at the BS on $[k-1, k]$ for RAT i:

$$P_{\text{BS}\to\text{MS}\,i,k} = [(k-1) * P_{\text{BS}\to\text{MS}\,i,k-1} + \delta_{\text{BS}\to\text{MS}\,i,k}]/k, \tag{11.10}$$

where $\delta_{\text{BS}\to\text{MS}\,i,k}$ is the mean transmit power needed per MT at the BS on $[k-1, k]$ and $i \in \{\text{RAT\#1}, \text{RAT\#2}\}$.

The mean FER experienced per MT on $[k-1, k]$ for i:

$$FER_{i,k} = [(k-1) * FER_{i,k-1} + \delta_{FER\,i,k}]/k, \quad (11.11)$$

where $\delta_{FER\,i,k}$ is the mean FER experienced per MT at the BS on $[k-1, k]$ and $i \in \{\text{RAT\#1}, \text{RAT\#2}\}$.

The inter-system handover decision is based on a combination of those both system and user indicators.

Given the previous formalism, finally, the different indicators to be monitored and assessed periodically within a defined time window $[(k-1) * T_{\text{OBSERVATION}}, k * T_{\text{OBSERVATION}}]$ are shown in Table 11.4.

Subsection 11.3.2.1.3 described the process in a macroscopic manner. Subsection 11.3.2.1.4 enumerated the different indicators concerning the user and the system to be monitored and assessed. The next subsection concerns the algorithm of the JRRM module.

11.3.2.2 Description of the JRRM Scheme

The global architecture of the JRRM scheme (initial phase and JRRM module) is defined in Figure 11.18, where only the important steps of the algorithm are denominated. The following will accurately describe the algorithm used before ordering an inter-system handover. The JRRM scheme can be divided into two steps: step 1 and step 2.

11.3.2.2.1 Step 1. The JRRM scheme principle is based on performance data gathered from network observations. Thus, the first step consists in processing those data. This step is considered as the initial phase of the JRRM scheme.

Notation. Let $Tr_{\text{native}\,i}$ ($i \in \{\text{RAT\#1}, \text{RAT\#2}\}$) be the actual traffic generated and gathered from experiments in a realistic network operating in i within a given time observation window (limited by $t_{\text{End}} - t_{\text{Init}}$), excluding additional traffic due to inter-system handover. Let $Tr_{\text{current}\,i}$ ($i \in \{\text{RAT\#1}, \text{RAT\#2}\}$) be the actual traffic generated and gathered from experiments in a current network operating in i within a given time observation window (limited by $t_{\text{End}} - t_{\text{Init}}$), including additional traffic due to inter-system handover.

Table 11.4 Summary of indicators to be monitored and assessed where $i \in \{\text{RAT\#1}, \text{RAT\#2}\}$

Type of assessment	Indicators
Native traffic	$\delta_{\text{native}\,i,k}$
Inter-system handover traffic	$N_{\text{ARRIVALS}\,i,k}$ $N_{\text{DEPARTURES}\,i,k}$ $\nu_{i,k}$
System QoS	$\delta_{\text{BS tot}\,i,k}$ $\delta_{\text{Available Codes}\,i,k}$
User QoS	$\delta_{\text{BS}\rightarrow\text{MS}\,i,k}$ $\delta_{\text{FER}\,i,k}$

Statistics gathered from the network performance observations. This initial phase aims at gathering satisfaction, blocking, dropping and bad quality call metrics in order to fully characterise the realistic overall system QoS when considering both modes and a large traffic dynamic range for each $Tr_{\text{native}\,i}$ ($i \in \{\text{RAT\#1}, \text{RAT\#2}\}$) gathered from network performance observations.

In this first step, the inter-system handover procedure is triggered when a user (in RAT#1 or RAT#2) experiences problems, as explicitly pointed out in Subsection 11.3.2.1. The following QoS metrics are measured in the network for RAT#1 and RAT#2 modes: the mean number of satisfied users $N_{\text{S},i}$ collected in the realistic network for i; the mean number of dropped users $N_{\text{D},i}$ collected in the realistic network for i; the mean number of blocked users $N_{\text{B},i}$ collected in the realistic network for i; the mean number of users who have experienced bad quality calls $N_{\text{BQC}\,i}$ collected in the realistic network for i; the total mean number of users $N_{\text{TOT},i}$ collected in the realistic network for i.

- All those parameters are indicators of the global QoS in both modes $i \in \{\text{RAT\#1}, \text{RAT\#2}\}$.
- Given $N_{\text{TOT},i}$, $N_{\text{S},i}$, $N_{\text{D},i}$, $N_{\text{B},i}$ and $N_{\text{BQC},i}$, other system QoS metrics can be also defined and additional statistics are processed as follows: $\%\text{S}_i = N_{\text{S},i}/N_{\text{TOT},i}$ is the percentage of satisfied users in the considered i, $\%\text{D}_i = N_{\text{D},i}/N_{\text{TOT},i}$ is the percentage of dropped users in the considered i, $\%\text{B}_i = N_{\text{B},i}/N_{\text{TOT},i}$ is the percentage of blocked users in the considered i, $\%\text{BQC}_i = N_{\text{BQC},i}/N_{\text{TOT},i}$ is the percentage of bad quality call users in the considered i.

At the end of this step, a set of performance curves is available for all i, $i \in \{\text{RAT\#1}, \text{RAT\#2}\}$, and for each QoS metric. Each curve displays the system performance vs. $Tr_{\text{current}\,i}$ ($i \in \{RAT\#1, RAT\#2\}$) for a given $Tr_{\text{native}\,j}$ ($j \in \{\text{RAT\#1}, \text{RAT\#2}\}$ and $j \neq i$).

Finally, at the end of step 1, the following curves are available (see Figure 11.19). These curves have been generated considering the TDD mode of the UMTS for RAT#1 and the FDD mode for RAT#2. This figure only represents the satisfaction metric $\%\text{S}_{\text{TDD}}$ for different $Tr_{\text{native FDD}}$. However, those kind of results are available for all metrics ($\%\text{S}_i$, $\%\text{B}_i$, $\%\text{D}_i$, $\%\text{BQC}_i$) for each $i \in \{\text{TDD}, \text{FDD}\}$.

Note: To make the formalism more tractable, in what follows, let $\%l_i$ such as $\%l_i = f(Tr_{\text{current}\,i})$ be the previously defined system QoS metrics for i with $i \in \{\text{RAT\#1}, \text{RAT\#2}\}$ and $l \in \{\text{S, B, D, BQC}\}$.

11.3.2.2.2 Step 2. Using the performance data obtained during step 1, step 2 aims to apply the JRRM algorithm to suitably decide whether the inter-system handover is triggered or not. This is achieved by the JRRM module. This step can itself be divided into several tasks, as displayed in Figure 11.20. It is assumed that the user is in i and tries to perform an inter-system handover from i to j, $i \neq j$ and $(i, j) \in \{\text{RAT\#1}, \text{RAT\#2}\}$.

(1) Description of the preliminary tests. Two simple tests are implemented at the beginning of the algorithm. Those tests (see Figure 11.21) are first performed to check basic requirements mandatory for performing an inter-system handover from i to j.

The first test aims to identify the multi-capability of the user's terminal. If the terminal has not the capabilities required by RAT#2, it is not possible to change its mode. It could be possible to download the required capabilities, but that is not considered here.

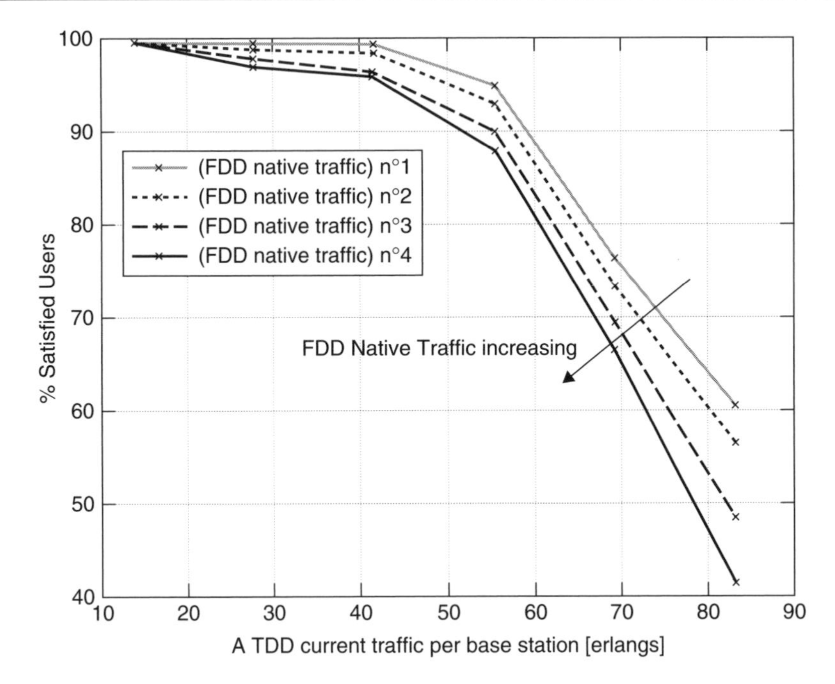

Figure 11.19 Set of TDD system performance curves for satisfaction metric $\%S_{TDD}$ and several FDD native traffic loadings $Tr_{\text{native FDD}}$

The second test is relative to the user speed. It is not necessary to study the possibility of an inter-system handover from RAT#1 to RAT#2 if RAT#2 is a WLAN system and if the speed of the user is a vehicular one. In that case, the mobility of the user will have a stronger impact on his communication. It would be the same if RAT#1 were the FDD mode of UMTS deployed in a macro environment while RAT#2 were the TDD mode deployed in a Manhattan environment. When a user is using the FDD system and is travelling with a vehicular speed through a town, in which a TDD Manhattan deployment may exist, a FDD → TDD inter-system handover might be forbidden for this user. The major reason is that he will make too many handovers in the TDD mode due to his speed, resulting in increasing the global interference level. That is why the speed is a critical parameter in this algorithm.

(2) Request on native and inter-system traffic. After validating the two tests, the algorithm requests the network (which centralises databases) for some information concerning each (i, j) $(i \neq j$ and $(i, j) \in \{\text{RAT\#1}, \text{RAT\#2}\})$:

- native traffic indicators, $(\delta_{\text{native}\,i,k}, \delta_{\text{native}\,j,k})$,
- inter-system traffic indicators, $(N_{\text{ARRIVALS}\,i,k}, N_{\text{DEPARTURES}\,i,k}, \nu_{i,k})$.

Those indicators are assessed (i.e. updated) and monitored automatically with a time cycle of $T_{\text{OBSERVATION}}$ in time unit. On the one hand, the period $T_{\text{OBSERVATION}}$ has to be very long to guarantee convergence in statistics since in this period, only variations are assessed and all system memory is included in the previous values $(k-1, k-2\ldots)$ owing to the recursive formalism. On the other hand, $T_{\text{OBSERVATION}}$ has to be small and

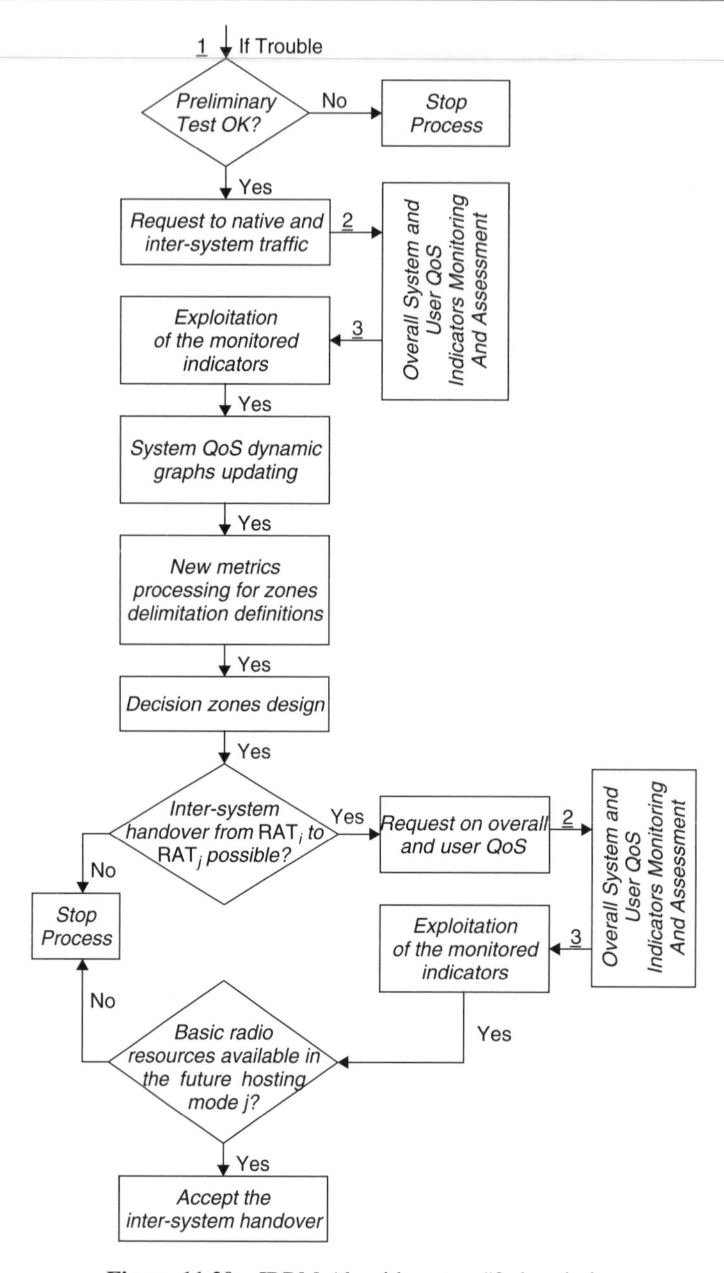

Figure 11.20 JRRM Algorithm step #2 description

also enable to update the indicators in a very small time observation window (i.e. with a high frequency) enabling the prediction of rapid traffic variations while having a good statistics convergence speed. It must be made possible to have a correct estimation of the instantaneous traffic even when the traffic load is time-varying. Therefore, a compromise has to be found in order to determine the most appropriate value for $T_{\mathrm{OBSERVATION}}$.

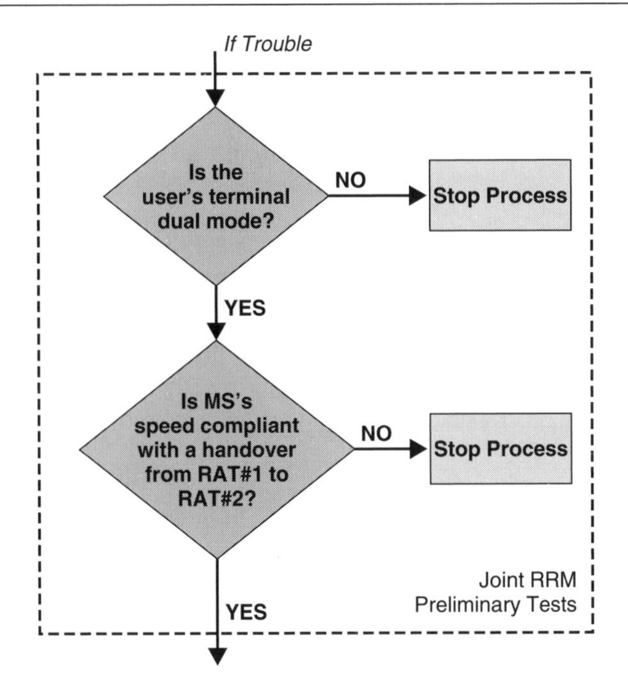

Figure 11.21 Preliminary tests description

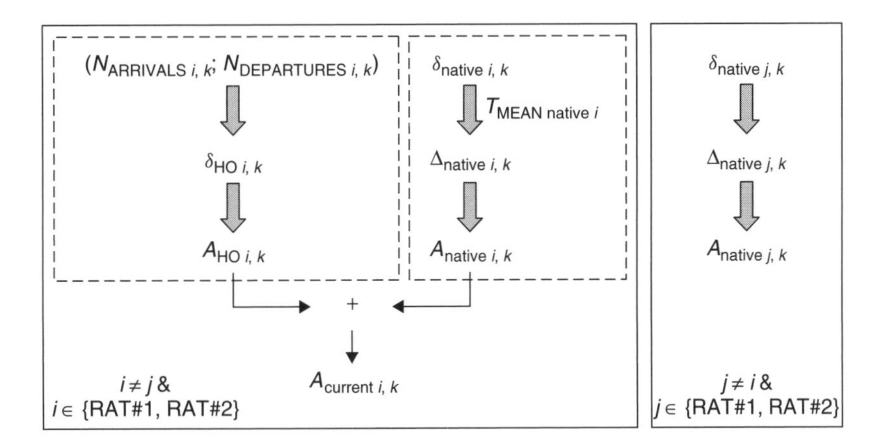

Figure 11.22 Exploitation of the monitored indicators

(3) Exploitation of the monitored indicators. Once the previous task is achieved, it is possible for each i, given $\delta_{\text{native}\,i,k}$, $N_{\text{ARRIVALS}\,i,k}$, $N_{\text{DEPARTURES}\,i,k}$, and $\nu_{i,k}$, to assess $A_{\text{current}\,i,k}$ and $A_{\text{native}\,j,k}$ with ($i \neq j$ and $(i, j) \in \{\text{RAT\#1, RAT\#2}\}$) as displayed in Figure 11.22.

(4) Dynamic updating of QoS characteristics. This task aims to update the selection of the performance curve $\%l_{\text{i}} = f(Tr_{\text{current}\,i})$ for each QoS l in i at $t = k$. Given the

availability of $A_{\text{native }j,k}$ for i ($i \neq j$ and $(i, j) \in$ {RAT#1, RAT#2}) achieved in the previous section, the large set of statistical performance curves $\%l_i = f(Tr_{\text{current }i})$ for several given $Tr_{\text{native }j}$ with $i \neq j$ and $(i, j) \in$ {RAT#1, RAT#2} and $l \in$ {S, B, D, BQC} obtained previously, it is possible to select the most relevant performance curve $\%l_i = f(Tr_{\text{current }i})$ relative to $A_{\text{native }j,k}$ for each $l \in$ {S, B, D, BQC} and $i \neq j$ and $(i, j) \in$ {RAT#1, RAT#2}.

In other words, this task consists in updating estimation on RAT#1 and RAT#2 performance at $t = k$ for the 2-tuple ($\%l_i$; $A_{\text{native }j,k}$) with $i \neq j$ and $(i, j) \in$ {RAT#1, RAT#2} and $l \in$ {S, B, D, BQC}. It is important to note that this estimation is performed before initiating any inter-system handover at $t = k$. This step is mandatory to obtain the different metric estimations to assess the potential disturbances the inter-system handover could create in both RAT#1 and RAT#2. This process is illustrated in Figure 11.23. The obtained values are strongly dependant on the real conditions in both RAT#1 and RAT#2.

(5) New metrics processing for definition of QoS zones. This step consists in defining new metrics useful for designing decision zones. In order to do so, a set of thresholds related to QoS indicators is defined to limit those decision zones. The current traffic associated to reach these thresholds is derived in the following method.

To make this task more tractable, let $L_{i,l,n}$ be the nth $Tr_{\text{current }i}$ for metric l and corresponding to a given metric performance threshold $\%Thr_{i,l,n}$, where $l \in$ {B, D, BQC}, $i \in$ {RAT#1, RAT#2}, and n is an integer, $n \in [1, N_{\text{init }i}]$, $N_{\text{init }i}$ being the maximum number of $\%Thr$. $\%Thr_{i,l,n}$ are QoS metrics matching a given quality of service for each previous metric $\%\text{B}_i$, $\%\text{D}_i$, $\%\text{BQC}_i$.

Let us also define $M_{i,n} = \Sigma_{l=1}^{3} \alpha_{i,l,n} \cdot L_{i,l,n}$ as new performance metrics, where

$$l = 0 \Leftrightarrow l = \text{S}; \quad l = l \Leftrightarrow l = \text{B}; \quad l = 2 \Leftrightarrow l = \text{D}; \quad l = 3 \Leftrightarrow l = \text{BQC}, \qquad (11.12)$$

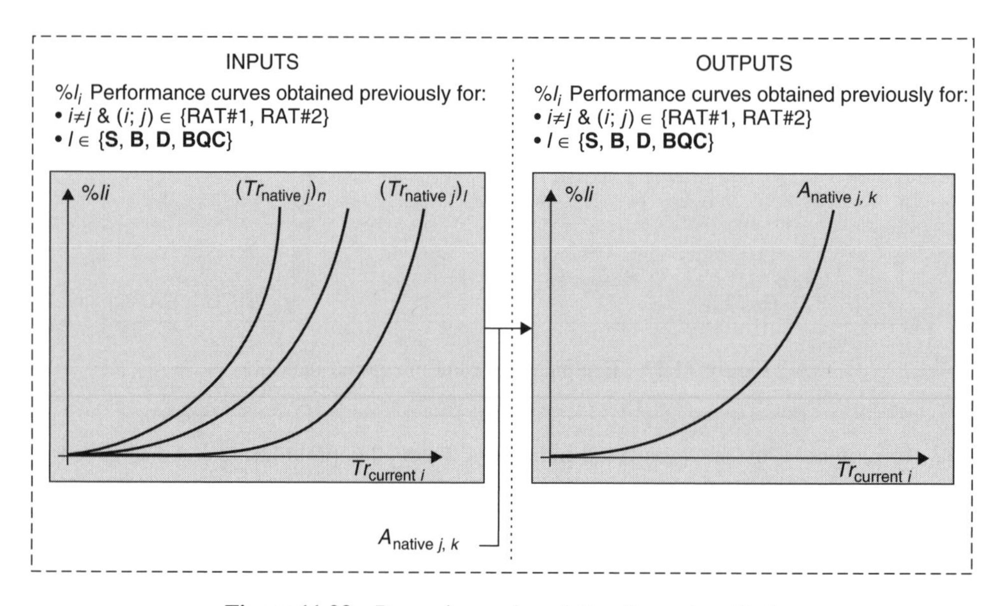

Figure 11.23 Dynamic graph updating for system QoS

$i \in \{\text{RAT\#1, RAT\#2}\}$ and $l \in \{\text{B, D, BQC}\}$, $\alpha_{i,l,n}$ is a positive real pondering weight such as $\Sigma_{l=1}^{3}\alpha_{i,l,n} = 1$ for a given n.

(6) $L_{i,l,n}$ Calculation. Given i and n, the $L_{i,l,n}$ calculation is based on $\%Thr_{i,l,n}$ as follows. Starting from $\%Thr_{i,l,n}$, $L_{i,l,n}$ is obtained by bijection on curves obtained in step 1 of the JRRM algorithm. This process is performed for all quality metrics, i.e. $l \in \{\text{B, D, BQC}\}$.

For instance, for a given i and n, $\%Thr_{i,l,n}$ are all equal $\forall l \in \{\text{B, D, BQC}\}$. This is illustrated in Figure 11.24. Of course, this is carried out for a given identified scenario specified by $A_{\text{current}\,i,k}$ and $A_{\text{native}\,j,k}$ with $i \neq j$ and $(i, j) \in \{\text{RAT\#1, RAT\#2}\}$.

For example, assuming that

- i is RAT#1,
- $N_{\text{init RAT\#1}} = 3$,
- $\%Thr_{\text{RAT\#1},1,1} = 1\%$,
- $\%Thr_{\text{RAT\#1},1,2} = 3\%$,
- $\%Thr_{\text{RAT\#1},1,3} = 20\%$,
- $\forall l \in \{\text{B, D, BQC}\}$,

the following values are obtained:

$$\%Thr_{\text{RAT\#1},1,1} = 1\% \Rightarrow L_{\text{RAT\#1,B},1}, L_{\text{RAT\#1,D},1}, L_{\text{RAT\#1,BQC},1}, \tag{11.13}$$

$$\%Thr_{\text{RAT\#1},1,2} = 3\% \Rightarrow L_{\text{RAT\#1,B},2}, L_{\text{RAT\#1,D},2}, L_{\text{RAT\#1,BQC},2}, \tag{11.14}$$

$$\%Thr_{\text{RAT\#1},1,3} = 20\% \Rightarrow L_{\text{RAT\#1,B},3}, L_{\text{RAT\#1,D},3}, L_{\text{RAT\#1,BQC},3}, \tag{11.15}$$

where $L_{\text{RAT\#1},l,1} < L_{\text{RAT\#1},l,2} < L_{\text{RAT\#1},l,3}$, $\forall l \in \{\text{B, D, BQC}\}$.

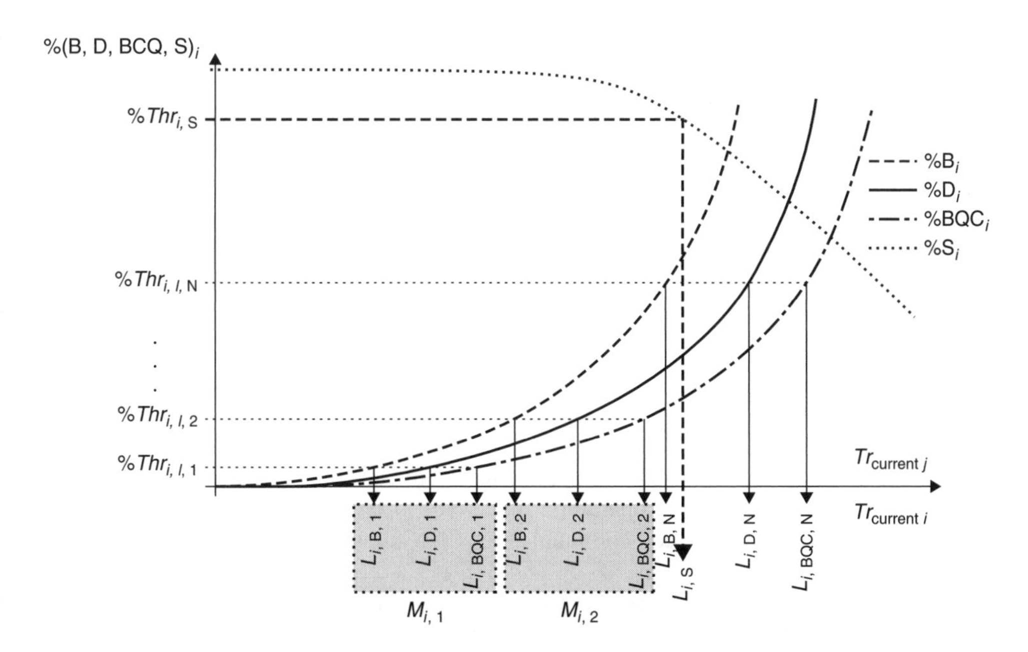

Figure 11.24 $L_{i,l,n}$ calculation for a given $Tr_{\text{current}\,i}$

A particular case of interest is when $l = \mathrm{S}$ since this performance metric provides the overall RAT#2 performance (respectively RAT#1) vs. $Tr_{\text{current RAT\#2}}$ (respectively $Tr_{\text{current RAT\#1}}$). It includes all other quality metrics (blocking, dropping and BQC). Therefore, we define the following metric specific to $l = \mathrm{S}$: let $L_{i,\mathrm{S}}$ be the $A_{\text{current}\,i}$ for the S metric, corresponding to a given metric performance threshold $\%Thr_{i,\mathrm{S}}$, where $i \in \{\text{RAT\#1}, \text{RAT\#2}\}$.

Note that this value is not indexed by n. The reason why is because $L_{i,\mathrm{S}}$ is an upper bound for all $L_{i,l,n}$ ($l \in$ {B, D, BQC}). In other words, all $L_{i,l,n}$ must be included within the window $[0; L_{i,\mathrm{S}}]$, warranting that mode i performance will be such that $\%\mathrm{S_i} \geq \%Thr_{i,\mathrm{S}}$.

Therefore, given $L_{i,\mathrm{S}}$, only $L_{i,l,n} \in [0; L_{i,\mathrm{S}}]$ are considered for the JRRM algorithm, where $i \in$ {B, D, BQC}, $i \in \{\text{RAT\#1}, \text{RAT\#2}\}$, and n is an integer, $n \in [1, \mathrm{N_i}]$, where $N_i = \{\max\,(n \in [1, N_{\text{init}\,i}])$ such that $L_{i,l,n} \in [0; L_{i,\mathrm{S}}]$, l and i being given$\}$, i.e. $N_{\mathrm{i}} \leq N_{\text{init}\,i}$. Given this constraint, N_{i} specifies the number of zones and consequently the number of decision zones. This number of zones is so $N_{\mathrm{i}} + 1$. Typically, the number of zones can depend on the nature of the service traffic (Speech, Web, Video) within the coverage area. Besides, it is important to note that N_i is not necessarily the same for both the RAT#1 and RAT#2 modes.

(7) $\alpha_{i,l,n}$ calculation. This step consists in determining $\{\alpha_{i,l,n}\}_{l=1,2,3}$ given $L_{i,l,n} \in [0; L_{i,\mathrm{S}}]$, for each $i \in \{\text{RAT\#1}, \text{RAT\#2}\}$ and for each $n \in [1; N_i]$. The way the $\alpha_{i,l,n}$ are determined for each $i \in \{\text{RAT\#1}, \text{RAT\#2}\}$ and for each n is as follows. $L_{i,l,n}$ are sorted by increasing values (for example, $L_{i,\mathrm{D},1} < L_{i,\mathrm{B},1} < L_{i,\mathrm{BQC},1}$ where $n = 1$). Let $(L_{i,l,n})_{\text{most constraining}}$ be the lowest $L_{i,l,n}$ value, i.e. the most constraining (the most constraining in the sense that the threshold is achieved for a low traffic load) value to ($L_{i,\mathrm{D},1}$ in the example) reach its corresponding target $\%Thr_{i,l,n}$ (in the example $\%Thr_{\mathrm{TDD,D},1}$). Let $(\alpha_{i,l,n})_{\text{most constraining}}$ be the weight corresponding to $(L_{i,l,n})_{\text{most constraining}}$. Therefore, the contribution of this value in $M_{i,n} = \Sigma_{l=1}^{3}\alpha_{i,l,n} \cdot L_{i,l,n}$ is provided with a higher weight value $(\alpha_{i,l,n})_{\text{most constraining}}$, i.e. with a higher likelihood.

Then, the remaining $\alpha_{i,l,n} \neq (\alpha_{i,l,n})_{\text{most constraining}}$ are processed such that the higher $L_{i,l,n}$ is, the lower its contribution is within $M_{i,n}$, i.e. the lower its likelihood is. One method to determine the remaining $\alpha_{i,l,n}$ is described below. For that, we define $a_{i,l,n} = L_{i,\mathrm{S}}/L_{i,l,n}$ and $s_{i,n} = \Sigma_l a_{i,l,n}$, where $l \in$ {B, D, BQC}. Each $\alpha_{i,l,n}(\neq (\alpha_{i,l,n})_{\text{most constraining}})$ is then assessed as $\alpha_{i,l,n} = a_{i,l,n}/s_{i,n}$, where $l \in$ {B, D, BQC} and $\Sigma_{l=1}^{3}\alpha_{i,l,n} = 1$. In the considered example,

$$\alpha_{\mathrm{TDD,D},1} = \frac{a_{\mathrm{TDD,D},1}}{a_{\mathrm{TDD,D},1} + a_{\mathrm{TDD,B},1} + a_{\mathrm{TDD,BQC},1}},$$

where $a_{\mathrm{TDD,D},1} = L_{\mathrm{TDD,S}}/L_{\mathrm{TDD,D},1}$. The same method is applied to $\alpha_{\mathrm{TDD,B},1}$ and $\alpha_{\mathrm{TDD,BQC},1}$.

Note: Other numerous methods can be applied to obtain each $\alpha_{i,l,n}(\neq (\alpha_{i,l,n})_{\text{most constraining}})$. The method can depend on the operator's objectives in terms of QoS. The operator can choose to emphasise ongoing call communications (i.e. avoiding dropping and bad quality calls) and therefore can emphasise $\alpha_{i,\mathrm{D},n}$ and $\alpha_{i,\mathrm{BQC},n}$.

(8) Decision zones design. After $L_{i,l,n}$ and $\alpha_{i,l,n}$ have been calculated for $i \in$ {RAT#1, RAT#2}, for $l \in$ {B, D, BQC} and for $n \in [1, N_i]$, it is then possible to assess $M_{i,n} = \Sigma_{l=1}^{3}\alpha_{i,l,n} \cdot L_{i,l,n}$ for each n and i. Given $M_{i,n}$, $i \in$ {RAT#1, RAT#2}, and $n \in [1, N_i]$, it is

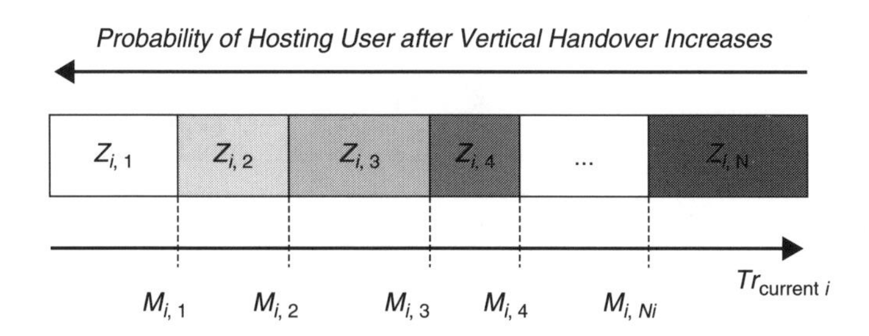

Figure 11.25 Decision zones design

possible to design zones. Finally, for each mode we obtain the N_i-tuples $(M_{i,1};\ldots;M_{i,n};\ldots;M_{i,N\mathrm{TDD}})$, $(M_{j,1};\ldots;M_{j,n};\ldots;M_{j,N\mathrm{FDD}})$ for $(i, j) \in \{\mathrm{RAT\#1}, \mathrm{RAT\#2}\}^2$.

A decision zone $Z_{i,n}$ is defined as a traffic interval delimited by $M_{i,n-1}$ and $M_{i,n}$. There are N_i decision zones, as illustrated in Figure 11.25. Those N_i zones are defined for a given mode in which the grey shaded rectangles are indicators of the QoS for a given expected load traffic to be supported in the case when the inter-system handover procedure were triggered. If the given traffic load expected after inter-system handover stays in the lower zone, than new incoming users from the other modes have a high probability of being satisfied.

While scanning from the left to the right on Figure 11.25, $Tr_{\mathrm{current}\,i}$ increases and the N_i successive designed decision zones refer gradually to traffic intervals in which the probability of being satisfied after inter-system handover is lower and lower. Of course, all traffic intervals are not necessarily of equal length, depending on previous steps.

(9) Is an inter-system handover from i to j possible? After obtaining the decision zones, with regard to the current traffic $Tr_{\mathrm{current}\,i}$ for $i \in \{\mathrm{RAT\#1}, \mathrm{RAT\#2}\}$ at $t = k$, the process of triggering an inter-system handover will continue or stop at this step. Given $A_{\mathrm{current}\,i,k}$ determined for all $i \in \{\mathrm{RAT\#1}, \mathrm{RAT\#2}\}$ at $t = k$, i is associated to $Z_{i,n}$ if $M_{i,n-1} < A_{\mathrm{current}\,i,k} \leq M_{i,n}$, i.e. $A_{\mathrm{current}\,i,k} \in Z_{i,n}$. Let $z_{i,k}$ be the decision zone associated to i at $t = k$.

Let us consider intersystem handover from i to j ($i \neq j$ and $(i, j) \in \{\mathrm{RAT\#1}, \mathrm{RAT\#2}\}$). To decide to continue or stop the process, we refer to the result of the decision function $f(z_{i,k}, z_{j,k})$, where $i \neq j$ and $(i, j) \in \{\mathrm{RAT\#1}, \mathrm{RAT\#2}\}$. This decision function indicates whether the switch from i to j can be envisaged or not, given the current traffic of both modes. If this function is equal to 1, than the JRRM module goes on or it stops here and no inter-system handover is processed. This function can be defined as $f(z_{i,k}, z_{\mathrm{j},k}) = \rho$, where $\rho = 0$ or $\rho = 1$ depending on the strategy used by the operator. This enables us to build up the decision rules, as shown in Table 11.5, for a handover from i to j, and $i \neq j$ and $(i, j) \in \{\mathrm{RAT\#1}, \mathrm{RAT\#2}\}$. If intersystem handover from j to i is considered, a new table is needed since this matrix is not symmetric. This task shows us whether inter-system handover is possible or not, but it does not trigger the inter-system handover yet. This is achieved in the next task.

(10) Are the required 'basic' radio resources available in the target mode? This tasks enables us to take the final decision. Let j be the hosting RAT in the following. The task refines the decision-taking to take account of additional metrics based on $P_{\mathrm{BS\,tot}\,j,k}$, $N_{\mathrm{Available\,Codes}\,j,k}$, $P_{\mathrm{BS}\to\mathrm{MS}\,j,k}$ and $FER_{j,k}$. The estimation of $P_{\mathrm{BS\,tot}\,j,k}$, $N_{\mathrm{Available\,Codes}\,j,k}$ and

Table 11.5 Decision rules $f(z_{i,k}, z_{j,k})$

$Z_{i,k}$ \ $Z_{j,k}$	$Z_{j,1}$	$Z_{j,2}$	$Z_{j,3}$	$Z_{j,4}$	...	$Z_{j,Ni}$
$Z_{i,1}$	1	1	1	0	...	0
$Z_{i,2}$	1	1	1	0	...	0
$Z_{i,3}$	1	1	1	0	...	0
$Z_{i,4}$	1	1	0	0	...	0
...	...	...	...	...	...	...
$Z_{i,Ni}$	1	1	1	0	...	0

$P_{\text{BS}\rightarrow\text{MS}\,j,k}$ enables us to minimise the risk of blocking even if $\%Thr_{j,\text{B},n}$ is already stringent in providing low blocking rates for a given j and all n. Let $P_{\text{BS tot threshold}\,j,k}$, $N_{\text{Available Codes threshold}\,j,k}$, $P_{\text{BS}\rightarrow\text{MS threshold}\,j,k}$ and $FER_{\text{threshold}\,j,k}$ be the respective thresholds of the indicators mentioned above.

The final inter-system handover decision from i to j is processed as follows: If $(P_{\text{BS tot}\,j,k} < P_{\text{BS tot threshold}\,j,k})$ and $(N_{\text{Available Codes}\,j,k} > N_{\text{Available Codes threshold}\,j,k})$ and $(P_{\text{BS}\rightarrow\text{MS}\,j,k} < P_{\text{BS}\rightarrow\text{MS threshold}\,j,k})$ and $(FER_{j,k} < FER_{\text{threshold}\,j,k})$, then the inter-system handover from i to j is triggered, otherwise the process is stopped.

Once the first inter-system handover has been achieved, the user cannot trigger another one before a given guard time T_{Guard} in order to avoid the 'ping-pong' effect. This usually will be the case for users travelling at high speed within a Manhattan network, for example.

11.3.3 Application to 3G Radio Access Technologies

11.3.3.1 Simulation Assumptions

The goal of the simulation scenarios is to introduce a basic JRRM scheme to enable 'smart' FDD-TDD inter-system handovers, and evaluate the potential gain of such a scheme. The first scenario (Figure 11.26) considers a circuit speech service and users evolving in a Manhattan environment, for both modes. Some values, for instance the time before triggering the JRRM scheme, or the threshold between two operating zones, are defined below.

A user experiencing dropping or blocking in his native mode is ordered to make an inter-system handover as late as possible. This means that the time threshold before triggering the JRRM scheme is 5 seconds (just before dropping the user). So, it gives the user only one attempt to obtain his required radio resources in the target mode. The main goal here is to prevent users from dropping and blocking and then to evaluate

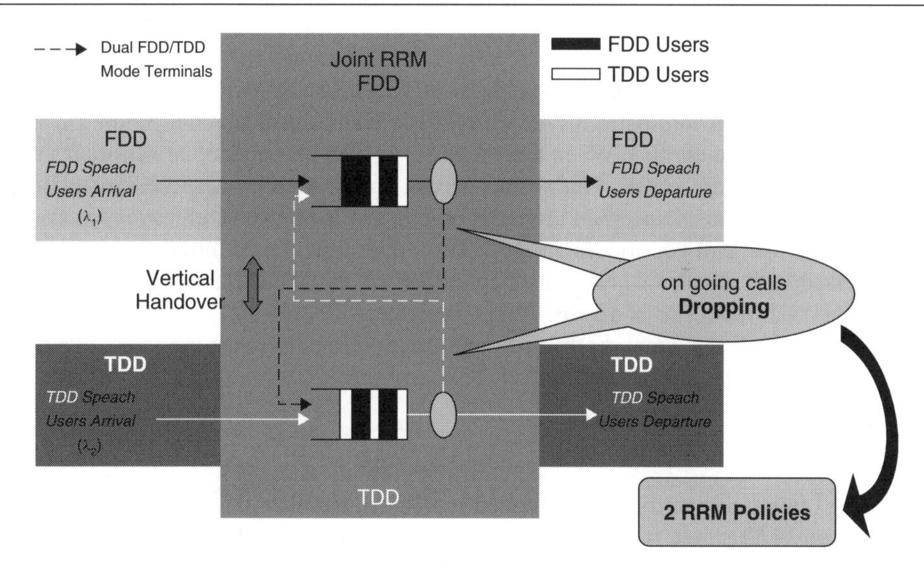

Figure 11.26 Vertical handover between the UMTS FDD and TDD modes in downlink

the performance of the two systems in terms of satisfied users. An improvement in the method then consists of decreasing the time threshold before triggering the inter-system handover process. Accordingly, the call quality of the user increases (his FER decreases). But this time the threshold must not be decreased too much so as to avoid too many inter-system handovers (the 'ping-pong' effect). One simulation has been performed for a time threshold of 5 s (RRM policy 1) and another one for a time threshold of 3 s (RRM policy 2).

To accept whether an inter-system handover is possible or not, only two operating zones are defined for this simulation scenario; the limit is fixed by the 95% satisfied users criterion. This criterion means that an inter-system handover is performed only if the target mode stays in the zone when its traffic is lower than $L_{S,95\%}$. The limit $L_{S,95\%}$ is given by the single-mode studies. The traffic value of the hosting mode is fixed and is not updated with the arrival and departure of inter-system handover users. To avoid the 'ping-pong' effect, a user can only execute one inter-system handover during his call.

The study is performed for a low loaded FDD system (call arrival rate $\lambda_{FDD} = 20$ call/s) and for variable traffic loads for the TDD system ($\lambda_{TDD} = 20$–70 call/s). Owing to inter-system handovers, the traffic load of both modes becomes variable. Performance gain is achieved by comparing the JRRM scheme (i.e. where the intersystem handover is possible) with the case where inter-system handover is not possible. The statistics are obtained considering the overall system (TDD + FDD).

A second scenario considers a mix of speech and web services. The web service is provided by the TDD mode and the speech service by the FDD mode. The same unbalanced traffic as for the first scenario is considered, but only one policy for dropping is considered (similar to RRM policy 1).

In the two scenarios, all terminals were considered to be dual mode ones. To highlight the impact of the penetration of reconfigurable terminals, a third scenario based on the first considers that only a percentage of terminals are dual mode.

11.3.3.2 Methodology

To assess the JRRM scheme performance, an event-driven methodology is used. Users arrive in the system according to a Poisson process. Each user is given a profile (in terms of service, position, mobility). When a user arrives, given his service, he negotiates radio resources in the access phase. This initial access phase is either a success or a failure depending on the strategy specified above. If he fails, he is blocked and leaves the system. Otherwise, the user call goes on. Inner loop power control and hard intra-system handover are processed individually. During this step, either the user ends his call successfully, or he fails due to dropping or bad FER.

Concerning the link budget, both intra- and inter-cell interference are considered. The main radio parameters are taken from refs [35] and [36].

11.3.3.3 Simulation Results

To compare and highlight the gain provided by JRRM, simulation studies on TDD single mode and on FDD single mode have been performed. For the TDD mode, some results can be found in ref. [38] and for FDD in ref. [37]. Figure 11.27 represents the percentage of satisfied users for the first scenario considering three different RRM policies: with no intersystem handover, with RRM policy 1 and with RRM policy 2. The implemented RRM policies using the JRRM scheme brought an overall system capacity gain in terms of the percentage of satisfied users. At a level of 95% of system satisfaction, this gain was about 7.9% for RRM policy 1 and 17.4% for RRM policy 2.

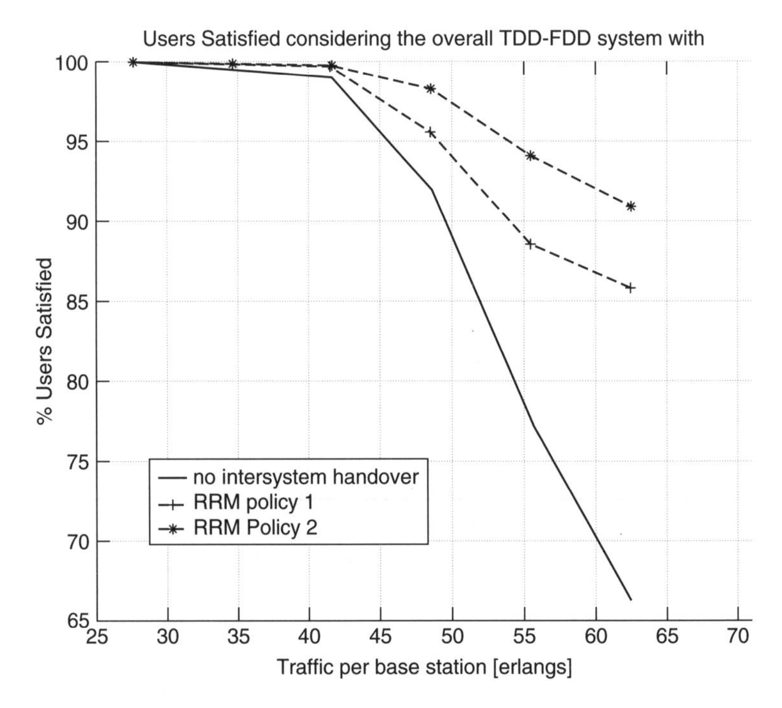

Figure 11.27 Capacity gain for the first scenario

In other words, for the same initial set of radio resources, more users are admitted and satisfied in the overall TDD/FDD network. With the introduction of those policies, no more users are dropped or blocked in the TDD/FDD systems. Accordingly, the QoS of users is also increased. Decreasing the dropping time threshold from 5 s to 3 s permits a significant reduction in the number of bad FER users. The levels of QoS improvement for blocking, dropping and bad FER are depicted in Figure 11.28. In other words, for the same offered traffic, each user's QoS is improved at those three levels. As expected, this QoS gain (compared with the case where no inter-system handover is performed) is drastically improved when the TDD system is highly loaded.

However, it is not conceivable to decrease the dropping time threshold too much, because in this case, too many users will potentially perform an inter-system handover, resulting in loading the other mode and degrading the overall QoS.

For the second scenario considering a mix of speech and web services, the capacity gain is all the more important, with a gain of 24%. This is mainly due to the statistic multiplexing between the circuit and bursty traffic.

Simulations were performed in order to highlight the importance of reconfigurable terminals enabling VHO between RATs. This simulation scenario is similar to the first scenario (speech only) using RRM policy 1, but here only some of the terminals are considered as dual mode.

Figure 11.29 illustrates the gain brought about by the JRRM for different penetration rates of dual mode terminals (in terms of capacity). If 25% of terminals are dual mode, than the gain brought by JRRM (in terms of capacity) is only about 1.6%, for 50% of terminals are dual mode, the gain is about 4.1%, and if all terminals are dual mode, the gain is about 8.3%.

Figure 11.30 shows the evolution of blocking and dropping probabilities in the (TDD + FDD) system. The main result that is there neither blocking nor dropping if all are dual mode terminals.

The introduction of the JRRM scheme has brought gains for the two scenarios (speech, speech + web) and has also highlighted the importance of the penetration rate of reconfigurable terminals on system performance. However, the scheme considered here is relatively basic. Higher gains are expected through enhanced strategies.

The JRRM performance has been assessed for the simple case of a 'fixed spectrum allocation' scheme (Subsection 11.2.2.2.1) supporting inter-system handover capability

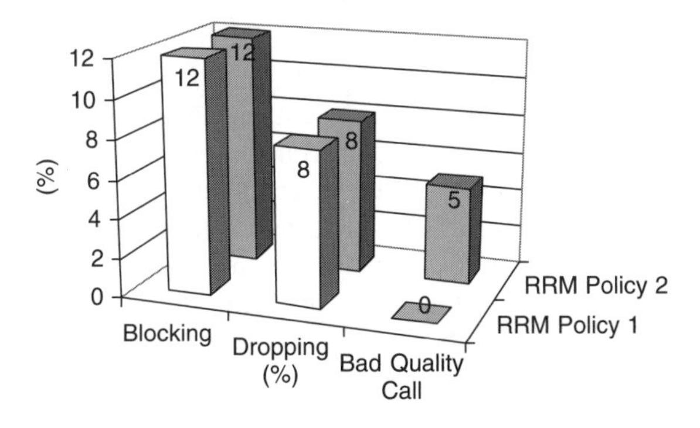

Figure 11.28 Overall TDD/FDD QoS improvement for RRM policies 1 and 2 (first scenario)

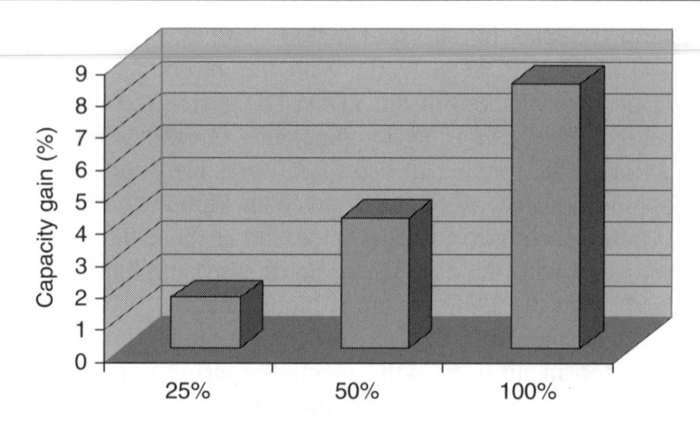

Figure 11.29 Influence of dual mode terminal penetration rate on the JRRM performance

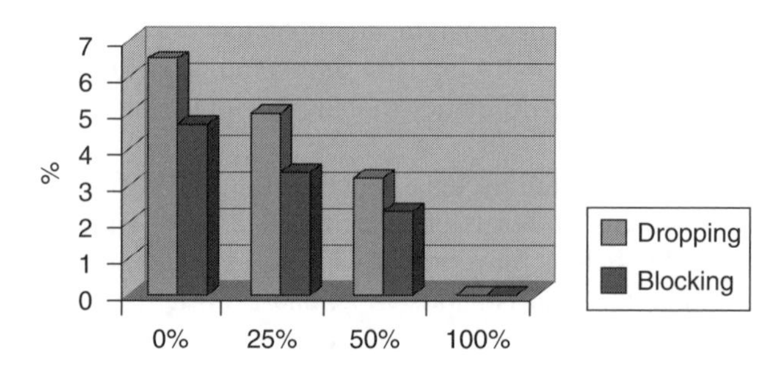

Figure 11.30 Blocking and dropping probabilities for different dual mode terminal penetration rates

between two radio access technologies. Hence, full benefit of JRRM will be all the more important with the introduction of more advanced schemes (described in Subsections 11.2.2.2.2 and 11.2.2.2.3) supporting further flexible spectrum allocation functionality including RATs' capability to be potentially shifted in frequency over the operational band of reconfigurable SDR equipment.

In addition to this, the context of reconfigurable radio introduces additional RRM challenges. The introduction of softly reconfigurable enabled terminals capability induces a new type of traffic: software download traffic. Future terminals could download a new air interface before performing inter-system handover or download new software in order to fix a bug or perform an upgrade. But the impact of software downloads on the QoS of regular users is a critical issue in evaluating the feasibility of over-the-air software downloading. Therefore, this impact has to be minimised by developing new RRM schemes. One potential solution envisages the use of broadcast/multicast to support the software download traffic in cellular networks. In Subsection 11.3.1.3, a cell grouping method allowing reducing neighbour cell interference when using broadcast/multicast was described. This requires investigation into an appropriate traffic model for this additional traffic in order to capture its main features and derive appropriate RRM rules (Figure 11.31).

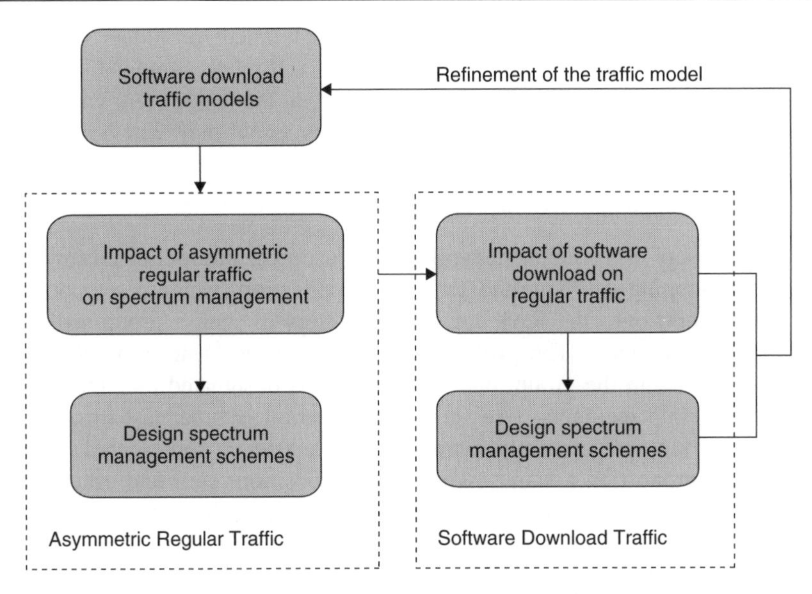

Figure 11.31 Software download challenges

11.3.4 Future Evolution in Heterogeneous Environments

As already detailed in the Vision SDR chapter of the BoV [39] and in ref. [40], the evolution of telecommunications in the next decade will be characterised by the convergence towards IP-based core networks and ubiquitous seamless access in a context of hierarchical and self-organising networks. Reconfigurable radio terminals and new appliances will be the key components of such a seamless network convergence. This section aims to describe what the evolution could be in terms of network management provided by reconfigurable radio technology enabling. The considered approach relies mainly on the exploitation of the concept of self-organising networks applied to reconfigurable networks (terminals and infrastructure) and spectrum sharing. Thus, a combination of *ad hoc* networks and RRM would be tackled. One way to investigate this is to optimally take into consideration the commonalities between users in terms of required applications and the terminal profile, as outlined in Chapter 1.

The previously mentioned commonalities between users rely on two major cases. (1) Whether the users under consideration have a common profile or, at least, have the same requirements (namely an unorganised group) and (2) whether a finite number of users act in a concerted way and can be grouped for the provision of new services (namely a concerted group). For both the unorganised and concerted cases, the creation, management and termination of a group of users is implied. Hence, a quite wide range of applications addressing user groups can be envisioned.

The following subsections discuss the above two groups.

11.3.4.1 Unorganised Group

What could be the new services provided to users?

One immediate application is to consider public events such as sport (i.e. an athletics competition in a stadium) or exhibitions, where information can be broadcast by networks to the users, if an agreement is made with specific content providers concerning the different possibilities to attend or take part in specific multimedia presentations, interactive shows, discussions, on-line games, etc. As explained, in this case, users will not know in advance who will be in the group they join, but they know they will be clustered with other users having the same kind of service requirements.

What could be the commonalities between users?

In this case, for users in a cellular network, radio resources are redirected towards a new RAT for better mapping the requested service or application (which is the common point between these users) onto the RAT capabilities. Users in such a group will also share common QoS requirements. The group set-up relies on the basic principle of a master and slaves node. Within the group, one of the nodes is designated as a master node and will handle the group members. User group information is broadcast in order to show what kinds of groups are currently available. The master is represented by the entity that initiated the group (i.e. a stand owner in an exposition, etc.) and asked in advance (a common agreement before the considered event between the operator and the master node) for dedicated resources. After receiving the information about the existing groups, if the user is interested in a group, he sends a message to his current cellular network notifying that he would like to join the identified group. In response, the network sends a message to this user containing all relevant information in order to properly handover from the cellular network occupied resources to the new dedicated resources. Those resources include the new frequency carrier of the new RAT, the application used, the master profile, etc. and, potentially, the RAT-specified software itself by downloading it if it is not yet embedded within the terminal. Once it has been successfully completed, the operator is aware that the considered user is no longer directly reachable except through the master. The user is also provided with the most appropriate RAT and resources to fulfil his tasks and is no longer consuming resources in the cellular network.

11.3.4.2 Concerted Group

What could be the new services provided to users?

Application fields in this situation address either private or professional contexts. Indeed, on the private side it can be envisaged that a group will use the terminal as push-to-talk transceivers, for instance when performing outside activities (e.g. trekking) or as WLAN transceivers (games, high-bit-rate interactive applications, etc.). For the professionals, the push-to-talk applications can be envisaged in the case of nomadic float and broadband access in a more local context such as meetings with interactions requesting a high bit rate (e.g. large files exchange) or on-line collaborative work (e.g. a video conference with several users disseminated physically around a site).

What could be the commonalities between users?

This case aims to fulfil the needs of a closed group of individuals in terms of applications and needed resources. Users know each other and have a clear vision of an identified application or service they want to do together. In contrast with the other case, users volunteer in order to initiate the group creation, but the principle to be redirected to the most suitable RAT in line with the profile of the users remain unchanged.

In that case the concerted group creation approach is slightly different. Here, future group members know each other and the master node is not known a priori, which is why the master node has to be designated by future members of the group before initiating the process. Once it has been done, the master node sends a message to the cellular network with the required information concerning the applications, requirements and profiles to set up. Based on this, the operator is able to decide if it is possible to dedicate a specific carrier frequency and, if so, determines the best configuration (RAT, available bit rate, etc.). After acceptance by the network, the dedicated resource is de-allocated and all necessary parameters are transmitted to the master, to reconfigure his terminal. All involved users send, at their own initiative, a message in an uncoordinated manner to the network to join the group. The network acknowledges all relevant information and notifies the master of the arrival of those users in the group. The master only keeps direct communication with the cellular network; the slaves do not exchange any information with the network but the master node acts as a relay between the native cellular network and the created *ad hoc* network.

One of the major outcomes of those combinations of technologies is not only to transfer user groups to a more suitable technology but also at the same time to unload what can be called the native cellular network. By initiating such a process it is possible for the operator to isolate some users that require a specific service or are responding to a specific profile, i.e. personalised services. In order to do so, the idea is to allocate resources to each of these groups in such a way that nobody outside of the group is able to use these resources. The most evident kind of resources would be, for instance, a frequency carrier (or several frequencies depending on the traffic to support) or time slots for which the group can use the most appropriate RAT, due to reconfigurable terminals (Figure 11.32).

The advantage is twofold, first users are isolated and it is also possible to maximise the offered QoS, and the RAT they are using is adapted to their application. Secondly, the operator by doing so release resources in the 'centralised' cellular network and is also able to accept more regular users and provide a higher capacity to them.

The spectrum allocated to an operator is limited and it seems obvious that the operator has to decide depending on constraints (traffic, load, coverage, etc.), on the balance between the resources dedicated to cellular access and those dedicated to user groups.

Considering both kinds of groups, a link with the 'external world' could be required if the group extremely interacts with other groups, for instance. This link will be typically dimensioned according to the needed bandwidth and the RAT could be chosen among cellular or broadcast technologies. The advantage of creating such *ad hoc* groups is that the cellular network considers master nodes as the only type of nodes for relaying messages to group members. The master node is a representative of the group members and the cellular network only allocates resources to this particular user (Figure 11.33).

The slave nodes do not have any direct contact with the cellular network. In the case of an incoming call, the network sends a message to the master node, which relays it. If the slave node accepts the call, he leaves the group and joins his network to perform this communication. When terminated, the user automatically rejoins his group. A user who is part of a group can notify his network that he will not accept incoming calls, or only from a pre-defined favourite list.

When a user decides to quit the group (whether private or public) the process would be to send a notification to the master node, which relays it to the cellular network and,

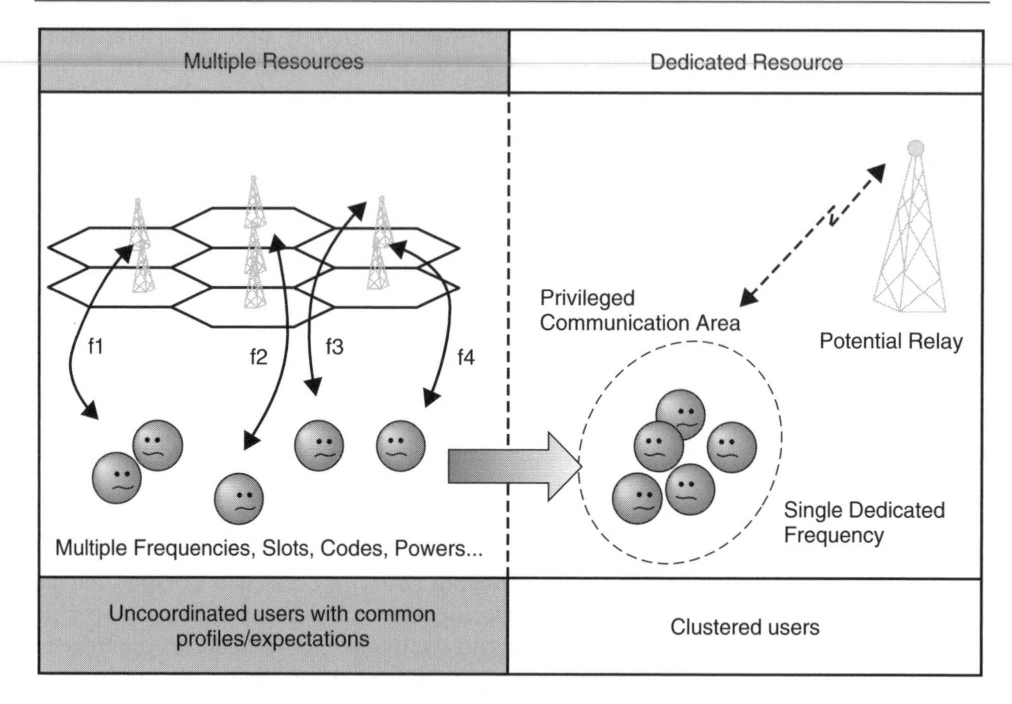

Figure 11.32 Concept of resource allocation for groups

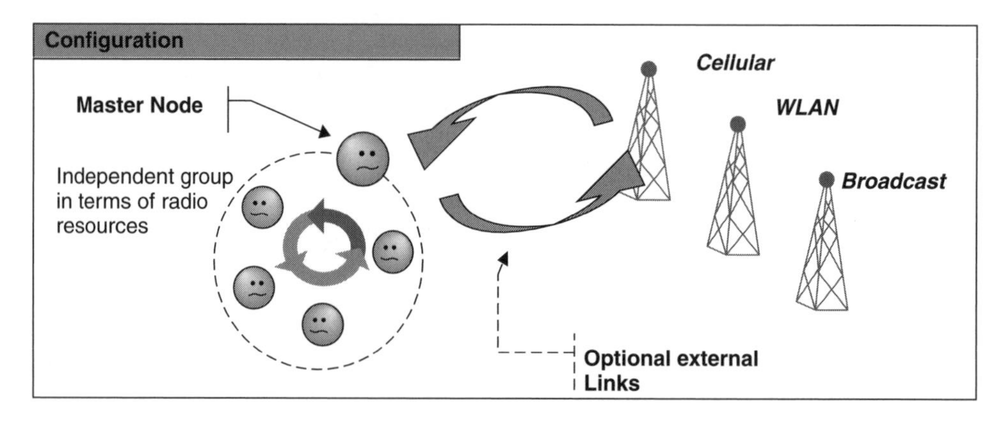

Figure 11.33 Global architecture

in the case that radio resources are available, an acknowledgement is sent to the user to perform the handover to the cellular frequency carrier and cellular RAT. Master node and operator must then update their databases.

Two options can be considered for the termination of the group. The master node is the only remaining member because previous slave users quitted and the master has to close the group to become a regular user again. On the contrary, the master node could initiate group termination when there are still slave users in the group. In this case a message is sent to the users as well as to the network and everybody must perform a handover to be attached to the cellular network.

The evolution described in this section provides the means in a cellular network to better respond to mobile users' expectations, especially when these latter are the same whether they know each other or not through intensive use of all RATs available in the area. It also becomes possible to offer a maximised QoS to the users' group by responding in a sharper way to what they expect, but also by releasing the native cellular network where users were when the group creation occurred. On operator side, the result would be a more efficient RRM offering a higher level of flexibility and always the possibility of reaching any user in a transparent manner through the group's master node.

References

[1] D. Calin, L. Elicegui and D. Grandblaise, 'Spectrum engineering options for software definable radio', Wireless World Research Forum (WWRF), 6–7 March, Munich, Germany, 2001.

[2] D. Grandblaise, L. Elicegui and D. Bourse, 'Radio resource management and flexible spectrum allocation for reconfigurable SDR terminals', Wireless World Research Forum (WWRF), SDR Group, 17–18 September, Stockholm, Sweden, 2001.

[3] Report of the 4th meeting of ITU-R working party WP-8F, Annex 4 to attachment 13 of the document 8F/268, February, 2001.

[4] 'Projection of the future usage of IMT 2000 systems by UMTS-Forum', Report No. 6 and ITU-R Report M (IMT.SPEC)'.

[5] *http://www.itu.int/home/*

[6] *http://www.cept.org/*

[7] *http://forum.europa.eu.int/Public/irc/enterprise/tcam/home*

[8] *http://europa.eu.int/comm/enterprise/rtte/*

[9] P. Bender and S. O'Fee, 'European regulation of software radio', in W. Tuttlebee, ed., *Software Defined Radio – Origins, Drivers and International Perspectives*, Wiley, 2002.

[10] *http://www.fcc.gov/*

[11] *http://www.ntia.doc.gov/osmhome/osmhome.html*

[12] *http://www.citel.oas.org/citel_i.asp*

[13] *http://www.arib.or.jp/english/index.html*

[14] *http://www.aptsec.org*

[15] *http://www.telec.or.jp/ENG/Index_e.htm*

[16] *www.IST-TRUST.org*

[17] *www.ist-drive.org*

[18] *www.IST-SCOUT.org*

[19] *www.ist-overdrive.org*

[20] J. Mitola III, 'Cognitive radio: An integrated agent architecture for software defined radio', Dissertation Thesis, Royal Institute of Technology (KTH), Sweden.

[21] R. Tönjes, K. Moessner, T. Lohmar and M. Wolf, 'OverDRiVE spectrum efficient multicast services to vehicles', *Proceedings of the IST Mobile and Wireless Telecommunications Summit 2002*, **June**, 473–477, 2002.

[22] P. Leaves, et al., 'Dynamic spectrum allocation in a multi-radio environment: Concept and algorithm', IEE 3G2001 Conference, March, pp. 53–57, 2001.

[23] J. Huschke and P. Leaves, 'Dynamic spectrum allocation including results of DSA performance simulations', DRiVE deliverable D09, January, 2002.

[24] P. Leaves, J. Huschke and R. Tafazolli, 'A summary of dynamic spectrum allocation results from DRiVE', *Proceedings of the IST Mobile and Wireless Telecommunications Summit 2002*, June, 245–250, 2002.

[25] P. Leaves and R. Tafazolli, 'A time-adaptive dynamic spectrum allocation scheme for a converged cellular and broadcast system', IEE Conference on Getting the Most Out of the Radio Spectrum, October, 2002.

[26] J. Huschke and S. Ghaheri-Niri, 'Guard band coordination of areas with differing spectrum allocation', IST Mobile Summit 2001, 10–12 September, Barcelona, 2001.

[27] ITU-8F White Paper, joint Motorola S.A/Siemens A.G Contribution to WG Vision Drafting Group 1 on IMT.Vis and Attachment 3.10: Technology Trends (Source: Doc. 8F/489).

[28] R. Tönjes, K. Moessner, T. Lohmar and M. Wolf, 'OverDRiVE spectrum efficient multicast services to vehicles', *Proceedings of the IST Mobile and Wireless Telecommunications Summit 2002*, June, 473–477, 2002.

[29] D. Grandblaise, D. Bourse, K. Moessner and P. Leaves, 'Dynamic spectrum allocation (DSA) and reconfigurability', SDR Forum, Software Defined Radio Technical Conference, 11–12 **November**, San Diego, USA, 2002.

[30] 'Reconfigurable SDR equipment and supporting networks reference models and architectures', Wireless World Research Forum (WWRF) SDR White Paper (version 2.0), Didier Bourse Ed., September 2002, *http://www.wireless-world-research.org*

[31] Mobile VCE's Reconfiguration Management Architecture RMA, 2002 *http://www.mobilevce.com*

[32] M. Beach, D. Bourse, R. Navarro, M. Dillinger, T. Farnham and T. Wiebke, 'Reconfigurable Terminals Beyond 3G and Supporting Network System Aspects', Wireless World Research Forum (WWRF), SDR WG, 17–18 September, Stockholm, Sweden, USA, 2001.

[33] *http://www.sdrforum.org/tech_comm/mobile_wg.html*

[34] Scenarios, Services, and Requirements, OverDRiVE Deliverable D03 to be published in September 2002, *http://www.ist-overdrive.org/*

[35] Access Networks – RF System Scenarios (Release 1999)', 3GPP – 3G TR 25.942 3.0.0 (2001-03), Technical Specification Group Radio.

[36] UMTS 30.03 version 3.2.0, 'Selection procedures for the choice of radio transmission technologies of the UMTS'.

[37] L. Elicegui, D. Grandblaise, N. Motte, D. Thomas and P. Booker, IST-1999-12070 TRUST, D4.4, 'Final report on spectrum sharing techniques and recommendations for future research', December, 2001.

[38] D. Calin and M. Areny, 'Impact of radio resource allocation policies on the TD-CDMA system performance: Evaluation of major critical parameters', *IEEE JSAC*, **19**(10), 1847–1859, 2001.

[39] Book of Vision Wireless World Research Forum (WWRF), December, 2001 *http://www.wireless-world-research.org*

[40] D. Grandblaise, L. Elicegui, N. Motte and D. Bourse, 'SDR applications in self-organising environment – evolution of radio resource management', Wireless World Research Forum (WWRF), SDR WG, 6–7 December, Paris, 2001.

12

Mode Identification and Monitoring of Available Air Interfaces

Georgios Vardoulias and Jafar Faroughi-Esfahani

Motorola Ltd

12.1 Problem Definition: Mode Monitoring and Identification of Air Interfaces

One can define three stages in any mobile phone's/terminal's activity time: (a) the first few seconds when the terminal is switched on and cannot make or accept any calls; (b) the period during which the terminal is able to communicate but there is no user activity; and finally (c) the period when the terminal is used in order to send/receive voice and data. This section is dedicated to stages (a) and (b). We will show that after keying in the PIN number, the terminal has to do a number of demanding tasks in a short time before we can actually use it. We will present the relevant GSM and UMTS procedures as an example of how things work today. This will be our guide in order to predict what the first few seconds of a Software Defined Radio (SDR) terminal will be like. Our intention is to present the problems, not the solutions. The software radio technology has many open (almost virgin) research areas and this is one of them.

As was pointed out in ref. [1],

> a software radio terminal does not just receive: It
>
> - Characterises the energy distribution in the channel and in adjacent channels
> - Recognises the mode of the incoming transmission
> - Adaptively nulls interferers
> - Estimates the dynamic properties of desired-signal multipath
> - Coherently combines desired-signal multipath
> - Adaptively equalises this ensemble
> - Trellis decodes the channel modulation
> - Corrects residual errors via forward error control (FEC) decoding to receive the signal with lowest possible BER.

Software Defined Radio: Architectures, Systems and Functions. Edited by M. Dillinger, K. Madani and N. Alonistioti
 ISBN: 0-470-85164-3

The above list is a set of demanding tasks that a software radio receiver will have to cope with. We will present the problems and requirements of the first two tasks of the previous list, either in terms of the Initial Mode[1] Identification (IMI – i.e. the search for all possible modes following the power on of the terminal) or in terms of the Alternative Mode Monitoring [AMM – i.e. monitoring other modes when the User Equipment (UE) is already using a certain mode]. IMI and AMM are similar tasks, ideally will use similar techniques, and hopefully the same hardware (in this case, the word 'hardware' can have a softer meaning and may be equivalent to specific software modules). IMI and AMM have different problems:

- *IMI:* The main problem is that the UE may have to scan large areas of the radio spectrum in order to locate and use a mode. In the worst case the UE may have to do that without any help from the network (blind IMI). On the other hand, there are no serious time constraints (given that a connection can be established in a reasonable time period, acceptable by the user).
- *AMM:* The UE is already connected to a mode and thus the network can (and probably will have to) assist the terminal in finding and evaluating all possible modes. The problem here is that there are severe time constraints. The terminal must switch from the current mode, take measurements in another mode and then revert to the current mode with no loss of data or synchronisation. Unless there is a dedicated, second transceiver, AMM is not at all a trivial task.

12.1.1 General Overview of the IMI and AMM

A SDR user equipment is by definition a multi-mode terminal that can detect, recognise and monitor any of the available Radio Access Technologies (RATs) (i.e. the existing such as GSM/IS-95 or proposed 3G systems such as UMTS/CDMA2000) and subsequently reconfigure itself according to the chosen RAT specifications. After the terminal is switched on, it must be able to discover, compare and monitor all the available RATs. Let us assume a user travelling within Europe. The terminal must be able to handle at least GSM and UMTS networks. We will restrict ourselves in those two RATs in order to give an introductory presentation of this subject, although the general principles are applicable for IS-95 and CDMA2000 networks.

It is important that a distinction between discovering a RAT and monitoring a RAT is made. In order to discover a RAT, the terminal has to perform some kind of energy detection. The amount of detected energy, its distribution across the spectrum and the correlation with some predefined functions are enough to indicate the presence of a RAT. For a SDR terminal the RAT detection requires a detector able to cope with both narrowband (GSM) and wideband spread spectrum (UMTS) systems. It does not require knowledge of any protocol. Monitoring, on the other hand, implies not only the ability to measure but also to recognise specific parameters with values corresponding to predefined actions. Thus, monitoring requires knowledge of the monitored RAT's protocol, or at least a part of it. For a software radio terminal, which theoretically is able to use any RAT, this is a serious problem, maybe more difficult than a multi-mode detector. If the terminal stores all possible protocols, then this would require a large amount of memory. Thus, it seems

[1] The terms 'Mode' and 'Radio Access Technology' are used interchangeably in the rest of this section.

that the combination of multi-mode detectors and flexible protocols is imperative for SDR terminals.

The initial mode identification procedure for GSM and UMTS (and also IS-95, CDMA2000) follows the same steps that are depicted in Figure 12.1. The first step is to select the Public Land Mobile Network (PLMN), i.e. the provider. This can be done manually or automatically by the UE as described in refs [2] and [3]. The next step is the cell selection and reselection [2, 3] in which the best cell is selected according to certain criteria. The UE will *camp on this cell in idle mode*, which means that the terminal will be able to receive system information from the PLMN and, when registered (step 3 in Figure 12.1: Location Registration or LR), will also be able to initiate a phone call and receive an incoming one. The cell selection step is very important for a terminal in any RAT, as only after its completion has the terminal any communication with the network. Thus, the minimum requirement for a SDR terminal, in order to be able to connect to a network, is to have the hardware and software that will enable it to camp on a cell in idle mode of at least one RAT. After that it should also be able to negotiate a channel for downloading any additional software if we assume over-the-air software download [5].

It is logical to assume that an advanced SDR terminal should be able to compare all the available RATs at any time and use the one that fulfils a number of quality-of-service (QoS) criteria, as well as the user preferences. This implies that the terminal should be able to monitor alternative RATs when connected to a given network. In the 3GPP specifications documents [6], there are four types of possible multi-mode 3G terminals defined according to their capability of concurrent reception and monitoring in more than one mode. Those four types are summarised in Figure 12.2. Type 1 can support only one mode, type 2 can support two modes, but not simultaneously, type 3 can support simultaneous reception (Rx) in two modes but not transmission (Tx), and type 4 can

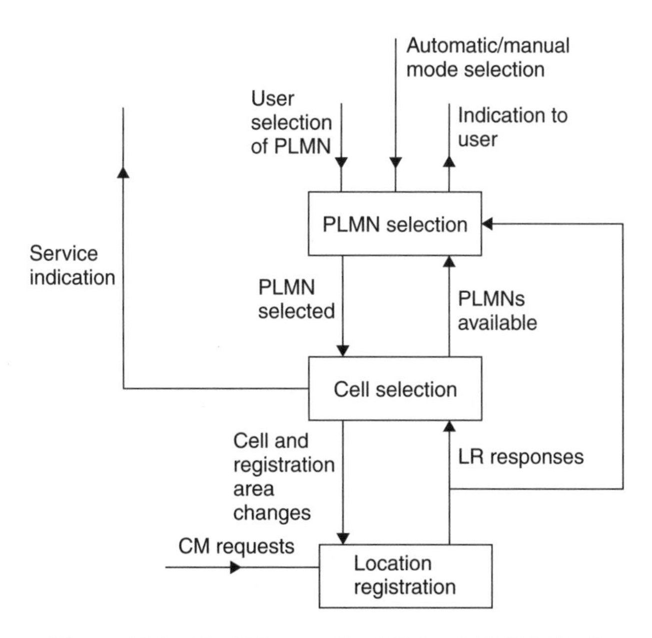

Figure 12.1 The IMI steps for GSM and UMTS [2–4]

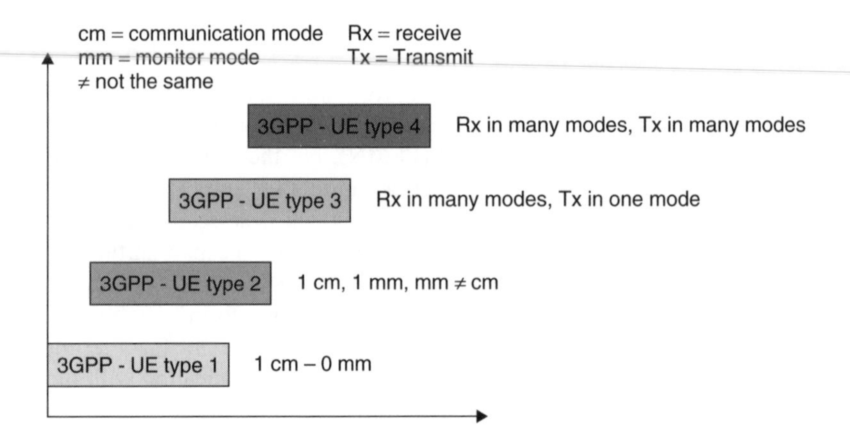

Figure 12.2 The four multi-mode 3G terminal types

support simultaneous Tx and Rx in two modes. Types 3 and 4 are not of interest for the UMTS system owing to their high complexity. Nevertheless, a SDR terminal must be a 3GPP type 4 equivalent.

The AMM capability investigation in TRUST can be subdivided into two approaches: (a) concurrent alternative mode monitoring and (b) non-current alternative mode monitoring. The former implies an architecture with multiple receive and transmit paths, a far more complex terminal (along the lines of the types 3 and 4 terminals in the 3GPP). With such a terminal, it is largely a trivial exercise to be able to monitor other frequencies and modes even whilst in communication in the current mode. Indeed, such a terminal could maintain communication sessions in as many modes as there are receive and transmit paths. Whilst this structure is viewed as 'trivial' in comparison to the alternative below, it is clear that there are issues here as well; for example, some consideration of interlocks, to ensure that a transmit stage does not 'deafen' a receive stage, is required.

The latter approach, utilising the reconfigurable nature of the terminal to the extreme, places some severe requirements on the underlying technology. In essence, the terminal must momentarily switch from the current mode, take measurements in one (or more!) alternative mode, and then revert to the current mode with no loss of data or synchronisation.

12.1.2 Time Considerations

The IMI stage time duration depends on the number of RATs that the terminal can support, the number of available networks in the surrounding area and the a priori information that the terminal has for the available networks.

The GSM All-Channel Scan process is a three-stage process. First, the received signal strength of all the channels is measured. Secondly, these measurements are sorted in order of strength. Finally, the strongest channels are singled out for further decoding if possible. Each cell broadcasts a single beacon signal, which will stand out amongst all the other channels emanating from that cell. The beacon signal is highly visible because it is continuously transmitted (unlike the other channels) and also because the transmitted data are specifically designed to create a perfect sine wave. The handset has to locate all these beacon signals (which are physically transmitted as 'frequency correction bursts' – FCBs), and measure the signal strength of each one. Having gathered the channel measurements, they can be sorted. The Mobile Station (MS) searches all RF channels in the system (124 for P-GSM, 174 for E-GSM, 194 for R-GSM, 374 for GSM1800 and 299 for PCS 1 900), takes readings of the received RF signal level on each RF channel, and calculates a certain parameter for each. The averaging is based on at least five measurement samples per RF carrier spread over 3–5 s, the measurement samples from the different RF carriers being spread evenly during this period. A multi-band MS searches all channels within its bands of operation as specified above. It is required that the overall process (including the network selection process) results in the terminal being able to select the correct cell and be able to respond to paging on that cell within 30 s of being switched on [7]. The UMTS procedure for the initial cell selection when there is no prior knowledge of which RF channels are UTRA carriers is introduced in ref. [3]. The UE scans all RF channels in the UTRA band to find a suitable cell. On each carrier, the UE searches first for the strongest cell and reads its system information, in order to find out which PLMNs are available. If the selected PLMN is found, then the search of the rest of carriers may be stopped. Note that the cell selection procedure consists of three stages, namely slot synchronisation, frame synchronisation and code-group identification, and finally scrambling-code identification [8, 9]. A generic framework for the calculation of the worst-case mean cell selection time, applicable in every RAT, is presented in Section 3.3.2.

For the AMM procedure using a single receiver common for communication and monitoring, the procedure that is depicted in Figure 12.3 must be followed.

The actual time available to perform any measurements is subject to the following calculation:

$$T_{\mathrm{m}} = T_{\mathrm{g}} - (T_{\mathrm{rct}} + T_{\mathrm{tct}} + T_{\mathrm{st}}) - (T_{\mathrm{rtc}} + T_{\mathrm{ttc}} + T_{\mathrm{sc}}), \tag{12.1}$$

where T_{m} = time available for the measurement, T_{g} = duration of gap in current mode, T_{rct} = time taken to reconfigure the terminal from current mode to target mode, T_{tct} = time taken to retune the frequency synthesiser from current frequency to target frequency,

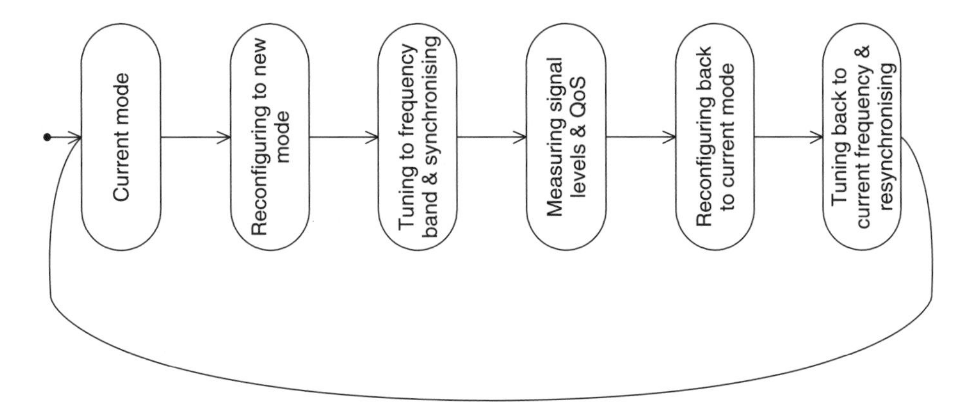

Figure 12.3 AMM stages

T_{st} = time taken to synchronise to the target system (including realigning the system clock, etc.), T_{rtc} = time taken to reconfigure the terminal from the target mode to the original ('current') mode, T_{ttc} = time taken to retune the frequency synthesiser from the target frequency to the original ('current') frequency, and T_{sc} = time taken to synchronise to the original ('current') system (including realigning the system clock, etc.). Note that equivalence of reconfiguration times, etc. cannot be assumed (e.g. the fact that it takes N milliseconds to reconfigure from mode A to mode B does not necessarily mean that it will take N milliseconds to reconfigure back again).

Taking into account the GSM/UMTS synchronisation times and the fact that the reconfiguration time cannot yet be estimated, it becomes obvious that, assuming a single transceiver, the monitoring of alternative RATs, while communicating, is a very difficult (if not impossible) task.

12.1.3 Solutions under Consideration

One of the most obvious and simple ways of allowing a multi-mode terminal to find the available RATs in a completely unknown environment is to point them out. In order to do that you need to define a *beacon* which will be known to and accessible by every terminal. Thus, the terminal will have to look only for the beacon instead of scanning the whole spectrum for a RAT. This beacon or *pilot channel* will transmit information about the existing services in a standardised format. The information could be anything, from the existence of a service to more advanced and useful tips, such as the quality and cost of the service. As different levels of service may be offered by different systems, it would be more efficient to be able to determine whether the level of service required is actually provided by a given system before attempting to register on it. The ability to discover information on the range of networks and services available by scanning a small set of frequencies would be very helpful. The idea of a pilot channel has attracted great industrial and academic interest. There were a series of proposals considered during the early scoping of the IMT-2000 system. Some of these proposals considered specifying a 'global radio control channel', to support multi-mode terminals in their quest for global roaming. Ultimately, these proposals were not pursued. An *ad hoc* subgroup of the ETSI SMG Plenary led by the UK DTI concluded that any such mechanism could be constructed as an overlay on top of the IMT-2000/UMTS system. As such, the global pilot channel did not affect the definition of UTRA [10]. The pilot channel was also an agenda item for the 1999 World Radio Conference (WRC99) [12]. It was recognised that a disparate set of frequencies and systems would most likely be deployed as the IMT-2000 system to allow for national and regional variations in frequency allocations and preferred technologies. Similarly, it recognised that GSM takes 'quite a long time ... to scan all available networks and choose one', even though the system works within a very limited range of frequencies, and that the sheer quantity of frequency bands and standards to search for may make the time taken to perform a brute-force scan prohibitive. It was felt at an early stage [13] that there might be scope to support reconfigurable terminals within this framework. A global pilot channel was considered as not only broadcasting information on available systems, but also transmitting tariff information and helping to 'boot' a reconfigurable terminal. An in-depth presentation of the subject can be found in refs [10]–[18].

Despite offering many advantages, the pilot channel concept is not regarded as a possible future solution. Issues such as the ownership of the channel, the involved costs and revenues and other economical and political implications will make it very difficult for the industry to accept and standardise a pilot channel. Thus, the TRUST project focused on alternative techniques.

As described above, the two major problems in IMI and AMM are: (a) the hardware and protocol stacks must be compatible with a large number of incompatible systems, and (b) there is very little time available, especially for monitoring other networks in parallel with an existing connection. Solutions for inter-monitoring between UMTS and GSM exist in the 3G specifications [19]. The monitoring ability of the terminal depends on the optimum use of every available time slot (GSM) or techniques such as the compressed mode of UMTS [20]. Nevertheless, the number of available slots is a problem and as the systems become more and more complicated, the amount of free slots will not allow any monitoring. Furthermore, although the compressed mode is part of the UMTS specifications, operators do not seem to like it, as it causes large performance degradation.

The time problem can be resolved if a second receiver chain, which can be used for all the monitoring tasks, is available. The disadvantages of this scheme are the higher power consumption and the possible cross impact of the two receptions. Nevertheless, the importance of alternative RAT monitoring dictates that a powerful and reliable solution must be provided. The suggested solution is depicted in Figure 12.4. It uses a dedicated receiver chain (but a single antenna) to perform mode monitoring independently of the data receiver chain which is responsible for the data/voice of the active connection. The monitoring receiver chain is controlled by a flexible protocol stack (the 'Boot Protocol' [25]) which is able to perform measurements in a large number of RATs (ideally all). Flexible protocols have been investigated mainly for computer network applications and CORBA environment applications. But, as the computer and wireless industry converge, it becomes obvious that new ways of effectively handling data from various sources, format and quality are necessary to be investigated [21–24]. The TRUST project has looked into adaptable protocols that will allow the initial connection and monitoring of any RAT. Note that the monitoring receiver chain must be a multi-RAT receiver. It will consist of a number of sub-receivers. This sounds demanding, but actually it is not as

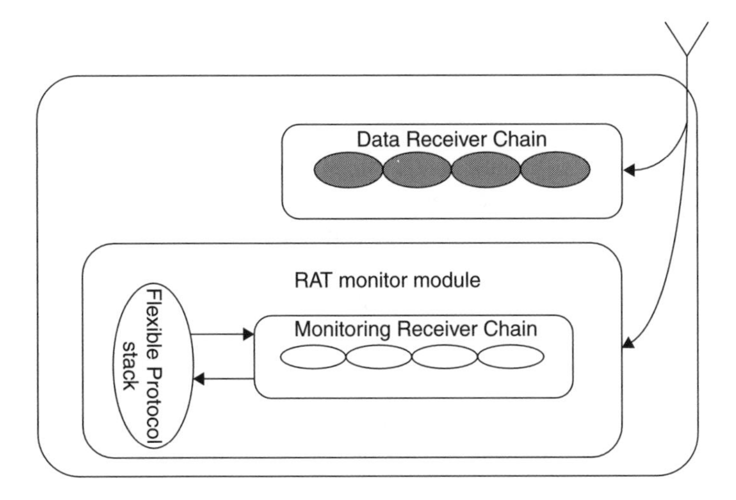

Figure 12.4 The RAT monitor module

difficult as it seems. In fact, two sub-receivers, one narrowband and one wideband, will be able to handle almost every RAT. A number of parameters might be necessary to change for each sub-receiver, but most of their modules will be re-usable. Similar two-chain receivers are currently available. The monitoring receiver chain can also be used for the initial RAT discovery. Thus, an integrated solution for both procedures (IMI and AMM) can be accomplished.

12.2 A Generic Framework for Calculation of the Worst-Case Mean Cell Selection Time in a Dynamic Spectrum Allocation Environment

SDR development is associated with challenging technological promises, such as seamless and transparent interoperability between different communication standards, enhanced roaming capabilities without changing terminal, the capability to select the most attractive network, the capability to download over the air application and core functional software. However, the full benefit of SDR technology will only be reached if new spectrum engineering practices are developed to maximise efficient and effective use of the scarce radio spectrum resource, taking into account the diversity of RATs, the multiplicity of operators (fully licensed or virtual) and the existence of regulator rules across the different regions. The spectrum allocation should evolve from the current static case to a flexible dynamic scheme (dynamic allocation and sharing of spectrum between operators). This evolution would require careful investigation but should be characterised by the progressive and regular introduction of some degrees of flexibility (limited spectrum sharing between RATs and/or operators).

The use of Dynamic Spectrum Allocation (DSA) algorithms will have serious implications on the time that is required for the terminal to scan the spectrum and locate an RAT carrier after it has been powered on. In a Static Spectrum Allocation (SSA) scenario (such as the existing 2G and 3G systems) the terminal sequentially searches a fixed list of frequencies and performs a number of measurements in order to determine which frequencies in the list are used as carriers and which of the available carriers is the best in terms of the received power. Even in the SSA case, the spectrum scan procedure can be time consuming if the terminal has no prior information about the most likely carriers. The prior information for GSM terminals is a list of the carriers that were used before the last switch-off. Usually, this kind of prior information is very helpful and the terminal can camp on the network in a few seconds. On the contrary, if the user is far away from his usual location (e.g. a different country), then the scan and camp-on procedure takes considerably longer because the terminal has no valid prior information. Things will be more difficult in a DSA environment, so special attention should be paid to the resulting mean cell selection time when designing DSA algorithms. For example, an extreme case of a DSA environment where every frequency can be a carrier of every RAT would require either very advanced spectrum scan algorithms and modules on-board the terminal (which may be non-practical) or dedicated pilot systems that would help the terminal to 'navigate' through the spectrum. For more realistic DSA algorithms things are simpler but attention to keep the cell search time within acceptable limits is still essential.

For the purposes of the following discussion, we define the mean cell selection time as the time required for scanning the spectrum, locating and identifying a valid carrier after the initial power on of the terminal. We will not take into account the duration of any information exchange between the terminal and the network after the choice of a carrier. The mean cell selection time depends on the DSA algorithm, the number of RATs that the terminal supports, the number of available RATs in the location area of the terminal, the number of carriers and the available a priori information about the location of the most suitable carriers.

12.2.1 Definitions and Assumptions

It is our intention to develop a framework for the cell selection procedure which is as generic as possible. Ideally, it must be suitable for all the cellular RATs and for both spectrum allocation schemes (static and dynamic). When we use the term DSA we assume a scheme according to which the possible carriers are fixed and known to the terminal but the mapping of RATs (e.g. GSM, UMTS) to these carriers is random, i.e. the same carrier can be used by different RATs or operators at different time instances depending on the traffic load, roaming agreements and other related parameters. The exact DSA algorithms can vary. For both the SSA and DSA schemes each carrier can be in two states:

- Void Carrier (VC): there is no associated RAT, i.e. carrier not in use.
- Occupied Carrier (OC): the carrier is used by a network of any RAT.

The status of a carrier (VC or OC) is not static, i.e. the same carrier can be OC at a time t_1 and VC at another time t_2.

Figure 12.5 depicts a generic procedure for the cell selection that can be used in any cellular system in a static or dynamic spectrum allocation. We define three basic steps (depicted as Steps 1, 2 and 3 in Figure 12.5), each one with a set of corresponding times and probabilities. The times are related to the duration of the steps and the probabilities are related to the reliability of each step, i.e. they indicate the probability of success at each step. The values of the times and probabilities depend on the cellular system, radio environment and the receiver architecture. The cell selection procedure starts by tuning the frequency synthesiser to a frequency F_i that may be used as a RAT carrier (depicted by the 'Tune to F_i' block in the flowchart). After the tuning, the terminal will perform some kind of energy detection in the selected frequency, i.e. it will make a decision whether the frequency is an OC or a VC, usually by performing energy/power measurements and comparing these values with a predetermined threshold. This procedure is depicted as Step 1 in the flowchart and the time required for the energy detection (also known as the dwell time) is T_1. The probabilities that describe the reliability of this step is the probability of detection, P_d, which is defined as the probability that an OC is actually detected and the probability of false detection, P_{fd}, which is the probability that a VC is mistaken for an OC (the threshold is exceeded due to noise). The implementation of this step (detection, algorithms, hardware) can vary significantly between different RATs, but the main objective is always the detection of a carrier for both SSA and DSA schemes. If the outcome of Step 1 is a VC decision, then the synthesiser will be re-tuned to the next possible carrier (the re-tuning is depicted in Figure 12.5 by the '$F_i = F_i + F_{step}$' block). The time that is required for the re-tuning of the frequency synthesiser is depicted as T_s. If the outcome of Step 1 is an OC decision, then the RAT identification step will take over. While the previous step is necessary for both SSA and DSA, this step is compulsory only in the DSA case as the RAT corresponding to the detected carrier is not determined by the frequency and thus it must be identified. There are a number of algorithms that can

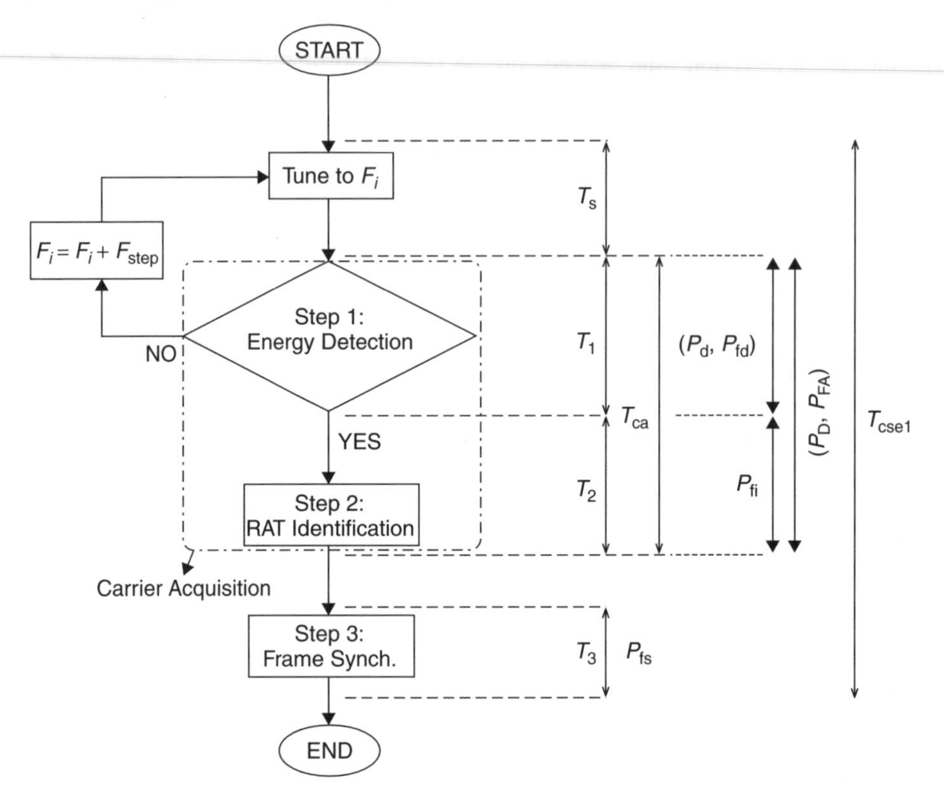

Figure 12.5 A generic flowchart of the cell selection procedure in a multi-RAT, DSA environment

be applied to the identification and classification of different RATs (e.g. [27–29]). On the other hand, it may be possible to combine the first two steps (energy detection and RAT identification) in order to avoid extra complexity and power consumption in the terminal. We will refer to the combined process as the carrier acquisition step. This combination is possible because of the differences between the various RATs. The energy detection of a CDMA signal requires different parameters (threshold, dwell time) than the energy detection of a narrowband (e.g. GSM) signal, even if we assume that the same detection algorithms and hardware are used. Thus, the algorithms or the parameters that are used in order to detect the carrier can also reveal the corresponding RAT. In Figure 12.5 the duration of the RAT identification is depicted as T_2 and the duration of the carrier acquisition procedure (energy detection and RAT identification) is $T_{ca} = T_1 + T_2$. The third and final step before the camping on of the terminal to a network in an idle mode is the frame synchronisation. The duration T_3, and the exact functionality and implementation of this step depend on the RAT (for the GSM/UMTS systems see refs [26] and [30]). In all cases, the main objective of the third step is that the terminal obtains slot and frame synchronisation. The exact terminology may be different from RAT to RAT but the point is that the terminal must find the boundaries of the frames that all cellular systems use in order to organise and transmit/receive their data. This step is common for both SSA and DSA schemes. The total cell selection time is defined as $T_{csel} = T_s + T_{ca} + T_3$. In order to make this framework as generic as possible we will treat the energy detection and RAT identification as two separate and independent procedures. Under this assumption, in a DSA scheme there are two possible outcomes of the acquisition procedure of a VC:

- The carrier is correctly declared to be void: VC correct detection.
- The carrier is incorrectly declared to be occupied: false alarm.

Similarly, there are four possible outcomes of the acquisition procedure of an OC:

- The carrier is detected, the corresponding RAT is correctly identified and the frame synchronisation is acquired (all steps successful): OC correct detection.
- The carrier is detected but the corresponding RAT is not identified correctly: false identification which is equivalent to a false alarm.
- The carrier is not detected at all: missed detection.
- The carrier is detected and identified but the frame synchronisation fails: equivalent to a missed detection.

For the RAT identification procedure we define the false identification probability P_{fi} as the probability that the RAT corresponding to a detected OC is falsely identified. For the frame synchronisation step, the probability of failed frame synchronisation P_{fs} is defined as the probability that the terminal fails to acquire frame synchronisation. Note that we regard a possible failure of the third step as equivalent to a missed detection, i.e. the terminal will tune to the next possible carrier and resume the search. For a DSA scheme, we define the probability of overall detection P_D as the probability that an OC is correctly detected and identified. This refers to the total carrier acquisition (Steps 1 and 2 combined). Furthermore, the probability of false alarm P_{FA} is defined as the probability that an OC is correctly detected but not identified, or that a VC is falsely declared as OC. If Steps 1 and 2 are independent we can write:

$$\begin{aligned} P_D &= P_d(1 - P_{fi}), \\ P_{FA} &= P_{fd} + P_d P_{fi}. \end{aligned} \tag{12.2}$$

The probabilities of overall detection and false alarm are particularly useful for a static spectrum allocation scheme or even for a dynamic spectrum allocation scheme with Steps 1 and 2 combined in a single operation (we will refer to this as a DSA-c scheme). In these cases, the probabilities of detection P_d, false detection P_{fd}, and false identification P_{fi} are meaningless since the energy detection and signal identification are not defined as separate processes. For the SSA and DSA-c schemes, the probability of overall detection P_{Do} is the probability that an OC is correctly detected during the carrier acquisition procedure (combined Steps 1 and 2) and the probability of false alarm P_{FAO} is the probability that a VC is mistaken for an OC by the detection procedure. The probabilities (P_D, P_{Do}) and also (P_{FA}, P_{FAO}) have the same name (overall detection and false alarm probabilities, respectively) because ultimately they describe the same things. Nevertheless, they are different quantities applicable to different spectrum allocation schemes [obviously equation (12.1) is valid for the DSA and not the SSA/DSA-c]. Table 12.1 summarises the results of the generic three-step cell selection procedure of Figure 12.7 for the DSA and SSA/DSA-c cases. For an OC the false alarm state is only defined in the DSA case as in

Table 12.1 A summary of the possible outcomes of the carrier detection procedure in DSA and SSA schemes

	OC			VC	
	Correct detection	False alarm	Missed detection	Correct detection	False alarm
DSA	Steps 1, 2, and 3 successful	Step 2 failed	Step 1 or Step 3 failed	Step 1 successful	Step 1 failed
SSA and DSA-c	Steps 1 and 3 successful	Not applicable	Step 1 or Step 3 failed	Step 1 successful	Step 1 failed

the SSA/DSA-c an OC can only be detected or missed, but not mistaken for a carrier of a different RAT.

12.2.2 Calculation of the Worst-Case Mean Cell Selection Time

For our calculations we will assume that there are ν possible carriers but only one is occupied (OC) by a single RAT during the acquisition procedure. This is obviously the worst case in terms of the mean cell selection time because there is only one available carrier that the terminal must find and use. The worst-case scenario in today's over-used spectrum is not a realistic one. However, worst-case studies are always useful for the design of any system. Moreover, the existence of only one carrier simplifies the mathematical analysis. This approach can be justified using the single H_1 hypothesis described below.

The cell selection procedure can be modelled as a Markov process, assuming that the search procedure for each carrier is independent. For the calculations we followed the *flow graph approach* (see refs [31] and [32]) as presented by Polydoros [33]. An alternative would be the *direct approach* by Jovanovic [34]. The direct approach is very useful if there is a priori information available for the position of the OC and if special search strategies are employed. Here, however, we assume that there is no available information and that the terminal examines all the possible frequencies in a serial fashion, i.e. no search strategy is employed. In this case, the flow graph approach is simpler. Both methods were originally presented in the context of the calculation of the mean PN code acquisition time (CDMA synchronisation) problem.

12.2.2.1 The Flow Graph Approach

Here, H_1 denotes the hypothesis that the terminal is tuned to the single OC and H_0 denotes the alternative hypothesis where the terminal is tuned to any one of the VC. We assume that the number of all the possible carriers is ν. The system has $\nu + 2$ states of which $\nu - 1$ states correspond to the carriers belonging to hypothesis H_0, while one state corresponds to the H_1. These ν stages are indexed in a circular arrangement, with the ith state ($i = 1, 2, \ldots, \nu - 1$) corresponding to the ith carrier position to the right of H_1. The two remaining states are the cell selection (CSEL) and false alarm (FA) states. Entry into the search process can occur at any one of the ν states, according to some a priori distribution (π_j, $j = 1, 2, \ldots, \nu$). Total uncertainty corresponds to a uniform distribution ($\pi_j = 1/\nu$, $j = 1, 2, \ldots, \nu$). Entry to the CSEL state can occur only by the single OC carrier, while entry to the FA state can occur from any of the VC carriers (step 1 failed). Entry to the FA state from an H_1 (OC carrier) state is only possible for a DSA scheme with independent energy detection and RAT identification steps. A segment of that circle, including the states numbered $\nu - 1$, ν, 1 and 2, is shown in Figure 12.6. For the SSA and DSA-c schemes a transition between the H_1 and FA states is not allowed (see Table 12.1). The corresponding flow graph is presented in Figure 12.9.

Let $p_{ij}(n)$ denote the probability that the Markov process will move from state i to state j in n steps and let z denote the unit-delay operator. If the unit delay specifically corresponds to τ time units, z is replaced by z^τ in the following. It has been shown in refs [31] and [32] that the state transition diagram can be mapped to its equivalent flow graph if each transition branch from i to j in the Markovian diagram is assigned a

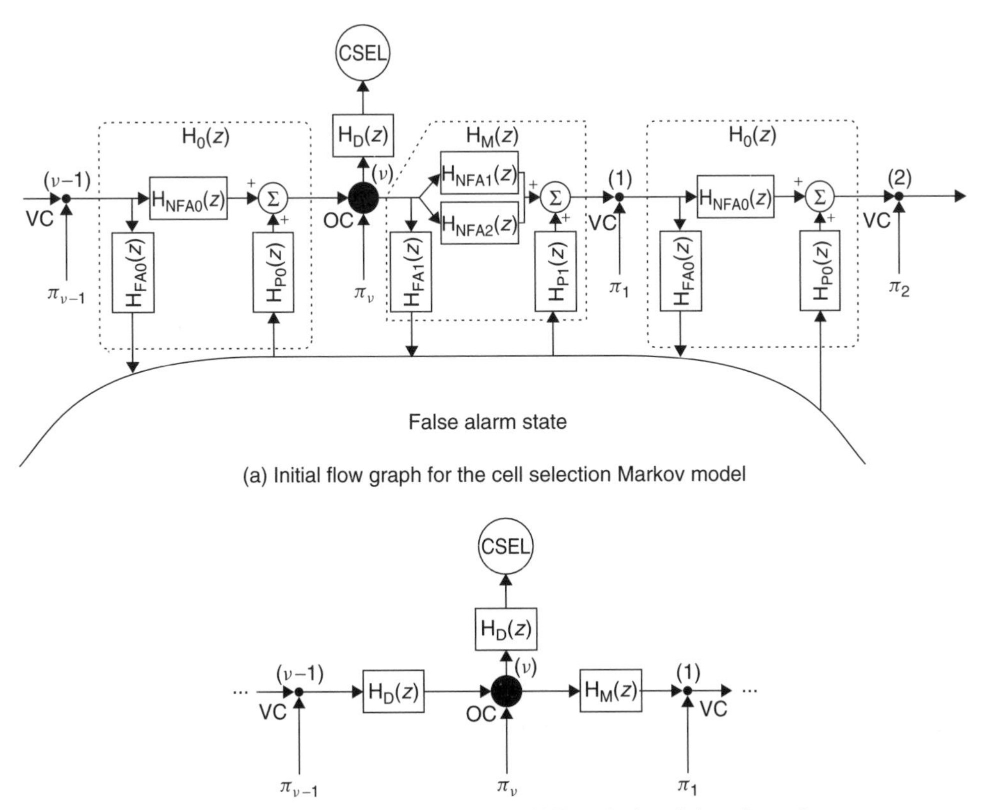

(a) Initial flow graph for the cell selection Markov model

(b) The flow graph after defining the total VC and missed detection gains

Figure 12.6 A portion of the flow graph diagram with an arbitrary a priori distribution (π_i) and various gains[2]

[2] The upper figure is the full flow graph and the lower is the simplified after defining the total VC and missed detection gains.

gain equal to $p_{ij}z$, where $p_{ij} = p_{ij}(1)$ is by definition the one-step transition probability and z represents the unit delay associated with that transition. The *generating function* is defined as:

$$P_{ij}(z) \stackrel{\text{def}}{=} \sum_{n=0}^{\infty} p_{ij}(n)z^n. \tag{12.3}$$

The generating function represents the transfer function from node i to node j on the flow graph and it is useful as it contains statistical information about the Markovian process. The generating function of a system can be derived using Mason's formula [31, 32]. In Figure 12.6(b), $\mathrm{H_D}(z)$ is the gain of the branch leading from node $\mathrm{H_1}$ (νth node) to the CSEL; $\mathrm{H_M}(z)$ is the gain of the branch connecting $\mathrm{H_1}$ with node 1 while $\mathrm{H_0}(z)$ is the gain of the branch connecting any other two successive nodes $(i, i+1)$, $i = 1, \ldots, \nu - 1$. Furthermore, the process can move between any two successive nodes $(i, i+1)$ with $i \neq \nu$ either without false alarm [in which case the gain is $\mathrm{H_{NFA0}}(z)$] or by first reaching the FA state [branch gain $\mathrm{H_{FA0}}(z)$], then pass from FA to node $i+1$ [branch gain $\mathrm{H_{P0}}(z)$] so that $\mathrm{H_0}(z) = \mathrm{H_{NFA0}}(z) + \mathrm{H_{FA0}}(z)\mathrm{H_{P0}}(z)$. Similarly, for the transition from $\mathrm{H_1}$ to $\mathrm{H_0}$ we have $\mathrm{H_M}(z) = \mathrm{H_{NFA1}}(z) + \mathrm{H_{NFA2}}(z) + \mathrm{H_{FA1}}(z)\mathrm{H_P}(z)$. It has been shown [33] that the generating function $P_{\mathrm{ACQ}} \stackrel{\text{def}}{=} \sum_{n=0}^{\infty} P_{\mathrm{ACQ}}(n)z^{\mathrm{n}}$ of Figure 12.6(b) is:

$$P_{\mathrm{ACQ}}(z) = \frac{\mathrm{H_D}(z)}{1 - \mathrm{H_M}(z)\mathrm{H_0^{\nu-1}}(z)} \sum_{t=1}^{\nu} \pi_i \mathrm{H_0^{\nu-t}}(z). \tag{12.4}$$

Using equation (12.4) we can calculate the mean cell selection time with:

$$\overline{T}_{\mathrm{csel}} = \left[\frac{\mathrm{d}P_{\mathrm{ACQ}}(z)}{\mathrm{d}z}\right]_{z=1}. \tag{12.5}$$

Equations (12.4) and (12.5) are very general and can provide a good statistical description of the acquisition process, as long as the Branch Gains are calculated and the a priori distribution π_i is determined. For the flow graph of Figure 12.6 the branch gains are:

$\mathrm{H_D}(z) = P_\mathrm{D}(1 - P_\mathrm{fs})z^{T_\mathrm{ca}+T_3}$, OC correct detection branch
$\mathrm{H_{FA1}}(z) = P_\mathrm{d}P_\mathrm{fi}z^{T_\mathrm{ca}}$, OC false identification branch
$\mathrm{H_{P1}}(z) = z^{T_\mathrm{p1}+T_\mathrm{s}}$, OC false identification recovery branch
$\mathrm{H_{NFA1}}(z) = (1 - P_\mathrm{d})z^{T_1+T_\mathrm{s}}$, OC missed detection branch
$\mathrm{H_{NFA2}}(z) = P_\mathrm{d}(1 - P_\mathrm{fi})P_\mathrm{fs}z^{T_\mathrm{ca}+T_3+T_\mathrm{s}}$, OC failed frame synchronisation branch
$\mathrm{H_M}(z) = \mathrm{H_{NFA1}}(z) + \mathrm{H_{NFA2}}(z) + \mathrm{H_{FA1}}(z)\mathrm{H_{P1}}(z)$, OC total missed detection branch
$\mathrm{H_{NFA0}}(z) = (1 - P_\mathrm{fd})z^{T_1+T_\mathrm{s}}$, VC correct detection
$\mathrm{H_{FA0}}(z) = P_\mathrm{fd}z^{T_1}$, VC false alarm branch
$\mathrm{H_{P0}}(z) = z^{T_\mathrm{p0}+T_\mathrm{s}}$, VC false alarm recovery branch
$\mathrm{H_0}(z) = (1 - P_\mathrm{fd})z^{T_1+T_\mathrm{s}} + P_\mathrm{fd}z^{T_1+T_\mathrm{p0}+T_\mathrm{s}}$, VC total branch,

where T_p0 and T_p1 are the penalty times after a false detection and identification, respectively, i.e. the time that is required in order to realise that a false detection/identification has occurred and resume the search. The penalty time appears in two branches, namely the false alarm recovery, T_p0, and the false identification recovery, T_p1, branch. This is because in both cases the terminal will finally realise that a wrong decision was made and it will resume the search. The penalty times depend on the configuration and architecture of the monitoring receiver so we cannot predict their values. Nevertheless, for computational purposes it is reasonable to model them as multiples of the duration of the step that follows the false alarm, i.e. $T_{p0} = k_0 T_2$ and $T_{p1} = k_1 T_3$. The gains $\mathrm{H_M}(z)$ (total missed detection branch) and $\mathrm{H_0}(z)$ (total VC branch) are used in order to transform the original flow graph [Figure 12.6(a)] to the simplified one [Figure 12.6(b)].

We assume total uncertainty about the position of the correct carrier which corresponds to a uniform a priori distribution, i.e. $\pi_j = 1/\nu$, $j = 1, 2, \ldots, n$. We also note that $\mathrm{H_0}(1) = 1$. Thus, from equations (12.3) and (12.4) we obtain:

$$\overline{T}_{\mathrm{csel}} = \frac{1}{\mathrm{H_D}(1)}\left[\mathrm{H_D}(1) + \mathrm{H_M}(1) + (\nu - 1)\mathrm{H_0}(1)\left(1 - \frac{\mathrm{H_D}(1)}{2}\right)\right], \tag{12.6}$$

where $\mathrm{H}'_x(1)$ is the value for $z = 1$ of the first derivative of the gain of the x branch. After substituting the branch gains in equation (12.5) we finally obtain:

$$\begin{aligned}\overline{T}_{\mathrm{csel}} = \frac{1}{P_{\mathrm{D}}(1 - P_{\mathrm{fs}})}\Bigg[&P_{\mathrm{D}}(1 - P_{\mathrm{fs}})(T_{\mathrm{ca}} + T_3) + (1 - P_{\mathrm{d}})(T_1 + T_{\mathrm{s}})\\ &+ P_{\mathrm{d}}P_{\mathrm{fd}}(T_{\mathrm{ca}} + T_{\mathrm{s}}T_{\mathrm{p1}}) + P_{\mathrm{d}}(1 - P_{\mathrm{fd}})P_{\mathrm{fs}}(T_{\mathrm{ca}} + T_3 + T_{\mathrm{s}})\\ &+ (\nu - 1)(T_1 + T_{\mathrm{s}} + P_{\mathrm{fd}}T_{\mathrm{p0}})\left(1 - \frac{P_{\mathrm{D}}(1 - P_{\mathrm{fs}})}{2}\right)\Bigg].\end{aligned} \tag{12.7}$$

Equation (12.6) gives the mean cell selection time for a single OC in ν possible carriers, as a function of the times and probabilities that were previously defined. Equation (12.6) is generic and independent of any particular RAT or carrier detection/identification algorithms. Different detection/identification algorithms will result in different probabilities (P_{FA}, P_{D}), while different RATs will result in different times (T_{ca}, T_3). Simulations were carried out in order to verify the validity of equation (12.6). The results show that theory and simulations are in excellent agreement (5% is the maximum difference between theory and simulations when $\nu = 12$; for $\nu = 124$ the results are identical). Figure 12.7 depicts the mean cell selection time as a function of the detection probability P_{d} for two sets of the false detection, false identification and failed frame synchronisation probabilities (P_{fd}, P_{fi}, P_{fs}). All the times are normalised to the energy detection time T_1. Equation (12.6) is valid if the energy detection and RAT identification are two independent procedures. For the SSA and DSA-c schemes the flow graph is different (Figure 12.8) because there cannot be a transition from the $\mathrm{H_1}$ to the false alarm state. The mean cell selection time for the SSA/DSA-c cases will be a function of the probabilities of overall detection, P_{Do}, and false alarm, P_{FAO}, as were defined above. The branch gains in this case are simpler:

$\mathrm{H_D}(z) = P_{\mathrm{Do}}(1 - P_{\mathrm{fs}})z^{T_{\mathrm{ca}}+T_3}$, OC correct detection branch
$\mathrm{H_{NFA1}}(z) = (1 - P_{\mathrm{Do}})z^{T_{\mathrm{ca}}+T_{\mathrm{s}}}$, OC missed detection branch
$\mathrm{H_{NFA2}}(z) = P_{\mathrm{Do}}P_{\mathrm{fs}}z^{T_{\mathrm{ca}}+T_3+T_{\mathrm{s}}}$, OC failed frame synchronisation branch
$\mathrm{H_M}(z) = \mathrm{H_{NFA1}}(z) + \mathrm{H_{NFA2}}(z)$, OC total missed detection branch
$\mathrm{H_{NFA}}(z) = (1 - P_{\mathrm{FAO}})z^{T_{\mathrm{ca}}+T_{\mathrm{s}}}$, VC no false alarm branch

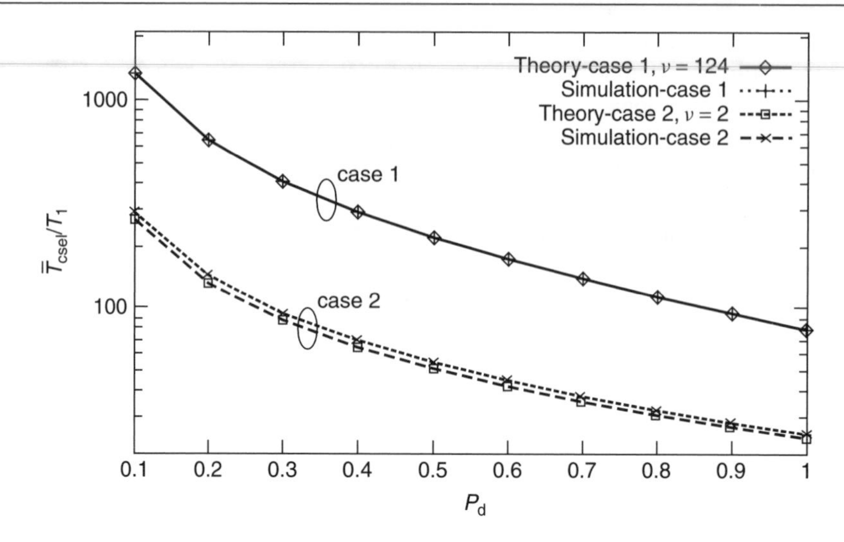

Figure 12.7 A comparison between theory and simulation results for two different sets of P_{fd}; P_{fi}; P_{fs}). For the simulations: $T_1 = 1$, $T_s = 0.01$, $T_2 = 2$, $T_2 = 4$ and $T_{p0} = T_{p1} = 10$

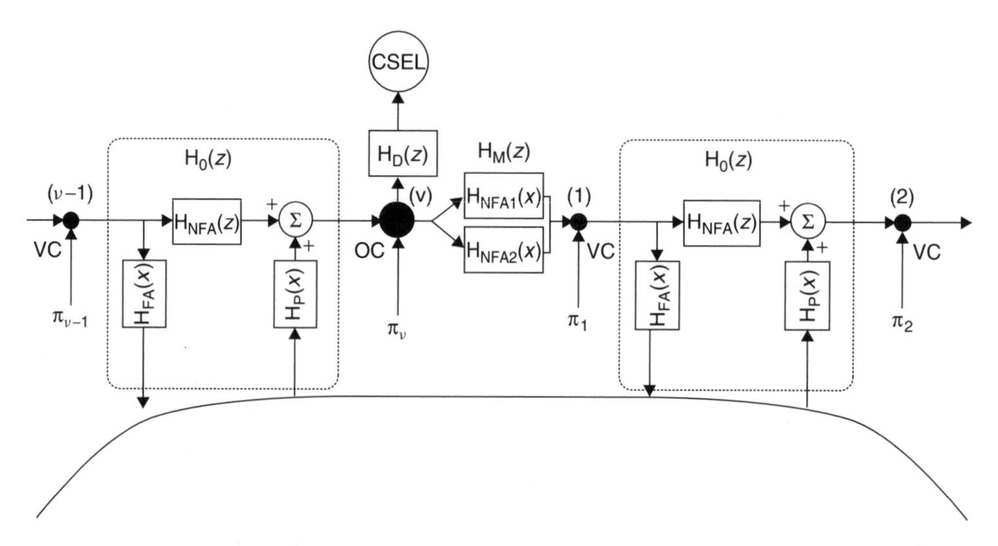

Figure 12.8 A portion of the flow graph diagram with an arbitrary a priori distribution (π_i) and for the SSA/DSA-c schemes

$H_{FA}(z) = P_{FAO} z^{T_{ca}}$, VC false alarm branch
$H_P(z) = z^{T_p + T_s}$, VC false alarm recovery branch
$H_0(z) = (1 - P_{FAO}) z^{T_{ca}+T_s} + P_{FAO} z^{T_{ca}+T_p+T_s}$, total H_0 branch.

The final (simplified) flow graph will be the same as the one in Figure 12.6(b). In this case, there is only one penalty time, T_p, which can be modelled as $T_p = kT_3$. After substituting the branch gains in equation (12.5) we obtain:

$$\overline{T}_{\text{csel}} = \frac{1}{P_{\text{DO}}(1 - P_{\text{fs}})} \Bigg[P_{\text{DO}}(1 - P_{\text{fs}})(T_{\text{ca}} + T_3) + (1 - P_{\text{DO}})(T_{\text{ca}} + T_{\text{s}}) \\ + P_{\text{DO}} P_{\text{fs}}(T_{\text{ca}} + T_3 + T_{\text{s}}) + (\nu - 1)(T_{\text{ca}} + T_{\text{s}} + P_{\text{FAO}} T_{\text{p}}) \left(1 - \frac{P_{\text{DO}}(1 - P_{\text{fs}})}{2}\right) \Bigg]. \tag{12.8}$$

12.2.3 A GSM/UMTS Based Example

In this section the GSM and UMTS cell selection procedure is modelled using the generic flow-chart of Figure 12.1. The purpose of this example is to demonstrate the applicability of our model in real cellular systems and to obtain a good estimation of the required cell selection time under various probabilities of detection/false alarm in a radio environment, where the terminal has no a priori information and help from the networks.

For our simulations we assumed a static spectrum allocation scheme based on this and the GSM-900 and UMTS-FDD systems. For the GSM we used a total number of carriers of $\nu = 124$ and for the UMTS-FDD, $\nu = 12$ UMTS (in reality, the number of possible UMTS carriers can be much larger due to the carrier raster [35, 36]). The mean cell selection time is derived for both systems. The worst-case mean time (one occupied carrier case) is calculated using the theory of the previous section and is compared with the mean cell selection time when there are more than one occupied carriers. The latter is derived using simulations. In the current GSM/UMTS networks the first step of the cell selection involves the detection of a number of carriers. The second step uses the strongest carrier that was found during the first step. This comparison of a number of carriers in terms of the received power is not modelled in our simulations. We assume that the second step uses the first carrier that has been found. However, this comparison can be included in our generic model by increasing T_1 and also taking into account that the detection probability would be different (increased) compared to the case that Step 2 is initiated immediately after the detection of the first carrier. In order to use the suggested model we must show the correspondence between the actual GSM/UMTS cell selection steps and the steps of the suggested generic procedure. In the next paragraph we will briefly describe the cell selection procedure in GSM-900 and UMTS-FDD in order to indicate the correspondence.

The GSM cell selection (or initial synchronisation) is described in (refs [26], [36], [38] and [39] and can be modelled as a three-stage procedure:

- *GSM-Step* 1*:* the terminal performs power measurements to get a list of *Nfreq1* frequencies with the highest received power. The duration of this step can be modeled as:

$$T_{\text{GSM1}} = \frac{Nfreq1 \cdot NmeasPerFreq \cdot FrameDuration}{NmeasPerFrame}, \tag{12.9}$$

where *Nfreq1* is the number of scanned carriers (for our simulations we use *Nfreq1* = 1 because we do not model the comparison campaign), *NmeasPerFreq* is the number of measurements per carrier to average the fading effect (5 is the number appearing in [37]), *FrameDuration* is the TDMA frame duration = 60/13 ms (8 GSM slots) and

NmeasPerFrame is the number of measurements per TDMA frame which depends on the terminal capabilities. We assume that $NmeasPerFrame = 5$.

- *GSM-Step* 2*:* Frequency Correction Bursts (FCB) detection. The terminal tunes to the frequency with the highest received power (from the previous step) until at least *Nfcb* FCB have been transmitted. The duration of this stage is modelled as:

$$T_{\mathrm{GSM2}} = Nfreq1 \cdot Nfcb \cdot 11 \cdot FrameDuration, \tag{12.10}$$

  where *FrameDuration* is the maximal timing distance between two FCB frames. For our simulations we used $Nfcb = 1$.
- *GSM-Step* 3*:* Synchronisation Burst (SB) detection. This step is not time consuming as it is often interleaved with FCB detection knowing that the SB frame follows the FCB frame.

The cell search procedure in UMTS-FDD is a three stage procedure whereby the UE determines the existence of a base station (Node-B), acquires frame synchronisation and identifies the code group to which the code used by the Node-B belongs, and finally identifies the scrambling code itself. A detailed description of the UMTS synchronisation procedure can be found in refs [9], [39]–[41]. A brief description will be given here:

- *UMTS-Step* 1*:* Slot synchronisation. The UE makes use of the primary synchronisation channel (primary SCH) in order to determine whether a Node-B is located in its locality. Typically, this is achieved by accumulating incoherently successive outputs of the primary synchronisation code correlator. The choice of the duration of the three UMTS steps in order to minimise the total acquisition time is not simple and does not have a unique, universal solution. For our simulations we used the values that are suggested in ref. [40] (for all three steps) because they are based on experimental data and not only simulations. Nevertheless, these values are just indicative and can vary in the real UMTS networks. The duration of the first step is $T_{\mathrm{UMTS1}} = 40$ ms, i.e. four UMTS frames.
- *UMTS-Step* 2*:* Frame and code-group synchronisation. During the second step, the UE uses the sSCH (secondary synchronisation channel) to acquire frame synchronisation and identify the code group of the base station found in the first step. This is done by correlating the received signal at the positions of the secondary code with all possible secondary synchronisation codes and identifying the maximum correlation value. Since the cyclic shifts of the sequences are unique the code group as well as the frame synchronisation is determined. The duration of this step is $T_{\mathrm{UMTS2}} = 30$ ms, i.e. three UMTS frames.
- *UMTS-Step* 3*:* Scrambling code identification. During this last step of the initial cell-search procedure, the UE determines the exact scrambling code used by the found base station. At this point, the common pilot channel (CPICH) is used and the scrambling code is identified through symbol-by-symbol correlation over the CPICH with all the codes of the code group identified in the second step. In order to reduce the probability of wrong/false acquisition, due to channel effects, the correlations have to be performed over a number of symbols. After the scrambling code has been acquired, the UE can read the broadcast channel (BCH) where system's information is available. The duration of this step is $T_{\mathrm{UMTS3}} = 10$ ms, i.e. one UMTS frame.

During the first cell selection step of both systems the carrier is discovered. Thus, this step is equivalent to the generic Step 1, i.e. the energy detection. The generic Step 2 (RAT identification) is not applicable here as both systems use a static spectrum allocation scheme. The GSM and UMTS frame synchronisation is acquired in both systems with the co-operation of their second and third step. Thus, the third generic step (frame synchronisation) is equivalent to the GSM/UMTS-Steps 2 and 3. Finally, the switching speed of the frequency synthesiser was set at $T_s = 100\ \mu s$ for both RATs. The correspondence between the GSM/UMTS and the generic steps and also their duration is summarised in Table 12.2.

12.2.4 Results and Discussion

For our simulations we used the values of Table 12.2 while the penalty time was set to $T_p = T_3 = 50.769$ ms for the GSM-900 and $T_p = T_3 = 20$ ms (i.e. two UMTS frames) for the UMTS-FDD case. The probabilities of overall detection, false alarm and failed frame synchronisation depend on the signal-to-noise ratio (SNR), channel model, the receiver architecture and parameters and also the times (T_1, T_2, T_3). They are also related to each other (e.g. in CDMA systems the lower is the probability of false alarm, the lower the detection probability would be). Nevertheless, for our simulations these probabilities were used as self-standing parameters and were set to arbitrary values. This point will be discussed further when necessary.

Figures 12.9 and 12.10 depict the mean cell selection time as a function of the overall detection probability, P_{Do}, for the GSM-900 and UMTS-FDD parameters, respectively, and different number of occupied carriers. The mean cell selection time is significantly improved when there are many OCs compared with the worst case (single OC). The improvement rate, though, drops rapidly as the number of OCs increases. In Figures 12.9 and 12.10 the OCs are uniformly distributed in the available spectrum. If most of the OCs are concentrated in a particular spectral area, then the mean cell selection time can be improved or increased depending on the starting point of the spectrum scan of the terminal. More precisely, if the spectral areas of large OC density coincide with the frequencies that the terminal examines first, then the cell selection time would be shorter. On the contrary, if the terminal starts the spectrum scan away from the large OC density areas the cell selection time will be larger. This is depicted in Figure 12.11. The terminal must scan the GSM-900 downlink spectrum [935.2, 959.8] MHz starting from the 935.2 MHz edge. We assumed 10 occupied carriers; when the OCs are concentrated

Table 12.2 GSM/UMTS cell selection equivalent stages and durations

	GSM-900	UMTS-FDD
T_s	100 μs	100 μs
Generic-step 1 equivalent duration	GSM-Step 1 $T_{GSM1} = 4.615$ ms	UMTS-Step 1 $T_{UMTS1} = 40$ ms
Generic-step 3 equivalent duration	GSM-Steps 2, 3 $T_{GSM2} = 50.77$ ms	UMTS-Steps 2, 3 $T_{UMTS2} + T_{UMTS3} = 40$ ms

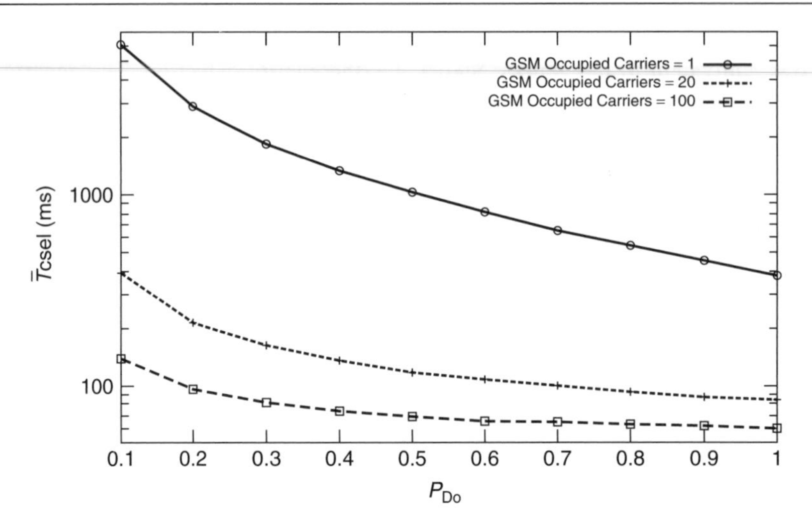

Figure 12.9 The mean cell selection time for three different OCs. The total number of carriers is $N = 124$. The duration of the steps is based on the GSM-900 values of Table 12.2. For the simulations we used ($P_{\text{FAO}} = 0.01$, $P_{\text{fs}} = 0.001$)

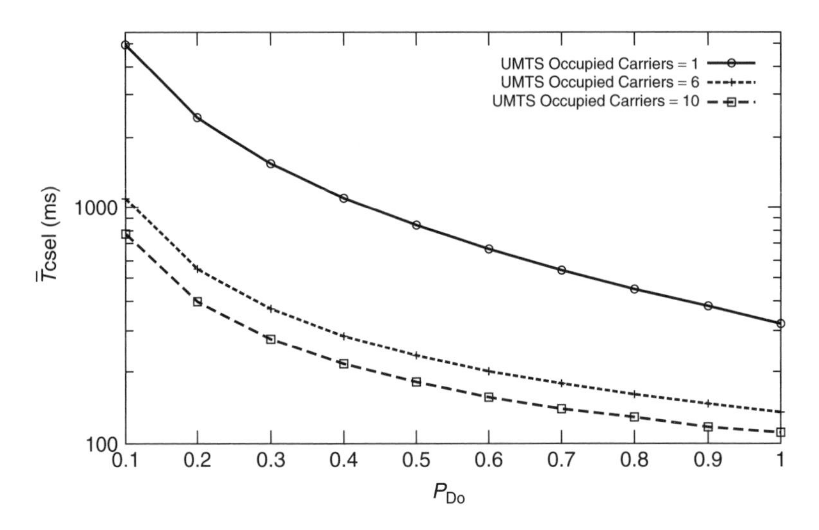

Figure 12.10 The mean cell selection time for three different OCs. The total number of carriers is $N = 12$. The duration of the steps is based on the UMTS-FDD values of Table 12.2. For the simulations we used ($P_{\text{FAO}} = 0.01$, $P_{\text{fs}} = 0.001$)

towards the 959.8 MHz edge of the spectrum (for our simulations we used the last 30 GSM carriers in [954.2, 959.8]) the cell selection time is significantly larger than when the carriers are uniformly distributed. When the carriers are closer to the 935.2 MHz edge (within the first 30 carriers in [935.2, 941.2]), the mean cell selection time is well below all other cases. These results show that it is very important for the terminal to have valid a priori information about the position of the available carriers in order to reduce the cell

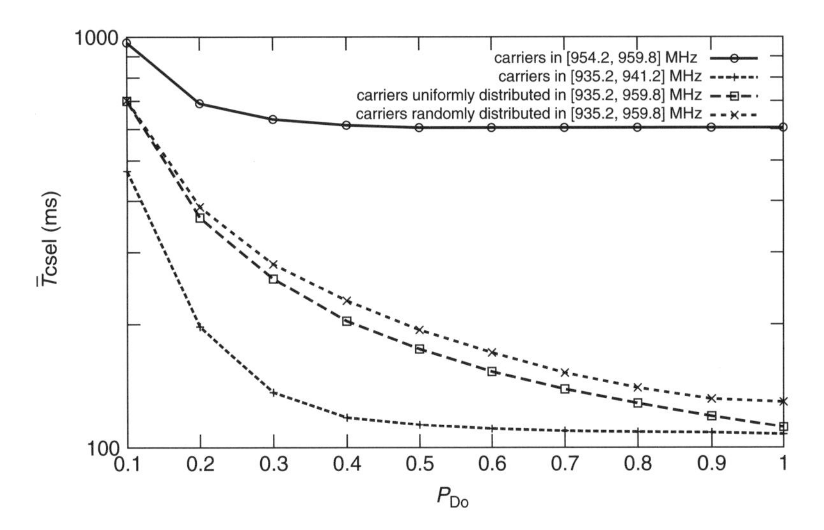

Figure 12.11 The effect of the carriers' distribution on the mean cell selection time. For the simulations we used ($P_{FAO} = 0.01$, $P_{fs} = 0.001$) and we assumed 10 OCs

selection time (and battery power drain). We also present the cell selection time for a totally random distribution of the carriers and we note that the mean cell selection time is very close to the case when the carriers are uniformly distributed.

The probability P_{Do} depends on the SNR and the duration T_1 and also certain parameters of the receiver which also affect the probabilities P_{FAO} and P_{fs}. Since we want to maintain this work independent of any channel conditions and specific receiver architectures we used P_{Do} as an independent variable. Given that (P_{FAO}, P_{fs}) and the times (T_1, T_3) are constant, one can view the detection probability as a metric for the SNR. If this framework is used to evaluate the cell search performance of a receiver in a specific RAT, then the relations of all the probabilities to the channel model, receiver architecture and times (T_1, T_2, T_3) must be taken into account (see, for example, refs [42] and [43]). In general, the probability of detection is proportional to the SNR and duration of the step. The duration of each step should be long enough to result in good detection probability but must not result in excessive cell selection times. It is also limited by the channel fading effects.

Figure 12.12 depicts the mean cell selection time as a function of the failed frame synchronisation probability for two different false alarm probabilities. The effect of a failed frame synchronisation on the cell selection time becomes important for $P_{fs} > 10^{-2}$, while a false alarm seems to have the same effects for all the probabilities. Obviously one expects that the effect of the false alarm will be proportional to the penalty time T_p. The probabilities of false alarm and failed frame synchronisation describe different steps and are expected to be uncorrelated.

A novel generic framework that describes the cell selection procedure in cellular networks was presented. Three steps, namely energy detection, RAT identification and frame synchronisation, are defined. Each step is characterised by its duration and the number of probabilities. The framework is independent of channel models, receiver architectures and spectrum allocation algorithms. Simple equations for the worst-case mean cell selection

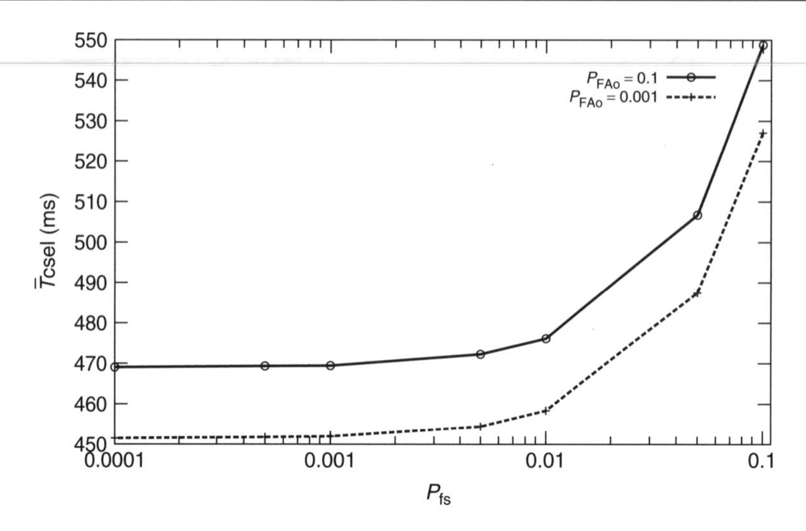

Figure 12.12 The mean cell selection time as a function of the failed frame synchronisation probability for two different false alarm probabilities. The duration of the steps is based on the UMTS-FDD values of table 12.2. The number of OCs is 1 and $P_{do} = 0.8$

times were derived using the flow graph approach for both the DSA and SSA/DSA-c spectrum allocation schemes. Using this framework the mean cell selection time for GSM-900 and UMTS-FDD was calculated under various probabilities. The improvement in the cell selection time offered by valid prior information has also been demonstrated. There are good indications that the framework can give accurate results if the correct values of times (T_s, T_1, T_2, T_3) are used. The framework can also be used in order to evaluate a receiver in terms of the cell selection procedure. In this case, the exact relations between the times and probabilities that were defined and the receiver parameters and channel models must be calculated.

References

[1] J. Mitola, 'The software radio architecture', *IEEE Communications Magazine*, **May**, 26–38, 1995.

[2] *ETSI GSM 03.22 version 7.3.0*, 1998, 'Functions related to mobile station in idle mode and group receive mode', *http://www.etsi.org*

[3] *3G TS 25.304 version 3.2.0*, 1999, 'UE procedures in idle mode and procedures for cell reselection in connected mode', *http://www.3gpp.org*

[4] *3G TS 23.122 version 3.2.0*, 1999, 'NAS functions related to mobile station (MS) in idle mode', *http://www.3gpp.org*

[5] J. Faroughi-Esfahani and G. Vardoulias, 'Locating and eliminating rogue software-reconfigurable terminals from networks', *3G2001 Mobile Communication Technologies Conference*, March, London, 2001.

[6] *3G TR 21.910 version 3.0.0*: 'Multi-mode UE issues', *http://www.3gpp.org*

[7] *ETSI GSM 05.08 version 7.3.0*, 1998, 'Radio subsystem link control', *http://www.etsi.org*

[8] *3G TS 25.214 version 3.2.0*, 2000, 'Physical layer procedures (FDD)', *http://www.3gpp.org*

[9] S. Kourtis, 'The initial synchronisation procedure in UMTS W-CDMA', *Proceedings of the 4th IEEE Multiaccess Mobility and Teletraffic Conference*, October, Venice, Italy, pp. 97–106, 1999.

[10] 'Consideration of a global pilot mechanism', *Tdoc SMG 879/97*, October, Budapest, Hungary, 1997.

[11] CEPT ERC Task Group 1 on UMTS, 'Liaison Statement to ETSI SMG, WRC-99 agenda item on global radio control channel', November, 1997.

[12] ITU Radiocommunication Study Groups, 'Discussion document on the global radio control channel', *Document 8-1/xx-E; TG1(98)55*, April, 1998.
[13] ITU Radiocommunication Study Groups, 'Proposal for a global radio control channel', *Document 8-1/xx-E, Draft 1.1, 3/4/98.*
[14] CEPT ERC TG1, 'Logical radio control channel for IMT2000', November, Helsinki, Finland, 1997.
[15] 'IMT 2000 control channels', Tdoc SMG 618/97, Nokia, August, 1997.
[16] 'Global pilot mechanism', Tdoc SMG2 UMTS 52/97, Lulea, April, 1997.
[17] C. Noblet and A.H. Aghvami, 'Software radio, the cutting edge of wireless communications', *IEE Colloqium: Personal Communications in the 21st Century*, 5–6 February, London, 1998.
[18] D. Hunold, A. Barreto, G. Fettweis and M. Meeking, 'Concept for universal access and connectivity in mobile radio networks', *PIMRC'2000*, London.
[19] *3G TS 22.129 version 3.2.0*, 1999, 'Handover requirements between UMTS and GSM or other radio systems', *http://www.3gpp.org*
[20] *3G TS 25.212*, 'Multiplexing and channel coding (FDD)', *http://www.3gpp.org*
[21] F. Vandermeulen, F. Steegmans, B. Vermeulen and S. Vermeulen, 'Dynamically configurable protocol stacks', *Multimedia and Expo, 2000. ICME 2000. IEEE International Conference*, Vol. 3, pp. 1391–1394, 2000.
[22] A. Fieger and M. Zitterbart, 'Transport protocols over wireless links', *Proceedings of the 2nd IEEE Symposium on Computers and Communications*, pp. 456–460, 1997.
[23] M. Zitterbart, B. Stiller and A. Tantawy, 'A model for flexible high-performance communication subsystem', *IEEE Journal on Selected Areas in Communications*, **11**(4), 507–518, 1993.
[24] W.W. Lu, 'Compact multidimensional broadband wireless: the convergence of wireless mobile and access', *IEEE Communications Magazine*, November, 119–123, 2000.
[25] M. Beach, D. Bourse, M. Dillinger, N. Drew, T. Farnham, J. Faroughi, R. Navarro, G. Vardoulias and T. Wiebke, 'Reconfigurable terminals beyond 3G: boot protocol', *Wireless World Research Forum – WG3-SDR. Available*: *http://www.wireless-world-research.org/*
[26] ETSI-GSM Technical Specifications, 'GSM 03.22 version 7.3.0 functions related to mobile station in idle mode and group receive mode' [online], 1998. *Available: http://www.etsi.org*
[27] E.E. Azzouz and A.K. Nandi, 'Procedure for automatic recognition of analogue and digital modulation', *IEE Procedures in Communication*, **143**(5), 259–266, 1996.
[28] H. Liu and K.C. Ho, 'Identification of CDMA signal and GSM signal using the wavelet transform', *Proceedings of Milcom'99*, **1**, 427–431, 1999.
[29] Z.S. Huse and S.S. Soliman, 'Signal classification using statistical moments', *IEEE Transactions in Communications*, **40**(5), 908–916, 1992.
[30] UMTS Technical Specifications, '3G TS 25.304 version 3.2.0 UE procedures in idle mode and procedures for cell reselection in connected mode' [online], 1999. Available: *http://www.3gpp.org*
[31] S.J. Mason, 'Feedback theory – some properties of signal flow graphs', *Proceedings of the IRE*, **41**(9), 1144–1156, 1953.
[32] S.J. Mason, 'Feedback theory – further properties of signal flow graphs', *Proceedings of the IRE*, **44**(7), 920–926, 1956.
[33] A. Polydoros and C.L. Weber, 'A unified approach to serial search spread spectrum code acquisition – Part I: general theory', *IEEE Transactions in Communications*, **32**(5), 542–549, 1984.
[34] V.M. Jovanovic, 'Analysis of strategies for serial search spread spectrum code acquisition – direct approach', *IEEE Transactions in Communications*, **36**(11), 1208–1220, 1988.
[35] UMTS Technical Specifications, '3G TS 25.101 version 3.2.2 UE radio transmission and reception (FDD)' [online], 2000 Available: *http://www.3gpp.org*
[36] Siemens 3GPP contribution, 'Definition of channel raster', 3GPP, TSG-RAN Working Group 4 (Radio) Meeting No. 3, 29–31 March, Tokyo, Japan, 1999.
[37] ETSI-GSM Technical Specifications, 'GSM 05.08 version 7.3.0 radio subsystem link control' [online], Available: *http://www.etsi.org*
[38] A. Mehrotra and L. Golding, 'Mobility and security management in the GSM system and some proposed future improvements', *Proceedings of the IEEE*, **86**(7), 1480–1497, 1998.
[39] Y.E. Wang and T. Ottosson, 'Cell search in W-CDMA', *IEEE Journal on Selected Areas in Communications*, **18**(8), 1470–1482, 2000.
[40] K. Higuchi, Y. Hanada, M. Sawahashi and F. Adachi, 'Experimental evaluation of 3-step cell search method in W-CDMA mobile radio', *Proceedings of the VTC 2000 – Spring*, Tokyo, 303–307, 2000.

[41] UMTS Technical Specifications, '3G TS 25.214 version 3.2.0. Physical layer procedures – FDD' [online], 2000, Available: *http://www.3gpp.org*

[42] E. Conte, A.D. Maio and C. Galdi, 'Signal detection in compound-Gaussian noise: Neyman-Pearson and CFAR detectors', *IEEE Transactions on Signal Processing*, **48**(2), 419–428, February, 2000.

[43] J.H. Iinatti, 'On the threshold setting principles in code acquisition of DS-SS signals', *IEEE Journal on Selected Areas in Communications*, **18**(1), 62–72, January, 2000.

Part V

Software and Hardware Reconfiguration

13

Reconfiguration of the Network Elements

Gyula Rabai and Sandor Imre
MTA, Budapest, Hungary

13.1 Introduction

This chapter is intended to provide fundamental technical information on the reconfiguration of network elements in Software Radio (SWR) systems, and Reconfigurable Radio Networks (RRNs) [1–4]. Smart network elements have the potential to revolutionise the field of wireless communication and are capable of providing adaptable services to end-users. To illustrate this, an overview of the reconfiguration technology used in the base stations and the mobile terminals is given here. During this overview we point out how it is possible to increase the IT effectiveness of these network elements, and why abstract modelling of the hardware is necessary. In addition, the role of local intelligence, and how it affects the performance of the systems, is explained.

The main goal of this chapter is to present the essential aspects and guidelines required to understand how reconfigurable network elements are working. Through the examination of the digital building blocks and the various hardware topologies, one can get an insight into how optimisation algorithms modify the behaviour of these elements. Various algorithms, which are applied in different scenarios, and the parameters involved in the optimisation of the reconfiguration of a mobile terminal and a wireless base station, are explained.

13.1.1 Reconfiguration of Base Stations and Mobile Terminals

The hardware of base stations and mobile terminals in a SWR network are built using reconfigurable hardware devices such as Digital Signal Processors (DSPs), Field Programmable Gate Arrays (FPGAs), digital interconnects and reconfigurable buses. These digital building blocks are partially configured at the time of design using complex design methodologies, e.g. platform-based design, and their configuration is changed from time to time during operation. When the configuration of the digital hardware is changed during

Software Defined Radio: Architectures, Systems and Functions. Edited by M. Dillinger, K. Madani and N. Alonistioti
 ISBN: 0-470-85164-3

operation, we say that reconfiguration is performed. The level of reconfiguration performed ranges from small parameter adjustments to changing the entire protocol stack. The trigger for a reconfiguration can be a local event or a remote network management decision.

As an example, one can think of a reconfigurable mobile terminal that changes from GSM into a CDMA IS-95 air interface standard. In this case some new software must be installed on the reconfigurable processors of the mobile terminal to support communication in the new environment. This includes new modulation, error detection and channel coding algorithms. If the installation of the new software (reconfiguration) is performed in a seamless way, then seamless roaming is possible. Another example is a GSM base station that detects an UTRA handset moving into its coverage area. In this case the base station can install an UTRA protocol processing chain in order to serve the handset.

To be able to perform such reconfigurations, the base stations and the mobile terminals must be equipped with special configuration management software, capable of adjusting the configuration of the hardware using different methods. First, the software implements a handle that makes it possible to identify each hardware device. Second, it defines a map to identify the hardware topology. The software also maintains a list of the different processing functions that could be set up on the hardware. With these tools and with logic built into the configuration software, runtime configuration management can be performed efficiently.

The next generation of network elements is expected to see a large increase in the data bit rate, a decrease in power consumption, and multiple protocol support with programmable parameters for multi-mode, multi-standard operation. This is further complicated by many variations within one single standard. For example, in GSM Phase 2+ the channel encoding schemes vary in many ways. For different modes different frame sizes are used with modified convolution polynomials and interleaving block sizes. The flexibility offered by a reconfigurable network element is therefore crucial if all mobile application requirements are to be fulfilled in the future. To improve the IT effectiveness of a radio terminal or a radio base station there are two main options:

1. One can introduce a new concept in hardware design, which is ready to adjust to the needs of future devices. This concept is based on the abstract modelling of the digital building blocks and the physical hardware topology. The main idea is to classify the digital building blocks and to propose several alternative configurations for the same task.
2. Another option is to introduce local intelligence decisions dynamically made by the network elements. These decisions use historical data and take into account several aspects to select a suitable configuration for the specific scenario. These are represented by algorithms, which are applied simultaneously most of the time. Since the decisions are based on theoretically proven algorithms, the probability that a well-performing configuration is created using this approach is high.

13.1.2 Abstract Modelling of Reconfigurable Devices

In designing a reconfigurable network, a layered approach in which the hardware and the operating software are organised into layers has many advantages. Each layer has specific functionality and provides services for the upper layers through a well-defined interface. The number of layers used in this approach can vary. We recommend that at

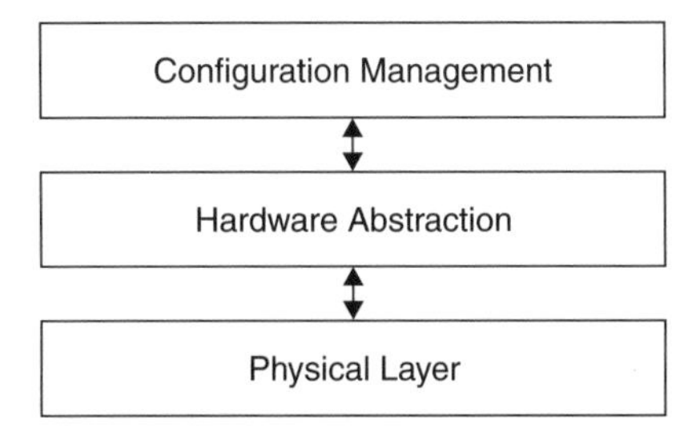

Figure 13.1 Layered structure for reconfiguring network elements

least three layers should be introduced, as presented in Figure 13.1. First, a physical layer is needed that makes it possible to connect the hardware devices to each other. A hardware abstraction layer is then required that provides a similar interface to operate each different device. Finally, a hardware management layer is necessary to manage the underlying hardware in an optimal way. Often the hardware management layer is referred to as the Resource Controller (RSC). From the controller's point of view, the most important layer of the platform is the hardware abstraction layer. This gives the different hardware devices a handle and makes it possible to support hardware configuration management at a higher level.

To be able to implement hardware abstraction, several design-time procedures must be carried out at runtime. When the configuration of the hardware is altered through this layer, then synchronisation, segmentation and reassembly are performed to keep the system running. It is possible that the different hardware devices cannot be reconfigured simultaneously. In this case, the hardware abstraction layer must perform queue management, channel assignment and scheduling in order to be able to present the different hardware elements as separate units. If more reconfiguration requests are handled simultaneously, then the hardware abstraction layer must work in multi-threaded mode and must be able to communicate with the upper layer asynchronously. Later in this chapter it is shown that choosing an appropriate hardware topology can help in creating this layer, and configuration can be managed from the upper layers by routing and configuration management strategies.

13.1.3 The Role of Local Intelligence in Reconfiguration

As mentioned earlier, making intelligent configuration management decisions at runtime can increase the IT effectiveness of a SWR network element. These decisions are based on information gathered during the operation, and are targeted at permitting or blocking reconfiguration attempts and at creating a configuration to comply with predefined aspects. In order to make correct decisions, the strengths and weaknesses of the network element need to be identified, and sufficient historical data are essential. In addition, the configuration management algorithms must take into account all the different aspects that affect the desired behaviour of the network element.

The information gathered to support locally made decisions contains data about hardware devices, usage patterns and the resource requirements of certain functionalities. If a configuration fits a desired usage pattern, then it is rated higher, but if it interferes with some configuration aspects, then it is rated lower. This rating mechanism is based on various reconfiguration strategies (reconfiguration based on comparison, resource recycling and flexible workload management) [2] and reconfiguration algorithms (hottest first, coldest first, maximum utilisation) [5]. The information generated can be used to make objective decisions to provide better functionality. For example, if a configuration attempt is not successful for any reason or a specific set-up has unwanted side-effects, such as decreased standby time in a handset, then with the modified ratings the probability that the same configuration will be installed in the future is going to be lower. On the other hand, if a certain configuration is performing well, then the intelligent configuration management mechanism will try to recreate the same configuration in a similar future situation.

One can conclude that smart network elements equipped with local intelligence make it possible to use the available resources of the mobile terminals and the base stations more effectively. The decisions made by a local intelligence unit are independent of the network management and are based on locally available information. The decisions are used in the mapping process, when the configuration platform installs functionality to the physical hardware devices. Local intelligence plays a vital part in the performance improvement of the network elements.

13.1.4 Performance Issues

In a mobile network, some services demand more resources than others, from the underlying hardware and from the communication infrastructure. If an application needs to deliver more information, then the communication network can be adjusted to assign more timeslots to the communication, thus allowing a wider bandwidth. This increased amount of data affects the network elements involved in the communication. In the mobile terminal and on the base station side additional processing capacity and additional internal connection resources are required in order to be assigned to the data flow; and the coding scheme must be adjusted to provide the desired quality of service. It is obvious that the performance of the mobile network relies heavily on the reconfigurable network elements. To be able to adapt to the changing performance requirements, the reconfigurable network elements need to have spare capacity available most of the time, or must be able to free allocated capacity when it is needed.

13.2 Classification and Rating of Reconfigurable Hardware

Performance and reconfiguration management decisions rely heavily on the information available about the different hardware devices of the reconfigurable network elements. This information partially comes from the classification of the hardware elements. Classification means that one can find out which digital hardware devices are fit for certain purposes, prior to operation. If this information is available, then the signal processing paths can be built more easily during the operation. In the following standards (DAMP, GSM, CDMA-IS95) the data flow in a processing channel consists of the parts [6] shown in Figure 13.2.

Each part performs data processing [7] and data modification in some way, and the modules following each other are connected in series. The data processing is handled

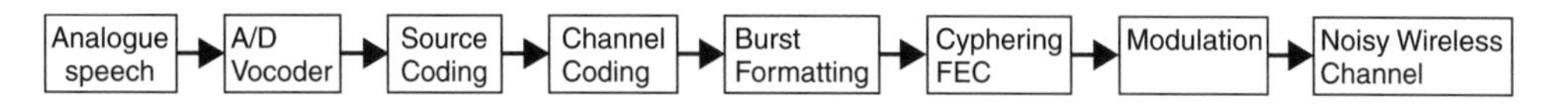

Figure 13.2 Data flow in speech transmission

by processing building blocks and the connections are managed by connection elements. These two main categories form the two major groups for device classification.

Taking a closer look at these categories from the software point of view, one can see that the processed data stream in SWR network elements typically employs block operations such as fast fourier transformations (FFTs), wavelet transformations, matrix multiplication, source coding, channel coding, scrambling and interleaving. These operations require different processing capacity and different input/output speed. In order to serve the stream, hardware devices with appropriate processing capacity and connections with sufficient capacity should be allocated to satisfy the demands. This means that information about device interconnection capacity is just as important as information about processing capacity.

13.2.1 Processing Elements

The processing elements are reconfigurable processors that modify on the data stream. Although these processing elements can be configured for different tasks, their internal architecture does not allow serving tasks at sufficient speed. In other words, different processors (e.g. DSPs, FPGAs, ASICs) have different characteristics and require optimal configuration for different tasks. For example, some can be better used for IF processing, some are more suitable for baseband processing, and some are faster for bitstream processing. The characteristics of the signal travelling through processors in these processing blocks are different. For example, the time between samples in the baseband are of the order of milliseconds to hundreds of microseconds and thus are better served with DSPs. In the IF stream, however, the time between samples is of the order of tens of microseconds to hundreds of nanoseconds, so an appropriate faster ASIC might be best fit for the job. In this scenario the input/output (I/O) data rates of the stream affect the decision on how the signal-processing path is installed on the reconfigurable hardware.

In general, there is more than one approach for reconfigurable processing elements classification. The classification can be done, for example, by identifying the processing capacity of the processor. In this case we assume that any functionality can be served by any processing unit, but at different speeds. Another approach for classifying the processors is to name the set of possible tasks they can serve. In this approach general-purpose microprocessors are the most suitable building blocks, because they can be used for all possible tasks, while task-specific circuits are located at the other end of the spectrum. The drawback with this approach is that using a general-purpose processor for all kinds of tasks (such as instruction fetching, decoding, execution, etc.) will result in higher overhead, and therefore the costs associated with running the processor. The best way is perhaps to assign a rating to every processing element for every possible task it can serve, as a basis for the maximum utilisation of resources.

13.2.2 Connection Elements

In a reconfigurable base station or a mobile terminal, devices used for connecting processing elements form a major part of the digital building blocks. These connections can be performed using buses, shared memories, IO channels, or any other connection elements. Each type of the different connection elements behaves differently. A shared memory, for example, does not require the two ends of the connection to be available at the same time; meanwhile a bus can only connect devices that are ready to transmit and receive information synchronously. It is possible to combine the functionality of connection devices to form a more efficient connection network. An example of this is shown in Figure 13.3, combining a communications bus with shared memory.

Classification of the connection elements can be done by different aspects, grouped by performance or by functionality. For classification purposes, it is advisable to treat the combined connection devices as a single connection element. To classify the connection elements, it is necessary to obtain information about how an individual connection performs in the reconfigurable hardware. The interconnect network of a SWR base station is a key component for providing flexibility. It enables the network element to create data paths dynamically by making it possible to define spatially programmed connections of processing elements. In this communication network a reconfiguration attempt initiated synchronously or asynchronously usually does not affect all the existing connections. When a new connection is created between two processing modules different aspects are taken into account. The primary aspect is that the connection device needs to be capable of providing the necessary bandwidth. Another aspect can be the power consumption. For example, the FPGA architectures are often trading flexibility against excess power consumption. For heterogeneous systems, some standard interconnect architectures are available. These architectures can be classified into two broad categories:

1. *Global interconnect networks.* These provide routes with identical costs for all connections between any pair of processing modules and functional entities.
2. *Hierarchical interconnect networks [8].* These are capable of exploiting the possibility of lower cost local connections and are capable of creating optimised routes.

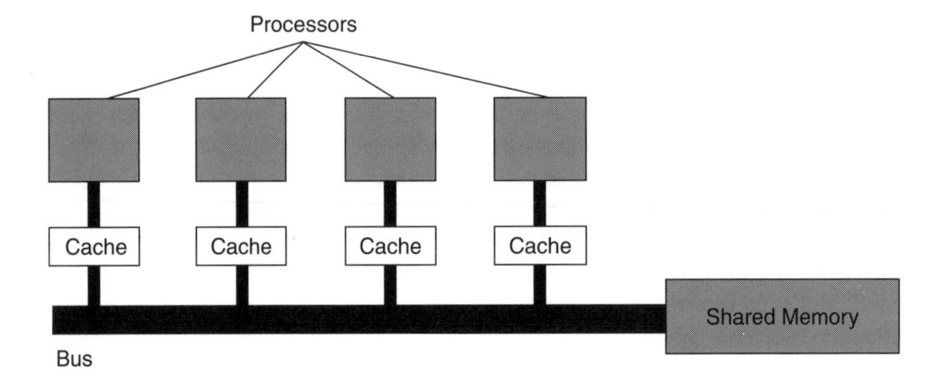

Figure 13.3 A Communications bus combined with shared memory

13.2.3 Global Interconnect Networks

In global interconnect networks we assume that any route between two processing elements has the same cost. Because of this, the primary (physical) hardware topology is irrelevant, since it does not matter which connection is used, and we only take into account the number of possible connections between two processing modules. In global interconnect networks a logical hardware topology is used. This logical network defines all the possible routes between the output ports and input ports of the processing entities. The optimal topology is made up of connection entities that make it possible to connect any two processing elements at any time. An example of this could be a bus shared between all the functional blocks. This shared bus makes it possible to route the data around the blocks in the desired way. Some typical global interconnect networks are discussed below.

13.2.3.1 Crossbar Interconnect Architecture

The crossbar interconnect architecture [9] allows simultaneous connections from any input port to any output port. The logical topology of such a network can be seen. This topology, shown in Figure 13.4, provides full connection flexibility but suffers from high costs. To implement such an architecture there are several possibilities. One technique is to have a global bus assigned to each input and then every output is connected to this bus through a switch.

13.2.3.2 Multi-Bus Architecture

As illustrated in Figure 13.5, in a multi-bus architecture the number of buses is reduced compared with the crossbar network. The maximum number of simultaneous interconnections in this case equals the number of buses.

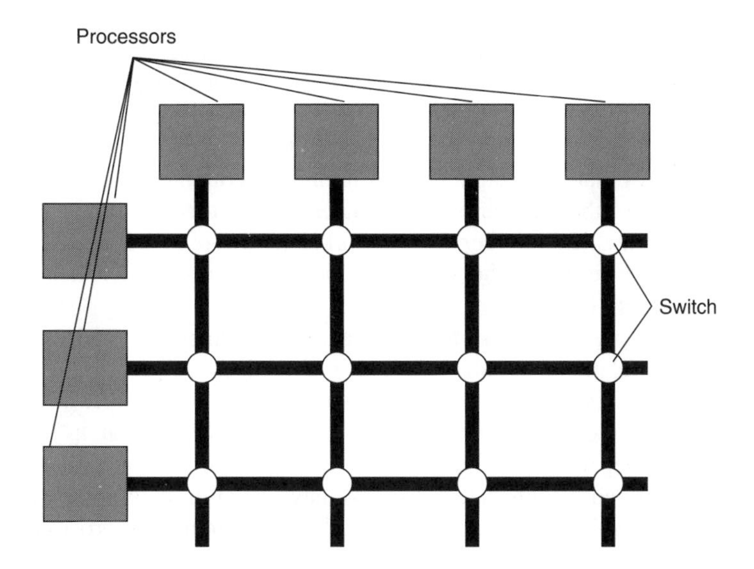

Figure 13.4 Crossbar interconnect architecture

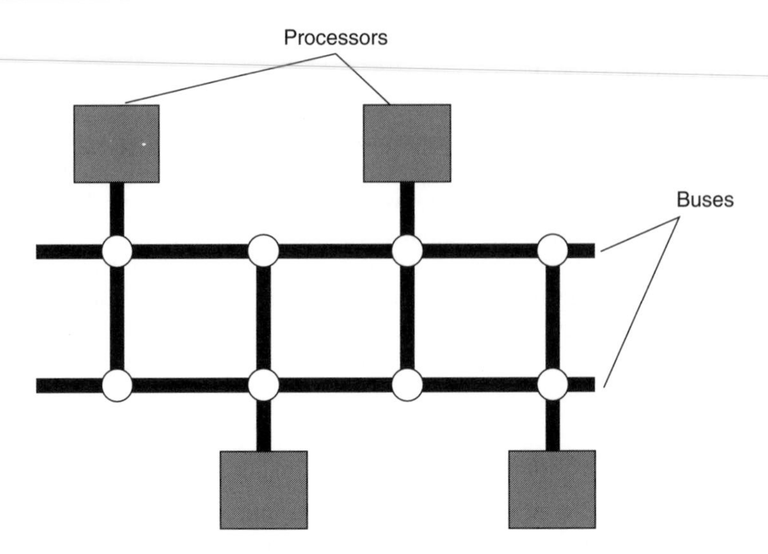

Figure 13.5 Multi-bus interconnect architecture

13.2.3.3 Multi-Stage Interconnect Architecture

Multi-stage interconnect networks make it possible to reduce the number of switches. These networks require complicated interconnect patterns between the switches; thus, implementation of them is not straightforward. A possible solution is to have a centralised multi-stage switch box with its inputs and outputs connected to the ports of the modules.

13.2.4 Hierarchical Interconnect Networks

In a hierarchical interconnect network each connection is described with an associated cost. The cost of the connection is considered to be higher if a connection takes more resources or for any reason it is not optimal for the intended hardware configuration. It is also feasible that there may be no possible routes between two processing elements, in which case the cost of that connection route is infinite. Hierarchical interconnect networks are better at describing the real hardware architecture than their ideal counterparts.

To construct a hierarchical interconnect network, a hardware model needs to be defined. This hardware model describes the system in an abstract way and it helps to find the optimal signal processing paths.

The model of the network is shown in Figure 13.6 where the nodes represent the resources and their relationships are reflected in the edges. A processing resource and a connection resource is related (connected with an edge) if a software installed on the processing resource can communicate with another software through that connection resource. In this model it is illegal to connect two 'processing' nodes directly, but two connection resources can be connected. To make navigation in the graph easier the nodes representing the connection resources and the nodes representing the processing elements are often marked with different colours. The building blocks of the graph are labelled, and the information contained in these labels depends on the task. For resource finding,

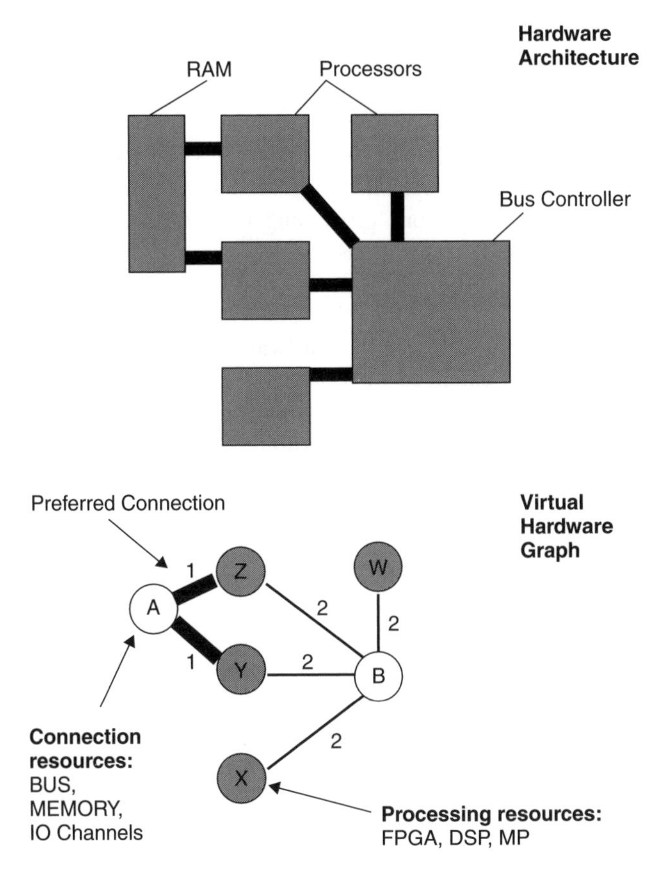

Figure 13.6 Hardware graph

the capacity information is used, and for resource selection purposes, the preference information is used. The information represented by the labels of the graph can change when the configuration of the hardware changes, or when circumstances change.

Figure 13.6 illustrates how the nodes of the graph can be assigned to processors and to connection resources. The nodes assigned to processing resources are labelled W, X, Y and Z. The nodes assigned to connection resources are labeled A and B. The edges illustrate which communication resource can be used for communication between the different processors. This graph illustrates the physical hardware architecture of a hierarchical interconnect network. It provides information about the capacity of the various resources and serves as a good starting point in the software mapping process. Connection resource A, for example, can only be used for communication between X and Y. The edges of the graph are labelled with numbers. These numbers help in selecting the preferred connection between the two resources. Connections with large number weights should not be used if other alternative lower weight routes are available.

On a typical reconfigurable hardware, processing elements can be used to serve more than one task. An FPGA can have multiple processes operating simultaneously and the shared memories or buses can serve many connections. Often it is needed to connect two software functions operating on the same processing resource. We call this an intra-device communication, and the connection between separate processors is called an inter-device connection. The intra-device connections sometimes are available using the internal resources of the processor (for example integrated memory), and sometimes require external resources. If the intra-device connections use the internal resources, then they are not represented by this graph. In this case different processing functions, operating in one device, are treated as one processing function and only define its external connections. If the intra-device connections can use external connection resources, then two edges are used to connect the processing device to the connection resource serving the communication.

To apply the above model to describe the hardware of reconfigurable network element, this model is required to be extended to be ready for intra-device connections that are handled internally in a processing unit. For this reason an additional layer is introduced. This additional layer contains a similar graph, with the difference that the graph represents the internal structure of the processing device. This internal structure is made up of smaller processing elements and internal connection resources. External connection resources are also present on this layer. These resources integrate the 'internal layer' into the upper external hardware architecture layer. A similar approach can be used to handle a large number of similar hardware units available in a reconfigurable base station. In a base station there are many boards containing several processing elements. When these boards are managed, they can be treated as a single processing unit and a bus (e.g. PCI bus) and can be treated as communication resources. Figure 13.7 shows the layered graph describing the hardware of the reconfigurable network elements.

When the signal-processing path is set up and decisions are made, the work is always done layer by layer. The preferred approach is bottom-up. In this case as many tasks as possible are installed on one processing resource using the graph located on the layer representing the processing resource. An upper layer is then used to connect the different functions installed on the different processing units.

13.2.5 Installing a New Configuration

The architecture graph represents the hardware map. When a new configuration is required on the network element's hardware, a new signal-processing path is required. The signal-processing path consists of a set of software functions installed on processing resources and a set of connections that are installed on connection resources. To install this processing chain one must navigate in the architecture graph. The starting point for this navigation is the wideband analogue-to-digital converter (ADC) for the receiver, and the wideband digital-to-analogue converter (DAC), for the transmitter. This ADC/DAC, connected to the RF front-end that is receiving or transmitting the signal, is the beginning of the digital signal processing path. The ADC/DAC is treated as a processing resource, even though it does not do too much processing on its own. It is, however, assigned to the processing devices class, since it is followed by a connection resource in the signal-processing chain. An architecture graph is illustrated in Figure 13.8, showing the available and allocated capacity for each resource. To find the processing path, one navigates from one resource to an other, and examines whether the given resource has sufficient capacity to serve the requirements.

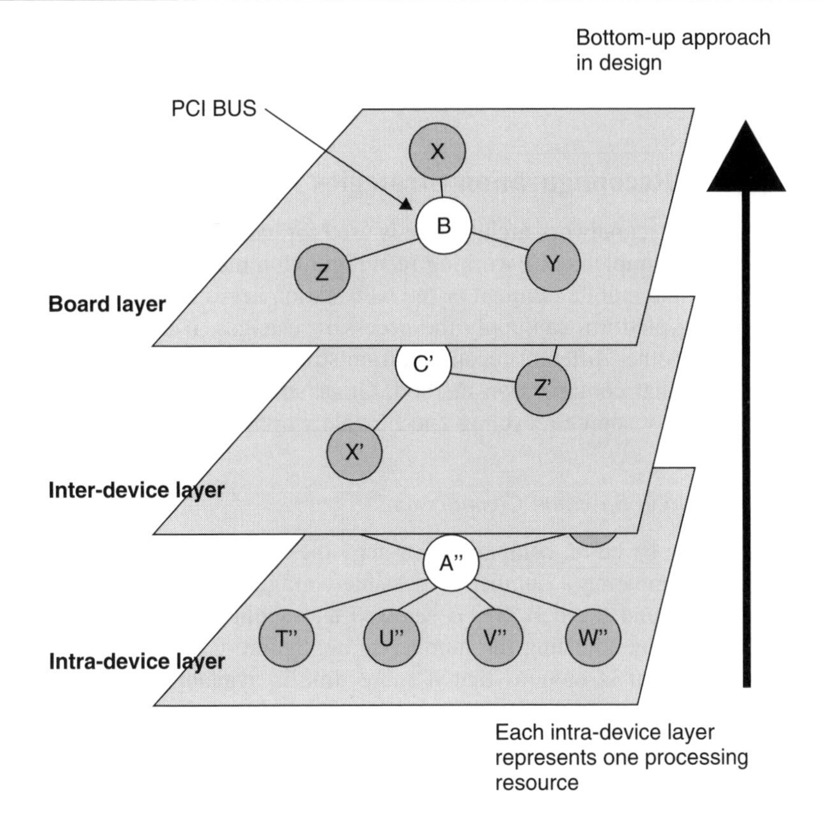

Figure 13.7 Layered hardware graph description for reconfigurable network elements

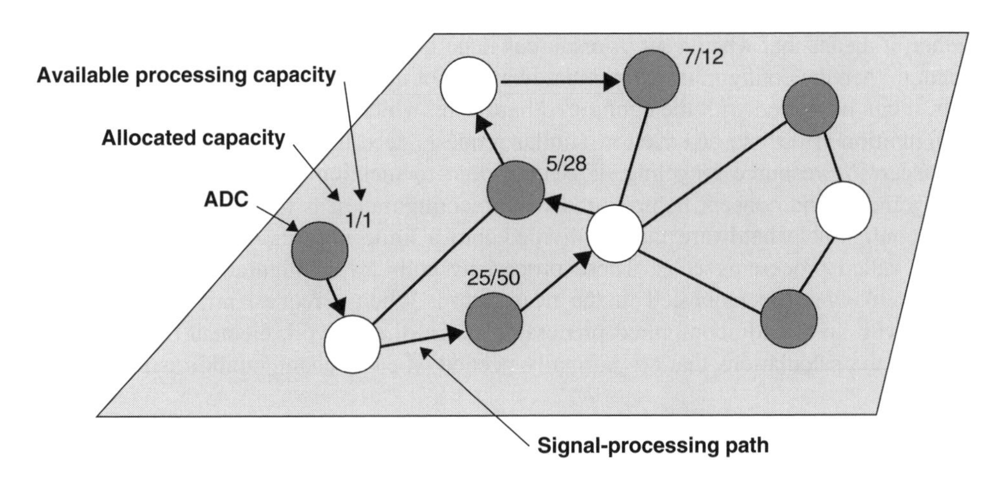

Figure 13.8 Reconfiguration using architecture graph

This path information is set up for connection and processing resources. For building the path, a decision is made at each resource to decide in which direction the signal should continue its way as part of a resource management strategy.

13.3 Applying Reconfiguration Strategies

Hierarchical interconnect network architecture is used for hardware abstraction. Using this model it is possible to implement a working reconfiguration management platform. When the configuration of the mobile terminal or the base station needs to be altered, the configuration management platform can apply the necessary changes. If this technique is used in combination with other different reconfiguration strategies, then it is possible to reach an efficient and optimal configuration method. Other strategies include: reconfiguration based on comparison, resource recycling and flexible workload management.

13.3.1 Reconfiguration Based on Comparison

In the reconfiguration based on comparison strategy the reconfiguration decision is based on examining and comparing a number of possible configurations. All possible configurations are compared and the best one is selected and applied. This type of strategy is highly scalable, since by adjusting the number of configurations examined, the decision time can be modified. It is obvious that if more time is available, then more possible configurations can be compared and a better decision can be made.

13.3.2 Resource Recycling

Resource recycling is a technology that allows multiple radio subscribers to share the same configured resources of a particular base station. The sharing of the resources does not mean that the resources are used for two different communications at the same time. Rather, it means that when a set of resources is no longer needed for a particular communication, then the configuration of the hardware is not modified. If a similar request arrives later, it can be served with the configured hardware, which already exists. This way much configuration effort can be saved, if similar requests need to be served frequently.

Concerning resource recycling, it is important to mention predictive reconfiguration management. The concept behind predictive reconfiguration is that a predefined area of the reconfigurable hardware map is divided into a finite number of smaller processing blocks called processing cells. These processing cells are configured to functionalities that are likely to be requested in the near future. When a request arrives that can be served with an already configured processing chain, the network element does not have to do all the calculations that are normally needed when a reconfiguration is performed.

13.3.3 Flexible Workload Management at the Physical Layer

When specific configurations are created in a reconfigurable network element, for example a base station, it is possible that over a period of time part of the hardware infrastructure experiences much greater usage than other parts of the hardware. If workload management is implemented, then it is possible to measure the load of certain parts of the hardware. If some parts are idle and others have more tasks, then load balancing can be performed. This way the overall performance of the base station can be improved, because no configuration processes will be delayed if the hardware has sufficient capacity somewhere else.

13.4 Optimised Reconfiguration

Previous studies have indicated that the locally made intelligent decisions in reconfigurations are most useful and can boost system performance. These decisions largely depend on the various aspects that should be taken into account. These aspects need to be treated simultaneously with variable weights. As circumstances change the resource management strategy must be updated, which means that the weight of the different aspects may change.

13.4.1 Optimisation Parameters and Algorithms

The maximum lifetime of the hardware is an important parameter in every system. If reconfiguration leads to wear and tear, then the frequently reconfigured devices could experience a lot of wear and tear and could fail to operate sooner than expected. In this case an algorithm for even utilisation is required. This algorithm extends the lifetime of the system by making sure that devices are used at pretty much the same frequency of utilisation.

Even utilisation is a key issue if fault tolerance is important. This is because if services are evenly spread over a large field of hardware, then only a few services will be affected if part of the hardware fails. There are several algorithms that accomplish even utilisation, such as random pick algorithms. Random pick algorithms create even utilisation, but they can waste valuable hardware resources. The performance of the network element can become very limited due to a badly chosen random pick algorithm. Maximum utilisation should always be taken into account, when only a limited number of resources are available, or a large number of services should be served. Device evaluation can be performed using the idle time of the hardware. The evaluation is normally carried out by assigning a number to the configurable device. This number describes the usefulness of the device. If a new chain needs to be installed, then the least valuable devices should be reserved. The maximum capacity algorithm describes a way to determine the number that describes how valuable a hardware element is.

In mobile handsets power consumption is one of the most important aspects. If a resource is not in use it can be powered down, thus saving energy. Since a processor can often serve more than one tasks, then when new functionality is needed it can be placed onto resources, which are already powered up. In such cases the main goal of the resource management is to only utilise the minimum amount of hardware.

13.4.2 Optimisation Algorithms

Appropriate configuration management algorithms are constructed according to the system configuration requirements [10]. Each algorithm is used to calculate the quality of a certain configuration according to a particular configuration parameter. In the reconfiguration strategy it is important to take into account all the different parameters simultaneously, using various weight factors. In other words, during each decision more than one parameter is considered, and each parameter is treated with different importance. This means that with the help of the optimisation algorithms one can construct and compare different configurations. This ability makes it possible to apply the reconfiguration based on the comparison management strategy described previously.

13.4.2.1 Maximum Capacity Algorithm

Another algorithm that can be used to maximize utilisation is the Maximum Capacity algorithm. The Maximum Capacity algorithm requires a lot of calculation. The goal of it is to find the 'least valuable' hardware object in the system. A device is considered 'more valuable' if it is suitable for more than one task, or if it is needed frequently. A number representing this value is calculated for each hardware resource after every configuration, taking into account all relevant parameters. In the hardware architecture graph, this means that the nodes are assigned numbers. The least valuable hardware device is used all the time, then the probability that more services can be served is higher. In order to maximise the utilisation, the resources with the lowest values are always selected and used first.

13.4.2.2 Hottest First Algorithm

Hottest First is a Last In First Out (LIFO) allocation. The goal is to allocate the resource that was last released, on the next resource allocation request. To implement this algorithm in the resource controller, timestamps are needed or a LIFO stack should be used. If a stack is used, then all the free hardware resources that can be reached from the current node in the graph are kept in the stack. When a resource is freed, it is pushed on the top of the stack. When the votes are distributed, then the device on the top of the stack is selected. The problem using this algorithm is that maintaining a separate stack for each hardware device in a reconfigurable network element can be a complex task itself.

13.4.2.3 Coldest First Algorithm

Coldest First is a First In First Out (FIFO) allocation. The configuration manager keeps the free hardware resources in a queue or maintains a timestamp for each device. When a resource allocation request is received, then the resource that is not allocated for maximum time is selected. If the implementation is done with a queue, then a resource allocation request is serviced by removing a resource from the head of the queue. A freed resource is returned to the free list by adding it to the tail of the queue.

This algorithm, like the Hottest First algorithm, also involves information storage at the nodes of the graph. Each node acts as a separate unit, which has information about the surrounding hardware. However, keeping this information up to date can be quite difficult. The main advantage of this algorithm is that it provides even utilisation of the hardware of the network element. This is useful, for example, in fault tolerance, lifetime improvement, or keeping the system temperature low.

13.4.2.4 Blind Monkey Algorithm

The basic idea behind the Blind Monkey (or random pick) algorithm is that when a resource selection is needed, then one asks a 'blind monkey' to randomly suggest a resource selection from the available free hardware resources that are related to the current node in the graph. Using this random selection process, an even utilisation is accomplished over a long period of time.

13.5 Specific Reconfiguration Requirements

It must be somewhat obvious that the reconfigurations of the base stations are handled differently to mobile terminal devices. This is because in a mobile terminal some configuration management parameters are more important than those in a base station. For example, for a mobile terminal low power consumption and maximum utilisation of the resources are always required, while in the base station fault tolerance and extendibility is more important. Below some special requirements for such reconfigurations are discussed.

13.5.1 Reconfiguring Base Stations

In a reconfigurable cellular wireless network, a base station is located in each cell and has a set of reconfigurable resources available to serve the mobile terminals active in that cell. Processing resources are allocated when a mobile unit becomes active in a cell and are released when the unit terminates the call or hands-off to another cell. In the base station, specific reconfiguration management aspects such as fault tolerance and predictive preconfiguration capability are most important. A reconfigurable base station is performing well if it is capable of focusing more processing resources on a particular subscriber, while minimising the impact of increased processing needs on the whole system. Also, it needs to be ready to serve new terminals that move into the coverage area. All this should be done with high reliability. The base station therefore needs to have the following features in its reconfiguration infrastructure:

- Additional processing capacity to serve the configuration management unit
- Redundancy
- Advanced error detection and error reporting capability

If additional processing capacity is assigned to the decision-making local intelligence unit, then more configuration parameters can be taken into account, and more calculations can be done. In this way, the overall decisions adapt better to the circumstances, and the service therefore improves. Redundancy and advanced error detection can also be added to the base station, because increased space requirements and increased power consumption are easily handled by the fixed hardware infrastructure. If these additional features are added to the system, then the possibility that a configuration will fail is greatly reduced, and the reliability of the network element will be increased. If error reporting is also added, then the base station is capable of notifying the network management unit of error conditions, so that the network management can take additional actions to minimise the effects of these errors.

13.5.2 Reconfiguring Mobile Terminals

To discuss the mobile terminal specific reconfiguration issues, first one must examine how the mobile terminals are used. In this case, one will notice a large variation of workloads between different periods in the usage pattern. This is because the surrounding scenarios are changing often and the communication rarely reflects continuous and moderate usage of the services. In this mobile terminal environment, the speed of reconfigurations at various levels are more important than in a fixed base station environment, and failing configuration attempts are more tolerable. To increase the speed of configuration modification in a mobile handset, different layers of reconfigurations can be used. Reconfiguration on an upper layer requires less effort and can be done more frequently, while lower layer reconfigurations can take more time. The following layers can be used:

Layer 1 – Parameter adjustment of already installed services
Layer 2 – Processing chain conversion
Layer 3 – New processing chain installation

The potential problem is that the time available for reconfiguration often is not sufficient to perform the necessary changes in a seamless way. For example, if the necessary reconfiguration cannot be performed in a roaming situation, then the ongoing conversation or data transmission can drop. To overcome this difficulty, we need to add additional functionality to our reconfiguration infrastructure to support parameter adjustments and partial reconfiguration, since sometimes even some small parameter changes are enough. Therefore using the suggested layered reconfigurations strategy, a mobile terminal will be able to adapt to different situations and requirements in a very flexible way.

13.6 Conclusions and Future Trends

In this chapter a systematic architectural study to develop reconfigurable network elements was introduced. These smart network elements are capable of adjusting to the changing user service application requirements better than their non-configurable counterparts. The reconfigurable network elements are built using reconfigurable processing devices and reconfigurable connection elements that are organised in a well-defined architecture. The processing devices provide a heterogeneous infrastructure with increased performance and flexibility, which is served by an interconnect architecture that supports sufficient concurrency within a configuration.

Optimisation parameters and algorithms for configuration management were described, which increase the efficiency of the dynamically applied configurations. These algorithms are easy to implement and can be used in base stations and mobile terminals. It was concluded that it is important to take into account the differences between the network elements, so that one can provide a way to exploit the possibilities offered by each system. If a mobile terminal is reconfigured it can be used in a network that supports a different communication standard, and the seamless roaming between different air interface standards becomes possible. If a base station is reconfigurable, then it can serve terminals of different standards simultaneously.

As the reconfigurable radio systems continue to expand into new application areas, the complexity of the systems and the diversity of hardware workloads are increasing. In addition to the 'classic' wireless voice communications transmitted as short packets in a digital radio environment we can now see workloads containing non-traditional data types, such as image, audio and video. Thus, reconfigurable radio network elements will have to deal with workloads with a wider range of resource demands and execution times in the future.

One also notices that users expect more functionality from their handsets, such as a larger screen, which consumes more power. However, they expect the standby time to increase at the same time. The number of channels is also increasing, hence base stations must minimise the power consumption per channel. Custom-designed circuits have power efficiency advantage over generic processor structures in some cases. However, it is expected that the reconfigurable network elements will be able to compete with non-configurable systems in the majority of future applications.

References

[1] J. Milota, 'Software radio architecture', in *Object-Oriented Approaches to Wireless Systems Engineering, Wireless Architectures for the 21st Century*, Addison-Wesley, 2000.

[2] J. Kovacs, S. Imre and P. Kacsuk, 'Resource management over the baseband subsystem of 4th Generation Wireless Networks in Software Radio Environment', International Symposium on Telecommunications, IST2001, 1–3, September, Teheran, Iran, 2001.

[3] T. Karran, G. Justo and K. Madani, 'Intelligent reconfiguration of large mobile networks using complex organic distributed architecture', VTC-Spring Conference, 6–9, May, Rhodes, Greece, 2001.

[4] K. Madani, et al., 'Configurable radio with advanced software technology (CAST) – initial concepts', IST Mobile Communications Summit 2000, 1–4, October, Galway, Ireland, pp. 139–144, 2000.

[5] T. Farnham, G. Clemo, R. Haines, E. Seidel, A. Benamar, S. Billington, N. Greco, N. Drew, T.H. Le, B. Arram and P. Mangold, 'Reconfiguration of future mobile terminals using software download', IST Mobile Communications Summit, 1–4, October, Galway, Ireland, 2000.

[6] S. Leung, A. Postula and A. Hemani, 'Customised reconfigurable block-based architecture for baseband data processing in telecommunication applications', The University of Queensland, Department of CSEE, St. Lucida, QLD 4072, Australia; Royal Institute of Technology, ESDLab, Kista, Sweden; International Conference on Chip Design Automation (ICDA2000), Beijing, China, 2000.

[7] T. Tuan, S.-F. Li and J. Rabaey, 'Reconfigurable platform design for wireless protocol processors', Berkeley Wireless Research Center, University of California at Berkeley, ICASSP2001, 7–11, May, Salt Lake City, Utah, USA, 2001.

[8] M. Eisenring and M. Platzner, 'Optimization of run-time reconfigurable embedded systems', Swiss Federal Institute of Technology, International Workshop on Field Programmable Logic and Applications FPL 2000, 28–30, August, Villach, Austria, pp. 565–574, 2000.

[9] H. Zhang, M. Wan, V. George and J. Rabaey, 'Interconnect architecture exploration for low-energy reconfigurable single-chip DSPs', University of California at Berkeley, IEEE Computer Society Annual Workshop on VLSI, 8–9, April, Orlando, Florida, USA, 1999.

[10] G. Rabai and S. Imre, 'Resource management in software radio architectures', 2nd Karlsruhe Workshop on Software Radio, March, 2002.

14

Management, Control and Data Interfaces

K. Moessner and S. Gultchev
University of Surrey and MVCE

14.1 Reconfigurable Elements

The classical approach for solving the various air interface standard problem has been that manufacturers were forced to produce their own legacy implementation of a radio platform. However, applying this rigid approach to software radios would lead to very restrictive software defined platforms that would merely allow the reconfiguration using manufacturer verified software. Such a legacy approach is a contradiction to an open platform and would deny the user the advantages of the original Software Defined Radio (SDR) ideas [1]. Instead, such a manufacturer-specific approach would result in numerous different radio processing platforms, each following a different modularisation approach. In addition, reconfiguration software would be product-specific instead of being generally useable.

Preceding chapters in this book have described the requirements for reconfigurable systems looking from the user, manufacturer, and operator viewpoints; they also outlined the functionality to be provided by networks to support reconfigurable terminals. But one may ask: What are the effects of reconfigurability on mobile networks? Which levels of the network are affected by reconfigurability? And how can reconfigurability be cast into a framework without limiting the degree of freedom it offers in the first place?

Reconfigurability influences all levels within a communication system. From the terminal viewpoint it affects all functional layers ranging from application down to the physical layer. From a radio viewpoint, reconfigurability impacts the whole radio environment. Looking at it from the network side, there is a need for supporting functionality as well as the requirement to actually reconfigure functionality within network nodes.

Putting these challenges and the anticipated open nature of SDR platforms into the context of reconfigurability requires making SDR terminals truly reconfigurable, by allowing the use of software that may be obtained from various sources and vendors. This requires a minimum framework consisting of open interfaces that provide access to the various

Software Defined Radio: Architectures, Systems and Functions. Edited by M. Dillinger, K. Madani and N. Alonistioti
 ISBN: 0-470-85164-3

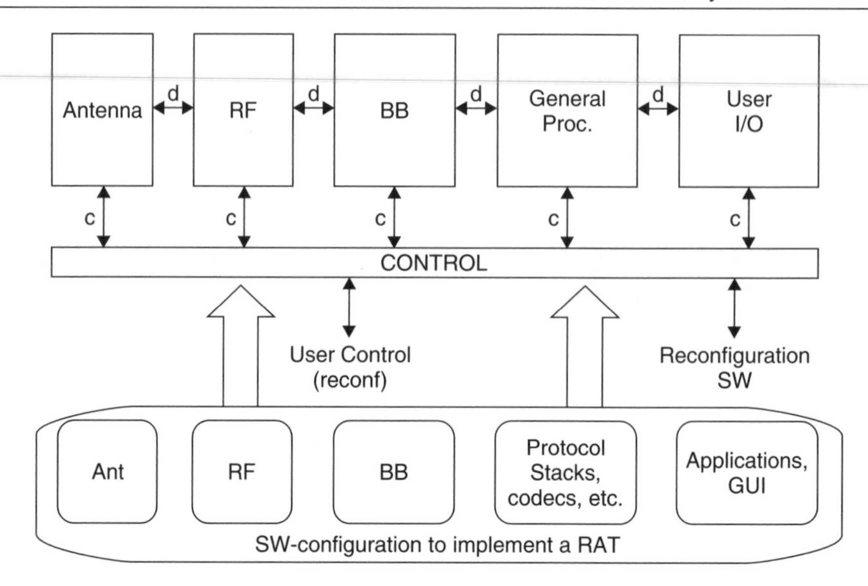

Figure 14.1 Basic SDR architecture

users, applications, and service profiles, as well as to the control and network support elements for the reconfiguration processes.

A structure taking these requirements into account has already been defined and propagated during the early days of software radio, when the SDR Forum (at that time called MMITS) advocated a structure derived from the SPEAKeasy [2] architecture. This initial architecture, albeit being not too detailed, already identified not only the various functional blocks and the software structure of a SDR, but it also recognises sets of interfaces between the various modules of a terminal, as shown in Figure 14.1.

The most important outcome of this architecture is that it initially divides and describes the modularisation of the various functional and logical blocks within a software radio. This chapter is built on this basic principle, it deepens the concept and describes the requirements, advantages and finally combines SDR terminal and network support technologies with reconfigurable network functionality by merging the interfaces of the various reconfigurable layers and entities within a reconfigurable mobile communication network in an open interface framework.

14.2 System and Reconfiguration Interfaces

The initial SDR architecture [3] (as derived from the SpeakEASY architecture) identifies not only the various functional blocks and the software structure of a SDR terminal but it also recognises sets of interfaces between the various modules within a terminal platform (i.e. the information/data (d) and control (C)) interfaces between the SW and HW platforms, and also between the HW platform and control infrastructure. Meanwhile there are a number of projects and organisations aiming towards the definition of software radio architectures (examples are listed in ref. [4]). Common to all these example architectures/structures is the separation between information and control paths, using the information path for the transport of the actually implemented radio system and the control path as a means for the configuration-related communication between the different radio building blocks. A further requirement that these architectures raise is that they need capable facilities and mechanisms to gather and analyse parameters and general system state information as a precondition for the execution of controlled and trustworthy standard compliant reconfiguration processes. Interfaces between the required functional blocks and system mechanisms are distributed across the mobile network. Applying the assumption that SDR terminals/equipment will be based on open platforms and programming interfaces, this will enable reconfigurable terminals to provide the possibility to use (soft-coded) radio implementations, which includes that they will be able to execute software elements that are obtained from various sources (i.e. this includes not only the terminal manufacturer and the network operator but also the third-party software provider).

14.2.1 System-Wide Control and Management Interfaces

The openness of this approach necessitates that reconfiguration and reconfiguration control/management cannot be confined or delegated to the terminal only but must be distributed between a number of nodes (i.e. the terminals and the network). In other words, if users (applications) are given the possibility to initiate reconfiguration of their terminal (i.e. the underlying radio subsystem) and when they are enabled to nominate the software modules that should be used for this reconfiguration and, furthermore, to apply this 'user-defined terminal functionality', then some network authority (e.g. the network operator or the service provider) will require the power and possibility to authorise or decline the reconfiguration of such a terminal. To achieve this, a control interface between the network and the terminal has to be installed. An example of this is the Mobile VCE Reconfiguration Management Architecture [5], which provides, as part of a mobile network-wide system, reconfiguration control space and mechanisms enabling verification of intended software reconfigurations. Reconfigurability of radios affects not only the terminals and the base stations, but the whole radio environment and access network. The effects of reconfigurability may even stretch into the core network. This implies that a 'system reconfiguration control space' literally covers the complete access networks rather than the reconfigurable terminal only.

Looking at a reconfigurable system from the perspective of the network operators, there are further reasons for the introduction of interfaces and mechanisms for reconfiguration control and management. Operators need reconfiguration management interfaces to properly manage efficient and non-interfering use of their limited radio resources (i.e. their licensed spectrum). On the one hand, operators need to ensure that the equipment used within their networks adheres to given radio standards, whilst on the other hand they will have considerable interest in control, support, trigger or even prevention of possible reconfigurations. They may want the means to download and install new algorithms that may achieve more efficient spectrum usage or any other kind of system software, enhancing the use and efficiency of a terminal within their network.

To provide the means for such a system-wide reconfiguration control and management, a reconfiguration control space (RCS) is required; this RCS must stretch across the interfaces within and the physical boundaries of the terminal into the access network. This is shown in Figure 14.2.

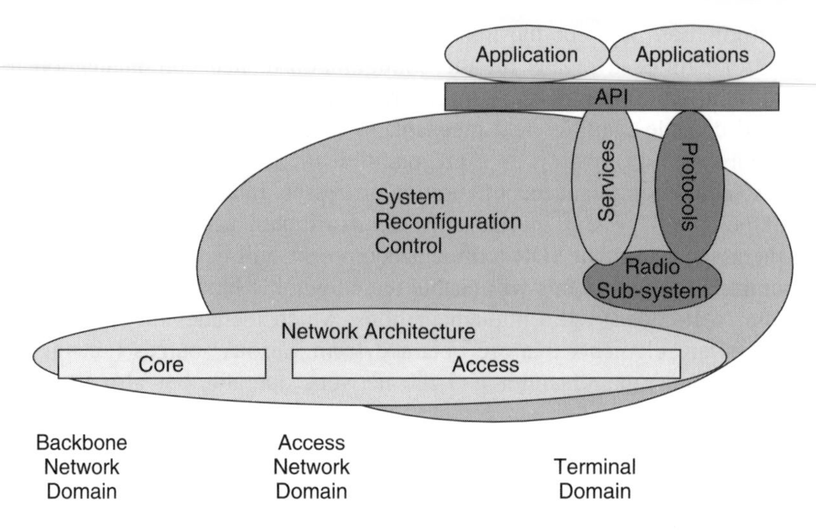

Figure 14.2 System reconfiguration control

14.2.2 Data and Reconfiguration Control Interfaces

The previous section describes the need for reconfigurable communication platforms to be reconfigurable on all system levels, ranging from reconfiguration supporting networks and reconfigurable base stations, to reconfigurability of the RF and BB modules in terminals. It is common understanding that there is currently no single platform and implementation technology capable of providing a homogeneous structure for SDR equipment [4]. Additionally, the anticipated flexibility of reconfigurable equipment requires modularisation of the various system elements and mechanisms capable of managing complete reconfiguration processes. Examples are the aforementioned RMA [5] and TRUST's Configuration Management Module (CMM) [6]. Taking the flexibility requirements into account, and at the same time aiming for open programmability to enable a 'PC-like' growth of application and system configuration software provision, requires an interface architecture for control of both reconfigurability and of data domains (Figure 14.3).

On the terminal side, a management unit (RMA, CMM, etc.) controls and steers the reconfiguration process whereby TRUST's 'Mode Switching' feature may be applied to

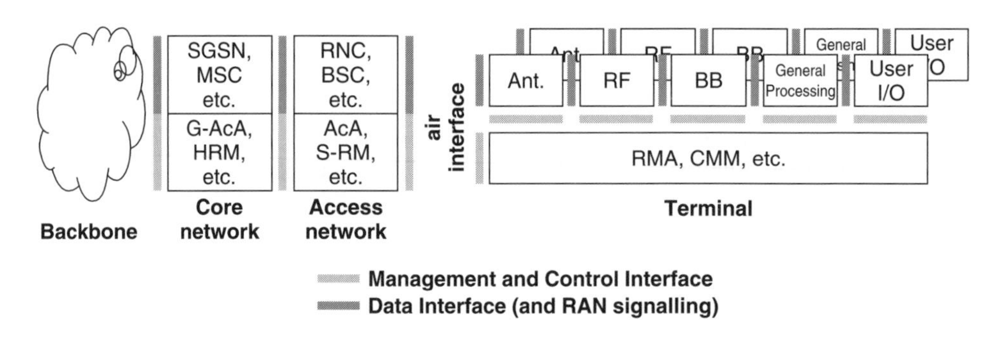

Figure 14.3 Interface architecture for reconfigurable systems and supporting network elements

monitor the radio environment and to support seamless HOs (see refs [6] and [7]). The principle of Mode Switching relies on the use of a complete switched shadow chain, whilst the MVCE's reconfiguration management controllers offer a solution that is targeted at on the fly reconfiguration of single modules within the transceiver chains.

Independent of the reconfiguration process used (mode switching or on the fly reconfiguration), a clearly defined and specified set of interfaces for both data and control of reconfigurable equipment needs to be developed. The following section describes the various aspects to be considered when defining interfaces for reconfigurable systems, followed by an example of reconfiguration management interfaces and their function within reconfigurable equipment.

14.3 Aspects of Interfaces in Reconfigurable Equipment

Reconfiguration of future commercial SDR terminals will be a rather complex task. Numerous parameters may influence reconfiguration procedures and programmability of terminals opening the possibility for any unwanted configurations. The ultimate goal is the complete description of the interfaces that will provide a common middleware encapsulation of the configurable hardware platform with provision of a high-level radio configuration interface to 'radio-application software'. This will facilitate the use of third-party configuration software by providing mechanisms that ensure conformance to a given Radio Access Technology (RAT) standard. This also includes 'standard test suites' to certify that the final instantiation of the configuration software implements the waveform, as well as the lower and the higher layers correctly and according to the standard [8].

In particular, development of a reconfigurable terminal model with clearly specified internal interfaces for communication is one of the main requirements for the specification of SDR equipment. This also includes descriptions for external protocol interfaces located between network and terminal, and requires application of a mechanism for secure reconfiguration software download. Therefore, reconfigurable communication system will require a new functional layer (i.e. a reconfiguration support plane) alongside the existing old signalling/user/management planes. Additional control between reconfigurable platforms, across the air interface, may also require the introduction of additional signalling channels as well as an expanded network infrastructure (i.e. eventually this may lead to the introduction of a universal/global communication control channel).

Such an additional functional plane that may be required for reconfigurable terminals and reconfiguration supporting networks will enable and support the secure and reliable reconfiguration of terminals and network nodes. It will also facilitate the download of trusted and secure reconfiguration software. Reconfiguration procedures may include all entities within a terminal, ranging from the application layer to the physical interface (i.e. the radio processing platform). Such a 'global' type of reconfigurability requires a strong reconfiguration management within the terminal but also the ability to interact with the responsible (and authorised) management units within the network environment. The architecture able to cope with these situations has to be distributed between the network and the terminals. We assumed an open programmable platform for the reconfigurable radio and developed distributed configuration management architecture to control this reconfigurable radio part [5].

These new terminals are the start of the trend towards dynamically reconfigurable base stations and terminals, which can support a multiplicity of access schemes. One single terminal will be able to operate in many different environments, potentially improving the efficiency of spectrum usage by dynamically adapting to the requirements of the radio environment. Open programmable radio platforms [9] will also enable third parties to provide multiple solutions for particular implementations (e.g. customised codec implementation) within such terminals. Assigning responsibility for terminals' standard conformance in such an environment in which the software from one vendor is running on a terminal from a different manufacture and using a different network operator while roaming on a foreign network, is not straightforward. The questions is: Who is responsible for approving reconfigurations in such environments? It requires robust and viable validation mechanisms that have to be recognised as authorised entities, to approve the deployment of configuration software in the terminal.

14.3.1 Abstraction and Simplification

Software radio terminals will be able to roam throughout the coverage areas of different networks. Therefore the terminals have to be fairly flexible, the degree of reconfigurability (i.e. how many parts of the terminal need to undergo a reconfiguration) depends upon how different these access schemes are. The alteration in the configuration may range from minor changes in the terminal settings (different frequency bands, i.e. GSM to PCN) to the complete reorganisation of all settings (i.e. GSM to IS-95). Requirements for software radio not only involve the reconfiguration of terminals, but also services such as software downloading, centralised user registration and centralised authentication. It also influences areas such as call establishment and maintenance across the borders of different access/traffic networks. Additionally, the network infrastructure should also support common billing, mobility management, authentication, and registration and signalling across the boundaries of different access systems. Furthermore, owing to the capability to adopt any legal (and illegal) configuration and potentially causing serious interference with other wireless systems, protection from illegal use is a serious issue to which the reconfiguration management entity must provide a solution.

14.3.2 Modularisation

The traditional telecommunication environments have limitations in the areas of service portability and fast service deployment. Using open interfaces is an approach to overcome these limitation and constraints by opening and making the networks accessible to third-party service providers. The most important factor is provision of access to network resources and network capabilities. Different solutions have been adopted to achieve this goal based on the principle of open Application Programming Interfaces (API). APIs are defined as sets of technology-independent interfaces in terms of procedures, events, parameters and their semantics and may be based on distributed computing concepts such as CORBA, JAVA, RMI, DCOM and other software technologies.

All these technologies are based upon the Object Oriented Technology (OOT) which provides the paradigm that supports reuse, rapid application development and many additional features such as inheritance, polymorphism and encapsulation. Systems designed using OOT allow the fragmentation of implementations into small functional modules and a series of cooperating components. Objects are self-contained software modules encapsulating functionality and attributes, and interactions between objects take place via clearly defined interfaces, whilst their functionality is hidden within the objects.

An open service architecture with open interfaces has much more flexibility in the underlying software technologies, including new software methodology extending the modularisation of software used today in the OOT. Thus, the different aspects of programs and underlying interfaces can be reflected much more cleanly in code as with the current approaches. Message-and event-driven-based programming complements the published user interfaces.

Reconfiguration is not only a matter for the air interface between terminals and base stations but also for the other network nodes along communication paths. Open systems and flexible configurations will eventually facilitate data transport with a minimum overhead across networks and the wireless link, but due to their potential of reconfigurability they are also prone to system misconceptions and data message collisions due to nonconformant states within the single network node. Therefore, reconfiguration has to be seen from different perspectives whereby reconfiguration management will have to deal with them. Reconfiguration management has to take care of the internal configuration but it also has to control the external relations of this node. This means that an external reconfiguration management needs to co-ordinate the states of all nodes along a communications path, whilst an internal counterpart has to control and manage the internal reconfiguration of the mobile station. Reconfiguration may occur in different variations/categories, which may also have different classes of effects on the communication link.

Open terminal platforms provide, through their programmability, the instrument to reconfigure a soft terminal in a way that they meet the network and application specific requirements of any mobile communication network. The concept of reconfiguration is the key to such flexibility and service adaptability.

14.3.3 Interface Hierarchies

While an open interface at the application level is the prerequisite for open service architecture, it is important that applications' requests be mapped appropriately onto the behaviour of the wireless network. Therefore, open APIs are needed also at low-level access to network elements (such as routers, switches, base stations, etc.) by defining appropriate interfaces. These interfaces then can be accessed by various entities such as protocols and agents, to offer advanced customisable and collaborative network services. This concept can be extended with mechanisms for distributing and executing code, which programmes the interfaces on behalf of individual applications, including those running on end systems. This 'new paradigm' is called active networking [10].

Reconfigurability of software radio terminals will not be confined to the physical layer alone; it will include reconfigurability of protocol stacks and application support environments. An ideal reconfigurable terminal should comprise a collection of radio protocol modules separated by open interfaces and controlled via a reconfiguration interface. Each radio module, for example, would offer its functionality, via the open interfaces, to the system and other modules, and could be reconfigured or exchanged via a reconfiguration interface.

Reconfigurability may be pursued in different ways: (1) using parameterised radio modules, (2) exchanging software components within a module, and (3) exchanging complete soft-terminal radio modules. Complete reconfigurability of the protocol stack, for example, calls for the introduction of flexible interfaces between protocol layers, i.e. Protocol Programming Interfaces (PPIs), that replace the rather static service access points. For example, OPtIMA enables reconfigurability through the introduction of programming interfaces between protocol strata, providing the possibility to write both single protocols or whole protocol stacks in a manner similar to the way applications are written in high-level programming languages. Java applications use different APIs, which are part of the class libraries with binding at runtime. This means that the functionality is outsourced to the API and the application simply defines the sequence and determines the parameters passed to methods within the APIs.

14.4 Reconfiguration Management Interfaces

Reconfigurable terminals will have to be viewed from more functional perspectives than the classical management, control and user planes, as defined for example in GSM. They will have to execute additional tasks that stretch beyond traffic transport, call control and connection management. There will be a variety of tiers each of which processes requests and tasks for new, value-added functionality; however, this new (reconfiguration) plane will still rely on the classical user traffic, control and network management planes and will still use their services. This means that the aforementioned reconfiguration support plane must be operated alongside the legacy planes. This additional functional plane, apart from providing basic connectivity, is also used to ensure secure and reliable reconfiguration of terminals and network nodes. Furthermore, it has to provide the means for a trusted and secure software download reconfiguration of different layers, ranging from the application to the physical layer. This 'global' type of reconfigurability brings the idea of a strong reconfiguration management within the terminal as well as a responsible management unit within the network.

When reconfigurable terminals are based on open platforms then the users have the possibility to implement and run within their terminals a variety of software elements that they may obtain from different sources. Therefore, reconfiguration control and management cannot be confined or delegated to the terminal only. This means that if users or even applications have the possibility to initiate a terminal reconfiguration to nominate the software to be installed, and to apply user-defined terminal functionality, that the network authorities (i.e. network operator or service provider) need to authorise the reconfiguration of the soft terminal. Additionally, since network operators are responsible for the efficient use of their limited radio resources (i.e. spectrum), they must have the means to ensure a complete adherence to approved standards. To achieve this, a distributed configuration management scheme is proposed. This structure delivers the tool for network providers to influence, authorise and control terminal reconfiguration. Reconfiguration management cannot be confined to the terminal alone, as there are interactions between different network entities and, in particular, between their configuration management units.

The following section describes this structure in more detail and introduces the concept of a set of reconfiguration interfaces and their general use as part of the reconfiguration

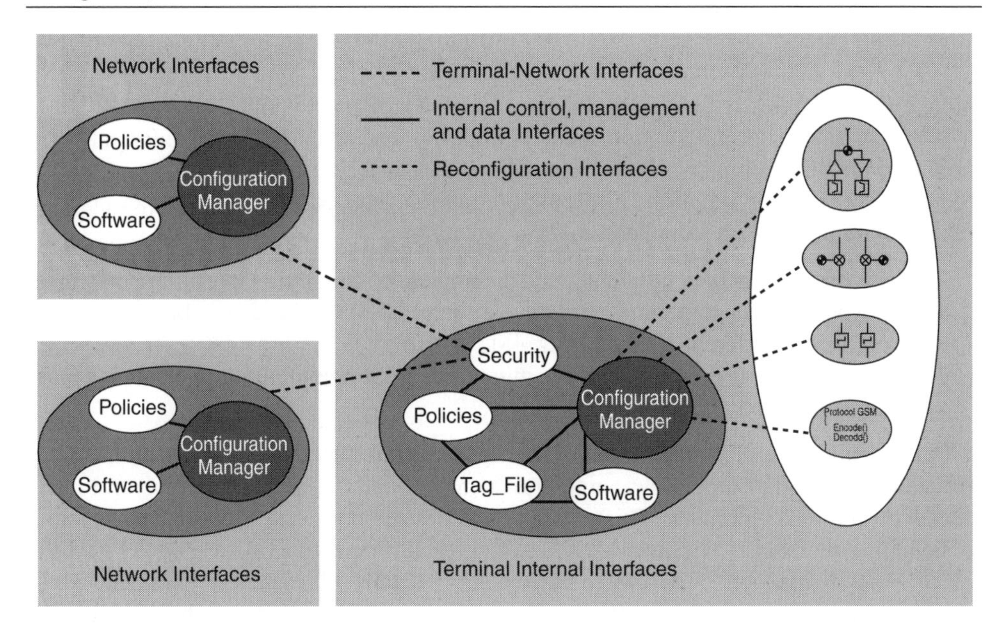

Figure 14.4 Reconfiguration support interfaces

management plane represented by the reconfiguration distributed system between the terminal and the network.

14.4.1 Terminal – Internal Interfaces

Situations requiring terminal or network reconfiguration may occur at any time, either caused by changing network conditions or alternating requirements of applications. This includes varying bandwidth, QoS provision or demands, etc. Therefore, it is inevitable that a configuration managing entity continuously oversees possible requests from the network and monitors the terminal conditions in its reconfiguration radio part. From the perspective of terminal reconfigurability, the aim of the internal reconfiguration management must be to control and manage the functionality of a network node before, during and after reconfiguration; and to facilitate compliance with transmission standards and regulations. Communication and signalling between the various parts of the terminal require the definition of protocols and interfaces between the various functional blocks involved in reconfiguration management. Internal reconfiguration encompasses a number of different reconfigurable modules, including hardware interface (such as antenna, filters, converters, etc.), protocol stack reconfiguration, protocol reconfiguration using active wireless networks, and application execution environment adaptation. The reconfiguration manager works alongside and interacts with all other parts of the reconfigurable terminal and other network nodes.

The above figure depicts the interaction between the reconfigurable radio platform and the reconfiguration management that is required to control the soft terminal settings. Additionally, the reconfiguration must adhere to given radio standards, assuming the open standard nature of SDR or soft terminals, and thus the possibility for third parties to deliver software for the various models within the soft terminal offers an infinite number of possible terminal configurations. Therefore, reconfiguration management has to ensure that the terminal complies with a given air interface standard and to control initial configuration, maintenance of configuration mode and partial and/or complete reconfiguration of mobile communication network nodes, and in particular the terminal and the base station. There are three major situations a reconfiguration manager has to handle; other possible reconfiguration needs are contained in these scenarios:

1. The network node is not configured, and a complete configuration has to be performed.
2. The network node is configured to a standard (e.g. GSM) and needs to be reconfigured to a completely different standard (e.g. IMT-2000).
3. The network node is configured and a minor or partial reconfiguration of one or more modules is required.

There should be a clear separation of management interfaces connected with the communication with other terminal reconfigurable entities and the management control interfaces with the radio modules. This structure enables the complete independent handling of signalling messages. The interfaces between the different functional modules and the reconfiguration manager hide the implementation details of each functional block. In addition, the classification of different types of message interfaces is handled by the configuration manager as 'notification' and 'dispatch' message interfaces. Notification messages do not issue messages to modules outside the reconfiguration manager, whilst dispatch messages are distributed to the specified connection point. The approach describes the decoupled and distributed structure of a manager. This distribution of the processing load between the internal entities decreases the complexity of the single function blocks within the reconfiguration terminal manager, and introduces more flexibility and decreases the individual message processing time.

The reconfiguration manager controls every configuration or reconfiguration process within the terminal; to pursue an actual reconfiguration process it has to manage the following:

- Installation of the necessary components within the its own process and within the radio module parts.
- Starting the so-called 'boot' start-up sequence of the terminal, when the terminal is being switched on for the first time.
- Handling external requests for reconfiguration from the network server.
- Observing notifications issued from the radio modules through their controllers, and handling internal requests for reconfiguration from them.
- Requesting a configuration procedure at the network side for authorising the configuration.
- Continuous monitoring and state control.
- Termination and removal of internal and external software components.

To implement these tasks the manager communicates, using various message sequence interfaces, with the other parts and modules within the management architecture. These sequences include:

- Request and procurement of software.
- Request and procurement of configuration policies.

- Creation of new tag-files (i.e. they are the 'blueprint' of a configuration).
- Establishment and termination of secure connections with the network.
- Installation of radio modules.
- Request procedure whether the user authorises a reconfiguration.
- Registration of the new configuration to the network server.

With regard to management interfaces, there are two cases of reconfiguration procedures that require the support of management interfaces. The first includes the functionality to be provided at the management interfaces, during the initial boot sequence of the terminal state as previously mentioned. The boot process includes initialisation of the internal reconfiguration management interface entities and creates the message handlers and a parameter matrix, specifying the system limits for each of the radio modules. Once all modules are activated the manager initiates the installation of the radio modules.

The second case requiring interfaces to be included in the reconfiguration management entities is the reconfiguration procedure where there is no trace of any previous configuration installed in the terminal. It may be assumed that this case will only occur in the manufacturing stage, when the required minimum radio modules are to be installed. For this reason, the first active reconfiguration procedure of the terminal has a minimum or no radio stack available. A possible way to install modules would be through wired connections or the use of a 'SIM card' and direct installation of the initial software modules and policies. This scenario starts the reconfiguration process with a mutual authentication procedure between the terminal and the vendor's software server (represented as a PC or 'SIM card'). The authentication procedure is performed in a similar manner as the usual case with the management entities from the network server. After completion of the authentication, the reconfiguration manager receives a request from the PC to install an initial set of configuration software. Once the system is ready to pursue the procedure and to install the required software, the manager responds by informing the other entities about the beginning of this procedure. In addition, following the notification, the procedures continue in a similar manner as in the later described download case (see Section 14.4.2). After this, the terminal needs to be reset and to reboot. Using the already installed software and policies, the reconfiguration manager will install the minimum possible terminal configuration, which suffices to provide a basic configuration of the radio part and enables future reconfigurations as described in the boot case above.

Reconfigurable terminals require this initial installation of a radio configuration; otherwise the terminal will not be able to connect to an air interface. The implementation of interfaces between the various functional modules of the reconfigurable system (using CORBA IDL) can be seen as 'hiding the implementation details of each functional block'. This structure forms a type of middleware platform enabling communication between the distributed objects within the reconfiguration management and the reconfigurable system. We assume an open programmable platform together with distributed configuration management, whereby a configuration management unit is resident within the reconfigurable or programmable terminal, whilst a reconfiguration and software download support server resides as counterpart within the access network. In general this distributed architecture has to follow the principle of the client–server communication model.

14.4.2 Terminal – Network Interfaces

Roaming across different air interface standards and the necessary reconfiguration of terminals requires the download of radio implementation software. To implement a configuration scenario in which the user may roam in vertical (within cellular systems) and horizontal directions (between different wireless technologies such as satellite and cellular), a number of requirements have to be met:

- Suitable reconfiguration control mechanisms need to be developed and employed.
- The infrastructure to ensure security and adherence to radio regulation needs to be specified and developed.
- The means to facilitate user and terminal mobility between different systems (i.e. to support vertical roaming) will have to be provided.

Several proposals have already been brought forward defining a 'Global Connectivity Channel' to support basic connectivity for SDR terminals and also to provide an infrastructure for OTA software download. Apart from ubiquitous connectivity for SDR terminals, such a channel may also be used to inter-connect the various legacy PLMNs and therein facilitate cross-system (vertical) mobility. Besides connectivity and mobility support, such a channel may be used as transport infrastructure for the signalling between the different parts of a reconfiguration management architecture, to connect terminal and network resident control authority facilitating secure software download and controlled reconfiguration of SDR terminals. A reconfiguration process, defined as the reconfiguration of a radio module controlled by the configuration manager, consists of three major steps that need to be completed before a terminal is able to apply its new configuration. These stages are:

1. Download of policies and rules and the requirement of a network provider, software provider, and manufacturer.
2. The creation of a tag-file which contains the 'blueprint' of the terminal.
3. The validation of the intended terminal configuration by the control part (i.e. the verification, within the network server, of the configuration described in the tag-file).

However, such a download to reconfigurable SDR terminals and the management of reconfiguration sequences within such terminals requires additional signalling and security capabilities. The signalling scheme to facilitate and implement these procedures relies on support from the network, and has to provide the secure transport of software, terminal, user, operator, and regulation-related reconfiguration information. Transactions take place between the network support node and the terminal. The complexity of such a system included in the reconfiguration-related functional plane will require implementation of these relationships using a set of standardised system interfaces. The relationships between the network and the reconfigurable terminal are becoming exceedingly complex and thus the need for reconfiguration control and management mechanisms with strong security enforcement is crucial.

Signalling mechanisms are required for software download, for the establishment of secure connections between the network and reconfigurable terminals, for the support of reconfiguration control mechanisms, as well as for inter-system roaming. Here the aspects of the different Radio Access Networks (RANs) have to be considered and signalling connections between the nodes of different networks have to be defined.

Installing a new configuration in an SDR terminal necessitates the availability of the required software. This software may be negotiated and downloaded via different channels, including dedicated traffic channels of the currently active RAT or a possible global control channel.

Independent of the download channel, there are signalling requirements to be met during a download sequence:

- Authorisation and authentication (i.e. in a mutual manner) and the establishment of a secure connection between terminal and network.
- Checking the availability of the required software within the network and procurement from other sources, if the software is not locally available.
- Acknowledgement and control of the download of software between network and network node.
- Download confirmation and termination of the secure connection.

An initial assumption for a reconfiguration process would be to have the required software module available within a software repository attached to the server. However, if the software is not available in the repository, then an extended search would need to be performed querying other network servers and their repositories to try to download the software module from a remote location. If a specific software module were required but was not available, then the reconfiguration process would have to be terminated. This directly corresponds to the interfaces required for completion of the software download.

By requesting a terminal to reconfigure to another RAT, the network (i.e. the operator or an standard authority) may issue an initial trigger for reconfiguration, including all three possible reconfiguration types ranging from sub-module, module to complete reconfiguration. The procedure requires support from the terminal, which has to install the radio modules, generates the 'blueprint' (i.e. a tag-file) of the new terminal configuration as well as software and configuration standard compliance confirmation. The network may also provide a tag-file for the intended configuration. This type of request is issued in the case that the network provider makes alterations or requires module updates, such as new software releases or bug fixes. In the case the terminal requests or triggers a reconfiguration, there are two possible procedures, which depend on the policies and security policy files used. These policy files are derived from the initial terminal hardware and user information or user profile; changes to policy files require confirmation from the user. In the case that the user has agreed to an 'open policy', the management entity responsible for the reconfiguration procedure becomes automatically authorised to proceed with the reconfiguration procedure without user intervention. If the affirmation is required, however, then the management entity will dispatch a message to the user for authorisation of the intended reconfiguration. Regardless of which approach is followed, once a reconfiguration is mutually authorised, the management entity within the terminal sends a notification message to the network and starts the actual reconfiguration sequence. If a reconfiguration has been triggered and authorised, and the terminal has ascertained that a software download is required, then it will use a secure connection (SC) to pursue the download.

Assuming the availability of the required software, which can be local or obtained from other repositories, the next step is to forward the module to the terminal. During the complete download, the software module becomes wrapped with security information, which takes place within the network server, and is then forwarded to the terminal where the security context is stripped off. After completion of the download and confirmation to the management entity (the reconfiguration manager module) the SC is terminated.

14.4.3 Network Interfaces

A single network server will represent the network control part (NCP) from the point of view of the terminal – connection to the network, authentication, download, secure protocol, exchange public key certificates, and virtual configuration procedure. The ultimate goal will be a user with a unique ID for location and identification purposes, which will allow the network server to distinguish each terminal and user.

In a similar manner to the reconfiguration manager of the terminal, it is responsible for the coordination of the control and data messages within the server and with the terminal requiring connection and information from the network. It also is responsible for the connections to other servers, for processing the user personal data and terminal information, and saving the user profile with the necessary billing information occurred based on any reconfiguration download performed and reconfiguration of the terminal.

The network server in general performs the following functions integrated into the network interfaces. It

- Monitors network nodes.
- Manages registration of configurations.
- Handles software downloads.
- Validates intended configurations ensuring standard compliance.
- Handles a secure connection protocol for an Alternating MAC Frame Utilisation Algorithm (AA).
- Ensures the privacy of user information.

14.5 Interface Framework for Reconfigurable Equipment

Flexibility on the one hand, and software definability and open programmability on the other, flag the grounds in which reconfigurable systems have to operate. Merging the various concepts such as the modularised radio and a reconfiguration management structure (e.g. the RMA) into a single framework leads to a framework for reconfigurable equipment. Based on the feature of complete reconfigurability, the only constant elements in this framework are the actual hardware processing platform and the various interfaces separating the different functional blocks. The overall principle of such a structure may be compared with a virtual machine, whereby in this reconfigurable communication architecture applications will access services, transport, signalling and reconfiguration means via sets of APIs whilst the underlying radio-execution environment (i.e. the equivalent to the virtual machine) performs the software-implemented radio tasks. Figure 14.5 shows this merged model as a framework of interfaces.

The framework consists of five major interfaces: at the top, there is an application-programming interface providing access to the lower, reconfigurable layers. In the case that the user application or network requests a reconfiguration, the reconfiguration control

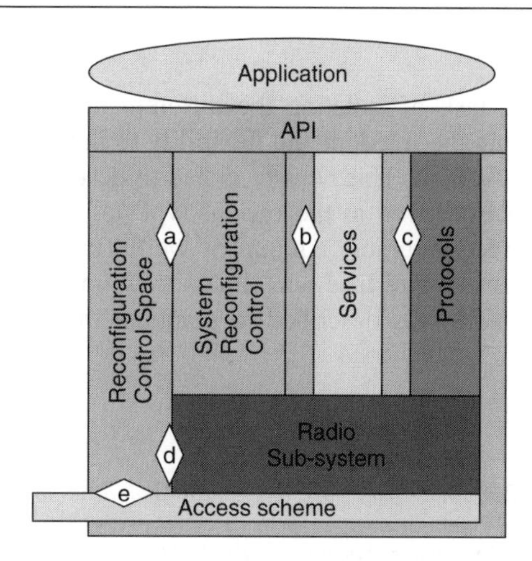

Figure 14.5 Interface framework for reconfigurable equipment

space (RCS) performs a reconfiguration of the radio-execution environment, initiated through the 'RCS API'.

At the core of the RCS is a distributed reconfiguration management architecture, such as those described in refs [5] and [6], that manages reconfiguration procedures, ensures standard adherence and provides the mechanisms to re-establish the old configuration in case the reconfiguration procedure were to fail. As already mentioned, reconfiguration management is not an action confined to the terminal; there will be interactions between different network entities and in particular between their configuration management units. This interaction ensures the adherence to given or agreed radio standards, hence there is the need for an 'outside' interface. Reconfiguration management has to ensure that the terminal complies with a given air interface standard, it has to control the initial configuration (using interfaces a, c and d), to maintain the configuration, and to revert the radio part of a terminal to the previous configuration if the reconfiguration process fails.

The other main section within the RCS structure deals with the radio sub-system (interface d); although many efforts are under way to investigate multiprocessor platforms providing the necessary processing capabilities and projects attempting development of software configurable radios, there is still a need to integrate the different technologies and platforms and to use these platforms, 'configurability' to make systems reconfigurable.

The upper layers of a reconfigurable radio [such as the protocols (interface c) and services (interface b)] are layered on top of the reconfigurable radio sub-system. The interfaces enable separation between reconfiguration management and the reconfigurable terminal parts; they facilitate independent download and installation of reconfiguration software to reconfigure the radio sub-system, the services and the protocols. See-able and accessible for users and application developers will only be a set of APIs, all other parts and the implementation of the reconfiguration mechanisms are encapsulated within the RCS.

Once truly reconfigurable processing platforms are available, developers of applications, services, protocols and even waveforms will require only the specification of a programming interface (API) to develop and write applications using the reconfigurable features of the underlying radio platform.

14.6 Conclusions

Interfaces have the basic task of isolating the different modules within reconfigurable equipment, thus introducing the freedom and flexibility to independently reconfigure each functional part within the system. This chapter aimed to describe both the system context of reconfiguration interfaces as well as the various types of interfaces required. Different aspects of interfaces in reconfigurable equipment were outlined and the importance of a reconfiguration management interface was illustrated. Finally, an interface framework for reconfigurable equipment was described to combine the various interfaces using a coherent model.

References

[1] J. Mitola III, *Software Radio Architecture: Object Oriented Approaches to Wireless Systems Engineering*, ISBN 0471384925, Wiley, 2000.
[2] M. Cummings, et al., *MMITS Standard Focuses on APIs And Download Features*, *Wireless Systems Design Magazine*, April 1998, Vol. 4 pp. 45–49, 1998.
[3] R. Lackey, et al., 'SPEAKeasy: The military software radio', *IEEE Communications Magazine*, **33**(5): 56–61, 1998.
[4] 'WWRF SIG on Reconfigurability, Reconfigurable SDR Equipment and Supporting Networks', White Paper, 2nd En, Presented at WWRF No. 7, 3–4 December, Eindhoven, Netherlands, 2002.
[5] K. Moessner, et al., 'The RMA: A framework for reconfiguration of SDR equipment', *IEICE Transactions on Communications*, **E85-B**(12), 2002, pp. 2573–2580.
[6] *www.IST-TRUST.org*, 2002.
[7] *www.IST-SCOUT.org*, 2002.
[8] M. Gudaitis and J. Mitola III, 'The Radio Virtual Machine', SDR Forum 21st General Meeting, 14–16 November, Mesa, AZ, 2000.
[9] K. Moessner, S. Vahid and R. Tafazolli, 'OPtIMA: An open protocol programming interface model and architecture for reconfiguration in soft-radios', 12th Tyrrhenian International Workshop On Digital Communication – Software Radio Technologies and Services, 13–16 September, Portoferraio – Island of Elba, Italy, 2000.
[10] WWRF, *Book of Visions for the Wireless World 2001*, *www.wireless-world-research.org*, 2001.

15

Reconfiguration Principles for Adaptive Baseband

David Lund
HW Communications Ltd

Mehul Mehta

15.1 Introduction

Reconfigurable networks and systems require many new support mechanisms to allow dynamic air interface operation to be a reality [1–37]. Amongst IST Framework V projects sponsored by the European Commission, the Configurable Radio with Advanced Software Technologies (CAST) project [1, 2], the Transparently Reconfigurable Ubiquitous Terminal (TRUST) project [3, 4], and others consider many such mechanisms. This chapter discusses results from both of these projects from the separate contexts of software, hardware and the boundary between the two. Traditionally, a mixed hardware/software system design requires consideration of both aspects separately. This separation is an essential requirement of Software Defined Radio (SDR) in order to maintain hardware independence and portability.

The following text forms only a summary of the concepts provided by both the CAST and TRUST projects. The reader is invited to study the detail of each project by referring to the public deliverables, which can be found on the project websites [2, 3], respectively.

The intention in this chapter is mainly to study the software architecture of the TRUST project and hardware concepts of the CAST project, and explore the link between the two projects in the context of reconfigurability.

15.1.1 The Software Element of the Baseband Sub-System

The physical layer signal processing blocks of a generalised radio terminal are considered, as shown in Figure 15.1. It consists of the following main parts:

Software Defined Radio: Architectures, Systems and Functions. Edited by M. Dillinger, K. Madani and N. Alonistioti
 ISBN: 0-470-85164-3

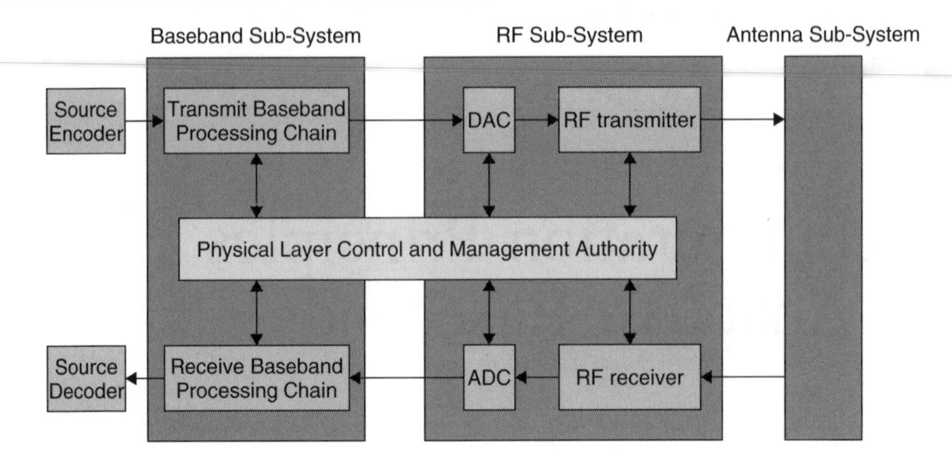

Figure 15.1 Typical signal processing blocks in the physical layer of a radio transceiver

- Source Coding and Decoding
- Baseband Sub-System
- RF Sub-System
- Antenna Sub-System

Given these typical segments of the physical layer, we concentrate primarily on the digital signal processing aspects that facilitate data and control channel processing.

The Baseband Sub-System consists of the digital signal processing elements of the transceiver, and is made up of constituent modules for the transmit chain and receive chain processing, respectively. Typical modules within the sub-system include: channel coding, modulation, signal detection algorithms, equalisers, etc.

In the context of SDR, the objective of baseband reconfiguration is to re-define the functionality, behaviour and performance of the baseband transceiver chains. This means that the digital signal processing algorithms making up these chains are to be re-defined so that their post-reconfiguration functionality is different [4, 5]. It is envisaged that baseband reconfigurations may involve a change in modulation, channel coding, etc. This would be necessary if the terminal were to be reconfigured to operate a different standard to that currently in use.

The air interface of a digital communication system normally requires a high degree of processing power in order to generate signals which can propagate through extremely hostile channels. The mobile channel is one such channel with many standards available that specify transmission of mobile data over an air interface to and from a digital network. Efficient processing of mobile channel data is particularly demanding for the 3G standard, UMTS Terrestrial Radio Access (UTRA). Coupling this high capacity processing with the need for reconfigurable processing requires the use of a combination of processing resources. Digital Signal Processors (DSPs) and Field Programmable Gate Arrays (FPGAs) form only a subset of the available configurable technology. Future devices may even contain a combination of these basic processing resources.

Reference [9], which appears earlier in this book series, gives a thorough review of available hardware together with some initial concepts.

15.2 Technical Challenges for an Adaptive Baseband

The following is a comprehensive list of the many technical challenges facing the designers and the developers of the reconfiguration of baseband chains [6]:

- *Transparent Reconfiguration.* It is well known that the baseband sub-system of a radio terminal consists of several inter-connected and inter-dependent processes. Now assuming that the radio terminal is software-definable, the problem is how to perform and administer reconfiguration without halting or terminating the incumbent baseband processes.
- *Selective Re-definition of Module(s).* Consider a baseband chain with constituent modules, each process performing a pre-defined functionality. These processes (modules) are inter-connected to one another and inter-dependent of one another to some degree.
- *Installation and Invocation of Software-definable Baseband Modules.* Given that each baseband module is defined in software, the problem is how to acquire the software. In addition, once the software is made available, the problem is how to install the software to change the functionality of a particular baseband module and then invoke that module to start processing data.
- *Micro- and Macro-level Process Management.* Assuming that the baseband chain is a collection of individual micro-processes (i.e. constituent modules), the problem is how to reconfigure one or more of these micro-processes within the macro-level baseband environment.
- *Analogue–Digital Signal Processing Divide.* With advances in modern digital processor technology, such as DSPs, FPGAs, ASICs and Reconfigurable Logic, the partitioning between analogue and digital signal processing in the context of SDR is a challenging topic of research.
- *Reconfiguration Complexity.* In order to administer a potential reconfiguration request, it would seem necessary to estimate the complexity for carrying out the reconfiguration. So, how does one estimate the complexity of a given baseband reconfiguration? This must include signalling overhead, hardware requirements, and algorithmic changes.
- *Levels of Reconfiguration.* Given that baseband reconfiguration may result in a complete re-definition of the whole transceiver or a selective upgrade of some of its modules, the overall software architecture must cater for both total and partial reconfigurations.
- *Degree of Intelligence.* Since the baseband sub-system must be reconfigurable through dynamic installation and invocation of software-defined modules, there needs to be an overall management and control authority for the whole sub-system. Consideration needs to be given as to how intelligent this authority ought to be.
- *Hardware Independent Architecture.* The main objective of reconfigurability is to design a *software* architecture, without the need to incorporate hardware limitations. This has been a major research topic in the CAST project.
- *Behaviour and Performance Authentication of Software.* It has been assumed that the functionality of any baseband module can be changed by downloading an appropriate software. This may be a collection of files, upgrade patches, parameters for re-setting the already available software, etc. However, it is not clear how one can guarantee that the reconfiguration of a given module will result in the desired change in its functionality, behaviour, and performance.

- *Reconfiguration Signalling.* In order to perform baseband reconfiguration, it is envisaged that there will be interactions between the baseband and other components of the terminal [7]. These interactions must conform to an agreed protocol, such that the order of priority and terminal working state are maintained in accordance with its mode of operation, compliant with the network, and is conformant with the user settings.
- *Software Repository and Access Methods.* Assuming that baseband software is downloaded and stored in a library on the terminal, how is it stored (e.g. database format, link-list format)? What are the access schemes for extracting the desired software?
- *Verifying the Reconfiguration.* Assuming that the baseband sub-system is reconfigurable by software installation and invocation, the problem is how to verify that the performed reconfiguration is reliable, and results in the desired baseband functionality, is acceptable by the network, and is compliant with the terminal's hardware characteristics and signalling structure.

The following section describes solutions to most of the above problems in the context of software. They are derived based on the work carried out in the TRUST project.

15.3 Adaptive Baseband Software

The software architecture of the reconfigurable baseband (R-BB) is based on object-oriented methodology [4, 5, 8]. Each module of the baseband transceiver chain is reconfigurable by instantiation of an appropriate class and/or re-initialisation with new parameters. It is assumed that the software (i.e. class) of each module (e.g. modulator, FEC decoder, etc.) is available (downloaded and stored in a baseband library), error free, and compatible. In order to reconfigure the baseband transparently, i.e. without interrupting the ongoing service(s), the baseband architecture must support dynamic creation and binding of new/modified modules. As a result, instantiation of downloaded classes must be administered through dynamic binding, whereby the required functionality of a given class is only made available at runtime, whilst the structure of the downloaded class is known a priori. The software download module of the terminal is responsible for ensuring that the correct software is downloaded. Parameterisation of baseband module classes allows different objects to be derived from the same downloaded class depending upon the agreed new reconfiguration. It is assumed that the available hardware is capable of meeting the processing and memory requirements of both foreground and background processes.

It is a requirement for the baseband reconfiguration to be facilitated without interrupting its current mode of operation. In light of this requirement, the proposed architecture offers two planes of processing as follows:

1. *Foreground Processing Plane.* This consists of the signal processing functionality (i.e. data and control channel processing) for the current mode of operation. For example, if the terminal is in GSM mode, then the foreground processing will consist of the GSM transceiver algorithms in a pre-defined modular processing chain.
2. *Background Processing Plane.* This consists of the signal processing functionality responsible for creating the next mode's transceiver chain, and the associate processing required to administer and manage a given reconfiguration from the current mode (e.g. GSM, to another standard).

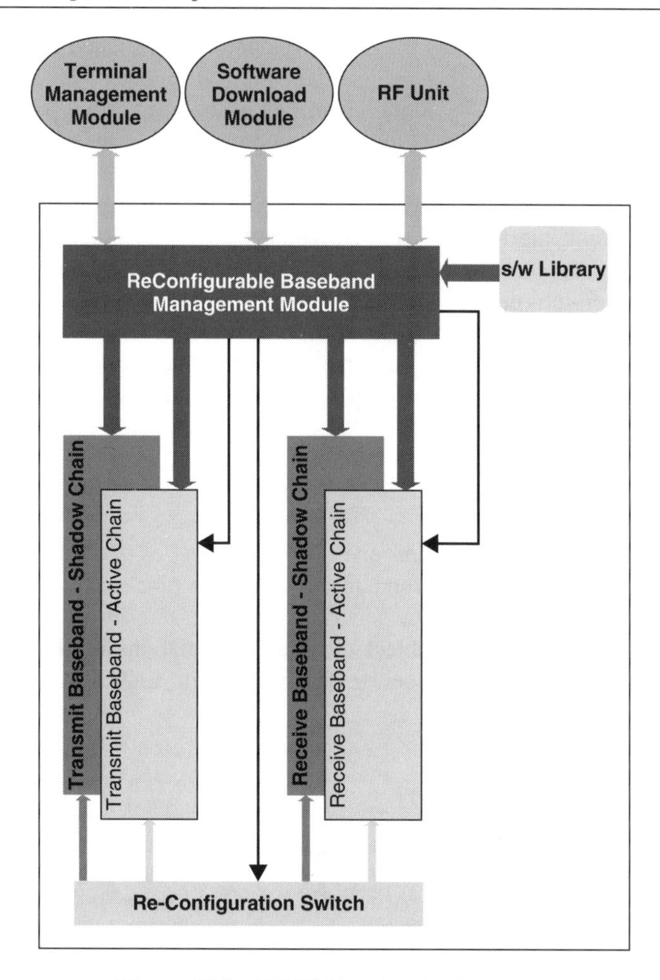

Figure 15.2 TRUST baseband sub-system

A schematic diagram of the R-BB sub-system is given in Figure 15.2. As shown in the figure, the sub-system consists of the following components:

- *Reconfigurable Baseband Management Module (RMM).* This is the overall controller of the R-BB sub-system. It is responsible for negotiating reconfiguration, creating active and shadow transceiver chains, and controlling the runtime behaviour of each module. The RMM also controls the RF sub-system through appropriate signalling, making it the overall physical layer authority.
- *Baseband Software Library.* This contains the active and shadow configuration maps of the baseband. These correspond to the active and shadow transceiver chains, respectively. A configuration map is a list of baseband modules (type, functionality, algorithmic identity, etc.), their associated interface definitions and inter-connections. It is the overall definition of the baseband, and is a piece of software in itself, which is downloaded when a new standard is to be implemented. In addition, the library will also store all the baseband module software that is currently in use and that used previously.

- *Active Transceiver Chain.* This is the currently operating baseband chain. It consists of transmit and receive modules that process data and control channels associated with the current mode of operation. The active chain lies in the foreground processing plane.
- *Shadow Transceiver Chain.* This is the post-reconfiguration baseband transceiver chain. It contains copies of modules that are unchanged from the active chain, and one or more new modules. The shadow transceiver chain complies with the agreed reconfiguration strategy. It does not interfere with the active chain and constituent processes are kept independent of corresponding ones in the active chain. The shadow chain lies in the background processing plane.
- *Reconfiguration Switch.* This is a conventional ON/OFF switch. It implements the ON/OFF signal from the RMM, in order to switch the shadow chain ON and the active chain OFF. The RMM will only issue the ON signal upon approval with the terminal's management entity.

15.4 Software Architecture

The active and shadow baseband transceiver chains consist of several Baseband Processing Cells (BPCs). The BPC is the fundamental building block of any transmit–receive baseband chain. The BPC class is a generic data-processing entity. The process class gives the functionality of a BPC object. Several instantiations of the BPC class yield the constituent modules of the transceiver baseband chain. Each BPC object consists of the following:

- A set of input ports, I
- A set of output ports, O
- RMM control signal interface, C
- A virtual process function, which implements an instance of the downloaded class

Consider a BPC object of identity A, i.e. BPC_A. (Figure 15.3). The algorithmic functionality of the BPC_A is defined by the *process* function, which in turn can be configured by its parameter(s), p. Note that p is a list of parameters of the leaf class used to instantiate

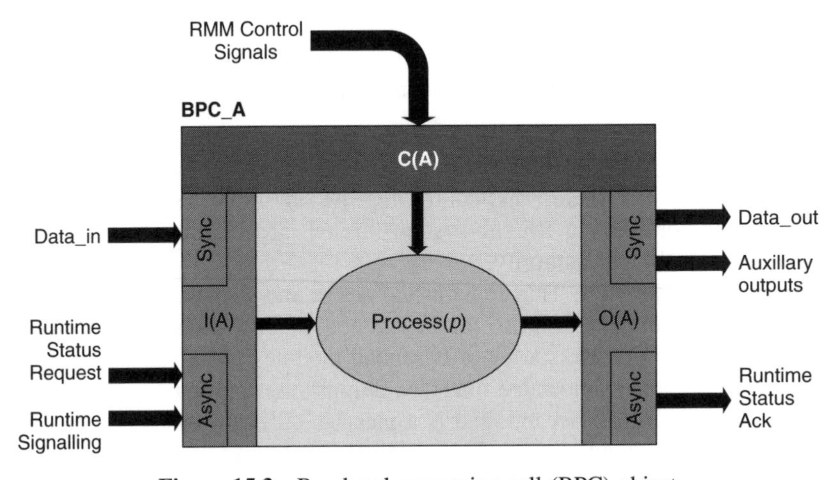

Figure 15.3 Baseband processing cell (BPC) object

the process function. As shown in Figure 15.3, the BPC_A contains input ports I(A), and output ports O(A). Each port has a unique name, type, and data rate. Both provide synchronous and asynchronous inputs and outputs, respectively. In addition, the BPC class includes a set of member methods, and they help define the state of the BPC object and the nature of the operation it is performing.

The following is a list of member methods of the BPC class:

- `initialise()` This method initialises the BPC object by instantiating the downloaded leaf class as a process of the object.
- `resume()` This method sets up the initialised BPC object in a 'get-ready' mode.
- `run()` The run method is invoked when the BPC object is required to process.
- `suspend()` This method will halt the running of the BPC object. It is akin to a pause, wherein data are no longer being processed.
- `restart()` This method will return the BPC object to its original state, i.e. initialise. This is necessary to clean up the BPC object and re-commence the object as if it were new.
- `reset()` This method will help re-instantiate the BPC object.
- `kill()` This method will destroy the process of the BPC object. It is invoked when the BPC object is no longer required. It provides an orderly shut down and destroy mechanism for removing the BPC object from the processing chain.
- `bpcStateQuery()` Each BPC object has its own state. This method is used to query the state of the process of the BPC. These methods control the states which the BPC can exist as shown in Figure 15.4.

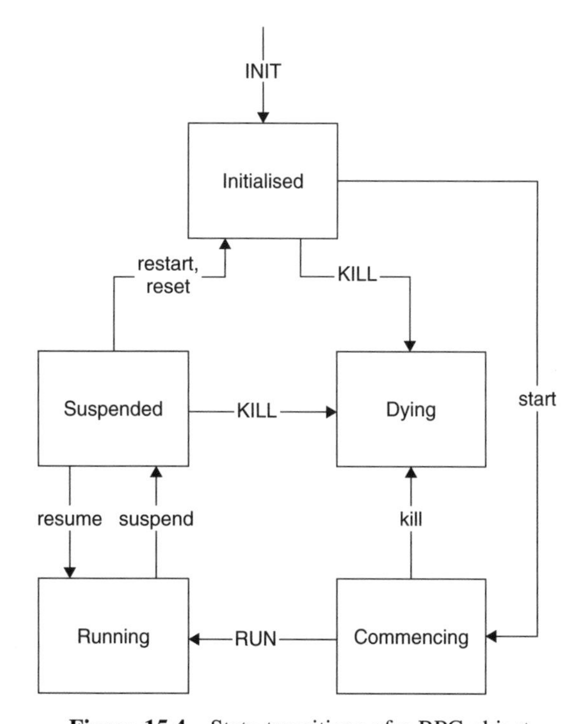

Figure 15.4 State transitions of a BPC object

15.4.1 Parameterisation of Baseband Modules

Algorithmic abstraction of process leaf classes can allow parameterisation of baseband modules. This allows different objects to be instantiated on demand by using the same downloaded leaf class. In other words, different versions of the same algorithm can be generated by instantiating the leaf classes with appropriate parameters. The full extent of valid parameters for a given leaf class will depend on the degree of abstraction, and the range of standards under which it can be used. Studies of common wireless communication standards have led us to derive example parameters for typical baseband modules, as shown in Table 15.1.

15.4.2 Implementation Properties

In addition to the algorithmic parameters, each leaf class has an associated property list. This is very important since the terminal's software download mechanism will need to ensure that the software is compatible in its form, fits the agreed configuration map (standard), and provides the desired functionality. Examples of typical properties are the following:

- Implementation details
  - Fixed or floating point
  - Hardware features, i.e. processor type, speed, memory requirements, etc.
  - Language, i.e. C, JAVA, VHDL, others
- Performance details
  - Power consumption for a given hardware
  - Benchmark results for qualitative appraisal
- Standard
  - Identity of the wireless standard that the software is compatible with
  - Parameter of the configuration map
- Vendor
  - Software manufacturer identity for compliance check with the terminal manufacturer
- Version
  - Version of software

Table 15.1 An example of parameterised baseband module classes

Baseband module	Example leaf class Id	Parameters
CRC parity check	CRCcheck class	crcPolynomial, $c(x)$ crcSize, p,
FEC encoder	ConvolutionalEncoder class	Constraint length, k Galois field polymonial, $G(p)$
Modulation	PhaseModulation class	BitsPerSymbol, n Angle, θ

15.4.3 Baseband Software Structure

In order to build the required transceiver chain, two types of primitive software are required:

1. *Configuration Map.* The Configuration Map (ConfigMap) defines the baseband transceiver in terms of module algorithm (functionality), inter-connections, and interfaces. It is the recipe for making baseband transceiver chains. The ConfigMap is defined in software and is download by the terminal. Each configuration map may include the following information:

- Identification and features of the communications standard, e.g. GSM, UMTS-FDD, UMTS-TDD, DECT, etc.
- Terminal class compliance.
- Terminal manufacturer compliance.
- List of all baseband modules with detailed information on the identity, type, and processing rate of each algorithm
- Inter-connections of the constituent modules
- Approval and authenticity signature to ensure ConfigMap is reliable and that it can be implemented with confidence
- Identity of the vendor
- Version
- Integrity check sum, to ensure that the downloaded ConfigMap software is error free

2. *Leaf Classes.* A Leaf Class is the downloaded baseband algorithm class software.

These two software-defined commodities are the key ingredients for the creation of any baseband processing chain. Given a ConfigMap and all the required leaf classes as defined in the ConfigMap, the RMM will use the available resources (processing power, memory, dynamic process management facilities, reconfiguration protocols) to implement the ConfigMap and create a shadow transceiver chain. Figure 15.5 illustrates a simple ConfigMap.

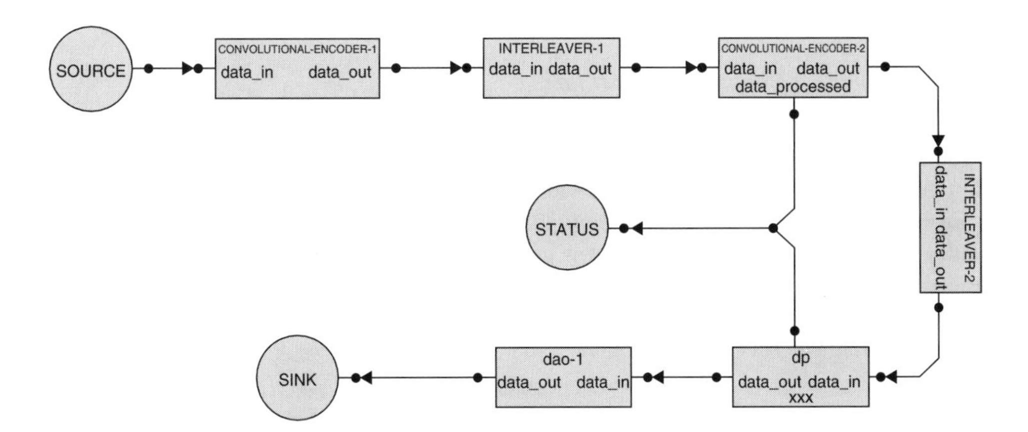

Figure 15.5 Baseband configuration example

15.5 Reconfigurable Hardware Abstraction

The previous section described three software primitives which are required in order to define a baseband chain, namely

- Baseband Processing Cell (BPC)
- Process Leaf class
- ConfigMap

This section describes a framework for the control of configurable processing resources in order to facilitate these three primitives. This is achieved by careful abstraction of hardware capability using Object Oriented methods. The system is known as 'Fizzware'. An open architecture is considered in Fizzware, to allow for extension to any air interface sets and any reconfigurable hardware subsystem.

The following description deals with physical hardware in terms of:

- Classification – description of the hardware and software components
- Control – method of controlling and reconfiguring the hardware components

Initially we define how to partition a hardware platform in order to distinguish between the different configurable devices. Hardware independence is provided by definition of the common control interfaces. A database structure is described allowing thorough description of hardware, functions (a term used interchangeably with BPCs), processes and chains (or ConfigMaps).

15.5.1 Partitioning of Hardware

A typical hardware system is usually defined at the time of design and the software is designed based on the hardware structure. A reconfigurable system takes additional software dynamically throughout its lifetime. To allow provision of information regarding the structure of the hardware for dynamic integration of the software, it is necessary to define a partitioning and classification of the hardware. The following description partitions hardware based upon the physical location of the configurable devices within the hardware system.

15.5.2 Defining a Configurable Partition

The purpose of this task is to isolate the reconfigurable hardware such that it is grouped together with its associated support hardware. This allows direct control of each individual configurable device at a resolution providing an efficient control of configuration.

Each partition has, at its centre, a configurable or reconfigurable device. The boundaries of the partition are defined such that it can incorporate all other resource 'it owns'. A device owns resources if they are accessible only by that device. However, if a particular resource is shared for some reason (e.g. a shared bus or RAM), then only one 'owning' device arbitrates to access the resource.

15.5.3 Connecting Configurable Partitions

Two partitions are connected together if a physical connection exists between them. These connections require description to define the structure of the system. Connections may be as small as individual wires or as large as complex buses (SCSI, PCI, etc.). These are termed Configurable Partition Connections.

15.5.4 Example Partitioning

The HW2000 FPGA development system described in ref. [10] consists of 4 Xilinx XCV600 FPGAs, field programmable interconnect and copious amounts of memory. This board is optimised for partial reconfigurability of FPGA and therefore serves as a good example of hardware partitioning. The example given in Figure 15.6 illustrates the HW2000 FPGA board plugged in to a PC. Partitions are defined as follows:

- *Partition A.* The PC and its CPU has a lot more functionality, which is not shown here. The PC CPU is specified as a reconfigurable resource capable of processing, as defined by some reconfiguration mechanism. This mechanism may actually exist as an executing process on the PC CPU so the partition only considers the PC resources freely available for configuration. The PC memory exists in this partition, as it is only accessible by the PC CPU. The PCI bus is included in this partition as its usage is arbitrated by the hardware (not shown) associated with the PC CPU. The HW2000 board has a region of shared memory, which is accessible by all FPGA devices and also by the PC (via the PCI bus). This is a member of partition A, as its resources are arbitrated indirectly by the PC CPU via the PCI bus hardware.
- *Partition B/D.* A Xilinx Virtex FPGA device forms the centre of both of these partitions. Some external memory is accessible only by these individual devices, and therefore the memory resides in the same partition.
- *Partition C.* The IQX cross-point device forms the heart of this partition. It owns no other hardware. A cross-point device allows reconfiguration of routing between its pins. This effectively allows redirection of signals between the FPGA devices it is connected to.

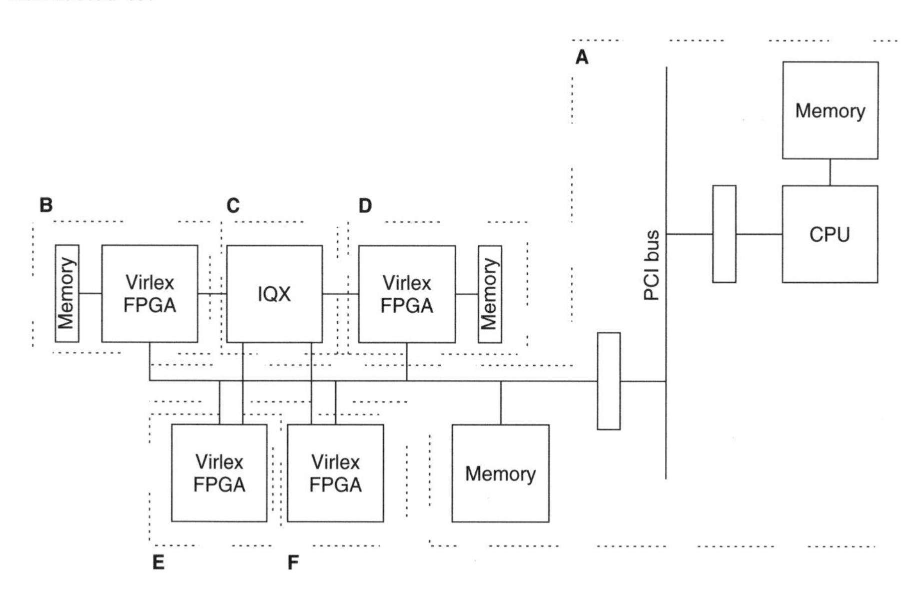

Figure 15.6 Partition of the HW2000 board installed in a PC

- *Partition E/F*. A Xilinx Virtex FPGA device forms the centre of both these partitions. No other hardware is owned in this diagram. On the real card, expansion connectors are provided which may allow ownership of other hardware connected to them. This simplified diagram assumes that nothing else is connected.

15.6 Hardware Independence

As described earlier, a key feature required to ease the operation and management of a reconfigurable system is that of hardware independence. In order to manage differing chains operating over differing classes and combinations of hardware it is desirable to be able to use functions from a library which can execute on many different hardware platforms.

A *process* is defined as the hardware-specific implementation of a *function*. There can exist many *processes* of a *function* and are identical in terms of the functionality, data interfaces and control interface. Data interfaces are defined logically so as to allow data transfer between other functions in a *chain*. Each process implements the function's functionality, logical data interfaces and parametable control in an equivalent physical manner dependent upon the target hardware.

Different hardware platforms have different types of processing resources. A DSP may have a memory map and a fixed level of processing ability. An FPGA, on the other hand, consists of arrays of different types of logic circuit and memory, which can be wired together. To provide a common treatment of configurable resources each device contains a quantity of configurable *space*. A function *process* may form a configurable *size* together with other special requirements. A *process* can be deployed to a particular *location* based on the resources available at that *location*. An XML description can be used to describe the specifics of the configurable *space* of a device and also the resources required by a *process*. The common quantity of space, size and location can provide the necessary common interface between configuration mechanisms of different devices, similar to memory maps related to microprocessors.

15.7 Database Structure

The Fizzware information structure, which indicates the relationship between chains, functions, processes and hardware, is shown in Figure 15.7. This diagram illustrates a database structure, which can store the core information required in order to describe the information for the deployment of functions, chains and hardware.

The following description defines standard database tables, which reference extensible mark-up language (XML) descriptions and ClassNames. Specific implementation for the CAST project uses an Oracle database and Java for control and XML interpretation. Many other implementation methods are possible. Descriptions of these database tables are given in the following sections.

15.7.1 Hardware Definition

15.7.1.1 Parts Table

Each hardware component is described electronically. Fields – *Type, PartNumber* and *Vendor* – give a direct link to any supporting information from the silicon vendor. The

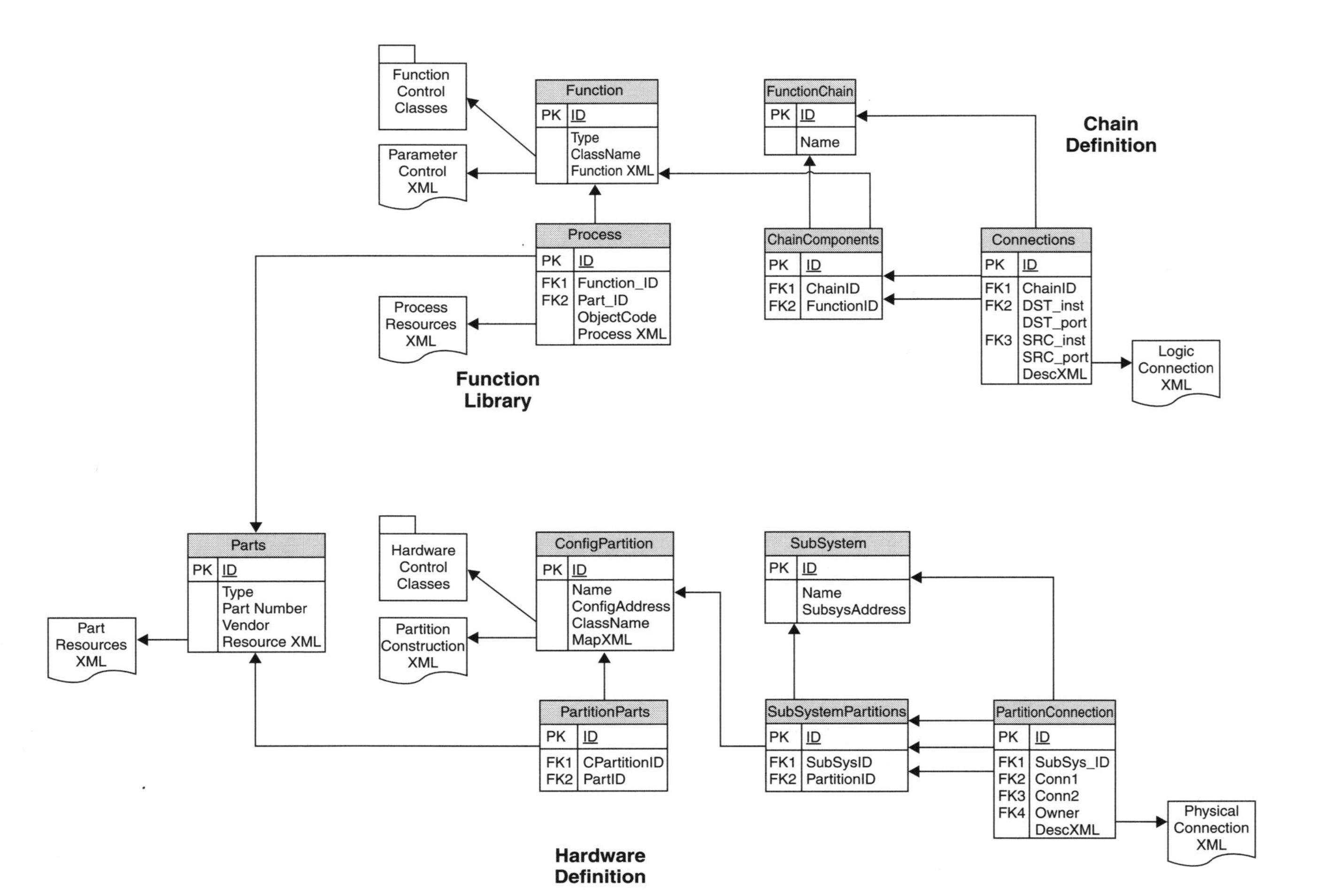

Figure 15.7 Fizzware information structure

ResourceXML field links to a more specific description of the hardware resources available on the device. This can effectively form an electronically readable datasheet.

15.7.1.2 SubSystem Table

A reconfigurable system is decomposed into subsystems. This may be in the form of separate PCBs on a shared bus (i.e. PCI) or any other medium which may separate hardware. A field – *SubSysAddress* – is used to provide a unique address for control of each subsystem.

ConfigPartition and PartitionParts Tables

Each subsystem is partitioned as described earlier and described in *ConfigPartition*. A unique address is given by the field *ConfigAddress*. A link is given via *ClassName*, which describes the name of a Java class, which can be used to control each partition. This is a class derived from the *HWResource* class described below.

The *PartitionParts* table describes each entity of *Parts*, which exist in each *ConfigPartition*. The *MapXML* field in *ConfigPartition* gives more detailed information regarding how each *PartitionParts* entity is physically connected within a *ConfigPartition* entity.

15.7.1.3 SubSystemPartitions Table

This describes which *ConfigPartition* entities exist within each *SubSystem* entity.

15.7.1.4 PartitionConnection Table

Connections between *ConfigPartition* entities are described. The fields *Conn1* and *Conn2* describe two *ConfigPartition* entities which are connected. *Owner* describes which partition is responsible for arbitrating the use of the connection, if any. *DescXML* gives more detailed information regarding the actual physical connections between *Parts* entities of the two different *ConfigPartition* entities.

15.7.2 Function Library

15.7.2.1 Function Table

This defines a *Type* of function available for deployment. The field *ClassName* refers to a Java class which can be use to control the parameters of the function once it has been configured to a device. This class is derived from the *Function* class described below. *FunctionXML* links to a more detailed description of the common parameterable control available to the configured function.

15.7.2.2 Process Table

This defines the device-specific implementation of a *Type* or *Function* entity. The field *PartID* links to the specific *Parts* entity. *ProcessXML* describes the resource requirements of the specific implementation. *ObjectCode* gives the device (*Parts* entity) specific executable code or configuration code for the actual configuration.

15.7.3 Chain Definition

15.7.3.1 FunctionChain Table

This is used to identify a whole processing chain for the configuration of a whole transmission mechanism.

15.7.3.2 ChainComponents Table

This identifies which *Function* entities exist within the *FunctionChain* entity.

15.7.3.3 Connections Table

This defines the logical connections that exist between the *Function* entities within a *FunctionChain* entity. The fields *DSTInst* and *SRCInst* indicate the two *Function* entities with *DSTPort* and *SRCPort* allowing multiple ports for each *Function* entity. *DescXML* provides a link to further information regarding the logical structure of the connection.

15.7.4 XML Descriptions

The XML links in the database to form a method for further description of hardware resources. The level of detail can be varied depending upon application. This can use basic descriptions of hardware resources, which simply count resources down to highly detailed description, whereby resource allocation mechanisms may even use analogue modelling techniques based on SPICE or IBIS models.

15.8 Control of Reconfigurable Hardware

This section describes the control mechanisms in the Fizzware architecture. Basic classes and interfaces are presented which form the major functionality for configuration and online control of functions. These allow common control of all different types of hardware and functions via a common API. Only the major classes and interfaces are presented here that form the basic Fizzware architecture. Further classes allow for detailed runtime control of the system and other minor functionality.

15.8.1 Class HWResource

The hardware resource class is the base class for all reconfigurable partitions in the system. *HWResource* objects are created using the *ClassName* information held in the *Config-Partition* database table. This class should also implement the Configurable Interface described below.

Methods

```
void Bind(Address resourceAddress);
```

Binds the resource class to a physical resource in the system. After a resource is created it must be bound to a Configurable Partition in order to specify what is being configured. The address is generated from information in the *SubSystem* and *ConfigPartition* partition tables.

```
HWLocation getLocation();
```

Obtains an object which can describe an independently configurable location on a resource (see below).

15.8.2 Class HWLocation

HWLocation represents an independently configurable location within the configurable partitions' configuration space. This class serves as a placeholder for the internal addressing scheme of the internal configuration. Specific implementation of this class can manage configuration space for the particular device.

Methods

```
void Set(...);
```

This sets the address for the target of the configuration. This class may also implement the Interface Configurable described below.

15.8.3 Interface Configurable

The configurable interface specifies which operations must be provided by any object representing a reconfigurable resource.

Methods

```
void PlaceFunction(Function function);
```

Configures the resource to run a function.

```
void Remove();
```

Removes the function from the resource.

```
ConfigueredFunction Get();
```

Gets a configured function object representing the function running on the configurable partition containing this location so that it may be controlled.

15.8.4 Class ConfiguredFunction

Represents a function configured on a resource. The operation of the function is manipulated through this class and its subclasses. Objects of this class would usually implement the controllable interface described below.

Methods

```
Configurable GetConfigurable();
```

Returns an object implementing the configurable interface which can be used to configure (more usefully unconfigure) the function.

15.8.5 Interface Controllable

This interface is implemented by the ConfiguredFunction class and its subclasses. It is used to perform basic control operations on a running function.

Methods

```
void Enable();
```

Activates the function.

```
void Disable();
```

Deactivates the function.

15.8.6 Class Function

This class serves as a base class for all the function classes on the system. It defines no methods of its own. Subclasses of function are created to represent configurable functions which are held in the *Function* database table. It is the joint responsibility of the classes derived from *Function* and the *HWLocation* class to locate a configurable process which is compatible with the partition type being configured.

15.8.7 Portability

The classes and interfaces, which are described above, form only the base classes. Further functionality may be added by simply extending these base classes. For example, classes implementing the controllable interface may also implement an interface formed from the methods which manipulate the state of the BPC style of function, as described earlier.

15.9 Example Use of the API

This example helps demonstrate how the API is used to configure a resource in a subsystem. The controller creates a HWResource object (FPGADevice) for the type of resource that is to be configured and binds it to the address of the resource in the subsystem by passing an address object to the resource objects `Bind()` method. The controller also creates an object derived from Function (Filter). Once the controller has determined how the function is to be placed on the resource it obtains the correct type of HWLocation object (FPGALocation) by calling the GetLocation function of the HWResource class. The controller may then configure the HWLocation object with the destination address

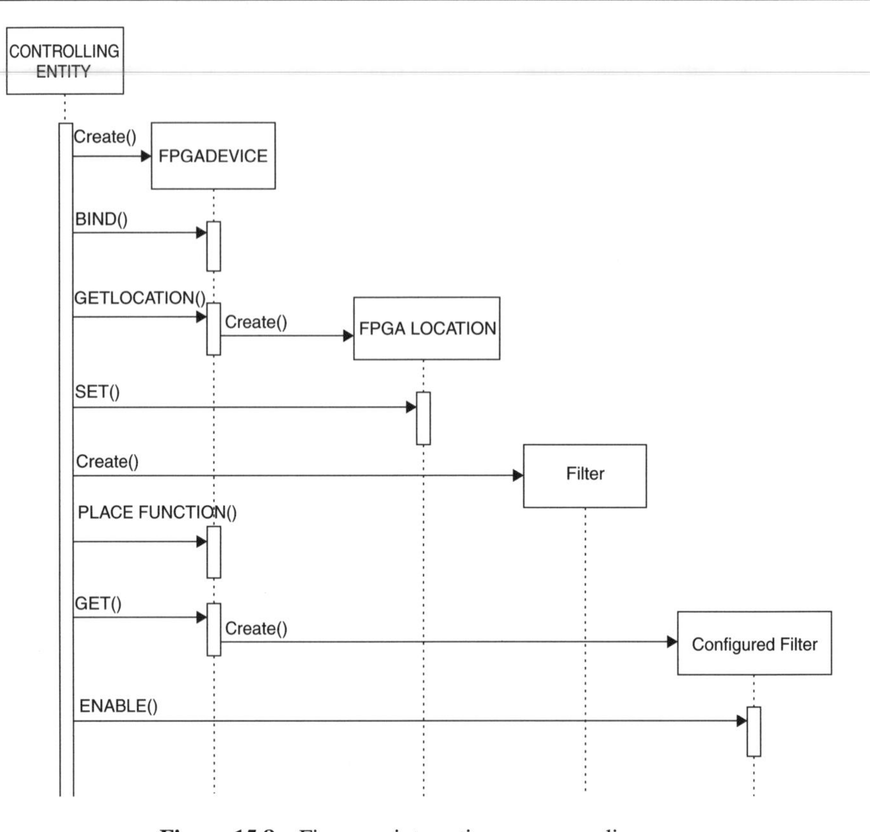

Figure 15.8 Fizzware interaction sequence diagram

for the configuration using the Set method. This address may be formed from the result of some hardware-specific resource allocation algorithm.

The RSC can then pass the function object to the AddFunction method of the HWLocation objects Configurable interface. The HWLocaton object and the Function object negotiate to determine the correct code to execute on that resource and the resource is configured with that code. The controller may then obtain a ConfiguredFunction (ConfiguredFilter) object by calling the `Get()` method of the Configurable interface. The ConfigueredFunction object can then be used to enable the function and perform other configuration operations. Figure 15.8 shows the Fizzware interaction sequence diagram.

15.9.1 Fizzware Structure

Figure 15.9 shows the Fizzware controlling architecture. The dotted line encloses the Fizzware framework. From the bottom the hardware itself is shown. Above the hardware, separate native drivers are shown each relating to a particular hardware partition. The Java machine forms the centre of the system and defines the barrier between the software and hardware. Resource drivers are formed from database entries and Java classes and each implement the common interface for device configuration. All of these resource drivers and several supporting runtime classes form the Fizzware API.

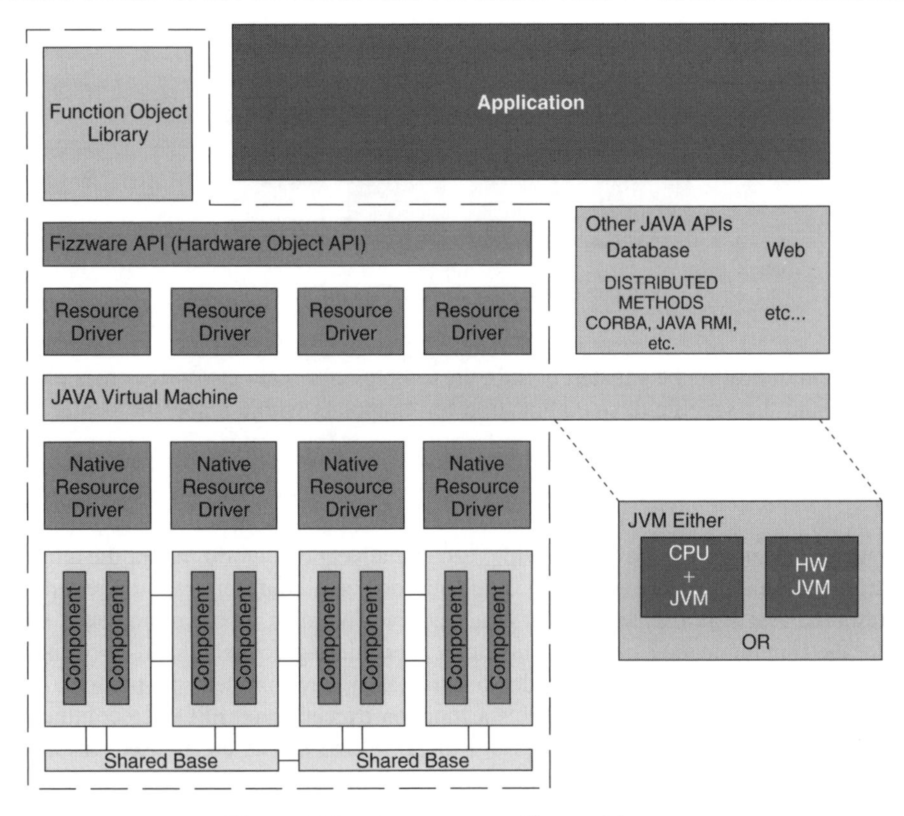

Figure 15.9 Fizzware controlling architecture

15.10 Runtime Support

Automation is paramount in relieving the designer from hardware-specific design tasks. The following describes how resource allocation can be used to automatically deploy function processes to hardware and connect them together to form a datapath.

15.10.1 Resource Allocation

Resource allocation can allow automatic deployment of function processes to hardware. A hardware subsystem may consist of many different types of reconfigurable device. Figure 15.10 illustrates a two-level resource allocation scheme.

Cross platform allocation may use heuristic approaches to measure loading of the different devices. Individual device allocation can decide how to specifically place the function on the device, including the actual location of its configuration bit stream or object code.

There are two distinct classes of function, which can be allocated to a particular resource. The first describes the functions, which implement the radio transmission standard. Cross platform allocation can decide on which devices to place functions. The second class of functions to be allocated is the support functions such as Endpoints. Although this task has not been considered in depth in the CAST project it is suggested that

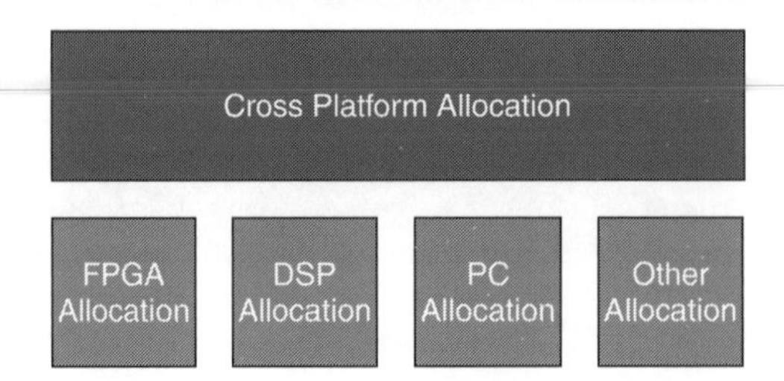

Figure 15.10 Resource allocation levels

the Endpoint allocation be situated both in the cross platform allocation for cross platform endpoints and the specific device allocation for endpoints within a specific device.

15.10.2 Connection Support

It is relatively well known how to provide resource allocation algorithms for the allocation of functions to hardware devices. For a specific device a count of resource requirement and resource available may be all that is necessary. The main problem, however, is how to allocate communication between functions. Considering connection of two functions executing on different devices provides a completely new problem. A particular route considered in the CAST project requires a function executing on FPGA to communicate with a function executing on a DSP. Both the DSP and FPGA are housed on different cards within a PC, therefore communication is as follows:

```
FPGA function ⇒ FPGA IO

FPGA IO ⇒ HW2000 PCI controller

HW2000 PCI Controller ⇒ PC native control

PC native control ⇒ DSP PCI controller

DSP PCI controller ⇒ DSP Operating System

DSP Operating System ⇒ DSP function
```

Each step in this path requires definition of an *Endpoint* at each end of the link. An Endpoint defines a routing entity which passes data from one environment to another. It uses a similar concept to the named pipe (indeed PC to PC function communication uses these).

Different Endpoints can be considered support functions which are required for every possible communication transfer in the system. Implemented natively, they should provide the most efficient mechanism for transferring data between different hardware environments. They are described in the same database as the Functions with hardware-specific processes. However, their choice is more likely via resource allocation-based on the placement of Functions and not specifically defined as a part of the Function chain.

15.11 Conclusion

Initially, this chapter began by describing some strict requirements for the construction of a reconfigurable baseband subsystem [4–6]. Most of these apply to both software and hardware aspects. One crucial aspect forms the most interesting of the initial requirements, namely hardware independence. The ability to decouple hardware from software and abstract away its physical structure is crucial for the success of future SDR systems. Most current treatment of SDR targets a purely software approach where compilation to microprocessor or DSP is relatively straightforward and compilers can be considered a simple form of hardware abstraction when set up carefully.

The advent of the FPGA and hybrid reconfigurable devices provide the high capacity processing required for today's communication systems, together with dynamic reconfigurability. More than a single compiler is required to support this diverse array of hardware.

This chapter presented several primitive components which can be used to fragment the hardware and software systems into separate domains. The function or Baseband Processing Cell (BPC) provides a common wrapper in the software world. This is composed of the hardware-dependant process which implement the real functionality. Chains or ConfigMaps define how the hardware-independent functions or BPCs can be connected together to form a communication transmission.

By allowing full automation of the deployment of individual function processes, the hardware can be hidden but described in detail. Enough detail can be given using an electronic form of datasheet to allow resource allocation algorithms the ability to automatically deploy processes and bind them together during the runtime.

The result of this hardware abstraction provides interfaces allowing hardware providers an automated description and allocation of resources. Software or IP providers have an interface to allow the provision of functions and processes to build the radio functionality. System operators and integrators have an interface to be able to deploy baseband chains without the need to understand any hardware-specific issues. The ability to build up libraries of hardware-independent functions allows faster future deployment of new and improved communication systems without the need to upgrade the fine technical detail every time.

References

[1] K. Madani, D. Lund and P. Patel, et al., 'Configurable radio with advanced software techniques (CAST) – initial concepts', IST Mobile Communications Summit, October, Galway, Ireland, 2000.

[2] CAST website – *www.cast5.freeserve.co.uk*

[3] TRUST website – *www.ist-trust.org*

[4] M. Mehta and M. Wesseling, 'Adaptive Baseband Sub-System for TRUST', *11th IEEE PIMRC 2000*, September 2000.

[5] M. Mehta and M. Wesseling, 'TRUST approach to Software Defined Radio – Baseband Considerations', *IST Mobile Communications Summit 2000*, October 2000.

[6] M. Mehta, et al, 'Reconfigurable Terminals: An Overview of Architectural Solutions', *IEEE Communications Magazine*, August 2001.

[7] L. Allmen, J. Garcia, J. Macleod and M. Mehta, 'Capability Characterisation of Re-configurable Terminals', *IST Mobile Communications Summit*, September 2001.

[8] M. Mehta, 'Re-configurable Baseband Architecture fro TRUST', *IEEE International Conference on 3rd Generation Wireless and Beyond*, June 2001.

[9] D. Lund and B. Honary, 'Baseband processing for SDR', in W. Tuttlebee, ed. *Software Defined Radio: Enabling Technologies*, Wiley, 2002.

[10] D. Lund, B. Honary and M. Darnell, 'A new development system for reconfigurable digital signal processing', IEE 3G2000, March, London, 2000.

[11] D. Lund and B. Honary, 'Design and maintenance of physical processing for reconfigurable radio systems', PIMRC2001, October, San Diego, USA, 2001.

[12] T. Karran, G. Justo and K. Madani, 'Intelligent reconfiguration of large mobile networks using complex organic distributed architecture', VTC-Spring Conference, May Rhodes, Greece, 2001.

[13] S. Imre, J. Kovács, P. Kacsuk, R. Ramos and K. Madani, 'Resource control and reconfiguration in software radio environment', 3GIS – International Symposium on 3rd Generation Information and Services, July, Athens, Greece, 2001.

[14] B. Honary, S. Colsell and D. Lund, 'Applications of FPGA, FPAA for future generations of mobile communications', Presentation at Reconfigurability Colloquium, June, Athens, Greece, 2002.

[15] J. Mitola, 'The software radio architecture', *IEEE Communications Magazine*, **33**, 26–38, 1995.

[16] A. Gatherer, T. Stetzler, M. McMahan and E. Auslander, 'DSP-based architectures for mobile communications: Past, present and future', *IEEE Communications Magazine*, **January**, 84–90, 2000.

[17] P. Kenington, 'Emerging technologies for software radio', *IEE Electronics and Communication Engineering Journal*, **11**(2), 69–83, 1999.

[18] M. Cummings and S. Heath, 'Mode switching and software download for software defined radio: The SDR forum approach', *IEEE Communications Magazine*, **37**(8), 104–106, 1999.

[19] C. Dick and F. Harris, 'Configurable logic for digital communications: Some signal processing perspectives', *IEEE Communications Magazine*, **37**(8), 107–111, 1999.

[20] D. Efstathiou, J. Fridman and Z. Zvonar, 'Recent developments for enabling technologies for software defined radio', *IEEE Communications Magazine*, **37**(8), 112–117, 1999.

[21] S. Srikanteswara, J. Reed, P. Athanas and R. Boyle, 'A soft radio architecture for reconfigurable platforms', *IEEE Communications Magazine*, **38**(2), 140–147, 2000.

[22] J. Mitola III and G. Maguire, 'Cognitive radio: Making software radios more personal', *IEEE Personal Communications Magazine*, **August**, 13–18, 1999.

[23] W. Tuttlebee, 'Software-defined radio: Facets of a developing technology', *IEEE Personal Communications Magazine*, **April**, 38–44, 1999.

[24] S. Kourtis, P. McAndrew and P. Tottle, *Software Radio 2G and 3G Inner Receiver Processing*, IEE Conference Press, pp. 6/1–6/7, 1999.

[25] B. Vyden, 'A software-implemented baseband processor for digital communications', IEEE Region 10 Conference, November, Tencon, pp. 31–35, 1992.

[26] C. Brown and K. Feher, 'Reconfigurable digital baseband modulation for wireless computer communications', IEEE Conference Publication, 0-7803-2492-7/95, 1995.

[27] K. Moessner, S. Vahid and R. Tafazolli, *A minimum air interface implementation for software radio based on distributed object technology*', IEEE ICPWC, pp. 369–373, 0-7803-4912-1/99, 1999.

[28] C. Noblet and A. Aghvami, *Assessing the Over-The-Air Software Download for Reconfigurable Terminal*, IEE Conference Press, pp. 6/1–6/6, 1998.

[29] P. Cook and W. Bonser, 'Architectural overview of the SPEAKeasy system', *IEEE JSAC*, **17**(4), 650–661, 1999.

[30] J. Gunn, K. Barron and W. Ruczczyk, 'A low-power DSP core-based software radio architecture', *IEEE JSAC*, **17**(4), 574–590, 1999.

[31] J. Pereira, 'Beyond software radio, towards reconfigurability across the whole system and across networks', IEEE Vehicular Technology Conference, pp. 2815–2818, 1999.

[32] P. Oreizy and R. Taylor, 'On the role of software architectures in runtime system reconfiguration', *IEE Proceedings on Software*, **145**(5), 137–145, 1998.

[33] K. Moessner and R. Tafazolli, 'Terminal reconfigurability – The software download aspect', IEE 3G Mobile Communication Technologies Conference, Conference Publication No. 471, pp. 326–330, 2000.

[34] SDR Forum, *http://www.sdrforum.org*

[35] European Commission Re-configurability Cluster, *http://www.cordis.lu/ist/ka4/mobile/reconfigurability2.htm*

[36] The Software Radio Home Page, *http://www-sop.inria.fr/rodeo/personnel/Thierry.Turletti/SoftwareRadio.html*

[37] Xilinx Website – *http://www.xilinx.com*

Index

Software Defined Radio: Architectures, Systems and Functions. Edited by M. Dillinger, K. Madani and N. Alonistioti

ISBN: 0-470-85164-3

RECEIVED
AUG 14 2003
ENGINEERING LIBRARY